ANSI FORTRAN IV

IBM System/360 Model 40 Console, Console Printer, and Keyboard. *Courtesy of International Business Machines Corporations.*

ANSI FORTRAN IV
A STRUCTURED PROGRAMMING APPROACH

J.W. PERRY COLE
UNITED STATES
AIR FORCE ACADEMY

wcb WM. C. BROWN
COMPANY PUBLISHERS
2460 Kerper Boulevard, Dubuque, Iowa 52001

Copyright © 1978 by Wm. C. Brown Company
Publishers

Library of Congress Catalog Card Number: 77-80094

ISBN 0—697—08125—7

All rights reserved. No part of this publication may be reproduced, stored in a retrieval system, or transmitted, in any form or by any means, electronic, mechanical, photocopying, recording, or otherwise, without the prior written permission of the publisher.

Fourth Printing, 1979

Printed in the United States of America

Contents

Preface xi

Acknowledgments xv

1
PROBLEM SOLVING WITH THE COMPUTER 1

1.1 Introduction to the Use of Computers 1
1.2 The Components of a Computer 2
1.3 Characteristics of Computers 4
1.4 How a Person Solves a Problem with a Computer 5
1.5 The Structured Programming Approach to Problem Solving 18
1.6 Review Questions 19
1.7 Flowcharting Exercises 20

2
INTRODUCTION TO THE FORTRAN IV LANGUAGE 22

2.1 Introduction to the History and Purpose of the Language 22
2.2 FORTRAN IV Source Statements 23
2.3 Basic Elements of the FORTRAN IV Language 24
2.4 FORTRAN IV Character Set and Keywords 25
2.5 Numeric Constants in FORTRAN IV 25
2.6 Variables in FORTRAN IV 28
2.7 Operators in FORTRAN IV 31
2.8 Coding FORTRAN IV Statements 32
2.9 Nonstandard Language Extensions 36
2.10 Review Questions 37
2.11 Exercises 38

3
DEVELOPING A BASIC FORTRAN IV PROGRAM 39

3.1 Introduction to Writing a FORTRAN Program 39
3.2 The Data Card 40
3.3 Problem Analysis and Flowcharting 41
3.4 The FORTRAN Program and Comment Cards 42
3.5 The READ Statement and Its Associated FORMAT Statement 43
3.6 The Assignment Statement 46
3.7 The WRITE Statements and Their Associated FORMAT Statements 47
3.8 The STOP and END Statements 49
3.9 Nonstandard Language Extensions (Format-free Input/Output Statements) 51
3.10 Review Questions 53
3.11 Exercises 53
3.12 Programming Problems 54

4
PROCESSING YOUR FORTRAN PROGRAM 57

4.1 Introduction to Processing a FORTRAN Job 57
4.2 Preparing Your Program for the Computer and Running the Job 57
4.3 Program Compilation Errors 61
4.4 Program Execution (Logic) Errors 64
4.5 Expanding the Basic Program to Include a Loop 64
4.6 The Unconditional GOTO and Logical IF Statements 67

v

- 4.7 Review Questions 71
- 4.8 Exercises 71
- 4.9 Programming Problems 72

5

MORE ON ARITHMETIC EXPRESSIONS AND THE ASSIGNMENT STATEMENT 73

- 5.1 Introduction to the Use of the Assignment Statement 73
- 5.2 Rules for Constructing Arithmetic Expressions 73
- 5.3 More on the Assignment Statement 77
- 5.4 Truncation Considerations with Numeric Values 79
- 5.5 Arithmetic Assignment Statements in Sequence 79
- 5.6 Counting and Accumulating Totals with the Assignment Statement 80
- 5.7 Nonstandard Language Extensions (Mixed Mode Expressions) 80
- 5.8 Review Questions 81
- 5.9 Exercises 82
- 5.10 Programming Problems 83

6

MORE ON READING/WRITING NUMERIC DATA 85

- 6.1 Introduction to the Use and Importance of Input/Output Statements 85
- 6.2 Rules for Using the FORMAT Statement and the I/O List of Variables 86
- 6.3 More on the Integer (I) and Floating Point (F) Format Codes 89
- 6.4 Reading Numeric Data 92
- 6.5 Writing Numeric Data 95
- 6.6 Using the Slash to Select Records with Reading and Writing Operations 97
- 6.7 A Sample Program to Prepare a Student Tuition Register 100
- 6.8 Nonstandard Language Extensions (END = option with the READ) 102
- 6.9 Review Questions 103
- 6.10 Exercises 105
- 6.11 Programming Problems 107

7

WRITING LITERAL DATA AND ADDITIONAL FORMAT SPECIFICATIONS 109

- 7.1 Introduction to Literal Constants and the Hollerith (H) Format Code 109
- 7.2 Writing Literal Data 110
- 7.3 Using the Slash to Write Multiple Headings 111
- 7.4 Repetition of Group Format Specifications 112
- 7.5 A Sample Program to Illustrate the Use of Headings 113
- 7.6 Nonstandard Language Extensions (T and ' Format Codes) 117
- 7.7 Review Questions 119
- 7.8 Exercises 119
- 7.9 Programming Problems 120

8

CONTROL AND DECISION-MAKING STATEMENTS 122

- 8.1 Introduction to the Use of Control Statements and Types of Program Loops 122
- 8.2 More on the Logical IF Statement 127
- 8.3 The Arithmetic IF Statement 130
- 8.4 The Computed GOTO Statement 131
- 8.5 A Sample FORTRAN Program to Illustrate Control Statements 132
- 8.6 Nonstandard Language Extensions 137
- 8.7 Review Questions 138
- 8.8 Exercises 138
- 8.9 Programming Problems 139

9

PROGRAM LOOPING AND THE DO STATEMENT 142

- 9.1 Introduction to Counter Loop Logic with the DO Statement 142
- 9.2 The DO Statement 142
- 9.3 Rules in Using the DO Statement 147
- 9.4 The CONTINUE Statement 150
- 9.5 Examples in Forming DO Loops 150
- 9.6 Nested Loops with Nested DO's 151
- 9.7 Rules in Using Nested DO's 152
- 9.8 A Sample FORTRAN Program to Illustrate DO Loops 156
- 9.9 Review Questions 161
- 9.10 Exercises 161
- 9.11 Programming Problems 163

10

SUBSCRIPT OPERATIONS AND ONE-DIMENSIONAL ARRAYS 166

- 10.1 Introduction to the Concepts of Arrays 166
- 10.2 The DIMENSION Statement 168
- 10.3 Subscripts 169
- 10.4 Input/Output Operations with One-dimensional Arrays 170
- 10.5 Manipulating One-dimensional Array Items 176
- 10.6 Sorting One-dimensional Array Items 180
- 10.7 Nonstandard Language Extensions 183
- 10.8 Review Questions 184
- 10.9 Exercises 185
- 10.10 Programming Problems 186

11

TWO-DIMENSIONAL AND THREE-DIMENSIONAL ARRAYS 189

- 11.1 Introduction to the Concept of Two-dimensional Arrays 189
- 11.2 Input/Output Operations with Two-dimensional Arrays 191
- 11.3 Manipulating Two-dimensional Array Items 194
- 11.4 Introduction to the Concept of Three-dimensional Arrays 198
- 11.5 Achieving Efficiency in Multidimensional Array Operations 199
- 11.6 Nonstandard Language Extensions 202
- 11.7 Review Questions 203
- 11.8 Exercises 203
- 11.9 Programming Problems 204

12

ALPHANUMERIC DATA AND COMPILE TIME SPECIFICATIONS 207

- 12.1 Introduction to Alphanumeric Data and Format Specifications 207
- 12.2 Comparing Alphanumeric Data 211
- 12.3 Introduction to Compile Time Specifications 212
- 12.4 The DATA Statement 213
- 12.5 The Explicit Type Statements 214
- 12.6 The EQUIVALENCE Statement 216
- 12.7 A Sample FORTRAN Program with Multiple Specification Statements 217
- 12.8 Nonstandard Language Extensions (IMPLICIT Statement) 219
- 12.9 Review Questions 221
- 12.10 Exercises 222
- 12.11 Programming Problems 223

13

SUBROUTINE SUBPROGRAMS AND THE COMMON STATEMENT 228

- 13.1 Introduction to the Use of Subroutine Subprograms 228
- 13.2 The SUBROUTINE, RETURN, and CALL Statements 232

- 13.3 Passing Arguments in a Parameter List 233
- 13.4 Sample Subroutine Subprograms 236
- 13.5 Flowcharting Subprograms and Job Setup of Subprograms 238
- 13.6 The COMMON Statement 240
- 13.7 BLOCKDATA Subprograms 246
- 13.8 Nonstandard Language Extensions (Multiple RETURN n Option) 247
- 13.9 Review Questions 248
- 13.10 Exercises 248
- 13.11 Programming Problems 249

14

STRUCTURED PROGRAMMING AND DEBUGGING TECHNIQUES 251

- 14.1 Introduction to the Concepts and Use of Structured Progamming 251
- 14.2 FORTRAN Implementation of the Six Control Structures 257
- 14.3 Top-down Program Design and Development 260
- 14.4 Techniques for Better Designed Programs 263
- 14.5 Debugging Techniques for Execution Time Errors 264
- 14.6 Review Questions 268

15

FUNCTION SUBPROGRAMS AND ADDITIONAL SUBPROGRAM STATEMENTS 269

- 15.1 Introduction to Function Subprograms 269
- 15.2 The FUNCTION Statement and a Sample Subprogram Using Arrays 273
- 15.3 FORTRAN Library-supplied FUNCTION Subprograms 274
- 15.4 The Statement Function Subprogram 277
- 15.5 The EXTERNAL Statement and Subprogram Names Passed as Arguments 280
- 15.6 Object Time Dimensions in Subprograms 283
- 15.7 Nonstandard Language Extensions (ENTRY statement) 283
- 15.8 Review Questions 285
- 15.9 Exercises 286
- 15.10 Programming Problems 287

16

ADDITIONAL FORTRAN CONSTANTS AND FORMAT SPECIFICATIONS 289

- 16.1 Introduction to the Additional Types of FORTRAN Constants and Data 289
- 16.2 Single-precision Exponential Type Constants and the E Format Code 289
- 16.3 Double-precision Constants and the D Format Code 292
- 16.4 Logical Constants and the Logical (L) Format Code 295
- 16.5 Complex Floating-point Data 297
- 16.6 The Generalized (G) Format Code 300
- 16.7 The P Scale Factor Specifications 302
- 16.8 Nonstandard Language Extensions (Hexadecimal and Octal Constants and Data) 303
- 16.9 Review Questions 306
- 16.10 Exercises 306
- 16.11 Programming Problems 308

17

MAGNETIC TAPE STATEMENTS AND OPERATIONS 310

- 17.1 Introduction to the Use and Concepts of Magnetic Tape 310
- 17.2 Formatted READ/WRITE Statements with Magnetic Tape Files 313
- 17.3 Unformatted READ/WRITE Statements with Magnetic Tape Files 314
- 17.4 The REWIND, ENDFILE, and BACKSPACE Statements 315
- 17.5 Use of Sequential Tape Files in Business Applications 318

- 17.6 Nonstandard Language Extensions and Additional Remarks 320
- 17.7 Review Questions 321
- 17.8 Exercises 322
- 17.9 Programming Problems 322

18

MAGNETIC DISK STATEMENTS AND OPERATIONS 324

- 18.1 Introduction to the Use and Concepts of Magnetic Disk 324
- 18.2 FORTRAN Statements Used with Sequential Magnetic Disk Files 326
- 18.3 Random File Processing on Magnetic Disk 328
- 18.4 The DEFINEFILE Statement 328
- 18.5 The Formatted and Unformatted Random READ Statements 330
- 18.6 The Formatted and Unformatted Random WRITE Statements 332
- 18.7 The FIND Statement 333
- 18.8 Nonstandard Language Extensions 334
- 18.9 Review Questions 336
- 18.10 Exercises 337
- 18.11 Programming Problems 338

19

ADDITIONAL FORTRAN STATEMENTS AND FEATURES 341

- 19.1 Introduction to Additional FORTRAN Statements and Features 341
- 19.2 The PAUSE Statement 341
- 19.3 The ASSIGN and Assigned GOTO Statements 342
- 19.4 Object Time Format Specifications 344
- 19.5 Additional Nonstandard FORTRAN Statements (NAMELIST, Specific READ, PRINT, and PUNCH) 346
- 19.6 Review Questions 351
- 19.7 Exercises 352
- 19.8 Programming Problems 352

20

FEATURES OF THE WATFIV COMPILER 354

- 20.1 Introduction to the Importance and Use of the WATFIV Compiler 354
- 20.2 WATFIV Extensions to the FORTRAN IV Language 355
- 20.3 How WATFIV Differs from the IBM FORTRAN IV "G" Compiler 362
- 20.4 WATFIV Job Control Statements 363
- 20.5 WATFIV-S Features 365

Appendix A: The 80-Column Punch Card 372

Appendix B: IBM 029 Keypunch Operations 375

Appendix C: Overview to Timesharing Operations 380

Appendix D: The Scientific Notation for Expressing Constants 384

Appendix E: Job Control Language (JCL) Statements for Various Computers 386

Appendix F: Comparison of FORTRAN Compilers 389

Answers to Review Questions and Odd-numbered Exercises 394

Index 413

Preface

This book is intended to serve as an introductory text for the beginning student who has little or no knowledge of computers or computer programming. The text uses a building-block approach to teaching FORTRAN. Only the basic elements and statements of the language are presented in the early chapters, to provide the information needed for getting the students started writing programs. The subjects, such as double-precision constants, literal constants, logical operators, and the like, are presented in later chapters as needed. Following this procedure, the student doesn't get bogged down with unnecessary terminology and concepts early in the course. Since the text material covers all the statements and features in most FORTRAN compilers, the text may also be used for a two-semester course, and serve as a reference book for professional programmers as well.

The text is intended primarily for use by business and social science students. However, it may also serve as an introductory text for math and engineering students alike. This book is written so as to enable the student to learn FORTRAN programming without an extensive knowledge of complicated mathematics. A knowledge of mathematics through the first year of high school algebra is sufficient math background. The emphasis is on the FORTRAN language. I believe that once the student has mastered the FORTRAN language and the programming techniques with low-level mathematics, one can apply this programming knowledge to more advanced mathematics and to other disciplines as effectively as his mastery of those disciplines permits him.

The FORTRAN language presented is ANSI Full FORTRAN as described in the X3.9-1966 specifications. This version of the FORTRAN language has been chosen for four main reasons. First, it is an official standard FORTRAN version recognized by computer manufacturers, the United States Government, programming professionals, and the data processing industry. Second, it contains a large subset of the FORTRAN statements and features found in most FORTRAN compilers. Third, virtually all modern computer systems include the ANSI FORTRAN specifications in their FORTRAN compiler, since the major computer vendors have agreed to make Standard FORTRAN IV available on their computers. Fourth, this version of the FORTRAN language learned by the students will be easily transferable to other computer systems and in the job market. The additional features found in the IBM 360-370 G and H level FORTRAN compilers, WATFIV, H6000 Y, B6700, and CDC 6000 compilers are also discussed in a separate section at the end of most chapters. Thus, the educational institutions using IBM FORTRAN G, WATFIV, and various other FORTRAN compilers should experience no difficulty in using this text.

A problem solving approach is presented and stressed throughout the book. This approach consists of how to define a problem solution and the orderly steps to be taken for developing that solution into a computer program. This is not to imply that a vast number of application problems with accompanying flowcharts are to be presented in each chapter, as some might expect. The emphasis is, rather, on the general approach to developing algorithms and not necessarily specific ones. Once the statements in the chapter are explained, a problem is presented to show the use of the statements and concepts discussed. The structured programming techniques used in program development are also illustrated and stressed throughout the text. A detailed discussion of top-down structured programming along with the basic control structures is presented in chapter 14. If students can arrive at the proper approach in the developing of solutions to assigned problems and writing programs, they will discover the computer to be a helpful and interesting tool with which to solve problems.

This textbook is intended to be conducive to a large measure of self-instructed learning on the part of the student. Most instructors have such a heavy load that they cannot administer to all student problems. Thus, the student searches frantically for someone who can assist him with his problem. This text goes into great detail in the use of the FORTRAN language, especially the more complex subjects such as input/output operations, DO loops, arrays, subprograms, magnetic tape, and magnetic disk. It is hoped, therefore, that the student can glean from this text more than enough to help answer his questions, as a result confining his approaches to his professor on only the more "impossible-to-find" errors.

The material is organized to enable the student to start writing FORTRAN programs as soon as possible. The student should be able to begin writing simple FORTRAN programs after about the third class meeting. The first chapter introduces the student to the computer—explains what computers are used for, how they function, and presents an overview of how one goes about solving a problem with the computer. The second chapter introduces the basic features and composition of the FORTRAN language. Chapter 3 develops a simple FORTRAN program; shows how to prepare a data card; and explains the READ, WRITE, FORMAT, STOP, END, and Assignment statements using a very simple problem. Then chapter 4 covers in detail the steps that a student must take to run a program on the computer and obtain the output results. Chapter 4 also expands the basic problem to include a loop and the use of the Logical IF and Unconditional GOTO statements. Thus, by the end of chapter 4 the student will have learned the simple features of the eight basic statements which are used in nearly every program.

Each chapter begins with an introductory section to provide an overview of the concepts and statements covered in the chapter. This same section primarily stresses the importance and use of the category of statements that are explained in the later sections. A section of review questions is included at the end of each chapter to encourage the student to test his understanding of the material. The answers to these questions are given at the end of the book.

Starting with chapter 3, each chapter (except 14 and 20) will have a section of exercises and a section of programming problems. The exercise sections include a large selection of detailed exercises by means of which the student can apply the information covered in the chapter. Answers to all the odd-numbered exercises are presented in the back of the text, enabling the student to check his work. Even-numbered exercises are extensions of the odd-numbered exercises, to be assigned as homework by the instructor. The programming problems require that simple-to-complex programs be written by the student for gaining experience in writing FORTRAN programs. Most of these problems are business oriented. However, some scientific problems have been included for the students who are proficient in math. Input and output specifications are included in the problems to provide standardization in large classes.

Chapters 3-13 and 15-19 each include a section titled "Nonstandard Language Extensions." In these sections the additional features found in other compilers (such as IBM, Honeywell, CDC, and Burroughs) are discussed. The instructor should review these sections and instruct the student to skip them if they do not apply to the FORTRAN compiler being used by the class. The entire content of chapter 20 is devoted to the extensions in WATFIV. Appendixes on the 80-column punch card, use of the 029 keypunch machine, timesharing operations, job control cards, comparison of FORTRAN compilers, and other subjects are also to be found at the back of the book.

Numerous coded routines are furnished in the various chapters to illustrate the FORTRAN statements and their application to problem solving. Each program example has been tested on the computer to verify its correctness. Numerous programming techniques, such as looping, counter logic, header card logic, trailer card logic, table searching, sorting, and file maintenance are explained to acquaint the student with commonly-used programming concepts. The presentation of these programming techniques is, in my estimation, just as important as the FORTRAN language itself.

I have used the basic information in this book in my FORTRAN IV programming course for many years. Handouts providing condensed subject information and extensive program examples were readily accepted by the students. It was their encouragement to put this information into published form that persuaded me to make the effort to develop this text. I sincerely hope that you, as instructors, and your students can benefit from this new text. All of the areas that I have found difficult for students to master have been given careful consideration. Any suggestions and comments on how to improve this text will be promptly answered by return mail.

This book is dedicated to my loving family, Judy, Jeff, and Jarrett, whose sacrifices have made my work possible.

Acknowledgments

The author is greatly indebted to so many people who have helped make this FORTRAN text possible. First of all, I would like to thank the students at San Antonio College who encouraged me to put my class notes into a published text, and who also made teaching a pleasure. I wish to extend my personal thanks to Alvin J. Stehling, Head of Data Processing, San Antonio College, and to other SAC professors, John R. Friedrich and Merle Vogt, each of whom were helpful in many ways and assisted me in the development of this text. Second, I extend a hearty thanks to Mrs. Kay V. Tracy and my wife, Judy, who so diligently typed the manuscript.

Third, I am deeply grateful to the many people who reviewed portions or all of the manuscript and made constructive suggestions. The quality of this text is a direct result of the outstanding efforts of these people who diligently made so many invaluable recommendations. The list is long and includes Professor Dennis M. Anderson, Purdue University at Fort Wayne; Professor Norman Wright, Brigham Young University; Professor E.G. Aseltine, University of Toronto; Dr. Charles Butler, Mississippi State University; Professor Mary C. Durkin, St. Petersburg Community College; Professor Gary Gleason, Pensacola; Professor C. B. Millhaam, Washington State University; Professor Martin Stacy, Tarrant County College; Professor Greg Ulferts, University of South Colorado; Professor Howard D. Weiner, North Virginia Community College; Captain James D. O'Rourke, United States Air Force; Major James H. Nolen, United States Air Force; Major Kenneth C. Hovis, United States Air Force; Mrs. Jan Powell and Mr. Robert Lay, San Antonio College; Lt. Col. Jerry B. Smith, United States Air Force Academy; Captain Ronald E. Joy, United States Air Force Academy; Major Kenneth L. Krause, United States Air Force Academy; and Lt. Col. Clifford J. Trimble, United States Air Force Academy.

An author seldom recognizes his publisher, but I would like to thank Wm. C. Brown Company Publishers and its staff for the outstanding cooperation, assistance, and kindness shown me during the development of this text. I would also like to thank IBM Corporation for their generosity in providing photographs and other illustrative material for use throughout the text.

IBM System/370 Model 165. *Courtesy of International Business Machines Corporation.*

Problem Solving with the Computer   1

1.1 Introduction to the Use of Computers

The use of computers for processing information has steadily increased since their introduction in the mid-1940s. Wherever we turn, we find punched cards and computer output products. Our bills, paychecks, savings bonds, and payment cards are often conveniently recorded in the form of the familiar punch card. People receive reports representing hundreds of different applications—each report neatly printed on printer listings.

The computer has become many things to many people. To some, it is a wonderful machine that has eliminated the drudgery of their work; to others, it is a brain that never makes a mistake and can perform almost as if it were human. Some feel that it is a stranger that has overcharged their account; others think that it is an idiot that always gets things mixed up and has no regard for personal feelings.

The computer is simply a machine capable of solving many different types of problems. Complex calculations which would take hours, days, or even weeks to perform by hand can be calculated on a computer in seconds. It can read data from hundreds of punched cards in a minute and print out information at the rate of hundreds of lines per minute. The computer can store millions of numbers and letters and provide almost instantaneous retrieval of this information.

A computer can perform simple to complex mathematical tasks, draw pictures, plot graphs, and even play games. It can sort data, search tables of information, and update files of data. It can make comparisons between data values and perform various actions based on the resulting decision.

But the computer cannot solve a problem by itself. The thought of a computer being a "brain" is far from the truth. We must remember that it is only a machine—a tool used to expand our capabilities. It cannot make any decisions on its own—it must be told explicitly what operations to perform. The computer is limited to applications having predetermined goals which can be defined by a series of explicit, logic operations. If we cannot define all the steps involved towards the solving of a problem, then the computer cannot solve that problem.

The tremendous speed of operations in the computer, and their accuracy, is fantastic. The computer will malfunction only if some component within it fails, and it can perform operations endlessly without getting tired. This is not to say that the products produced by the computer are always perfect. We often find errors in output listings, reports, and billings. But the errors really originate in the program that instructed the computer what operations to perform or the data that was given to it. If we instruct it incorrectly, it does not question what we might have wanted. If we give it bad data to process, the computer accepts it, manipulates it at a fantastic speed, and outputs the results. According to the acronym GIGO (garbage in, garbage out), we can get valid results out only if we put correct data in the computer and instruct it correctly in the desired operations.

Computers are used to solve a wide variety of problems. From computing the census to putting a man on the moon, from computing payrolls to monitoring hospital patients, computers are used in nearly every realm of business and scientific applications. Our airline reservation systems today operate with computer systems to provide us with instant information in making flight reservations. Accounting, manufacturing,

1

inventory control, processing control, banking, and many other areas of business applications use computers. The computer is assisting the police in criminology as well as in traffic regulation. Research and education are greatly aided by the role of computers. Problem solving by the computer has indeed become an everyday way of life.

1.2 The Components of a Computer

A computer in the simplest organization of components is made up of five basic units. Each unit is constructed from transistors, integrated circuits, wires, motors, and/or switches. These parts are used to form the ***central processing unit (CPU)***. The CPU consists of three major units:

1. The **control unit**—for coordinating and directing the operations of the computer.
2. The **arithmetic-logic unit**—for performing calculations and decision making.
3. The **memory unit**—for storing data and computer instructions.

Auxiliary devices such as card readers, typewriters, line printers, card punches, magnetic tape drives, and magnetic disk units are connected to the CPU by electrical cables. These auxiliary devices form the last two units of the five basic components of a computer and are known as:

4. **Input devices**—to read data and transfer it to the memory unit.
5. **Output devices**—to take data from memory and write it upon some output medium.

These input and output devices are connected to the CPU to form a system of machines called a ***computer system*** or simply a ***computer***. The collection of the various physical units and equipment is referred to as ***hardware***. The auxiliary input/output devices are also known as ***peripheral equipment*** since they are normally arranged in a periphery around the computer operator console.

The relationship between these components is illustrated in figure 1.1. The solid lines represent the flow of information through the computer system. The broken lines show the relationship of control between the units.

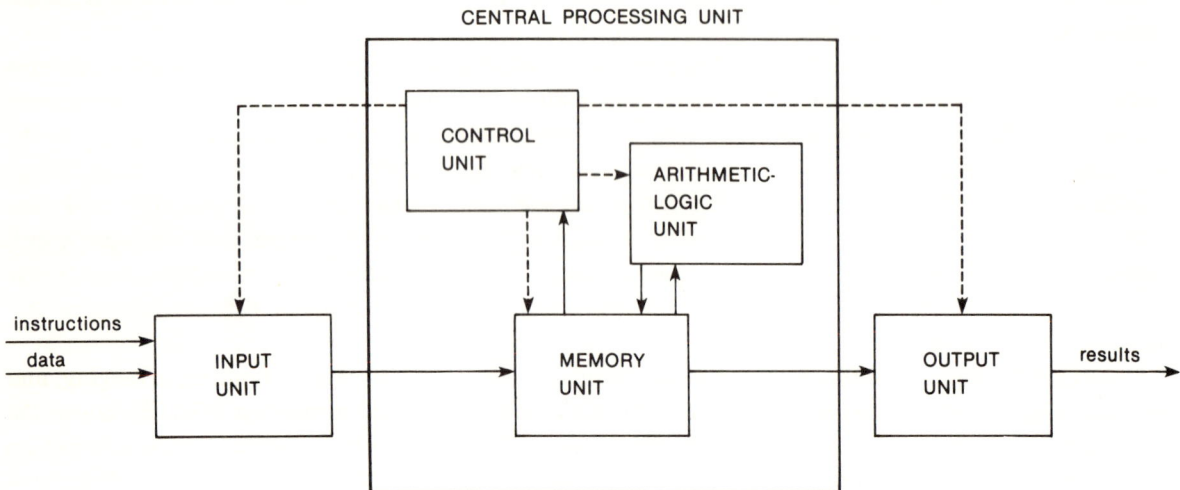

Figure 1.1. The Five Basic Components of a Computer

Let us examine each of these basic components in more detail.

Control Unit. The *control unit* serves two main purposes. First, it is the control hub of operations and directs the operations of all the other units. It informs the input devices when to transfer data to memory, and the output devices when to take data from memory and output it to their respective medium. It directs the arithmetic-logic unit in all of its operations. The second function of the control unit is to "fetch" (obtain), decode, and execute each of the instructions supplied for a problem solution. This unit might be compared to the traffic cop who directs the flow of traffic along a busy intersection.

Arithmetic-logic Unit. The *arithmetic-logic unit (ALU)* performs the basic arithmetic operations of adding, subtracting, multiplying, and dividing. This unit also performs the logical and decision-making operations such as comparing two values and determining the resulting condition such as equal to, less than, or greater than. The values to be manipulated in this unit are brought in from memory and the result of the manipulation is sent back to the memory unit.

Memory Unit. One of the advantages of a computer is its ability to store information and retrieve it in later use. The *memory unit* allows the computer to store the data read by the input devices or items calculated in a program. Input data must always be transferred to the memory unit before it can be processed by the computer. The memory unit is often referred to as *primary* or *internal storage* since the data is immediately accessible by the control unit.

The memory unit is broken down into many individual locations which may be thought of as mail boxes. Each location, however, can store only one piece of information at a time. The value stored at a particular memory location is called the *contents* of that location. New information may be stored at a location, but destroys the data value previously contained at that location.

Each storage location in memory has an *address* by which the location may be referenced. A programmer simply refers to a location by a name (symbolic address) such as X, HOLD, NUMBER, etc. Of course, the computer uses machine (numeric) addresses, but suffice it to say that the computer can always obtain the proper address when we use a symbolic name (called a variable). When we refer to an address, we are referring to the contents of a storage location or the value stored at the address of that location. (We are not referring to the address value, itself.)

Input Devices. The *input devices* allow us to read program instructions for the computer to execute and data pertaining to an application for the computer to manipulate. The data is recorded on a specific medium such as punched cards and transferred to the memory unit upon a READ command from the computer. There are many types of input devices available today. Each has its own advantages and disadvantages. The devices chosen depend largely upon the application and processing environment.

The *card reader* is probably the most commonly used input device in the student environment. It is used to read data punched in the punch card. The type of punch card most commonly used today is the *80-column Hollerith card* (named after its founder, Dr. Herman Hollerith), known simply as the punch card. The Hollerith card is a rectangular pasteboard of high-quality paper divided into 80 vertical columns, each capable of containing one character. A character may be an alphabetic letter, a numeric digit, or a special character such as a semicolon, asterisk, comma, etc. (See **Appendix A** for a complete description of the 80-column punch card if you are unfamiliar with this input medium.)

Many other devices exist today that provide input to the computer. Magnetic tape units read data recorded as magnetized spots on a long ribbon of tape much like our home tape recorders. Magnetic disk units read data also recorded as magnetized spots on a thin metal platter covered with magnetic oxide similar to our long-playing phonograph records. Other input devices include paper-tape readers, magnetic ink character readers, optical character readers, remote typewriters, and remote visual display terminals.

Output Devices. After data has been manipulated in such various ways as computing, arranging, and summarizing, we wish to output it in some form. An output command (normally a WRITE) causes the computer to transfer data from the memory unit to the specified output device which records the data on its respective medium. The type of output device used is specified in the "WRITE" command.

The *line printer* is probably the best known and most frequently used output device. This device prints program listings and output results on a paper sheet in readable form. Most printers have the capability for printing 132 characters of data per line.

Many other types of output devices exist to record output data. Output results may be punched into punch cards or into a strip of paper tape by a card punch and a paper-tape punch, respectively. Magnetic tape and disk devices may record output data. The remote typewriter and visual display units are highly popular output devices.

Now that the basic components of a computer have been explained, you are probably wondering what a computer or card reader and line printer look like. Figure 1.2 illustrates some of the modern-day com-

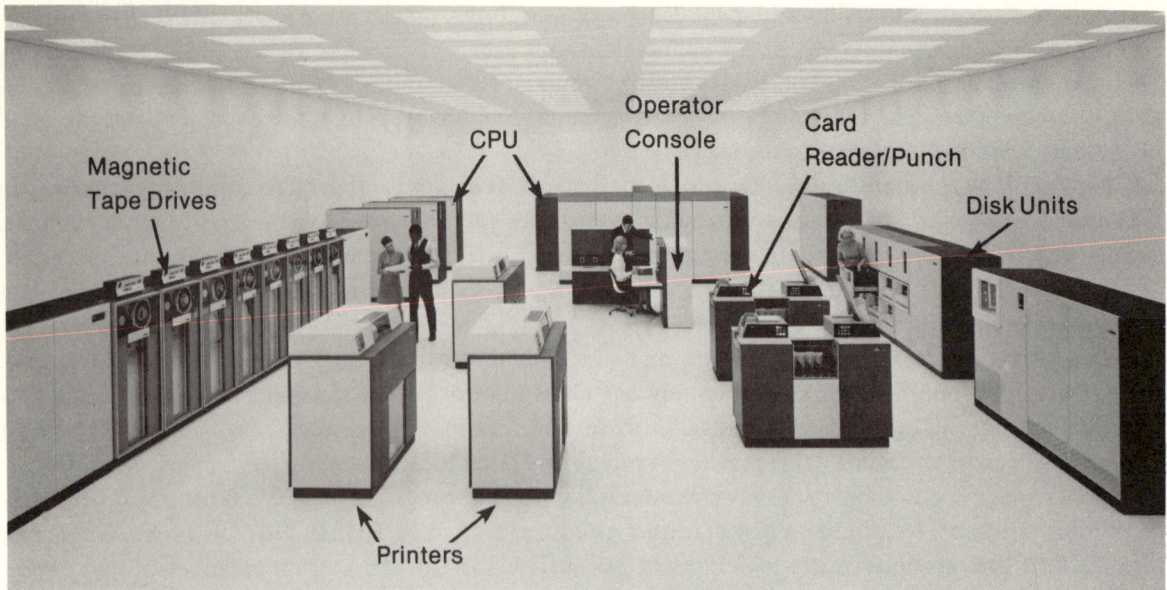

Figure 1.2. IBM System/370 Model 165 with Hardware Components of a Computer System. *Courtesy of International Business Machines Corporation.*

puter hardware found in a large IBM 370 data processing installation. The computer and peripherals in your college data processing center may not be as elaborate as those in the system shown in figure 1.2.

Computers and their related pieces of hardware are very expensive. Large computers cost in the millions of dollars. Your college may, therefore, rent time on a computer located away from the campus. So the only parts of a computer system you may see are the card reader and line printer to provide the input and output operations. If you have never seen a complete computer system with all its elaborate devices, I suggest that you ask your instructor to arrange for you to take a tour of a computer center.

The computer is indeed a marvelous machine which synchronizes its operations with great sensitivity and precision. But remember, a person must instruct it in every operation it performs, whether it be an input operation, a calculation, a logical evaluation, or an output operation. A computer must be told to read if it needs values contained in a punch card. Just because the computer has two values placed into memory does not cause the computer to calculate their sum. A command must be given to add the two numbers together. Likewise, the sum will not be output unless the computer receives a command to write out the result. The computer does not "think," but merely obeys the commands given to it in a sequence of instructions contained in a program developed by a human. This, then, is our goal in learning FORTRAN—to be able to supply the commands needed to instruct the computer in solving a problem.

The following section discusses the stored program concept and how a series of instructions are used by the control unit to direct the operations of the computer.

1.3 Characteristics of Computers

Some computers are built as *special-purpose* computers to solve one kind of problem such as air-traffic control or controlling a metalworking machine. Other computers are *general-purpose* computers that can be used to solve a wide variety of problems. It is the general-purpose computer that is used for the many applications in the business and scientific fields.

There are two basic types of computers relative to the form of the data which is processed. *Analog computers* process data that occurs in a continuous flow such as temperature, voltage, and pressure. The output from analog computers are often displayed on a cathode ray tube or plotting devices to monitor the results. Analog computers are frequently found in engineering and scientific applications where the form of input data occurs in a continuous signal. Common forms of analog devices you are probably familiar with are the slide rule, the gas gauge in a car, and the pressure gauge on a pressure cooker, since these are all measuring devices.

Digital computers operate on discrete (individual) quantities represented by numbers and letters. Digital computers are useful in business applications where separate quantities such as an account number, payment, receipt, etc., are processed. The scientific, engineering, and mathematical fields use digital computers for applications that involve numerical manipulations. The hand calculator is an excellent example of a digital computer on a very small scale.

We shall devote our attention in this text exclusively to general-purpose, digital computers.

A general-purpose, digital computer manipulates information under the direction of a written program stored in its memory unit. The *stored program* is simply a set of instructions provided by a *programmer* (one who writes the program), and directs the computer in all its operations. Instructions in the program tell the computer when to read data, when to perform required computations and logical operations, and when to display the output results. The programs to direct the computer in its operations and in the solving of application problems are also commonly referred to as *software.*

A program must be written for each problem to be solved by the computer. The information that the computer uses in the solution is comprised of all the data items required in the problem. The data normally consists of those related items pertaining to the activity, as, for example, student registration and grades, payroll for a company, parts in an inventory control system, or customer accounts in an accounts receivable system.

We shall now turn our attention to the activities that are required to enable us to interact with the computer. We shall review the steps one normally goes through in solving a problem on the computer.

1.4 How a Person Solves a Problem with a Computer

One may imagine that solving a problem on the computer is a very simple process. This is often true, especially with simple problems. But the work can be time-consuming and complicated. An orderly sequence of steps is followed to reach the desired problem solution.

There are five main steps normally followed to produce a working computer program and supporting documentation. These five steps are described in both general and programming terms below:

Step	General Terms	Programming Terms
1.	Analyzing and defining the problem	Problem analysis and preparing job specifications
2.	Planning the problem solution	Program design and flowcharting
3.	Describing the problem solution	Writing (coding) the program
4.	Implementing the problem solution	Program conversion to an input medium. Program testing. Program implementation
5.	Documenting the problem solution	Program documentation

Each of these steps will be discussed sequentially in the five following sections in order to answer the various questions surrounding the mystery of how one goes about solving a problem with the computer.

1.4.1 Analyzing and Defining the Problem

The first step in a problem solution is to recognize the problem and to define the various aspects surrounding it. The problem must be thoroughly analyzed and the best methods selected for the solution. If the problem is not a recurring one, the cost of solution by the computer may be far more expensive than if it is solved manually. Because of its speed, accuracy, and economy, however, the computer may be by far the best tool for solving the problem. We will assume that it is.

As with most problems the *scientific approach* to problem solving is recommended for properly analyzing the problem and for developing a solution. The steps normally followed are:

    a) Define the problem requirements
    b) Identify the variables (items)
    c) Identify the relationships between the variables
    d) Develop a model or formula using the variables
    e) Compute the solution
    f) Check the results

In the first step (a) we analyze the various factors concerning the given problem and determine what needs to be done. Your instructor has in many ways defined the problem given to the students. But the student must study the problem to understand what all is involved and needed. When the student achieves the proper understanding of the problem requirements he can make rapid progress with the other steps. If a proper understanding of the problem has not been achieved, one will flounder in the other steps and become highly frustrated. Time spent in adequately analyzing a problem usually saves valuable time in the long run.

Once the problem has been defined, the variables involved should become known, step (b). We need to know the variables we must have to begin work, those variables we must calculate, and finally those variables we need to develop as output values. In programming we need to know the form (whole numbers, fractional numbers, or alphabetic data) of each of the input items and their location on an input record. Likewise, we must define the form of the variables to be output and their location on each output record. (Note: A punch card is one input record and a printed line is one output record.) To sum it up, we can develop our calculations and decision-making requirements if we know our output requirements and input values.

In step (c) we need to know the relationship involved between the variables. Certain variables might be used to compute new variables. By knowing the relationship of the variables we can proceed to develop our model for our solution.

In step (d) we develop a model involving the variables. The model may simply be a single formula to compute our solution. Or the model may be a series of computations to arrive at our answer such as the computation of an employee's pay. The relationship of the variables and the model used is often inherent in the problem requirement. If we wish to compute simple interest, our model simply becomes the formula—$i = p\ r\ t$—for calculating simple interest. ($i = p\ r\ t$ means that simple interest is computed by multiplying principal times rate times a period of time.)

In steps (e) and (f) we take some simple numbers and compute the solution according to our model. If the results are correct, we can proceed to develop our solution into a computer program. For a computer solution is developed in much the same way as if we do it by hand. If the results are not correct, we must modify our model or perhaps develop an entirely new one.

For simple problems, many of the above steps will be almost automatically performed or combined. For complex problems, the above steps provide an orderly approach to the problem requirements and solution. Our next big step then is to organize our approach into some logical description that we can use in writing our program.

1.4.2 Planning the Problem Solution

After analyzing the problem specifications, the programmer carefully develops a plan to follow in the program development. *Program logic* involves the actions and decisions included in the solution; it also includes performing the necessary operations at the proper time. A program logic chart, better known as a *program flowchart,* is developed by the programmer to illustrate the various algorithms needed to perform the required actions. An *algorithm* is simply a series of operations followed to accomplish a task. For example, a cake recipe is an algorithm for preparing a cake. The recipe contains all the steps required to bake a cake. Also note that the steps in the recipe must be performed in the correct sequence in order to produce a properly baked cake.

The *flowchart* is a pictorial representation of the processes and operations that take place in the computer program. One might do as well by writing a narrative outline of the steps involved in program logic; but, as the saying goes, one picture is worth a thousand words. By using different symbols to represent various logical operations, the person reviewing the flowchart can more quickly and easily determine the function of the program (see fig. 1.3).

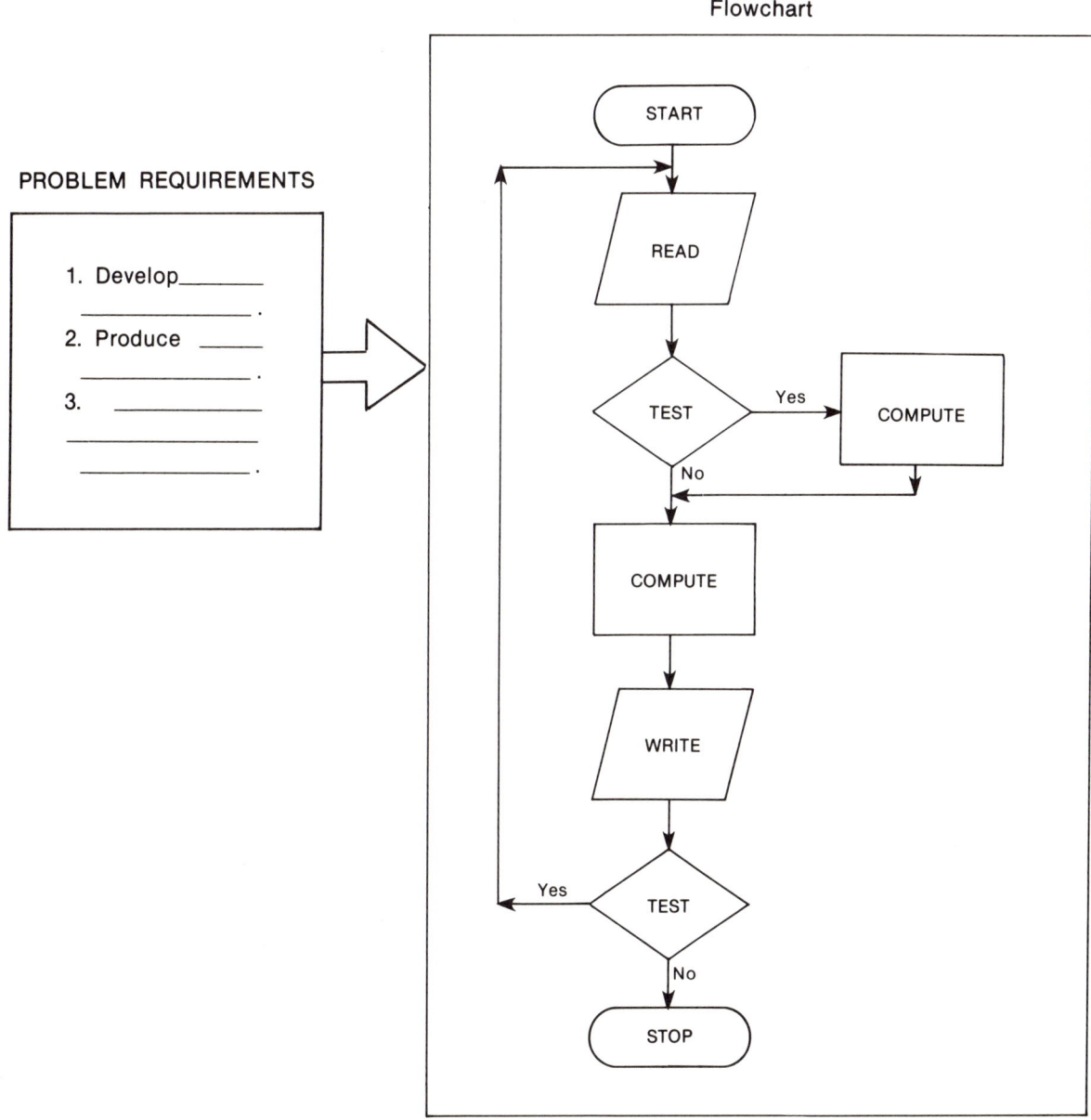

Figure 1.3. Developing a Flowchart to Depict the Algorithm Used for the Problem Solution

8   Problem Solving with the Computer

The program flowchart is a very important tool in programming. It is used to plan the logic or sequence of operations in a computer program. Just as a carpenter uses a blueprint to build a house, the programmer constructs a flowchart to ensure that all operations are considered and performed in the correct sequence. If the logic of the problem is very simple, a flowchart may not be needed. For most problems, however, a flowchart will prove invaluable.

The program flowchart consists of various symbols to show the operations performed by the computer. These symbols are used with explanations within the diagrams to depict the reading of data, arithmetic calculations, logical operations, decisions, and the writing of the output results. The order in which these symbols appear relates to the sequence of instructions in the program. Thus, a flowchart illustrates the logical order of computer operations to be performed by a program.

Program Flowcharting Symbols

Seven standard symbols with which one may construct program flowcharts are presented. Standard symbols are recommended so that all programmers will recognize the type of operation indicated by each symbol and will have a better understanding of the documentation.

Various symbols are chosen to represent different computer operations. The shapes of the symbols are important since the shape depicts the type of the desired operation. Size of the symbols has no significance. Short messages are written inside each symbol to provide further meaning concerning the operation. Such information might include the names of the input/output files, what data is tested for decisions, the arithmetic calculation to be computed, and so on. If a message is too long to fit within a symbol, don't panic. Draw a larger symbol or have a part of the message extend outside the symbol. An annotation symbol also is available for the insertion of long messages. Each symbol is connected by a line with an arrow to show the proper direction of flow in the logic.

The seven standard symbols to be used in constructing basic program flowcharts follow:

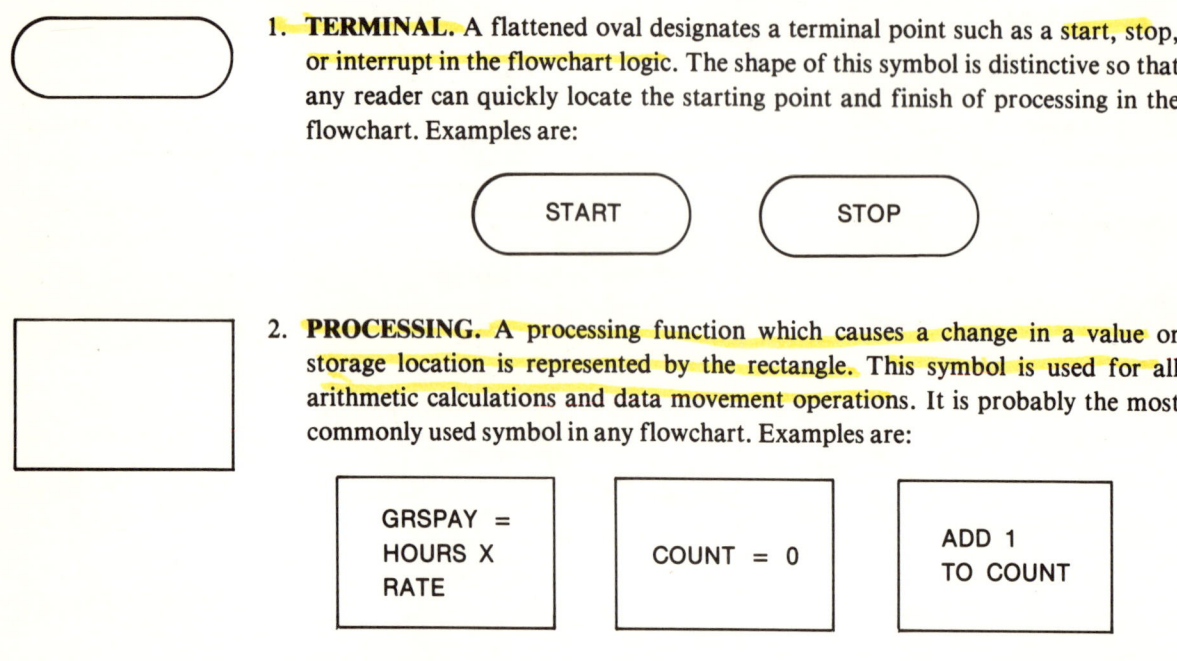

1. **TERMINAL.** A flattened oval designates a terminal point such as a start, stop, or interrupt in the flowchart logic. The shape of this symbol is distinctive so that any reader can quickly locate the starting point and finish of processing in the flowchart. Examples are:

2. **PROCESSING.** A processing function which causes a change in a value or storage location is represented by the rectangle. This symbol is used for all arithmetic calculations and data movement operations. It is probably the most commonly used symbol in any flowchart. Examples are:

3. **INPUT/OUTPUT.** The parallelogram depicts a general input/output (I/O) function. All I/O operations such as the reading and writing of data are represented by this symbol. Examples are:

Problem Solving with the Computer   9

Some programmers prefer to use the system symbols to indicate the specific form of input or output medium being used. For example:

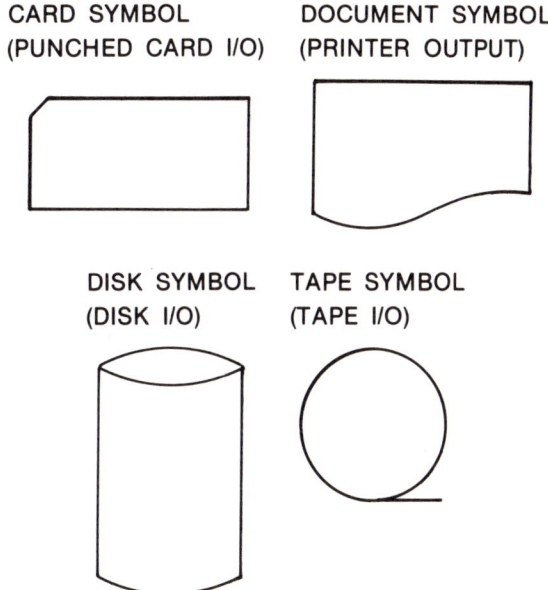

These four are not standard program flowcharting symbols. Your instructor may, however, require you to use them since they are widely used and show specifically what input or output medium is used.

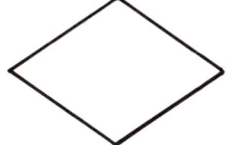 4. **DECISION.** The diamond illustrates a decision function that is concerned with alternate routes in a program. The decision operation is based on a test that determines which alternative path of logic to follow. The question to be answered (i.e., the test) is usually written inside the symbol. The possible answers are shown as alternative flowlines proceeding from the diamond. Three examples are:

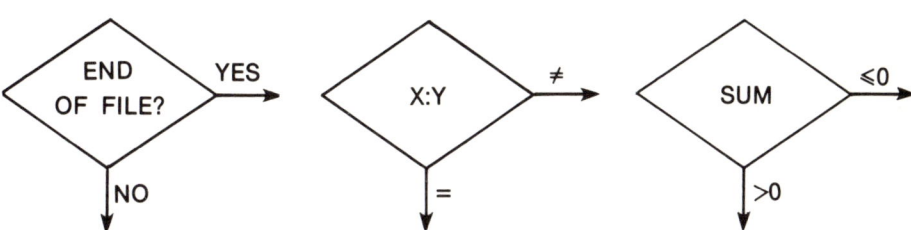

Each exit must be clearly marked to identify the results of the condition for each path. The entire condition may be stated inside the diamond with the results stated on each exit. Or part of the condition may be stated inside the symbol with the remainder given at each exit. The results may be given for each exit in abbreviated form. For example, Y and N may be used for yes and no, respectively. Mathematical operators also may be used. These include the $=, \neq, <, \leq, >,$ and $\geq$ symbols.

10   Problem Solving with the Computer

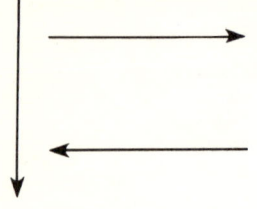

5. **FLOWLINES.** Lines with arrowheads link the symbols together. They also show the sequence of operations and direction of data flow. The normal sequence of logic is read from the top of the page to the bottom and from left to right. An example of connecting two symbols is:

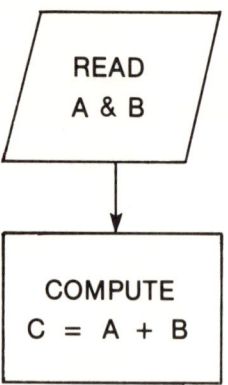

6. **LOGIC CONNECTOR.** A small circle is used as an exit to or entry from another part of the flowchart to indicate the continuation in the flow of logic. This symbol is used: (1) to flow the logic to another page to which it is impossible to draw a flowline—the symbol should then be divided in half and the page reference included; (2) to avoid drawing excessively long flowlines (for example, from the bottom of a page to the top of the page); and (3) to keep flowlines from crossing. Put a letter in the exit connector and its corresponding entry connector. One normally starts with the letter A and proceeds through the alphabet. Use double letters if the single letters have been used up. Some programmers like to use numbers in place of letters. (In fact, some prefer to put their statement number in the symbol.) Examples are:

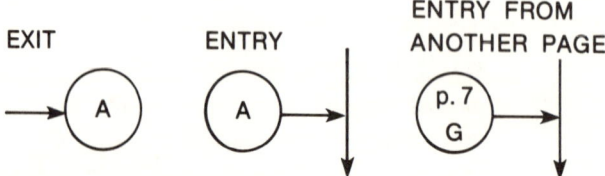

A flowline with an arrow pointing to the connector means an *exit* is to be made from the logic at this point and a jump made to the location with the matching entry symbol. The *entry symbol* has an arrow pointing away from the connector and into a flowline.

Some programmers prefer to use the *off-page connector* when the flow of logic continues on a different page. Examples of the off-page connector are:

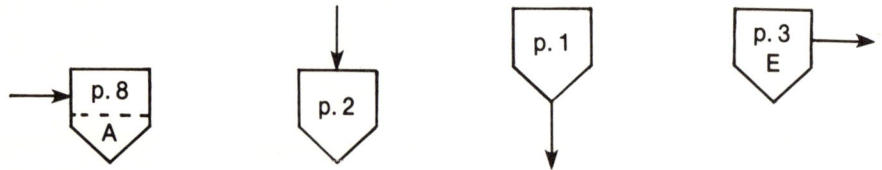

where the page number is included in the symbol. This symbol is especially useful when we wish to continue the logic onto the next page. Here again, we have a symbol *not* recognized as a standard flowcharting symbol. It is widely used, however, and may be required by your instructor.

Problem Solving with the Computer   11

7. **ANNOTATION.** The three-sided rectangle with a connecting broken line is used to add comments in a flowchart. This symbol allows one to include further explanations to an operation. The broken line links to the other symbol to which the annotation refers. For example:

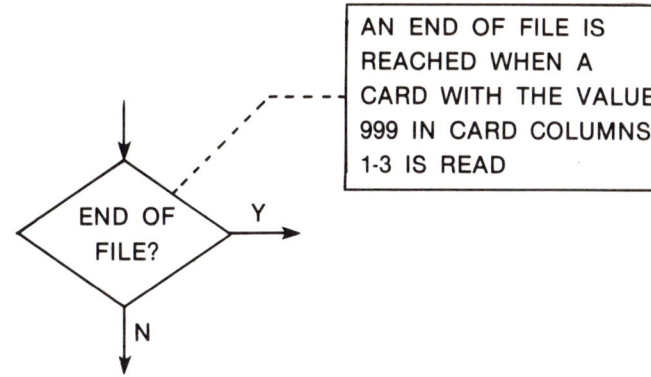

Two additional symbols will be introduced in later chapters when we need to use them in the more advanced concepts. Let us first review the rules on how to put the symbols together to form a flowchart before illustrating a complete flowchart. Remember, the sequence of the symbols in our flowchart correspond to the order of our steps in the problem solution.

Rules for Constructing Program Flowcharts

The rules to be followed when constructing program flowcharts are:

Rule 1: Use a standard set of symbols. This improves readability and interpretation of flowcharts.
Rule 2: The title of the flowchart (program name) should appear on every page. Each page should be numbered sequentially.
Rule 3: The steps constituting a flowchart should start at the top of the page and flow down and towards the right. Several columns of the flowchart logic may be drawn on each page to conserve space.
Rule 4: The messages used to describe each step should be in common English terms insofar as possible, and not in "machine language."
Rule 5: In each step, all the writing should be clearly written within the symbol (to the extent possible). But do not use too many abbreviations. Put yourself in the place of a person not informed of the problem—would he understand it? Use the annotation symbol, if needed.
Rule 6: Symbols are connected by flowlines with arrows to show the lines of logical flow.
Rule 7: Input/output and process symbols should have exactly one entry line and one exit line.
Rule 8: Decision symbols will have one entry line and two or more exit lines.
Rule 9: Do not cross over flowlines if avoidable. Use connectors. If a line must be crossed, one line should be "half-mooned" over the other to show that no intersection is intended. For example:

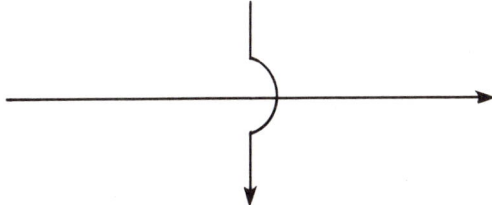

Most programmers do not like to see flowlines of this kind. Thus, connectors are preferable by far.

Rule 10: Long flowlines (normally over four inches) should have embedded arrows to clearly delineate the flow of logic. For example:

$$\longleftarrow \longleftarrow \longleftarrow \longleftarrow$$

Again, this may be disturbing to some; therefore, use connectors.

Rule 11: Normally separate symbols by at least one-half inch.
Rule 12: Allow at least one-inch margin on the left margin of the page in case the pages are bound.

I am aware that these rules are probably too much at this time for you to fully understand, since you have not yet drawn your first flowchart. Therefore, it is suggested that you digest those you can now and then come back to them after you have had some practice in drawing flowcharts.

One may wonder how much detail to put in a flowchart. It depends upon the intended use. Generally, there are three recognized levels of flowcharting: a *general level, semidetailed,* and *detailed.* The *general level* provides an overview to what the problem requires. That is, only the main tasks are identified and in the proper sequence. This is a summary level approach that you might give to management to let them know what the problem consists of. They are not interested in details so don't get wrapped up in them at this point. For example, a general level flowchart to compute one's pay might have only three identified operations: (1) compute gross pay, (2) compute deductions, and (3) compute net pay.

The second and third levels of flowcharting are those used by programmers to write their programs. The level used depends upon the complexity of a routine (task) in the problem. When in doubt as to how to accomplish a routine, go into more detail. Too much detail is better than not enough.

It is heartily recommended that your first flowchart not contain any detail, but show only the main tasks needed in the program. In that way, you will have a better idea of what the problem looks like and not get all bogged down in details.

To sum it up, a flowchart shows *how* to solve a problem. It should be complete, clear, accurate, and neat. Someone who has no knowledge of the problem being solved should be able to prepare a working computer program from the flowchart.

Several Flowchart Examples

We will consider two problems to illustrate flowcharting. Our first problem is simply to read a card record with three values representing principal, rate, and time (in years). The three values will be used to compute simple interest according to the formula $i = p\ r\ t$. We then wish to print the three input values and computed interest. Our flowchart might look like figure 1.4.

Our second problem is the calculation for a group of students' grades. Our first operation is to print some column headings to identify our columns of output values. A student's identification number and five test scores will next be read off a punched card. We wish to compute the grade average by adding the five test scores and dividing the sum by five. The input values along with the sum and average will be printed for each student. We want to do these operations for every student's record, so we must loop back to read the next card until all records have been read. When an end-of-file condition (no more data cards are present) is reached, program execution is to be terminated. Figure 1.5 illustrates the flowcharted problem.

Learning to draw flowcharts which represent the solution logic is a difficult hurdle for the beginning programming student to overcome. Perfection comes from practice and experience. All too many students draw his or her flowchart after the computer program has been written and the correct results obtained. Considerable time and frustration can be saved if a student gets into the habit of developing a "rough" flowchart to represent the logic operations required in the problem solution before coding the program. Once a flowchart has been developed, the writing of the computer program becomes much easier.

1.4.3 Describing the Problem Solution

After the program flowchart has been completed, the next step is to write the computer program. The term *program* should not be a new one to us. When we attend a play, we are given a program which outlines for

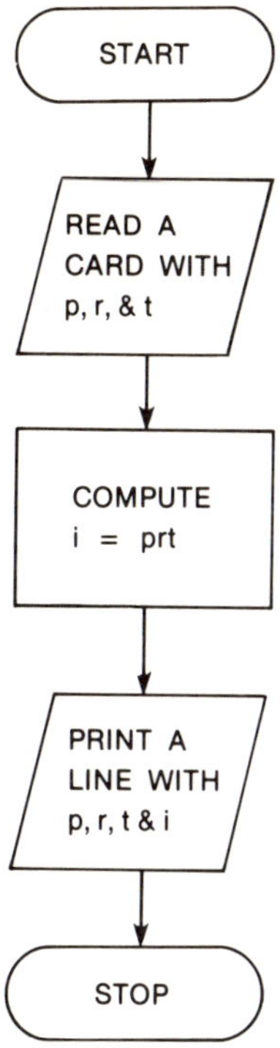

Figure 1.4. Example of a Flowchart to Compute Simple Interest

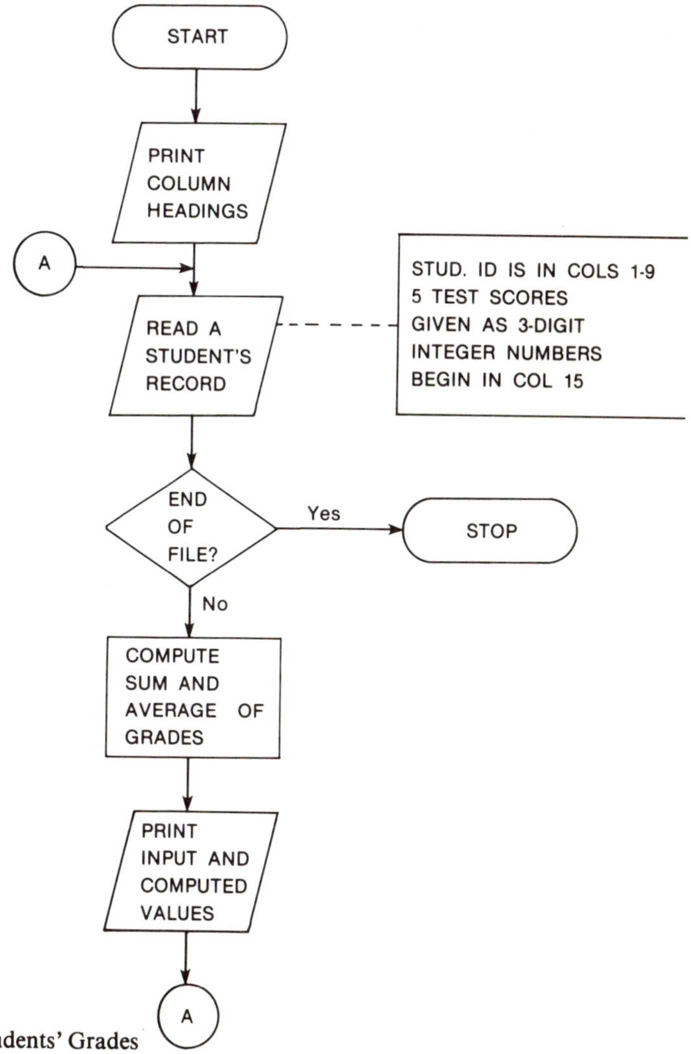

Figure 1.5. Flowchart to Compute a Group of Students' Grades

us the series of events that are to take place. Similarly, the computer program specifies the series of operations that must be accomplished in the problem solution.

Following the symbols illustrated in the flowchart, the programmer *codes* (writes) the instructions according to the rules of the programming language that correspond to each symbol (see fig. 1.6). In writing the program the programmer must know the record formats of both the input and output items. A *record format* specifies what items are read or written, their location on the record, and what form they will take. The instructions in the program are normally coded on specially designed coding forms to aid in the later transformation to an input medium.

Coding the Program from a Flowchart

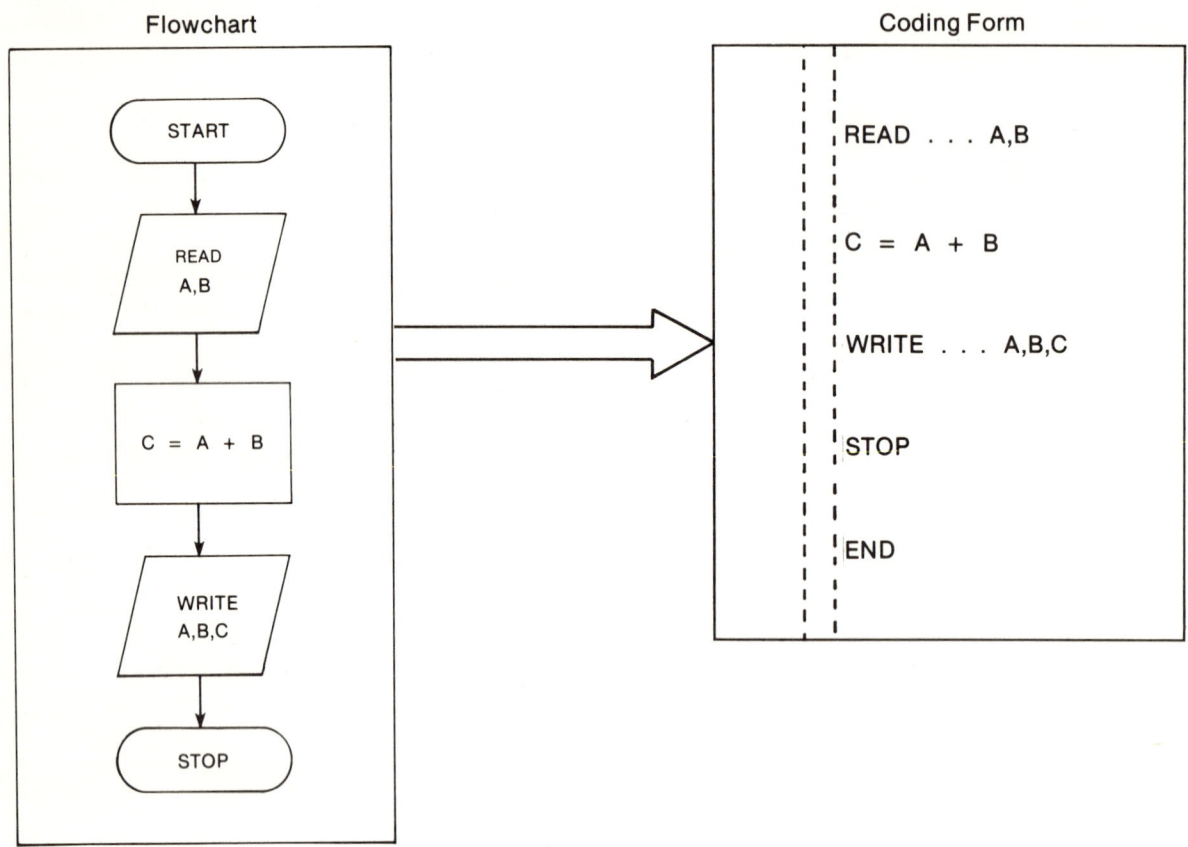

Figure 1.6. Coding a Source Program from the Flowchart

The computer instructions written by the programmer relate to the particular computer language used. For many years people wrote instructions for the computer in a *machine language*—the actual language the computer understands. Commands and memory addresses were all specified in numeric digits. Writing programs in machine language proved very time-consuming and lacked the flexibility needed for correcting program errors. Furthermore, a different machine language had to be learned for each type of computer. Thus, high-level, symbolic languages (such as FORTRAN) were developed to shorten the time required to write programs and to provide greater flexibility.

Rarely is a machine language used today to code programs. The high-level, symbolic languages are normally used. These languages are most widely used inasmuch as they are more specifically designed to deal with business, scientific, engineering, and other application problems. Two of the most common high-level languages used are **FOR**mula **TRAN**slation (FORTRAN) for science- and engineering-oriented problems, and **CO**mmon **B**usiness **O**riented **L**anguage (COBOL) for business-oriented applications.

Programs written in a high-level language have to be translated into the needed machine language required by the computer. A program called a *compiler* (a language translator) is written for the high-level languages (such as FORTRAN and COBOL) to convert the respective *source language* into the computer's *machine language.* This process of converting the source language written by a programmer into machine language is called *compilation* or *compiling*. Compilation of a source program written in a high-level, compiler language is all transparent to the programmer. The programmer merely tells the computer what compiler language (such as FORTRAN) the source program is written in, and the program is converted into machine language instructions.

After we have written our program onto coding sheets, we are ready to proceed to our next step. Our coded source program is now ready to be converted to some input medium to be run on the computer.

1.4.4 Implementing the Problem Solution

Converting a Source Program to an Input Medium

Our coded program is next typed onto punch cards so that they may be read by the computer. Each line on the coding form is transferred to a separate punch card (see fig. 1.7). The resulting deck of cards is commonly referred to as a *program deck* or a *source deck*. The source deck is now ready to be combined with a few computer *job control cards* (to specify to the computer what devices and resources will be used by the job) and the data cards to be processed by the computer. The data cards are placed behind the program cards to be read as called for in the program.

Converting the Coded Program to Punched Cards

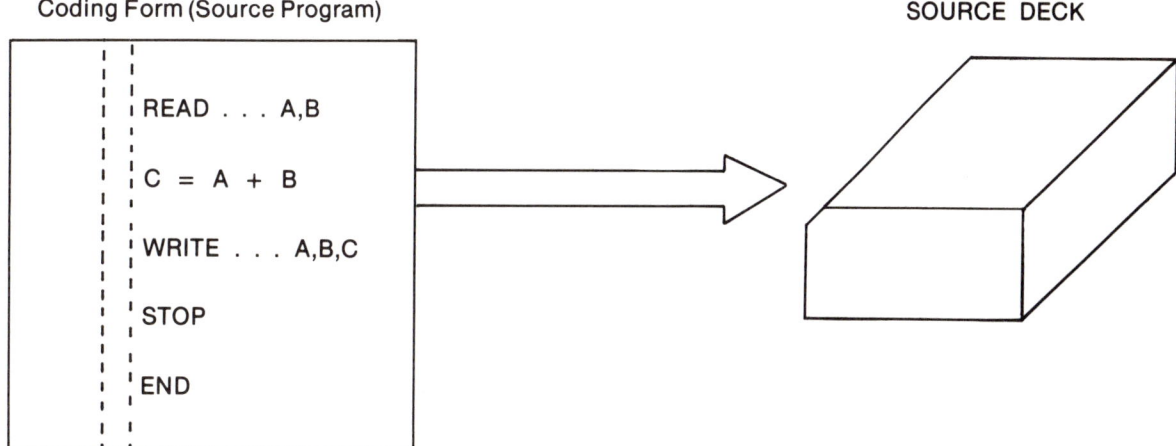

Figure 1.7. Converting a Coded Program onto Punched Cards Input Medium

Program Testing

After the coded program has been transferred to an input medium, it is ready to be read into the computer (see fig 1.8). The appropriate compiler is called in by the job control language cards that specify the needed computer resources. To convert a FORTRAN program to machine language, a *FORTRAN compiler* is used. The compiler reads the source program, one statement at a time, and generates the proper machine language instructions from the source statements.

A computer language requires that certain rules and specifications be followed in writing a program. Each instruction must be written in a predetermined form and structure. If an instruction coded by the programmer does not meet all the language rules, a *syntax* (construction) *error* will occur. A diagnostic message will be produced by the compiler to describe the error in the instruction. The programmer uses these compilation error messages to tell him what mistakes were made so that they can be corrected.

The compiler reads all the instructions in the program. A printer listing of the source program, along with any compilation error messages, will be produced by the computer. If an error is encountered in program compilation, no machine language code is generated and no program execution can occur.

Source Program Compilation and Execution

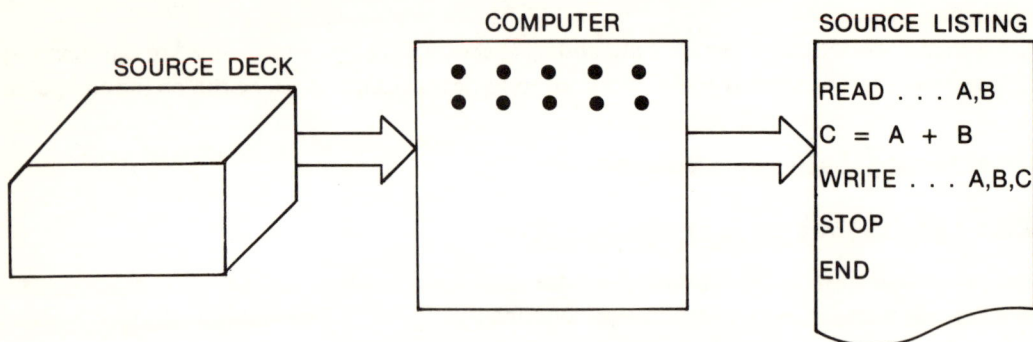

Figure 1.8. Running a Source Program on the Computer

The programmer must correct all the compilation errors and resubmit the program for another compilation. This process continues until all compilation errors are corrected. Whenever a "clean compile" (no compile errors) is accomplished, the machine language code to solve the problem is produced. This resulting code is known as *object code* or an *object program* which can now be loaded for execution.

Human errors in programming occur quite frequently and increase in number with increased size and complexity of a program. It is indeed a rare event for a program to compile and produce the correct output results on the first attempt. Just because a "clean compile" has been achieved does not mean there are no errors in the program. All syntax errors may have been removed, but logical errors may exist in the program nevertheless. So the program must next be run with the data cards to test the correctness of the output results. *Logical errors* include such errors as performing the wrong operation at some point in the program, or omitting a needed instruction. Logical errors normally result from the programmer's failure to properly understand the problem specifications, or from an oversight in the required logic. Most logical errors can be avoided if you spend a bit more time in the planning phase. A little extra time spent in the careful analysis and design of your solution will save more time later on, in the testing phase. After the flowchart has been drawn, step through the entire logic with some simple pieces of data. If all seems to work without any hitches, then proceed to run the solution on the computer.

It is during the *testing and debugging phase* that all known errors in a program are found and eliminated. *Testing* is the process of executing the program using the input data to verify the output results. *Debugging* is the process of locating and correcting any program errors which occur. The testing and debugging step may be a very lengthy and time-consuming phase. Generally, the larger the program, the longer period of time required to debug the errors. Valuable time and money may be saved as a result of proper problem analysis, accurate program development, and systematic debugging throughout the program testing.

A few output answers are compared to manual calculations answers to verify correctness (see fig 1.9). Several answers that take different legs or routes in the program logic should be checked to make sure that all parts of the program are working. Simple test data may be used in checking out the program if many complex calculations are made with decimal numbers.

If any discrepancies are found, the programmer must locate and correct the error. A *patch* (change) is made to the program. It is recompiled and tested to see if all *bugs* (logic errors) have been removed. This cycle of testing and debugging a program recurs until the program produces correct output results.

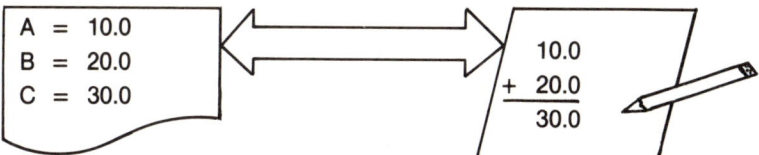

Figure 1.9. Checking the Output Results for Correctness

After a program is executed (processed) by the computer, it is no longer available for another execution. That is, the computer keeps track of all the programs in memory or in the input queue (holding area) awaiting execution. When the computer finishes with a program, it is removed from the list of programs awaiting execution. In order for you to reexecute your program, it must be read in again from your cards.

Program Implementation

Once all known errors in the program have been removed, the program is ready for a production run. In a production run "live" (actual) data is used with the program. The output results are carefully checked to ensure their validity. If the output products are correct, the program will be scheduled and run as a normal production job. If the output results are not correct, program testing will be resumed and continued until all corrections have been made.

When the program has been accepted for production processing, the object code is stored on magnetic tape or disk. In some cases, the object code may be punched into cards, but this is rarely done anymore. That is, the source program is not recompiled during each production run. There is no need to recompile the source program and produce a new object program unless some change is made to the program.

1.4.5 Documenting the Problem Solution

The activity of program documentation is an important consideration in all steps of developing a working program. Even though listed last, program documentation should be a part of, and accomplished within each of the preceding four steps—not put off until the end. Program documentation is a must for production programming environments where the turnover in the programming staff is a continual process. Programmers are promoted, fired, or simply switched to other projects. Program documentation provides the continuity in the *maintenance* (upkeep) of operational programs when the original programmer leaves.

Program documentation consists of all the paperwork which explains how the program works. This paperwork allows one unfamiliar with the program to make changes in it. These changes are called *program modifications* and may be corrections or even new features added to the program.

The following items are normally included as program documentation to provide an explanation to the program function and operation.

1. Problem *specifications* (requirements)
2. Input records description and layout
3. Output records description and layout
4. Flowchart
5. Source program listing
6. Sample test runs with output results
7. Job control cards required to run the program
8. Narrative descriptions of complex algorithms used within the program

Your instructor may require many of the above items to be turned in with your computer program solution to each assigned problem.

A second type of program documentation available is one that provides invaluable aid in following program logic. *Internal program documentation* is the insertion of comments in the source program itself.

Nearly all programming languages allow the programmer to include remarks which explain what function an instruction or group of instructions perform in the program. You should make liberal use of internal comments in your programs to provide yourself and your instructor with a clearer explanation of your program logic.

The student should realize that the process of developing a computer program to solve a given problem may be a very time-consuming and lengthy process. Most assigned problems will require days for the development of a working computer program. The student is, therefore, encouraged to start working on a problem immediately instead of waiting until the last day to begin.

The final section in this chapter briefly discusses a revolutionary concept in problem solving with the computer. The structured approach to programming has proven how valuable it can be for saving numerous manhours and dollars in programming projects. This approach is widely becoming the modern-day method to problem solving with the computer.

1.5 The Structured Programming Approach to Problem Solving

The *structured programming* approach to problem solving with the computer is concerned with the way one goes about writing the program to solve a problem. All too often we encounter a program containing a mass of instructions that is almost impossible to understand and to follow. The objective of structured programming is to provide a structured and orderly approach to writing a program such that the resulting instructions provide a clear, concise approach to the problem solution. The purpose of this section is to provide a general overview on this subject. The technical aspects of structured programming in FORTRAN will be covered in a later chapter.

A good approach to defining the aspects of a problem and programming is the *top-down approach*, which means that one starts with the general logic and works down to the details. Do not become involved in getting to the "nitty-gritty" calculations and decision making right away. Get the big picture! Find out the main tasks to be achieved before you start writing the program. So many programmers today jump into the middle of the problem and start writing the program before they understand the control logic involved. They are so anxious to get started in coding some of the specific routines that they fail to see how all routines will fit together.

The traditional approach to programming has been the *bottom-up approach.* Using this approach, a programmer starts by writing one of the detailed routines in a program to achieve a required task. Then he writes another one of the required routines and ties it in with the previous one. This procedure is continued until the overall problem is solved. The resulting program was often a mass of confusion to anyone other than the original programmer. Numerous transfer instructions made the program very difficult to follow due to the fact that one had to jump about so much in the program to follow the solution logic.

One way to make programs easier to write and understand is to reduce the number of branch/transfer instructions in a program. One should strive to have as few branches as possible in his program, so that the logic may be followed in a simple sequential (top down) fashion. This orderly approach provides a clear sequence of logic and control in a program and prevents one from having to follow so many branch statements that he soon loses his understanding of what is happening in the logic. Some programming languages provide commands by means of which it is possible to write complex programs without a single branch statement. This is rarely the case in the FORTRAN language because of the commands available. Nevertheless, students should strive to keep the number of branch statements to a minimum to provide a clearer path of logic in program development.

The most important aspect of structured programming is this top-down approach. Approach the problem from this standpoint: what are the main tasks to achieve? Then divide each of these main tasks into subtasks. Divide the subtasks into smaller parts if all of the logic is not yet apparent. These divisions of logic into subtasks can be regarded as levels of control. Each of the tasks and subtasks should be logically independent (not interconnected) of the other tasks and subtasks at the same level. Let us take the ordinary telephone call and break it up into the different control tasks to illustrate this concept. Figure 1.10 illustrates three levels at which we can view a problem/task.

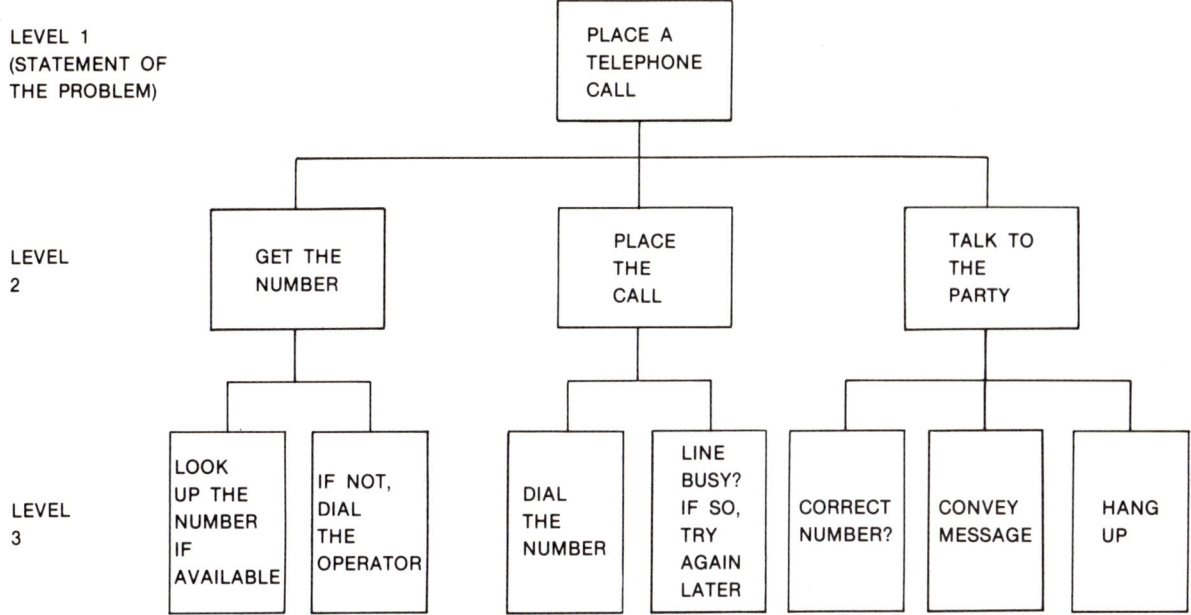

Figure 1.10. An Example Showing the Different Levels of Problem Definition

The idea behind structured programming (as illustrated in figure 1.10) is, then, to view the problem at different levels of definition. Each lower level breaks out the tasks into more finite steps. If the problem is approached in a left-to-right order of defined tasks, we can provide a highly organized sequence of control logic to the problem solution. If a problem is defined in this manner, we will usually have a program more clearly written, in the shortest time, and with fewer errors. What more valuable technique could a student learn to assist him in solving his problem on the computer!

Some of you may ask why you are taking this course. Why do *you* need to learn about computers? The computer is, in many respects, the most important development in the twentieth century. Much of our way of life today depends upon computers. So, no matter what your field of specialization, you will be affected by the computer.

The computer is a tool to help you. It can mean the difference between success and failure in many circumstances. Even if you do not plan to ever instruct the computer again after this course, you should gain an appreciation for what the computer can do for you and mankind. The computer can extend your productive capabilities far beyond what you have ever imagined—if you will only learn to use it. Learn to use it well and it can become one of your most important *tools* in solving problems. You are encouraged to strive to get the most you can from this course on programming. Who knows, this may be one of the most important courses you take in your college studies! Good luck and happy programming!

1.6 REVIEW QUESTIONS

1. How do people feel and react towards computers?
2. What are two limitations of computers?
3. List three main advantages of computers.
4. List three application areas or uses of computers.
5. Name the five basic components of a computer.
6. What is the most common type of punch card used today?
7. What is the purpose of output devices?
8. Name five output devices.
9. What is the maximum number of characters that can be printed on a print line by a line printer?

10. Define a storage address.
11. What is the purpose of the arithmetic-logic unit?
12. The central processing unit consists of what three units?
13. What are the two basic functions of the control unit?
14. What is the stored program concept and why is it so important?
15. Do programs remain in the computer memory after they have been executed?
16. List the five steps followed in solving a problem with a computer.
17. Explain the various steps followed in analyzing a problem with the scientific approach.
18. What is the importance of flowcharting a problem before attempting to write the program?
19. Why is the shape of a flowcharting symbol so important?
20. List the seven standard programming symbols used in flowcharting and their shapes.
21. A set of instructions to direct the computer in its operations is called a _____ .
22. A _____ is used to convert a source program written in a high-level language to machine code.
23. Name two compiler languages and give their main area of use.
24. _____ _____ cards are used to tell the computer what resources the job needs and how to process the job.
25. Explain the process of testing and debugging a computer program.
26. What happens to a stored program after its execution is finished?
27. What is the main idea in structured programming?
28. Draw a structured approach down to three levels to depict the steps involved in computing a person's pay check. Hint: Show gross pay, deductions, and net pay at level two.

1.7 FLOWCHARTING EXERCISES

1. Draw a program flowchart to read a single data card with the values, RATE and TIME; compute DISTANCE according to the formula d = r t; and print the three values on a printer output line.
2. Draw a flowchart to read three records with the values of A, B, and C, respectively. Use these three values to solve the equation: Y = A + B − C. Print the three input values and the computed value of Y.
3. Construct a program flowchart to read an inventory card file containing the value of unit-cost along with other related values; prepare a printer listing of all input records; accumulate the overall unit-cost of all items; and print the overall unit-cost for all items after listing the file.
4. Develop a program flowchart to read ten values, each on a different record, and compute their average value. (Note: An average is found by dividing the total of the ten values by 10.) Print the ten values, their sum and average on different lines.
5. Draw a flowchart to read three values off a card record, determine which value is the largest of the three, and print the largest value.
6. Draw a flowchart to compute the square, square root, cube, and cube root of the integers from 1 through 10. Print each integer and its related computed values on double-spaced lines. Print appropriate column headings for each column of values.
7. Draw a flowchart to solve the quadratic equation: $ax^2 + c = 0$. Read the values of A and C. Use these values to solve the equation according to the solution of X = ± $\sqrt{-c/a}$. If the computed value of −c/a is negative, write an error message to

the effect that an imaginary root has been calculated. Otherwise, print the positive and negative roots.

8. Draw a flowchart to read and compute the sum of fifty numbers. Print each value as it is read. Print the accumulated sum of the fifty numbers at the end of the program. The student should establish a counter to keep count of the number of data records read.

9. Draw a flowchart to read the values X, Y, and Z, respectively, off three input records. Print each input value. Add X and Y. If the results are negative, return to read another set of values. If the results are zero or positive, add Z; print the sum of the three values; and return to read another set of values.

10. Draw a flowchart to read the input values of A and B from a single data card and solve the equation: C = 2A + 3B. Write the input values and the computed value of C. Make the program repetitive to handle multiple sets of input records.

Introduction to the FORTRAN IV Language

2

2.1 Introduction to the History and Purpose of the Language

FORTRAN is the name given to a high-level, problem-oriented compiler language primarily designed to solve problems that can be easily represented by mathematical formulas. The acronym, *FORTRAN*, stands for *FORmula TRANslation* or *FORmula TRANslator*.

Development of the first FORTRAN compiler was begun in 1954 under the direction of John W. Backus. Mr. Backus, who worked in programming research at IBM, was convinced that a programming language could be constructed that would enable computers to produce their own machine code and thereby reduce the errors and time required to program the problems being put on computers. During the late 1940s and early 1950s, most programs were written in a machine language that required much coding and debugging time in contrast to the high-level languages of today.

Three years and 25,000 lines of machine instructions later, the first FORTRAN compiler had been developed so that engineers could easily understand and code their own problems. Today, FORTRAN is solidly established as the language of thousands of persons who communicate with the computer. Since the original FORTRAN, many improvements have been added during the course of the FORTRAN language evolution. *FORTRAN IV,* the latest version, is a highly flexible, powerful language that provides even more options for the convenience of the programmer.

FORTRAN compilers have been written for nearly all computer systems of all major computer vendors. Many hardware vendors provide multiple FORTRAN compilers for even the same computer system. The wide use of FORTRAN has prompted the adoption of a standard FORTRAN language. One standard FORTRAN compiler is known as *ANSI* (American National Standard Institute) *FORTRAN* version X3.9-1966. The University of Waterloo in Ontario, Canada, developed an extremely fast FORTRAN "in-memory" compiler, primarily for use in academic institutions. Their first compiler was called *WATFOR* (**WAT**erloo **FOR**tran). In recent years, the University of Waterloo developed a follow-up to the WATFOR compiler and named it *WATFIV* (the one after WATFOR). The name, WATFIV, is quite fitting; it may be thought of as **WAT**erloo Fortran **IV**.

Each FORTRAN compiler is written for a specific computer and must convert the FORTRAN statements of a program into the machine language code of that machine. FORTRAN programs are, however, generally independent of the computer on which they will be run. As long as the FORTRAN statements in a program meet all the syntax requirements of a FORTRAN compiler, the program can be compiled into the object language of the specific computer used. The FORTRAN programmer, therefore, does not need a detailed knowledge of the computer system on which his programs will be run. A general knowledge of the FORTRAN language along with the proper job control statements to allow interface with the computer will be sufficient to process most FORTRAN programs. The ANSI FORTRAN language is taught in this text inasmuch as programs written under these specifications will run on virtually all modern-day computers.

Let us now look at the type of commands that make up a FORTRAN program.

2.2 FORTRAN IV Source Statements

Source statements are the instructions written by a programmer. A *FORTRAN source program* consists of a set of source statements from which the FORTRAN compiler generates machine language instructions, establishes storage locations, and supplies information needed to process the program. Each FORTRAN statement is either an executable statement or a nonexecutable statement. An *executable statement* is one that generates machine instructions which are executed by the computer. *Nonexecutable statements* are those which specify instructions to the compiler; they do not generate machine code. Thus, they are normally referred to as *specification* statements. Nonexecutable statements may establish arrays (tables of information) at compile time, specify the type of data belonging to a storage location, and indicate the end of a source program.

These source statements have been categorized according to the operations they perform. The six major categories of statements are listed to provide you with an idea of the type of operations that you can perform in a program. The first four categories contain the basic statements you will use in your first few programs. Most of the statements contained in categories five and six will not be used until you are well along in the course.

1. **Input/Output Statements (executable):** These statements are used to control input/output devices and to transfer data between memory and an input or output medium. Input/output statements include such instructions as READ and WRITE.
2. **Assignment Statement (executable):** This statement simply assigns a value to a storage location or performs calculations and assigns a new value to a storage location (variable). The Assignment statement is not identified by a keyword, but uses the **equal sign** symbol (=) following a variable. The Assignment statement is used to provide the formulas we need in the problem solution.
3. **Logical and Control Statements (executable):** These statements enable the programmer to control the order of execution of the statements in the program and to terminate program execution. Statements included in this category are the IF, GOTO, STOP, and DO.
4. **FORMAT Specification Statement (nonexecutable):** This statement is used in conjunction with formatted input/output statements and specifies the location and form in which the data appears in a data record. The FORMAT statement is identified by the keyword FORMAT.
5. **Specification Statements (nonexecutable):** These statements provide the FORTRAN compiler with information that is needed at compile time. Some statements initialize storage locations with specified values, describe the attributes of certain storage locations, establish arrays, and supply other pertinent data about the program. Keywords identifying some of the specification statements are DATA, DIMENSION, DOUBLEPRECISION, and END.
6. **Subprogram Statements:** These statements identify subprograms, specify arguments, and include control statements that allow the execution of subprograms. Nonexecutable statements include the FUNCTION and SUBROUTINE statements which identify the type of subprogram used. Executable statements include the CALL and RETURN keywords.

Following is a simple FORTRAN program which reads two values and computes their sum (fig. 2.1). The category of each source statement is denoted within parentheses.

```
      READ (5,990) A, B                    (Input statement)
  990 FORMAT (F5.2, F5.2)                  (FORMAT specification statement)
      SUM = A + B                          (Assignment statement)
      WRITE (6,81) A,B,SUM                 (Output statement)
   81 FORMAT (1X,F5.2,3X,F5.2,3X,F6.2)     (FORMAT specification statement)
      STOP                                 (Control statement)
      END                                  (Specification statement)
```

Figure 2.1. A FORTRAN Program to Sum Two Numbers

See how simple it is to write a FORTRAN program. Don't worry about the construction of each statement now. This will be explained in the next chapter. First, we want to look at the elements in the FORTRAN language and see how we use these elements to form statements.

2.3 Basic Elements of the FORTRAN IV Language

Learning to program in FORTRAN is very much like learning a foreign language. A *character set* like an alphabet is available from which to choose characters to form elements of the language. Certain *words* (called *keywords*) are used to denote the operations to be performed by the computer much like verbs we choose to denote various actions. *Identifiers* (called *variables*) are established by the programmer from the character set to denote storage locations which we wish to use in various operations. *Numbers* or *constants* are formed to represent certain values we also may need in the operations. *Mathematical operators* are available in the character set to allow us to perform mathematical computations with values stored in memory and numerical constants. Other operators may be used when comparison of data is desired. We combine these keywords, variables, numeric constants, and/or operators to form a FORTRAN statement analagous to a sentence in our conversational language. The entire collection of FORTRAN statements composes the FORTRAN program.

Consider the following FORTRAN source statements:

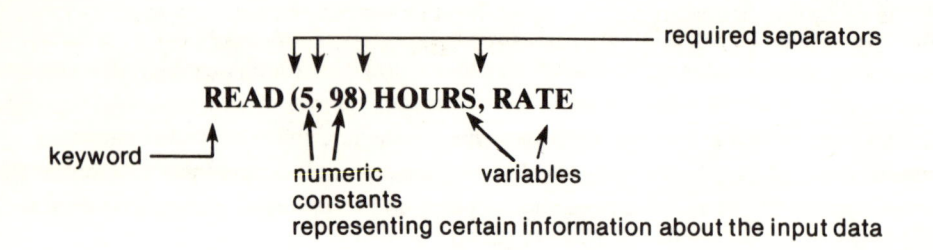

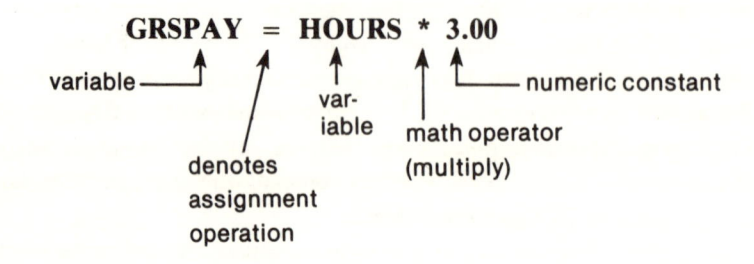

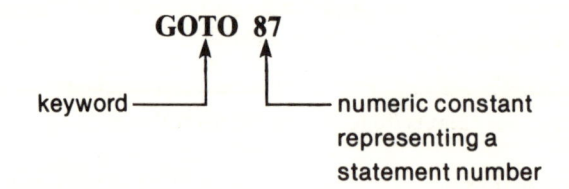

```
          READ (5, 98) HOURS, RATE
```
- keyword: READ
- required separators: (,)
- numeric constants: 5, 98
- variables: HOURS, RATE
- representing certain information about the input data

```
          GRSPAY = HOURS * 3.00
```
- variable: GRSPAY
- denotes assignment operation: =
- variable: HOURS
- math operator (multiply): *
- numeric constant: 3.00

```
          GOTO 87
```
- keyword: GOTO
- numeric constant representing a statement number: 87

The first statement allows us to read two values from a card record and transfer them to the storage locations identified by the names "HOURS" and "RATE." The second statement computes a value from the expression (HOURS times 3.00) to the right of the equal sign (=) and assigns it to the memory location identified as "GRSPAY." The third statement directs the computer to transfer control to the statement with number 87 assigned to it.

Let's take a closer look at these individual elements before forming complete FORTRAN statements.

2.4 FORTRAN IV Character Set and Keywords

Character Set

Forty-seven different characters make up the FORTRAN IV character set. There are 26 alphabetic characters, 10 numeric digits, and 11 special characters. These characters, grouped with respect to their category, are:

Alphabetic (26)	*Numeric (10)*	*Special (11)*
A thru Z	0 thru 9	. , ( ) = + − * / $ ∅
		(where ∅ represents a blank)

The alphabetic letters are used primarily in forming keywords and variables. The numeric digits are used in forming numeric constants and may also be used in variable names. The special characters are used as mathematical operators and various delimiters (separators) as required by certain FORTRAN statements.

Keywords

Keywords denote what operation is to be performed by the computer with respect to the FORTRAN statement. With the exception of the arithmetic Assignment statement, the keyword appears as the **first part** of a FORTRAN statement. The various keywords available in ANSI FORTRAN IV are presented in figure 2.2.

ASSIGN	END	PAUSE
	ENDFILE	
BACKSPACE	EQUIVALENCE	READ
BLOCKDATA	EXTERNAL	REAL
		RETURN
CALL	FORMAT	REWIND
COMMON	FUNCTION	
COMPLEX		STOP
CONTINUE	GOTO	SUBROUTINE
	IF	WRITE
DATA	INTEGER	
DIMENSION		
DO	LOGICAL	
DOUBLEPRECISION		

Figure 2.2. ANSI FORTRAN IV Keywords

A keyword must be spelled exactly as it is given in figure 2.2. Imbedded blanks do not matter. In fact, the keywords GOTO, BLOCKDATA, DOUBLEPRECISION, and ENDFILE are often coded as two words (e.g., GO TO, BLOCK DATA, etc.).

2.5 Numeric Constants in FORTRAN IV

A *numeric constant* is a known numeric value that always represents a fixed quantity in a source program. Therefore, numeric constants represent explicit numbers which the programmer wishes to use in his program. A numeric constant is composed of one or more decimal digits and may contain a decimal point and a sign. The absence of a decimal point implies a whole number; whereas an included decimal point implies a real (fractional) value. The constant may represent a positive or negative value depending upon an attached sign. If the sign is a minus sign, then the constant is considered to be negative. A plus sign or the absence of a sign implies a positive quantity. Whenever a sign is used, it must precede the first digit.

Embedded commas are *never* permitted in a numeric constant in FORTRAN. Some examples of FORTRAN constants are:

1	0	+3.0
+12	0.	+162.07
4708296	0.0	−55.
−386341	8.569	−10.3875

Numeric constants, however, are specifically recognized in FORTRAN as either being an integer-type constant or a real-type (floating point) constant. These two types of numbers must be handled differently in the computer and care must be taken to assure that the proper type is used in a FORTRAN statement. It is important, therefore, for you to have a good understanding of these two different types of numbers.

Integer Constants

An integer constant is a *whole number* written without a decimal point. It is formed from the digits 0 through 9. A sign is permitted in the high-order (leftmost) position of the number. A minus sign is mandatory for negative numbers. A zero may be written with a preceding sign, but it has no effect on the value zero. That is, zero is always zero; the difference between a positive or negative zero is not represented in the computer.

The maximum size of an integer constant allowed varies from computer to computer. The size of the computer word (length of a storage location) on each computer determines the limits of an integer constant. The smallest integer value that may be represented on the IBM 360 and 370 computers is -2147483648 (i.e., -2^{31}). The largest integer value that may be represented is 2147483647 (i.e., $2^{31} - 1$). It is not necessary to memorize these numbers since few integer numbers in practice exceed this magnitude. It is important to realize that different types of computers have specific magnitude (both positive and negative) limitations. You should keep in mind only the general limitations and look up the specific limitations if you ever write a constant that you think approaches those limitations.

Integer constants are often referred to as *fixed-point numbers* because the decimal point is always assumed to be located after the rightmost digit in the number. The assumed position of the decimal point is fixed; thus, only whole numbers are always represented. Both of these terms are used interchangeably. Integer constants should always be used when *counting*, since only whole numbers are involved. For example, the number of students in a class is an integer value, so is the number of cards in a program deck. An integer constant, therefore, always implies that there can be no fractional part associated with the number.

Examples of valid integer constants are:

$$123$$
$$-36$$
$$+4289$$
$$0$$

Examples of invalid integer constants are:

37.5	(not a whole number)
86−	(sign must precede number)
1,240	(contains an embedded comma)
3286946175	(exceeds largest value permitted)

Real Constants

A real constant is a string of decimal digits, which contains a decimal point. A real quantity may contain an appended sign as with integer constants. Real constants are more commonly called *floating-point numbers* to reflect the fact that the position of the decimal point may be moved (or floated) in the number. The terms real and floating-point will be used interchangeably in this text.

Real constants are thought of as *measuring numbers,* since many fractional positions can be included in a floating-point number. When using a measuring number, one is concerned with the *magnitude* (or range) and the *precision* (or accuracy) of the quantity. If we write 1.7 inches, 32.0 inches, and 568.2 inches, we are concerned with successively increasing magnitudes. If we measure the diameter of a shaft, we may come up with 3.0, 3.04, or 3.038 inches, depending upon whether an estimate, a ruler, or a micrometer was used. A micrometer would provide us with a more precise measurement value.

Both larger magnitudes and greater precision are expressed by using more digits in the numbers. There must be a limit, however, to the number of digits in the quantity handled by the computer, since this quantity factor has an important bearing on computer cost. A concept was devised to represent real numbers in the computer with a fractional quantity and an exponential part, the same as with the scientific notation for expressing numbers. This, then, is why we use the term floating-point: the exponential part actually denotes the placement of the decimal point.

In the scientific notation, the precision (number of digits) of the number is expressed in either integer or real form. The magnitude of the number is expressed as a power of ten by which the number is multiplied to obtain the desired value. For example:

$$1.86 \times 10^5 \quad \text{(value of 186,000)}$$
$$93 \times 10^6 \quad \text{(value of 93,000,000)}$$
$$.1 \times 10^{-2} \quad \text{(value of .001)}$$

The precision of most quantities can be expressed by less than six digits. Based on this consideration, many computers have been designed to allow internal floating-point calculations with numbers having a maximum of seven decimal digits and an exponent. Externally, a floating-point number may be constructed with or without the exponent. The exponent, when used, is expressed as a signed or unsigned power of ten by which the number is multiplied to obtain its true value. The exponent can range from a power of -78 to $+75$ on most IBM computers. Thus, we can see that extremely small and large numbers can be represented in floating-point form. We will not, for the time being, concern ourselves with expressing a real constant with an exponent. The everyday, common form for expressing real numbers will be used.

The use of one to seven digits to represent a real constant is known as *single-precision mode*. A need may occur for a precision of more than seven digits. Thus, most computers also permit an extension of up to roughly sixteen digits for increased precision. This form of extended digits is called *double-precision mode,* which is discussed much later in the text.

A single-precision, floating-point number is normally formed as a string of decimal digits with a decimal point in the way most of us represent numbers with fractions in our everyday life. There is no restriction on the number of digits that may be written with a real constant; however, no more than seven (sometimes eight) significant digits will be retained by the computer. Thus, there is no need to write more than eight digits in a real constant. A whole number may or may not be present with the fraction. Or a fractional quantity may or may not be present with the whole number portion. A decimal point, however, is always included with the number. A real constant must *not* contain embedded commas.

Valid examples of single-precision, floating-point (real) numbers in decimal form are:

$$-12.75 \quad\quad +4.$$
$$264.8 \quad\quad -.0029$$
$$.567 \quad\quad 1.$$
$$-0. \quad\quad 3481.026$$

Invalid examples of single-precision, floating-point numbers in decimal form are:

    57   (no decimal point)
  3,860.21   (contains embedded comma)
    .589+   (sign follows number)
  26*5.3   (contains a special character)

We will now turn our attention to the subject of variables, which are used to contain the types of constants we have just discussed.

2.6 Variables in FORTRAN IV

The FORTRAN programmer selects symbolic names in accordance with specific rules—primarily to denote quantities in memory. A name, thereby, identifies the value stored at a specific storage location. Since the contents of these locations may vary within the program, we call these quantities variables. A *variable* is, therefore, a symbolic reference to the value that occupies that storage location. A variable can have only one value assigned to it at any given time in program execution. The latest value assigned to a variable at a point in program execution is called the current value of the variable. Whereas the computer refers to a value at a memory location by a machine address, a FORTRAN programmer refers to the value at a location by its assigned variable. A variable is also referred to as a *variable name*.

The concept of variables and memory may be thought of as residential mailboxes with which we are all familiar. Memory is divided into storage locations as though it were divided into many "boxes," as indicated by the following:

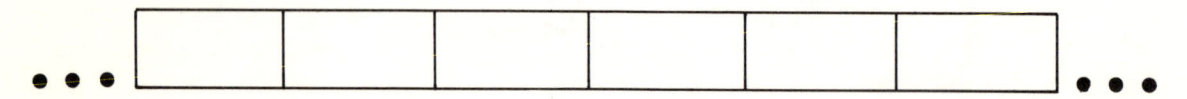

Each box or storage location is capable of storing a single value. Thus, a particular box may contain any value we place in it, such as

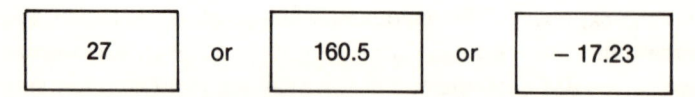

    27   or   160.5   or   −17.23

The computer refers to a value in a box by the machine address assigned to it just as the postman uses a street number to locate our mailbox. In FORTRAN and other programming languages, however, we assign a variable name to a box to refer to its value, just as our mailboxes are identified with our names. For example:

variable name	HOURS	RATE	GRSPAY
contents	40.0	5.50	220.00

Thus, variable names are our way of telling the computer which box (or value) we are using.

We can do two things with a box or storage location by referring to its assigned variable name.

1. We can put a value into it by reading an input value off a card and storing it there, or we can assign to the box a value by means of an Assignment statement. For example:

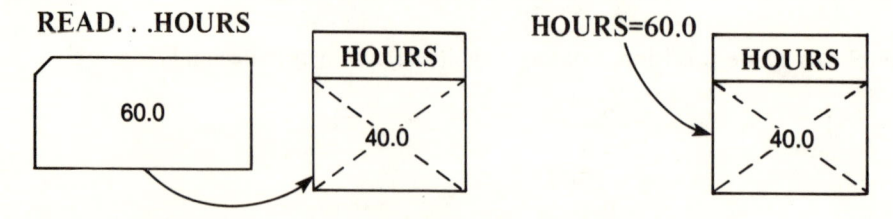

READ...HOURS   HOURS=60.0

  60.0 → HOURS (40.0)   HOURS (40.0)

Introduction to the FORTRAN IV Language   29

In this example, the old value of 40.0 in the variable HOURS is destroyed and replaced with the new value of 60.0 from the execution of the READ or Assignment statement. When a new value is stored into a box, the value currently stored in the box is destroyed, since a box can contain only one value at any given time. The value in a box is replaced only when we instruct the computer to put a new value in the box.

2. We can use the value within a box in a statement to compute other values or to print the value on the line printer. For example:

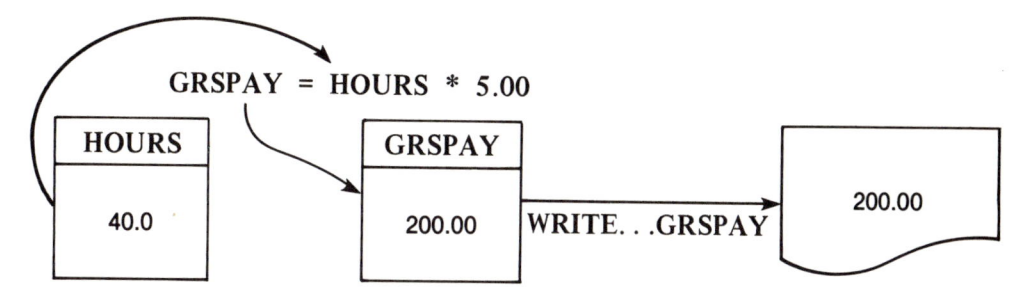

```
            GRSPAY = HOURS * 5.00
   HOURS           GRSPAY
                                                        200.00
    40.0            200.00     WRITE...GRSPAY
```

In the above example, the value in HOURS is used in the formula to compute GRSPAY. The contents of GRSPAY is printed on the line printer with the WRITE statement. When a value is used in a box, it is not destroyed.

Thus, variable names will be used in all programs to store values at the indicated boxes and to refer to these values for later computations and/or output. The concept of variables, therefore, is fundamental to programming and must be understood before a program can be written.

Variable names are formed from alphabetic and numeric characters (A-Z, 0-9). No special characters are allowed. A variable name may be a single alphabetic letter or it may be formed by any combination of alphabetic and numeric characters up to a maximum of *six*. The first character of any variable, however, must always be alphabetic (A-Z). The remaining characters are optional and may be either alphabetic or numeric. Examples of valid variable names are:

```
        COUNT          C5R2
        SUM            MONEY
        K              N12345
```

Examples of invalid variable names are:

```
        COUNTER      (exceeds 6 characters)
        A*B          (contains a special character)
        R2.5         (contains a special character)
        5FEET        (first character is not alphabetic)
        SUM−2        (contains a special character)
```

The choice of variable names is entirely up to the programmer. For aid in documentation, the programmer should assign variables with a mnemonic (memory aid) value. This enables one to better understand what the variables stand for and what is being accomplished in the program. For example, use:

GRSPAY = HOURS * RATE   (note: the * denotes multiplication)

and not:

A = B * C

to indicate the calculation of one's gross pay.

You may begin to wonder why we use variables and not just constants in our programs. The use of variables provides the flexibility we need in programming. We can read different input values into a variable, yet use the same formula to calculate the results. For example:

READ (5,99) ID, HOURS, RATE
GRSPAY = HOURS * RATE
WRITE (6,98) ID, GRSPAY

In the READ statement we are reading an employee's identification number, hours worked, and pay rate from a data card. The statement:

GRSPAY = HOURS * RATE

computes the employee's gross pay by multiplying (indicated by *) hours times rate and storing (indicated by the =) the results at the variable named GRSPAY. The WRITE statement will print the employee's ID and computed gross pay on the line printer.

Sure, we could have written:

GRSPAY = 80.0 * 4.50

and computed the gross pay for one employee. When we use constants, however, we are "locked in" to these numbers. What if we needed to compute the gross pay for 1,000 employees. Would we have 1,000 such computational statements in which only constants were used? Certainly not! By using variables we can read different values into storage locations and compute a result using the current value in a specific variable. Then we can repeat these steps for another employee, read the hours, and rate values from another data card for a different employee. Now the variables HOURS and RATE contain new values, but the same formula

GRSPAY = HOURS * RATE

is still valid to compute the next employee's gross pay. Thus, what may seem like a gigantic problem to solve on the computer becomes quite simple when we use variables. We are getting ahead of ourselves, but I believe this explanation is in order for enabling you to understand the use and importance of variables.

The FORTRAN language also distinguishes between the types of variables used in a program. FORTRAN variables are identified as being either type integer or type real just like constants. Each type relates to the form of the numeric value stored at the memory location.

Integer Variables

Integer variables represent a whole number in memory. All integer variables must begin with one of the letters I, J, K, L, M, or N to inform the system that it is representing (contains) an integer value. As a memory aid, remember I-N for integer variables. Examples of valid integer variables are:

KTR
N
IHOLD2
JSUB

Real Variables

Real variables represent numbers which may include a fractional value. For example, your grade point average and paycheck amount are both real quantities. The fractional value may be zero, but the capability is there for the fractional part to contain any value so desired. All real variables must begin with one of the letters A-H and O-Z (all letters other than I-N), to inform the system that it is representing a real (floating-point) value. Examples of real variables are:

RADIUS
SUM
A37
X
HLD3T

The assignment of names to the integer and real variables appearing in the program is entirely under the control of the programmer. Special care, however, must be taken to observe the rule for distinguishing between the names of integer and real variables. Every different combination of letters and digits represent a different name to the compiler. For example, the variable B7 is not the same as BSEVEN. A common error made by beginning programmers is to misspell the name used as a variable and thereby not getting the correct results. To the compiler each differently spelled name represents a unique variable. For example, the names RATE and RATES represent two different variables to the compiler.

2.7 Operators in FORTRAN IV

Arithmetic Operators

Arithmetic operations are represented in FORTRAN by symbols. Symbols must be used, since the use of English words as operators would be confused with variables. The five arithmetic operators allowed in FORTRAN, and their operation, are:

```
    +    addition
    −    subtraction
    *    multiplication
    /    division
    **   exponentiation (raising to a power)
```

These operators might be used with variables and numeric constants to compute a value as follows:

$$\text{PAYNET} = \text{HOURS} * 4.50 + \text{BONUS} - \text{DEDUCT}$$

Hours are multiplied by 4.50; the value at BONUS is next added to the product; and, last, deductions are subtracted from the total. The resulting computed value is stored at the variable, PAYNET.

Arithmetic computations that involve more than one calculation are performed according to a hierarchy or precedence of the arithmetic operators. This hierarchy is the same as that in algebra, and is as follows:

Precedence	*Arithmetic Operator*
1st	**
2nd	* and /
3rd	+ and −

For those operators on the same level of precedence, the calculation will be performed in a left to right order. Parentheses may be used to override this hierarchy and to enclose those calculations to be performed first. For example:

$$A = (B + C) / D - E * F$$

The order of calculations for this statement is:

(1) B is added to C because of the parentheses
(2) this sum is divided by D
(3) E is multiplied by F (because multiplication is a higher precedence than subtraction)
(4) the product in (3) is subtracted from the previously derived quotient in (2)
(5) the result is stored at the variable A

Assignment Operator

The equals sign is used only in the arithmetic Assignment statement. It is used to assign (store) the single resulting value from the arithmetic expression (given to the right of the =) to the variable located on the left side of the = symbol. For example,

$$B = 10.5$$
$$A = B + 7.2$$

In the first statement, B is assigned the value of 10.5. In the second statement, A gets assigned the value of 17.7 (the sum of 10.5 and 7.2).

To get across the operation performed by the assignment operator, many people use different terms for the = symbol. Some people refer to it as a replacement operator. Thus, our second statement in the preceding paragraph would read: A is replaced by the sum of B + 7.2. An easy-to-understand term used by others for the = symbol is "gets the value of." Thus, the second statement in the prior paragraph would read: A gets the value of B + 7.2.

Relational Operators

Another type of operators known as relational operators is used for the comparison of values contained in variables or represented as constants. Thus, relational operators provide the capability of testing whether a certain condition is met and for decision making. Let's not concern ourselves with the details of relational operators for now. They will be discussed when we need to use them.

Next we want to look at the procedure for writing these statements in a format that the compiler can understand.

2.8 Coding FORTRAN IV Statements

A FORTRAN program is coded on a special form called a ***coding sheet.*** (See figure 2.3 for a sample FORTRAN coding form.) Each line on the coding sheet represents an 80-column punch card. (See figure 2.4 for the frequently-used FORTRAN program card.) Each designated column on the coding form represents the same relative card column in the punch card. For example, column 7 on the coding sheet represents card column 7 on the punch card. Each FORTRAN statement is coded on one or more lines on the coding form.

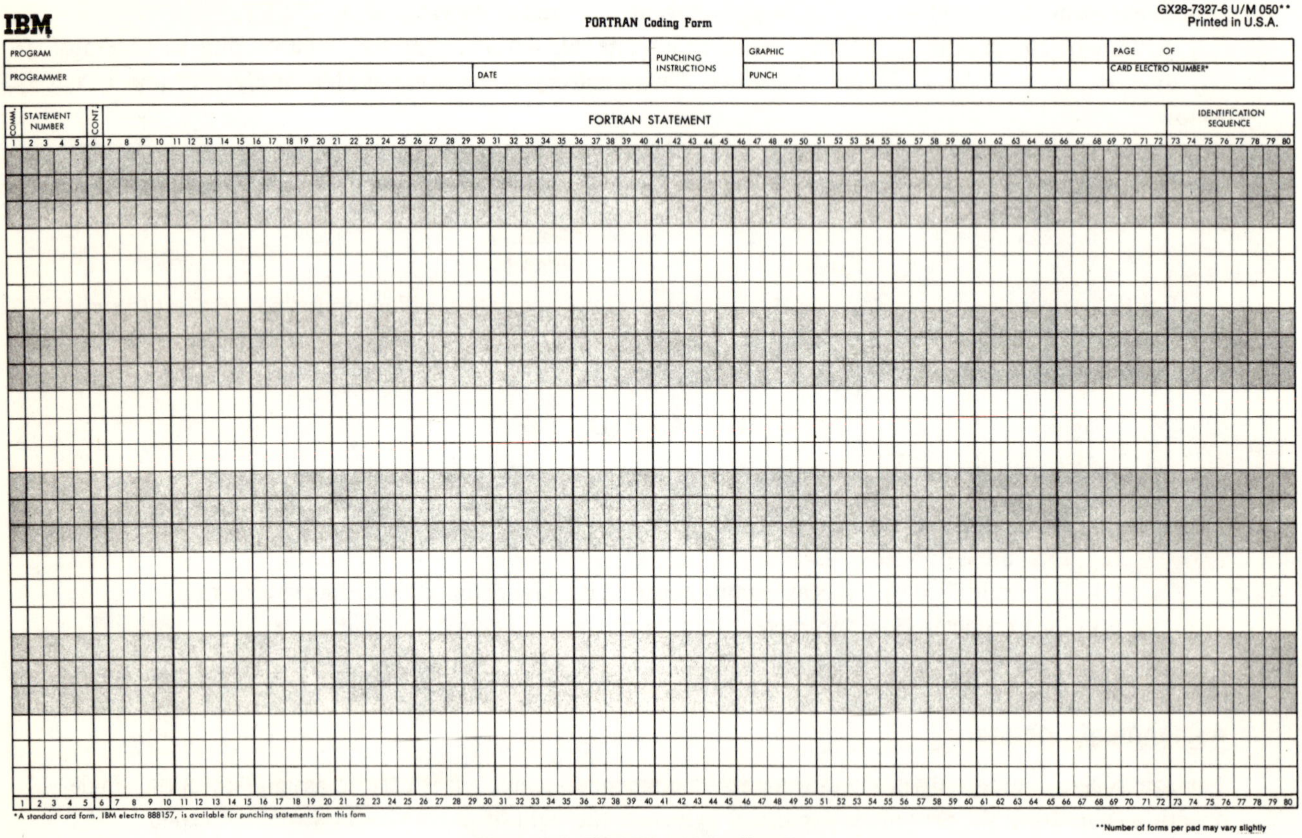
IBM
FORTRAN Coding Form
GX28-7327-6 U/M 050**
Printed in U.S.A.
PROGRAM
PROGRAMMER
DATE
PUNCHING INSTRUCTIONS
GRAPHIC
PUNCH
PAGE OF
CARD ELECTRO NUMBER*
COMM. STATEMENT NUMBER CONT. FORTRAN STATEMENT IDENTIFICATION SEQUENCE
1 2 3 4 5 6 7 8 9 10 11 12 13 14 15 16 17 18 19 20 21 22 23 24 25 26 27 28 29 30 31 32 33 34 35 36 37 38 39 40 41 42 43 44 45 46 47 48 49 50 51 52 53 54 55 56 57 58 59 60 61 62 63 64 65 66 67 68 69 70 71 72 73 74 75 76 77 78 79 80
*A standard card form, IBM electro 888157, is available for punching statements from this form
**Number of forms per pad may vary slightly

Figure 2.3. FORTRAN Coding Form

Introduction to the FORTRAN IV Language   33

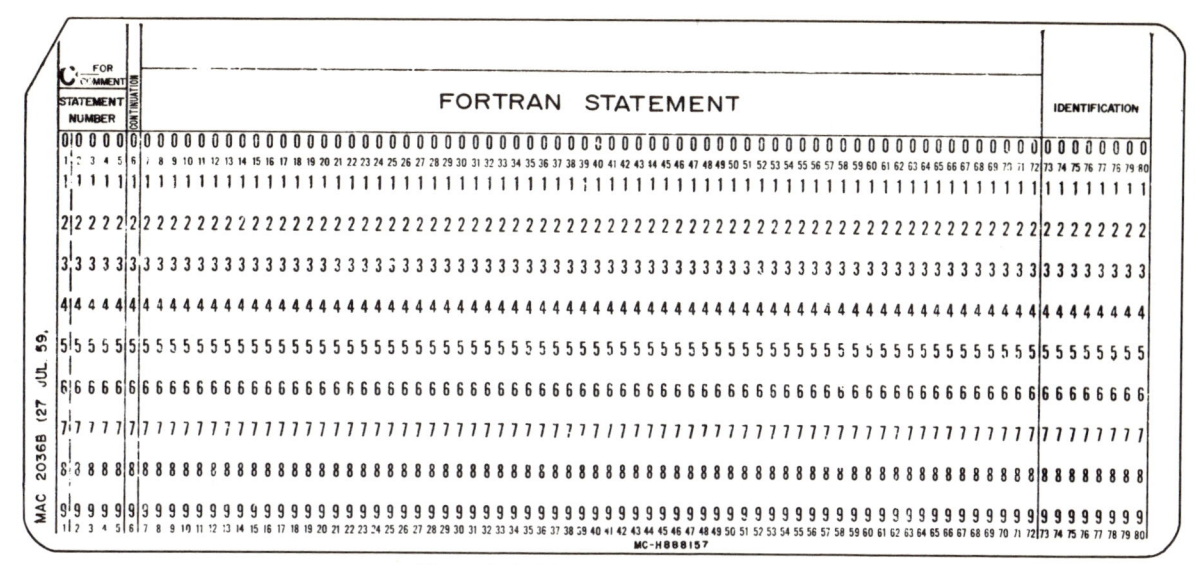

Figure 2.4. FORTRAN Program Card

FORTRAN statements are broken down into four different areas or fields. FORTRAN coding sheets have these specific fields marked off for easy reference. These four fields are:

    Columns   1-5   Statement number
    Column   6   Continuation
    Columns   7-72   FORTRAN statement
    Columns   73-80   Program identification and/or sequence number

Statement Number

Columns 1 through 5 are reserved for a statement number. This statement number serves as a statement label or tag which allows one to refer to a FORTRAN statement. This permits the program to skip various statements and to change the execution sequence of statements.

All statement numbers must be one to five numeric digits and may range from 1 to 99999. Statement numbers may be left-justified (begin in column 1), right-justified (end in column 5), or punched anywhere between columns 1 through 5. Imbedded blanks and leading zeroes in statement numbers will be ignored.

Statement numbers are not required for all statements. Only those statements which are referenced by other statements need be assigned a statement number. However, one may elect to assign a statement number to all statements as an aid in keeping the statements in the desired order. This is not recommended because of longer compile time of the program.

Statement numbers need not be assigned to statements in a program in any specific sequence. All statement numbers assigned within a program, however, must be unique. *No* two statements may be assigned the same number; otherwise, a compile error of duplicate labels will occur.

Continuation

Column 6 is reserved for the continuation column. Since FORTRAN statements may exceed one card, a provision must be made to indicate that succeeding cards are a continuation of the FORTRAN statement and not a new statement. Any character other than a zero or blank in column 6 indicates a continued FORTRAN statement. Up to nineteen continuation cards are allowed for each FORTRAN statement. In the first card of a statement, column 6 must be blank or zero. Statement numbers are forbidden on continuation cards.

FORTRAN Statement

The FORTRAN statement is punched in columns 7 through 72. A FORTRAN statement must not be coded before column 7; however, it is not mandatory to begin in column 7. Blanks within the FORTRAN statement are ignored completely (except for one special case which we will mention later).

Program Identification/Sequence Number

Columns 73 through 80 are reserved for program identification (ID) and/or sequence number. These columns are ignored by the compiler and are primarily for the benefit of the programmer. Any information punched in these columns will be printed on the source listing. If the statements in a program were shuffled (from having been dropped, etc.), they could easily be rearranged in the correct order by sorting on a mechanical sorter or even by hand. A program ID would prove beneficial if several FORTRAN programs were shuffled together.

An example of a combination program ID and sequence number would be PAY00100. Sequence numbers should increase by tens or even in increments of 100 to allow for the addition of new statements in the program. This field is normally not used by students to conserve time in punching a program. Production programs, because of their large size, are normally coded with a program identification and/or sequence numbers.

Figure 2.5 illustrates each of these four areas on the punch card.

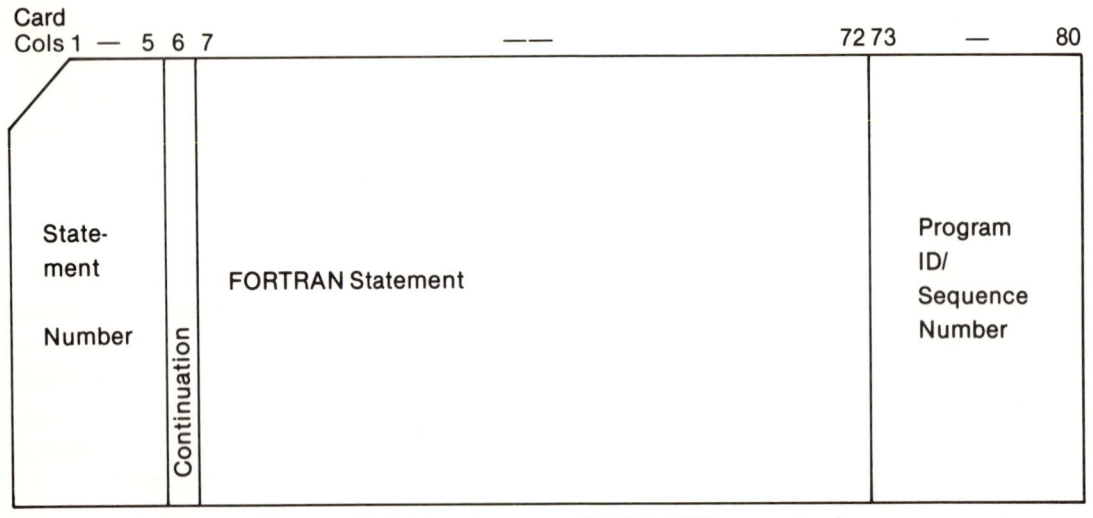

Figure 2.5. The Four Areas on a FORTRAN Program Card

Even though there are specially designed FORTRAN program cards, most data processing installations use the regular 80-column punch card for FORTRAN programs.

Comment Card

The character C in column 1 indicates a comment card. Comment cards are used to insert remarks into a program. Any comment or information may be coded in columns 2 through 80. There is no limit to the number of comment cards allowed in a program. If a comment is entered onto more than one consecutive card, each card must have a C in column 1.

Comment cards are not processed by the FORTRAN compiler. The comment cards are simply printed on the source listings. Any character in the FORTRAN character set may be included in the comment card. Comment cards may be placed anywhere in a program except between continuation cards.

Introduction to the FORTRAN IV Language   35

The FORTRAN program presented in figure 2.1 to read and compute the sum of two values is coded on a FORTRAN coding form to illustrate the process of coding FORTRAN statements. Figure 2.6 shows the coded program on a FORTRAN coding sheet.

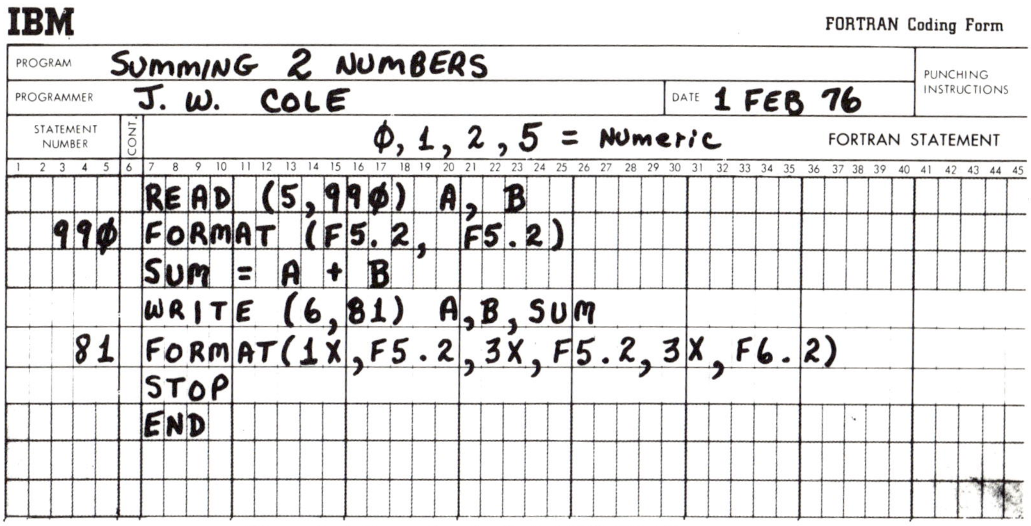

```
IBM                                                    FORTRAN Coding Form
PROGRAM      SUMMING 2 NUMBERS                         PUNCHING
PROGRAMMER   J. W. COLE                DATE 1 FEB 76   INSTRUCTIONS
STATEMENT          Ø, 1, 2, 5 = Numeric                FORTRAN STATEMENT
NUMBER

        READ (5,990) A, B
  990   FORMAT (F5.2, F5.2)
        SUM = A + B
        WRITE (6,81) A,B,SUM
   81   FORMAT(1X,F5.2,3X,F5.2,3X,F6.2)
        STOP
        END
```

Figure 2.6. Example of a Coded FORTRAN Program

In coding a FORTRAN program, only capital letters are used. Lower case letters do not exist in FORTRAN. Special care should be made to print each character with maximum legibility on the coding sheet if the program is to be keypunched by someone else. Certain characters are very difficult to distinguish. It is recommended that you use the following coding conventions in order to provide an accurate, easy-to-read program, and data, to the keypunch personnel.

Numeric Digit : *Print*	*Alphabetic Letter* : *Print*
zero Ø	O O
one 1	I I
two 2	Z Z
five 5	S S
eight 8	B B

When any one of these specific symbols is printed, especially the numeric ones, you should indicate their form in the punching instruction blocks provided on the coding sheet. For example, if you slash your zeroes, you might include a note that says: Ø = numeric, or Ø = zero. Some installations, however, may require the programmer to slash the letter "O" instead of the numeric zero. The zero will be slashed in coded programs on coding sheets within this text.

You may include spaces in FORTRAN statements at any column you wish. The FORTRAN compiler ignores all spaces in FORTRAN statements and reads the statement as if no blanks were present. However, spaces are important and counted, in literal constants (a special character string). Spaces should be included in statements whenever it makes the statements more readable. For example, one might code:

GRSPAY = HOURS * RATE + 50.0

FORTRAN will read this statement as:

GRSPAY=HOURS*RATE+50.0

The inserted blanks between the variables, constants, and operators, however, make the statement more readable. It is recommended that blanks be used only as separators after keywords, and between variables, constants, operators, and special characters (such as a comma or parentheses), to improve readability. Do not insert blanks in the middle of variables and constants, even though most FORTRAN compilers will ignore them!

The basic elements in the FORTRAN language have now been explained to help us in writing individual FORTRAN statements. Chapter 3 will require us to develop a basic FORTRAN program that gets us started into that fantastic and amazing field called **programming.**

2.9 Nonstandard Language Extensions

Different computer manufacturers' FORTRAN compilers permit extensions to the features discussed in this chapter. On the other hand, some FORTRAN compilers are more restrictive and limit these features. These extensions and limitations are given for each of the features that may differ, in the following paragraphs.

Character Set. Two additional characters are included in the IBM 360-370 G level and WATFIV compilers. The dollar sign ($) is used as an alphabetic letter in addition to A-Z. Thus, twenty-seven alphabetic letters with which to begin variable names are possible. The single quote mark (') represented by an 5/8 punch is used to indicate a "literal" string of characters. The ampersand (&) is also allowed and has a special meaning.

On Honeywell 6000 FORTRAN Y and B6700 FORTRAN the double quote (") is used to indicate a literal. The $ is permitted on Honeywell 6000 and CDC 6000-7000 compilers to permit and separate multiple statements on a card. The ";" is used on Burroughs and WATFIV compilers to permit and separate multiple statements on a card.

Keywords. The IBM 360-370 G level and WATFIV compilers have additional FORTRAN statements denoted by the keywords DEFINEFILE, ENTRY, FIND, IMPLICIT, NAMELIST, PRINT, and PUNCH. Consult **Appendix F** for the statements (keywords) allowed in various compilers.

Variables. Some basic FORTRAN compilers like ANSI Basic FORTRAN and IBM Basic FORTRAN IV limit the number of characters in a variable to five. The CDC 6000-7000 compiler permit up to seven characters in their variable names. The Honeywell 6000 Y compiler permits up to eight characters.

Statement Numbers. The ANSI Basic FORTRAN compiler limits its statement numbers to four digits. IBM Basic FORTRAN IV permits five digits, but limits the range to 32767. The Univac 1107 permits statement numbers from 1-32767. In WATFIV a warning message is issued for all statements that have statement numbers assigned to them and are not referenced in the program.

Continuation Cards. The IBM E level and ANSI Basic FORTRAN compilers limit the number of continuation cards to five. The WATFIV and the B6700 compilers permit an unlimited number of continuation cards.

FORTRAN Statement. The CDC 6000-7000, Honeywell 6000 FORTRAN Y, B6700, and WATFIV compilers permit multiple FORTRAN statements on a card. The CDC 6000-7000 compiler uses a $ symbol to indicate the end of a statement. The B6700, Honeywell FORTRAN Y, and WATFIV compilers use a ; symbol. Although the use of multiple statements per card produces a smaller source deck, the use of multiple statements in a card is strongly discouraged. The requirement of transportability to another FORTRAN compiler can result in many hours of added hard work caused by having to reconvert the multiple statements back to single statements on a card. The use of multiple statements in a card also increases the amount of debugging time and effort that must be given to a program.

Comment Card. The CDC 6000-7000 and Honeywell 6000 compilers permit an asterisk (*) in addition to the character C to be used in column one to designate a comment card. The character C is always recommended to provide comment card compatibility among FORTRAN compilers. Most FORTRAN compilers other than ANSI permit all characters available on a computer to be used in the comment instead of restricting them to the FORTRAN character set. IBM compilers restrict the number of consecutive comment cards to forty-nine. On most other compilers the number of consecutive comment cards is unlimited.

2.10 REVIEW QUESTIONS

1. From what two words is the acronym FORTRAN derived? *Formula Translation*
2. The first FORTRAN compiler was completed in what year? *1957*
3. Name the three categories of executable statements. *Input/Output, Assignment, Logical and Control Statements*
4. How many characters are in the FORTRAN character set? *47*
5. What is the purpose of FORTRAN keywords? *denote the operations to be performed by the comp. with respect to the Fortran statement*
6. With the exception of the Assignment statement, where do keywords appear in a FORTRAN statement? *As the first part of a Fortran statement*
7. Define a numeric constant in FORTRAN. *is a known numeric value that always represents a fixed quantity in a source program*
8. What are the two types of numeric constants that may be formed in FORTRAN? *integer and real*
9. An integer constant represents a *whole* number.
10. A real constant is always considered to be in *decimal-form* form.
11. What is the maximum number of decimal digits that can be represented in a single-precision real constant on most computers? *7*
12. Define a variable in FORTRAN. *a symbolic reference to the value that occupies a storage location*
13. Variables may be formed of *1* to *6* characters.
14. The first character of a variable must be *A-Z* while the remaining characters (if used) may be *A-Z* or *0-9*.
15. An integer variable must begin with one of the letters *I* through *N*.
16. A real variable may begin with what letters? *A-H, O-Z*
17. What are the five arithmetic operators used in FORTRAN? *+, -, *, /, ***
18. List three terms that may be used to express the function of the assignment operator. *assign, replace or store*
19. FORTRAN instructions may be referenced by labels called statement numbers assigned to each instruction. (True/False) *F*
20. Statement numbers used to identify FORTRAN statements are punched in card columns *1* through *5*.
21. Statement numbers assigned to a FORTRAN statement for reference purposes may range from *1* to *99999*.
22. Statement numbers may be assigned without regard to order or sequence, but no two statements may be assigned the same statement number in the same program. (True/False) *T*
23. Card column *6* is reserved for the statement continuation column.
24. Any character other than *0* or *b* may be used to indicate a continuation of a FORTRAN statement.
25. The FORTRAN statement itself begins in card column *7* and may extend through *72*.
26. Card columns *73* through *80* are used for statement sequencing and/or program identification, and are not checked by the compiler.
27. Comment cards may be entered in a program to insert comments/remarks and are indicated by placing a *C* in card column *1*.
28. Comment cards may be placed anywhere in a FORTRAN source program except between continuation cards. (True/False) *T*
29. What are the alphabetic characters that are often confused with the numeric digits 0, 1, 2, 5, and 8? *0 with O   1 with I   2 w/Z   5 w/S   8 w/B*
30. What is the rule regarding spaces in FORTRAN statements? *you may include space in any col. Fortran compil. ignores all spaces However, spaces are imp. & counted in literal constants. Spaces should be included in statements whenever it makes the st. more readable*

2.11 EXERCISES

1. Identify the following variables as to type integer or real:

    1. JOE — I
    2. RATE — R
    3. PI — R
    4. S — R
    5. TIME — R
    6. C5A — R
    7. F16 — R
    8. M5T — I
    9. LOOP — I
    10. HOLD — R

2. Identify the following variables as to type integer or real:

    1. KTR — I
    2. CTR2 — R
    3. I — I
    4. NSUM — I
    5. MOUSE — I
    6. ITEMP — I
    7. A — R
    8. X21 — R
    9. J578 — I
    10. SA5T — R

3. Explain why the following variables are invalid:

    1. COMPUTER — too many char
    2. 3B — beg. w/num. char.
    3. COST.9 — spec. char
    4. G86R29B — too many char
    5. A*B — spec. char
    6. T5000– — "

4. Explain why the following variables are invalid:

    1. SUMOFXANDY — too many char
    2. NET-PY — spec. char.
    3. A123456 — too many char
    4. X+7 — spec. char.
    5. 86TYM — begins with a numeric char.
    6. UT**2 — spec. char

5. Identify the following constants as to type integer or real:

    1. 12.3 — real
    2. 15 — I
    3. +.56 — R
    4. 0 — I
    5. 0. — R
    6. –79 — I
    7. –8.7569 — R
    8. +3274 — I
    9. 4296.002 — R
    10. 29036 — R

6. Identify the following constants as to type integer or real:

    1. 0.0 — R
    2. 38.462 — R
    3. 149 — I
    4. 00 — I
    5. –.003 — R
    6. +47.03 — R
    7. +0. — R
    8. –2 — I
    9. .000001 — R
    10. 29.0 — I

7. Explain why the following constants are invalid:

    1. 4,296 — comma
    2. 25– — sign follows no.
    3. 4896102457 — too large
    4. 4.56. — two decim.
    5. .–237 — sign follows dec.
    6. 2*5 — contains a special char.

8. Explain why the following constants are invalid:

    1. 46**3 — special character
    2. 0–23.4 — the 0
    3. 86,279 — comma
    4. 5132861004 — too large
    5. .289– — sign follows no.
    6. 2,357+ — " and comma

Developing a Basic FORTRAN IV Program

3.1 Introduction to Writing a FORTRAN Program

It would be nice if we could simply talk to the computer and instruct it how to solve our problem. Basically that is what we do in writing a program. However, the language we use must be in a form that the computer understands and must contain only those commands provided in the language. FORTRAN is one such language that has been developed for the computer to understand. One must organize these commands into a proper sequence to produce the correct results.

The directions that a computer is to follow are supplied by a series of instructions called a *program*. Each instruction is normally a command to the control unit to direct the operations of the various components of the computer. In a problem-oriented compiler language like FORTRAN, one writes the instructions following very much the same sequence he would use if he were solving the problem by hand. For example, if we manually computed the average of a student's test scores, we would: first, obtain the scores; second, add the scores together to obtain their sum; and then, divide the sum by the number of test scores to obtain their average.

These steps in our hand-calculated average might be represented in a block diagram form as follows:

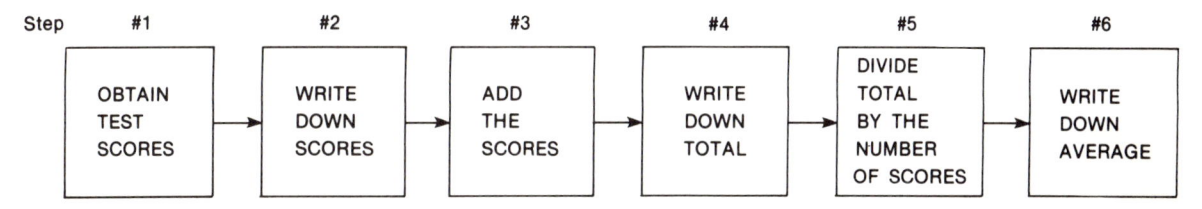

Step	#1	#2	#3	#4	#5	#6
	OBTAIN TEST SCORES	WRITE DOWN SCORES	ADD THE SCORES	WRITE DOWN TOTAL	DIVIDE TOTAL BY THE NUMBER OF SCORES	WRITE DOWN AVERAGE

This diagram represents the algorithm or logic we must follow to solve the problem of obtaining our grade average. *Logic* is doing the right things in the proper sequence. That is, some operations must be performed ahead of others to achieve the correct results. For example, we cannot compute an average until we have computed the total of the test scores; we cannot compute the total until we have obtained the test scores.

We would follow basically the same procedure if we instructed the computer to solve the same problem with a computer program. To supply the test scores to the computer they are generally read into memory. We then compute the required total and average. Finally we write the test scores and computed values to the line printer. A block diagram to illustrate these operations/instructions performed by the computer are as follows:

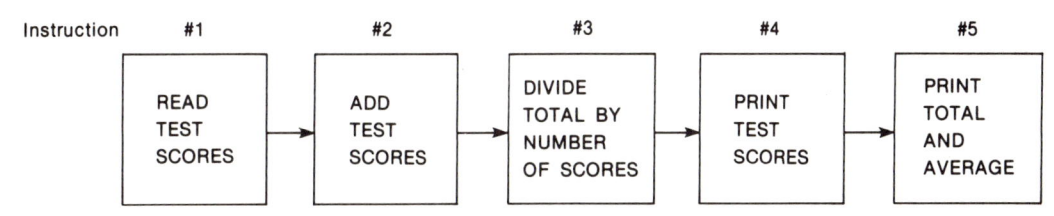

Instruction	#1	#2	#3	#4	#5
	READ TEST SCORES	ADD TEST SCORES	DIVIDE TOTAL BY NUMBER OF SCORES	PRINT TEST SCORES	PRINT TOTAL AND AVERAGE

We could have followed exactly the same steps that we established for our hand-computed solution, but with computers we usually read a set of values, process them in the required manner, and lastly write the desired values. An overview diagram of the general logic for most computer programs thus becomes:

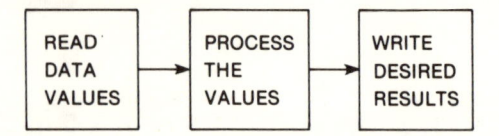

```
┌──────────┐     ┌──────────┐     ┌──────────┐
│  READ    │     │ PROCESS  │     │  WRITE   │
│  DATA    │ ──► │  THE     │ ──► │ DESIRED  │
│  VALUES  │     │  VALUES  │     │ RESULTS  │
└──────────┘     └──────────┘     └──────────┘
```

We should see from our second previous block diagram that the computer begins execution at the first instruction in our program. Instruction #1 is first executed, then instruction #2, then instruction #3, and so on. The implicit order in which the computer executes the instructions in a program is in a sequential order starting at the first one.

Control statements may be added in programs to change this sequential order when we need to do so. For example, if we had multiple sets of data to process, we might instruct the computer to go back to instruction #1 to reexecute the instructions in our program for a new set of data. For example:

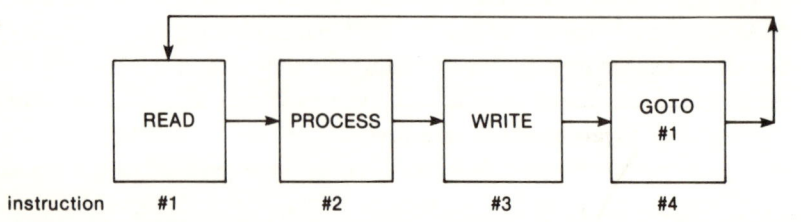

```
              ┌──────────────────────────────────────────┐
              │                                          │
              ▼                                          │
         ┌───────┐   ┌─────────┐   ┌───────┐   ┌──────┐  │
         │ READ  │──►│ PROCESS │──►│ WRITE │──►│ GOTO │──┘
         │       │   │         │   │       │   │  #1  │
         └───────┘   └─────────┘   └───────┘   └──────┘
instruction  #1         #2            #3         #4
```

A flow of execution (control) such as this series of instructions is said to be a *loop*. The program would be reexecuted for each set of data until all our sets of data are processed. With a control instruction like the GOTO we can process many sets of data with the same set of instructions. Thus the power of the computer to solve a problem with many sets of data is practically unlimited.

To sum up the way the computer knows how to process the instructions in a program we say that it begins at our first instruction and proceeds in a sequential manner (top to bottom) until it encounters a control instruction which directs it otherwise. To start with, however, we will concern ourselves with programs that contain no control instructions that cause a loop—just instructions that are executed only once in a top-to-bottom sequence.

Let us develop a FORTRAN program that will indeed compute a student's grade average from three test scores. The three test scores will be read from a data card. After computing the total and average we will write the three input values and the two computed values to the line printer.

First, it is necessary that we understand how our data card will be punched with the input values to be read.

3.2 The Data Card

As mentioned earlier, the commonly used punch card is divided into 80 vertical columns, each column capable of containing one character. We will use nine columns—three columns for each test score, assuming we use an integer number from 0 to 100 to represent each test score. You may wonder why we reserve three card columns for each score since some may be less than 100. The rule is to reserve as many card columns for an input item (field) as its largest possible value.

Let us assume three test scores of 100, 76, and 83, respectively. We could punch our three fields of data in any card columns we wish; we do not have to begin in card column 1. We do not have to punch the fields adjacent to each other; we could leave some blank columns in between each field. The computer does not care since we will instruct it where to look for the three values. Therefore, let us punch the three fields next to each other beginning in card column 1. Test score one would be punched in columns 1 through 3, test score two in columns 4 through 6, and test score three in columns 7 through 9, as shown in figure 3.1.

High-order zeroes do not have to be punched in a field; but we must insure that the least significant digit (last position) of the score is punched in the rightmost column of the field.

Figure 3.1. Data Card Containing Three Test Scores

The *keypunch machine* is used to punch data cards as well as program instructions. The characters are printed along the top of the card at the same time the holes are punched. The printing of the characters is for our benefit in reading the punched data; the computer is concerned only with reading the holes. *Appendix B* explains the operation of the keypunch machine for those unfamiliar with its use.

Let us now proceed to develop our FORTRAN program.

3.3 Problem Analysis and Flowcharting

The problem to compute the average from the three test scores is simple enough to grasp without extensive analysis. We will name the three test scores to be read as NTEST1, NTEST2, and NTEST3. (The beginning letter, N, denotes that these are integer variables.) To allow us to develop our input statements, we need to know the layout of our input records. Our input values are punched in a card record as follows:

Field Name	Card Columns
NTEST1	1-3
NTEST2	4-6
NTEST3	7-9

We will use these three variables which represent the test score values to compute our desired output values.

We need to decide what output values we desire to have printed to help us determine what calculations are needed. Let us print our three test scores on a line starting on a new page. Then let us print their sum and average on a new double-spaced line. Hence we see that we need to compute two values—sum and average. Let us give the computed values for sum and average the variable names TOTAL and AVE, respectively. We could calculate the average in one FORTRAN statement; but if we need both the values of sum and average for output values, we must use two separate statements.

A *printer layout form* is available to programmers for use in describing the format and location of printer output results. Each cell on the print layout form corresponds to a print position on the printer output line. Our developed print layout requirements are illustrated in figure 3.2. We use N's (or 9's) to represent numeric fields. An X (or A) is used to represent alphanumeric output fields. We do not write the ac-

42 Developing a Basic FORTRAN IV Program

NAME: **J. W. Cole**

FORM TITLE: **Output fields for sample program**

PRINT LAYOUT

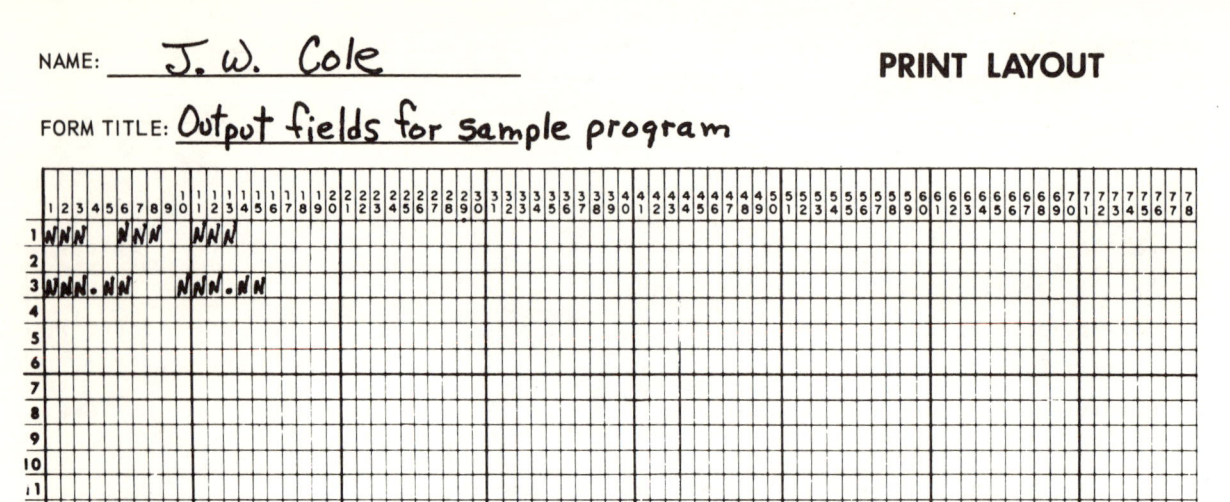

Line 1: `NNN   NNN   NNN`
Line 3: `NNN.NN   NNN.NN`

Figure 3.2. Printer Layout Sheet Showing Output Fields

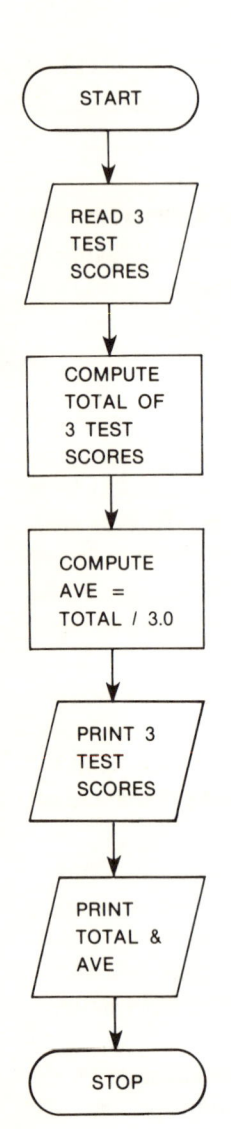

tual output values on the print layout sheet for three main reasons. First, we are concerned only with the placement of the output field on the print line. Secondly, we would not want to write the output values for dozens or perhaps hundreds of output values. For repeated fields in an output column we simply draw a line down the column to show that the fields are repeated. Third, we do not know what the values of some output fields will be since they are calculated by the program. Decimal points and special characters are written at their desired locations. From the output requirements and print layout forms, the programmer can develop his output statements in FORTRAN.

To get us in the habit of flowcharting our plan of solution, we should next construct our flowchart. Our developed logic is presented in figure 3.3.

Along with our input record layout and our print layout specifications, we can use our developed flowchart as a guide in writing our FORTRAN program.

3.4 The FORTRAN Program and Comment Cards

A FORTRAN program consists of a series of FORTRAN statements. Each FORTRAN statement is an instruction to the computer to perform specified actions. We can follow our flowchart and write the FORTRAN program directly from the developed logic. Of course, we must write the actions in the FORTRAN language and provide additional information such as contained in the input and printer record layout specifications. Each FORTRAN statement is coded on the FORTRAN coding sheet to facilitate punching the statements onto a punch card. Our coded FORTRAN program is illustrated in figure 3.4.

We shall examine each of these statements in the program to determine their construction and use.

Flowchart:
- START
- READ 3 TEST SCORES
- COMPUTE TOTAL OF 3 TEST SCORES
- COMPUTE AVE = TOTAL / 3.0
- PRINT 3 TEST SCORES
- PRINT TOTAL & AVE
- STOP

Figure 3.3. Flowchart of Problem to Compute Average of Three Test Scores

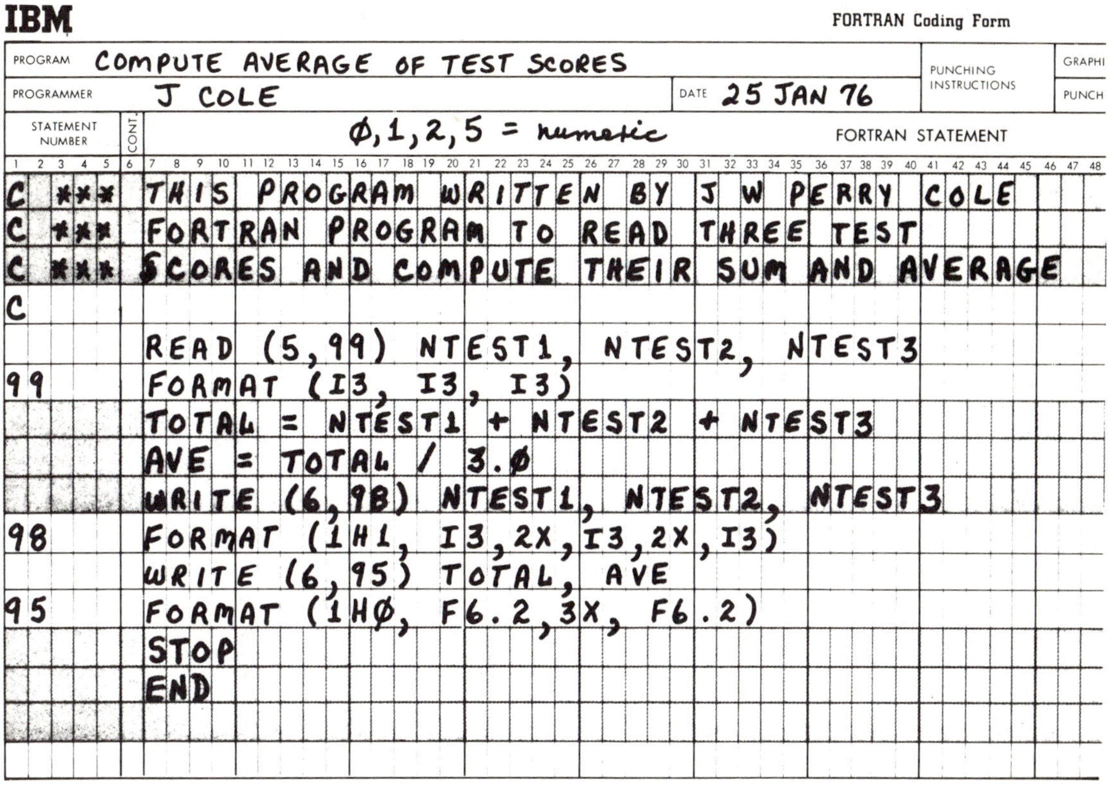

IBM FORTRAN Coding Form

PROGRAM: COMPUTE AVERAGE OF TEST SCORES
PROGRAMMER: J COLE DATE: 25 JAN 76
PUNCHING INSTRUCTIONS — GRAPHI — PUNCH

STATEMENT NUMBER | CONT. | FORTRAN STATEMENT

Ø, 1, 2, 5 = numeric

```
C ***  THIS PROGRAM WRITTEN BY J W PERRY COLE
C ***  FORTRAN PROGRAM TO READ THREE TEST
C ***  SCORES AND COMPUTE THEIR SUM AND AVERAGE
C
       READ (5,99) NTEST1, NTEST2, NTEST3
99     FORMAT (I3, I3, I3)
       TOTAL = NTEST1 + NTEST2 + NTEST3
       AVE = TOTAL / 3.Ø
       WRITE (6,98) NTEST1, NTEST2, NTEST3
98     FORMAT (1H1, I3,2X,I3,2X,I3)
       WRITE (6,95) TOTAL, AVE
95     FORMAT (1HØ, F6.2,3X, F6.2)
       STOP
       END
```

Figure 3.4. Coded FORTRAN Program to Compute Tests Average

Comment Cards

The first four lines represent comment cards to provide remarks about our program. The first comment card includes the name of the person who wrote the program. The inclusion of a comment card at the beginning of a program to identify the author and owner can be very valuable. This comment card can help identify the owner of the program in case the beginning job card gets lost and/or the job deck gets misplaced in the computer center. I highly recommend that you punch your name and class section in the first comment card for these reasons. The next two comment cards provide remarks about the nature and purpose of the program. The fourth comment card without any remarks is used to separate the comment cards from the start of the program, and thus improves readability.

Comment cards are especially useful in programs that include complex logic or a unique way of solving a problem. A routine may take considerable time to understand—even to the person who originally wrote it when he comes back later to make a change. Thus, comment cards prove highly beneficial by facilitating an understanding of the problem and its logic. The student is encouraged to make liberal use of comment cards to provide valuable internal program documentation.

3.5 The READ Statement and Its Associated FORMAT Statement

The next statement in the program is the READ statement. The READ statement begins with the keyword READ and is a command for the computer to read a data record and store the data items into memory.

The computer needs to know much more than simply to read data. It must be told what input device to use since we have the choice of reading from punched cards, magnetic disk, and many other media. The computer must also know where to look for the values on the input record and what form they are in, e.g., integers (whole numbers), decimal numbers with decimal points (real numbers), or even alphabetic data. And finally, the computer must know how many data items are to be read and at what memory locations to store them. All this information must be provided in the READ statement and its accompanying FORMAT statement.

This first number inside the parentheses on the READ statement denotes the *file unit* or input device to be used. The 5 specifies that the data will be read from a punch card by the card reader. (Other numbers specify different input devices.) The second number inside the parentheses, 99, relates to a FORMAT statement that must be used in conjunction with the READ statement in order to specify the form of the input data.

The names following the closing parenthesis are names by which we wish to refer to the data items being read. These variable names are actually the symbolic addresses of the storage locations at which the values will be placed. When we refer to that variable name later in the program, the computer will always use the value stored at that location. We chose the three variable names NTEST1, NTEST2, and NTEST3 to refer to our three test scores, respectively. We might have used some other names such as I, J, and K. The computer doesn't concern itself with what choices we make as long as we obey the rules of the FORTRAN language. We should select names that provide some meaning to the data items we are using.

The parentheses and commas are necessary delimiters (separators). A comma must be used to separate each variable. A comma might be thought of as saying another variable name follows in the list of variables.

The general form for the syntactical construction of the READ statement is given in figure 3.5.

READ Statement—General Form

READ (unit, n) list

where:

- **unit** — is an unsigned integer constant or variable used as a file reference number.
- **n** — is the statement number of the associated FORMAT statement which describes the attributes of the input data.
- **list** — is an I/O list which may include one or more variables separated by commas.

Figure 3.5. General Notation of the READ Statement

Examples of the READ statement are:

```
  READ (5,98) ITEMNR, NQTY, COST
3 READ (5,187) NRSSAN, HRS, PAYRAT
  READ (INFILE,63) X, Y
```

The syntax construction for the statements in FORTRAN is described by a general form notation to indicate the required parts of the statement and the elements that must be supplied by the programmer. The *general form notation* is a symbolic representation of the rules for constructing a FORTRAN statement. It is from the general form of a statement that the programmer determines what keyword and elements of the FORTRAN language are required and permitted in writing a specific statement.

Throughout this text, the general form notation is given for each new FORTRAN statement. All words in capital letters and all delimiters in a FORTRAN statement must be coded exactly (spelling, location, etc.) as they appear in the general form notation. All words in lower case letters represent elements which are formed and supplied by the programmer. An explanation of each programmer-supplied element is included in the diagram to further explain various aspects of these elements. Optional elements are also indicated. No indication is made as to which statements may be assigned a statement number, since any executable statement may be given a statement number. Each general form notation is enclosed in a boxed diagram to call attention to the statement's general form.

Some compilers permit the use of format-free READ statements which do not require an associated FORMAT statement. That is, the data fields may be punched in any card columns and separated by a comma or space to indicate the end of the field. Thus, no FORMAT statement is needed. Your instructor will tell you about this type of READ statement if it is available on your compiler. (A discussion of format-free input/output statements is included in section 3.9.) Many compilers, however, do not have format-free input/output statements. Thus, a FORMAT statement must be used to supply the needed information about our input data fields as referenced by the READ.

The FORMAT Statement for the READ

The FORMAT statement is used in conjunction with a READ or WRITE statement to describe the form and location of the data on the input or output record. A FORMAT statement is always assigned a statement number that is referenced in the READ or WRITE statement to denote which FORMAT statement is to be used. A unique statement number must be assigned to each FORMAT statement. The keyword FORMAT identifies the FORMAT statement. A FORMAT statement is usually placed after the READ or WRITE statement; however it may be placed at other locations in the program, such as at the beginning for quick reference.

Following the keyword is a set of parentheses which contain format specifications. A comma is used to separate each format specification if more than one is given. A *format code* describes how each individual data field is to be read or written. This input information is obtained from the input record layout.

The two most common format code types are the *Integer (I)* to denote whole numbers and *Floating-point (F)* to denote real (decimal) numbers. Following the format code letter is a value (called the width part) to specify how many positions are included in the field. For example, the first FORMAT statement (assigned statement number 99) has three integer format codes of I3. This implies that we are reading three data fields of three-digit integer numbers, respectively. Remember, each input or output field must have a respective format code to describe that field. Each format code in the format statement corresponds to the variables in the READ (or WRITE) statement on a one-to-one, left-to-right basis. These format codes also correspond to the data fields on the input card on a one-to-one, left-to-right basis. See figure 3.6.

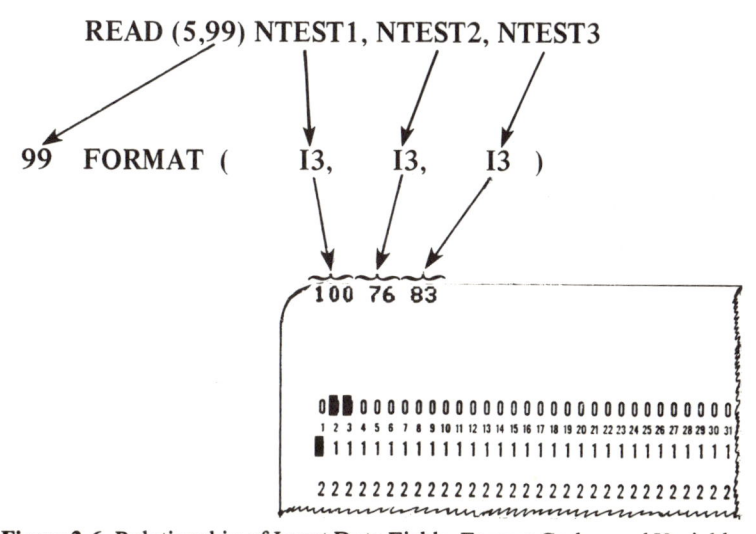

Figure 3.6. Relationship of Input Data Fields, Format Codes, and Variables

The computer assumes the fields to begin in column 1 of an input card unless told otherwise. The width of the field is added to the first position of the field to locate the position of the next field (i.e., column 1 plus a field width of three says the next field will begin in column 4).

If input fields are adjacent to each other and use an identical format code specification, a repetition factor may be used as a shortcut notation in specifying the format codes. The repetition factor is an unsigned integer number (from 1 to 255) which is placed in front of the format code to indicate how many times it is to be repeated. In our FORMAT statement we could have specified 3I3, which means to repeat the format code I3 three times. Remember, for one to use the repetition factor the input fields must be next to each other with no blank columns intervening between the fields.

Figure 3.7 gives the general form of the FORMAT statement.

FORMAT Statement—General Form

---

    nnnnn FORMAT (fc1, fc2, ..., fcn)

where:

nnnnn — represents the 1-5 digit statement number assigned to the FORMAT statement for reference by an I/O statement.

fc1, fc2, ..., fcn — represent the format codes that may be included to describe the form of each data field transferred to or from memory. Each format code is separated by a comma.

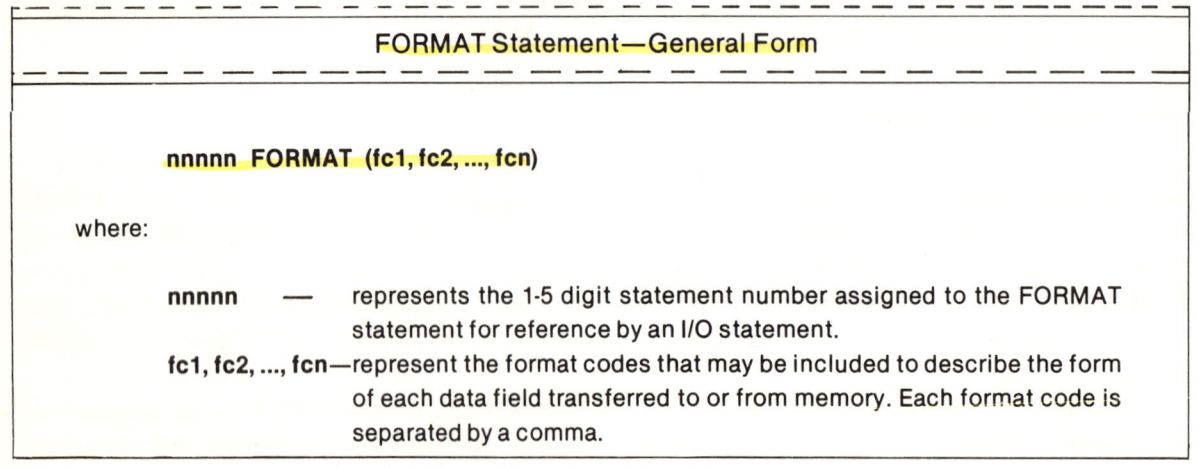

Figure 3.7. General Notation of the FORMAT Statement

Examples of the FORMAT statement are:

    98 FORMAT (I2, I4)
    76 FORMAT (3I3, F5.2, I5)

3.6 The Assignment Statement

Now that the input values have been read into memory and stored under the variable names NTEST1, NTEST2, and NTEST3 we can perform various calculations with them. The next two FORTRAN statements are known as Assignment statements. The function of the Assignment statement is to evaluate the expression to the right of the equal sign (=), perform any necessary calculations, and assign the one resulting value to the variable name to the left of the equal sign. Thus, the Assignment statement is used to perform needed computations and the assignment of the resulting value to a variable. The Assignment statement is often used to assign a single numeric constant to a variable or to move the value assigned to a variable to another variable (location). The equal sign (=) does not mean "to make equal to," but rather "to assign a value to," or "replace the value of" the variable to the left of the equal sign with a new value.

The first Assignment statement causes the three test scores (referred to as NTEST1, NTEST2, and NTEST3) to be added together and the sum placed in the variable named TOTAL. The second Assignment statement then causes this sum at the variable TOTAL to be divided by 3.0 and the results assigned to the variable AVE. Note that the variable names TOTAL and AVE are type real so they can contain fractions.

Figure 3.8 shows the general form of the Assignment statement.

Developing a Basic FORTRAN IV Program   47

Assignment Statement—General Form
---

        variable = expression

where:

        variable   — represents any type variable name.
        expression — may be a single constant, variable, or complex combination of arithmetic operations.

Figure 3.8. General Notation of the Assignment Statement

Examples of the Assignment statement are:

```
  2 KOUNT = 0
    PI = 3.1416
    NEWK = K
 31 XSQD = X * X
    RESULT = 4. * (R+S) / T - 6.5
```

3.7 The WRITE Statements and Their Associated FORMAT Statements

The WRITE statement in FORTRAN causes the computer to write data stored in memory to some output device such as the line printer, magnetic disk, etc. The printer is designated as file unit 6 in FORTRAN on most computer systems. Thus, the 6 within parentheses designates the printer as our output device. The second number within the parentheses specifies the related FORMAT statement to be used with the WRITE command. We use 98 to correspond to the FORMAT statement we have established to describe how our first line of output will be formatted.

The variable names which follow the set of parentheses denote the items to be accessed in memory and to be written on a line of output. The first WRITE statement says to write the three test scores as our first line of output. The values will be written in the order in which they are specified in the WRITE statement.

The second WRITE statement says to write out the computed values of TOTAL and AVE. The values will be written on a new line since a separate WRITE statement is used. That is, each new WRITE statement implies a new output record is to be written. The FORMAT statement number 95 is used to describe the output fields in the second WRITE statement.

Figure 3.9 on the next page gives the general form of the WRITE statement.

Examples of the WRITE statement are:

```
    WRITE (6,97) ITEMNR, HRS, PAYRAT, PAY
 30 WRITE (6,83) X, XSQD
```

The FORMAT Statements with the WRITE

A FORMAT statement is supplied for each WRITE statement to describe the form and location of each output field on an output line. A format code must be supplied for each output item. In addition, an output FORMAT statement for the printer must include a descriptor to specify how the paper is to be moved

---

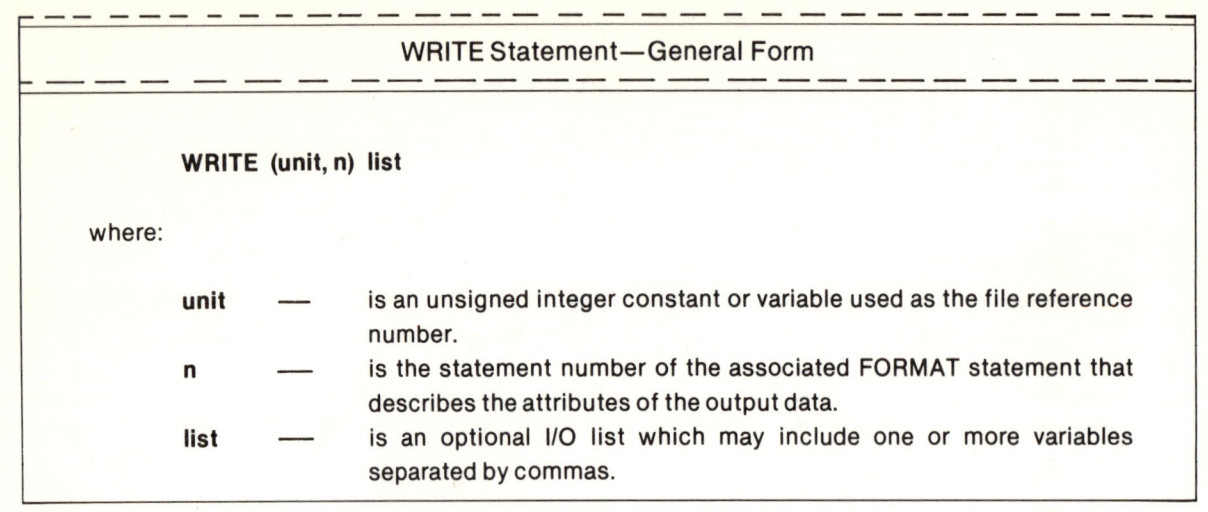

```
                    WRITE Statement—General Form
---

        WRITE (unit, n) list

where:

    unit    —    is an unsigned integer constant or variable used as the file reference
                 number.
    n       —    is the statement number of the associated FORMAT statement that
                 describes the attributes of the output data.
    list    —    is an optional I/O list which may include one or more variables
                 separated by commas.
```

Figure 3.9. General Notation of the WRITE Statement

(called carriage control operations). This descriptor (a *carriage control specification*) must be specified at the beginning of the FORMAT statement before any format codes for the output fields. A carriage control specification of **1H1** causes the printer to start printing at the top of a new page (page eject). We would use a **1H0** or a **1H∅** (∅ represents a blank) to cause the printer to (respectively) double-space or single-space the print line. (A carriage control specification of 1X will also provide single-spacing.) A **1H+** means to suppress spacing, i.e., no paper movement.

Only the FORMAT statements for the WRITEs need a carriage control specification; those for the READ statements do not. The carriage control specification is never printed, but is "stripped off" by the compiler and used to provide the needed character which dictates how the paper is to be moved. Note that the vertical movement of the printer paper is forward only. There is no way we can instruct the printer to roll the paper backwards.

The second FORMAT statement (number 98) corresponds to the first WRITE statement. The carriage control specification of 1H1 inside the parentheses causes our output to begin at the top of a new page. The first format code of I3 causes the first test score (NTEST1) to be printed in the first three print positions. The next format code, 2X, is a new one. The X represents the *blank format code* which is used to specify blank positions on an input or output record. This format code does not equate to any variable in memory. The X format code merely tells the computer to skip positions or look further down the record. The number before the X format code specifies how many blank positions (or columns) to skip. Thus the format code 2X tells the computer to skip two positions from the last position at which it was situated.

The same rule for input format codes apply to output format codes. The first variable in the WRITE statement corresponds to the first format code in the output FORMAT statement which specifies that the form and location of the first value will be printed (NTEST1 in positions 1-3). The 2X causes two positions to be skipped to prevent the first output number from running together with the second. The second format code I3 causes the second test score (NTEST2) to be printed in positions 6-8. The next 2X causes two more positions to be skipped to prevent the second number from rumnning together with the third. The last I3 causes the third test score (NTEST3) to be printed in print positions 11-13. The relationship of the output variables, format codes, and output results is illustrated in figure 3.10.

The third FORMAT statement (number 95) is for the second WRITE statement. We wish to write the values of TOTAL and AVE double-spaced below the previous output line. The carriage control specifications of 1H0 causes the double-spacing of the print line. The value of TOTAL will be written as an F6.2 specification. The format code F says a decimal field with a decimal point will be written on output. The first digit following the F denotes the total width of the field which includes the decimal point. The decimal

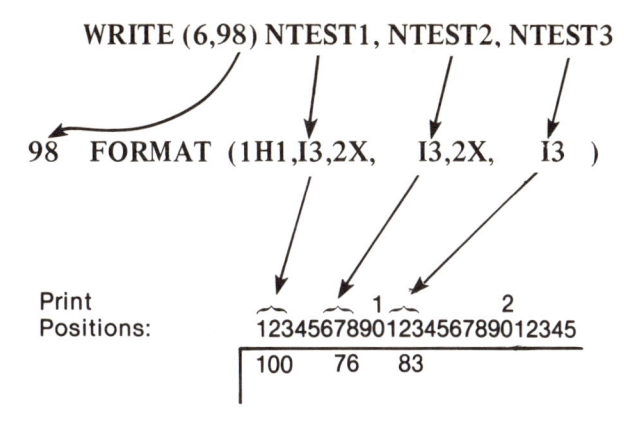

```
         WRITE (6,98) NTEST1, NTEST2, NTEST3

    98   FORMAT (1H1,I3,2X,   I3,2X,   I3  )
```

Print Positions:
```
              1         2
     12345678901234567890l2345
     100   76   83
```

Figure 3.10. Relationship of Output Data Fields, Format Codes, and Variables

point is a necessary separator in the format code. The last digit tells how many digits will follow the decimal point. Thus the format code F6.2 says to use a six-position output field that includes a decimal point and two fractional digits plus a maximum of a three-digit integer. The next format code 3X causes three positions to be skipped so that the next output value will not be run together. The format code F6.2, which includes space for three integer digits, a decimal point, and two fractional digits, is used to output the value of AVE. The F format code is also used to describe real *input* values.

Our lines of ouput with their respective values would appear as:

Print Positions:
```
              1         2
     12345678901234567890
     100   76   83
     259.00     86.33
```

Output fields are always right-justified (aligned from the right) in the output field description (format code). If our output value does not occupy the entire field width specifications, blanks will be inserted in the high-order (left) positions. For example, the value 76 written under a format code of I3 would appear on output as a ƀ76 (where the ƀ is a blank). If the output value ever takes more positions than the field width given in the format code, asterisks will be printed for the output results by most compilers. For example, the value 143 written under an I2 format code would cause two asterisks to be printed as output. (The output results would not be 43 as some might think.) Or the value 259.00 written under an F5.2 format code will cause five asterisks to be printed. As many asterisks are written as the field width specifies. (Note: Some compilers use different symbols, such as ##, to perform the same function as the asterisks.)

3.8 The STOP and END Statements

The STOP Statement

Once the computer begins the execution of a program, it will continue until (1) the computer or computer operator terminates the job, (2) no more data is available to be read, or (3) an instruction is reached that directs the computer to cease processing the program. The first two types of program termination give us what is called an *abnormal termination*. We can cause a *normal termination* in our program by providing a control statement to signal the computer to stop processing. The control statement most frequently used is the STOP statement.

The STOP statement tells the computer to stop the processing of our program. Program execution is terminated, we are given our output results, and the computer begins processing another person's job. The general form of the STOP statement is given in figure 3.11.

50   Developing a Basic FORTRAN IV Program

```
┌─────────────────────────────────────────────────────────────────┐
│                  STOP Statement—General Form                    │
├─────────────────────────────────────────────────────────────────┤
│                                                                 │
│         STOP                                                    │
│                                                                 │
│   or                                                            │
│                                                                 │
│         STOP  n                                                 │
│                                                                 │
│   where:                                                        │
│                                                                 │
│         n      —     is an integer octal constant of 1 to 5 digits. │
│                                                                 │
└─────────────────────────────────────────────────────────────────┘
```

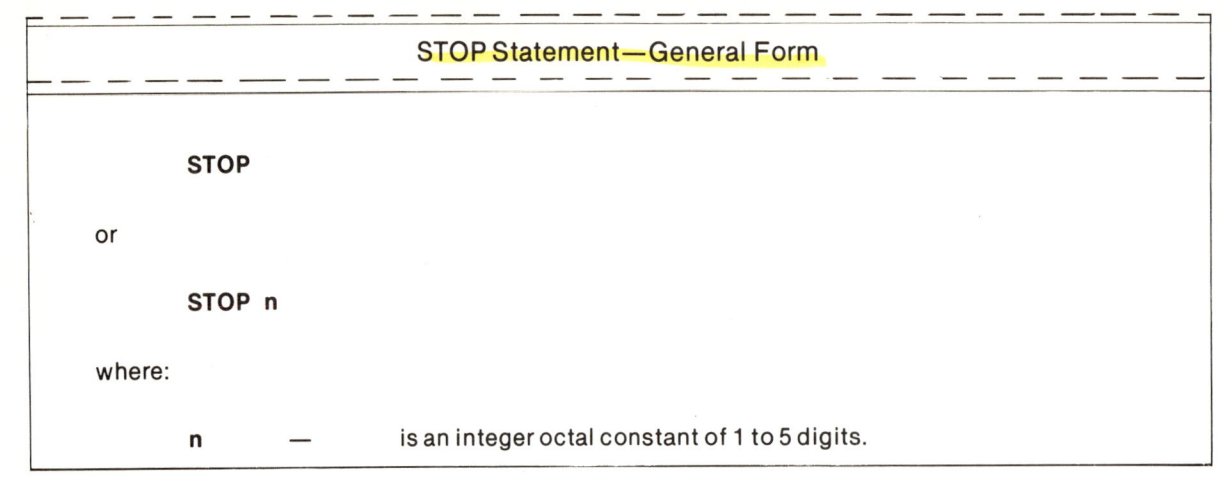

Figure 3.11. General Notation of the STOP Statement

Examples of the STOP statement are:

76 STOP
STOP
STOP 317

The STOP n option of the STOP statement consists of the word STOP followed by an integer octal constant of up to five digits, e.g., STOP 266. It has the same effect as the regular STOP statement; but in addition to terminating the program, the word STOP followed by the octal constant is displayed on the computer operator's console typewriter. If multiple STOP statements have been included in a program, the displayed number provides a record as to the STOP statement which terminated the program.

It is permissible to have multiple STOP statements in a program if conditional testing or branching is used. When a single STOP statement is used in a program, it is normally placed at the end prior to the END statement. This placement, however, is not mandatory. It may be placed anywhere in the program as long as it is the last statement executed in the program logic.

In some FORTRAN compilers another statement available by which to terminate program execution is the CALL EXIT. The general form of this statement is simply: CALL EXIT. This statement is actually a call to a system program which performs program wrap-up and termination. It is used in the same manner as the STOP statement.

The END Statement

The last statement in every FORTRAN program must be the END statement. This statement simply says this is the end of the program. The END statement does not end processing as one might be lead to believe; the STOP statement causes this action. The END statement is nonexecutable; it is a specification statement used to inform the compiler that no more FORTRAN source statements follow in the program. Thus, it is a necessary statement that must be written at the end of all FORTRAN programs.

The END statement consists solely of the word END. It must not have a statement number assigned to it; if it does, a compilation error will occur. Only one END statement is allowed in each FORTRAN program. Figure 3.12 gives the general form of the END statement.

Many beginning students are confused as to the order in which the FORTRAN statements are processed. The computer starts with the first statement and executes the statements in a sequential order.

END Statement—General Form

END

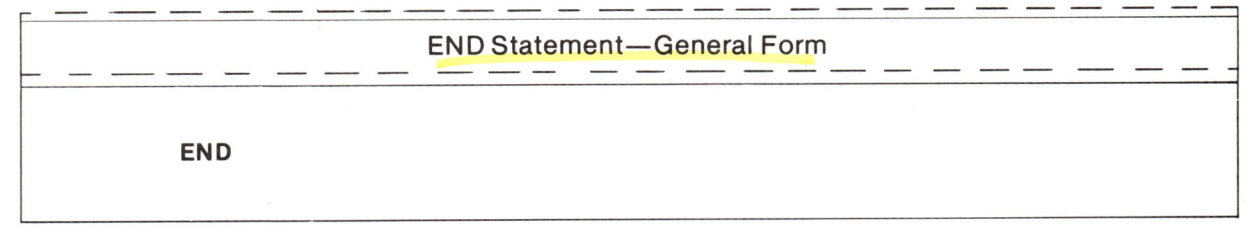

Figure 3.12. General Notation of the END Statement

Thus, the order in which the statements are arranged within the program is very important. You could not have placed the Assignment statement to compute the average of the test scores before the statement to compute the total. Had you done so, the computed AVE would not have been correct. The computer executes one statement at a time in sequence. When it computed AVE from

$$AVE = TOTAL / 3.0$$

it does not look around or check to see if TOTAL had been calculated already. The computer assumes that you know what you are doing and have already given values to all of the necessary variables for any calculation. If any value has never been defined, the results are, of course, incorrect. Thus, the statements to be executed must be in the proper sequence; otherwise our processing does not match the required logical order of the solution to the problem.

The six basic statements that we have just discussed are normally used in every FORTRAN program. The discussions in this chapter were intended to give you the basic features of these statements to let you start writing simple FORTRAN programs as soon as possible. Chapters 5 through 7 will present a more detailed explanation with illustrations and rules on how to use the Assignment, READ, WRITE, and FORMAT statements for more complicated problems.

The best way to learn to program in FORTRAN is to write FORTRAN programs. Sure, you will make mistakes, but so do the most experienced programmers. It is a very rare occurrence for a seasoned programmer to write a fairly lengthly program without a bug (error). The more experience you acquire in writing programs, the easier it becomes to develop and write programs.

Programming is an art and not a science as is the field of mathematics. The art of learning to program is analogous to the art of learning to drive a car or to swim. We can talk about driving (or swimming) all day and discuss elaborate techniques to use. But the true test comes when we get behind the wheel (or get into the swimming pool). It is the actual practice that pays off. We may swerve around the road or flounder in a pool at first. Experience is our best teacher. It is with each new attempt that we incorporate the information learned from our previous experiences and become more proficient. So in programming we each apply our own techniques and style learned from "doing."

Even though we have written a basic FORTRAN program, there are additional steps that must be completed before one can get results from the computer. First, you must keypunch the source program statements onto punched cards for entry into the computer. Various control statements must be included with our source deck and data cards to identify our job and specify what we want done. Chapter 4 provides a detailed explanation of how you go about accomplishing these steps in order to run your program on the computer.

3.9 Nonstandard Language Extensions (Format-free Input/Output Statements)

The Waterloo WATFOR and WATFIV compilers provide format-free forms of input and output statements, easy to use in reading and writing data, for the beginning student. Thus, the student does not have to worry about the confusing data-field specifications for the format information of input and output records. The student can, therefore, concentrate more on the basic principles of programming concepts.

Format-free Input

The two general forms of the format-free READ statements are:

READ, list

and

READ (5,*) list

where **list** is a list of variables, each separated by commas. The * in the second form is necessary to indicate free-field input to the compiler. Examples are:

READ, NTEST1, NTEST2, NTEST3
READ (5,*) NTEST1, NTEST2, NTEST3

The data fields for format-free input are punched in one or more punched cards in the form of proper type constants. One or more blanks or a comma must be used to separate each data field on the card so that the compiler can isolate each value. The input values do not have to appear in any specific card columns, since no format specifications are given. For example:

```
100    76    83,    123.56,
```

Format-free Output

The two forms of the format-free output statements are:

PRINT, list

and

WRITE (6,*) list

where **list** is a list of one or more variables, each separated by a comma. The * is a necessary symbol to indicate free-field output. For example,

PRINT, NTEST1, NTEST2, NTEST3
WRITE (6,*) TOTAL, AVE

The output items are printed according to their specific type. The items are evenly spaced across the printer page with a maximum of ten items per line. Real values are always represented in an exponential form. Carriage control operations are automatically performed by the computer.

Other compilers also provide this format-free input/output capability. For example, the Burroughs 6700 provides free-field READ and WRITE statements which uses a slash (/) in place of a format statement number to indicate free-field data. The general form of these statements are:

READ (5,/) list

and

WRITE (6,/) list

Developing a Basic FORTRAN IV Program   53

The use of format-free input/output statements are highly encouraged by many instructors. The student has enough trouble learning FORTRAN without getting bogged down with format specifications at first. Some compilers, like ANSI and IBM, do not have this capability, thus the student must use the general READ and WRITE statements for the input/output operations.

File Unit Numbers

The file unit number which designates the type of device a data file is read from or written to varies by computer system and the compiler. For example, IBM Disk Operating Systems (DOS) use file unit 1 to indicate the card reader and file unit 3 to indicate the printer. Most of the other FORTRAN compilers use 5 and 6 to designate the card reader and line printer, respectively.

3.10 REVIEW QUESTIONS

1. What purpose does a print layout form serve? *for use in describing the format and location of printed output results*
2. Why is internal program documentation so important?
3. The FORTRAN statement used with READ and WRITE statements to describe the form and location of data items is the *format* statement.
4. What is the purpose of each format code in the FORMAT statement? *Each format code corresps to the variables in READ on a 1-1, L→R basis and also corresp to data fields on the input card.*
5. Many calculations can be performed in a single Assignment statement. (True/False) *T*
6. What is the purpose of the carriage control specification in printer output FORMAT statements? *instructs the comp. how the paper is to be moved for WRITE STATEMENTS*
7. If a data item in memory is in integer form, it may be written using an integer variable and the floating-point (F) format code. (True/False)
8. The FORTRAN statement used to terminate execution of the object program is the *STOP*.
9. More than one STOP statement is permitted in a FORTRAN program. (True/False) *T*
10. At least one STOP statement must be placed at the *physical* end of the program. (True/False) *F — only one STOP when logic is fini.*
11. The END statement is simply a command to the FORTRAN compiler to mark the end of a source program. (True/False) *T*
12. The END statement may also terminate execution of the object program. (True/False) *F*
13. Only one END statement is allowed in a program and must be the last physical statement. (True/False) *T*

3.11 EXERCISES

1. Find the errors in the following statements: *comp. would not take this as error would use HOURSRATE as 1 var.*
   a) READ (5, 10) HOURS RATE *no comma betw. HOURS and RATE*
   b) 10 FORMAT (F5.2, 3X, F4.1 *needs an end parenthesis* (*X means omit or skip*)
   c) GPAY = HOURS ⊗ RATE *should be ***
   d) WRITE (5, 33) HOURS, RATE, GPAY *refers to reader / code no. for printer is 6.*

2. Find the errors in the following statements:
   a) READ (6;20) HOURS, RATE *wrong code, should be 5 — The semicolon betw. 6 & 20 should be a comma*
   b) 20 FORMAT (F5.2; 3X; F4.1)) *no semicolons, only commas*
   c) GRSPAY = (HOURS)(PAY) *should be specified (operation)*
   d) WRITE (6, 88) HOURS, RATE, GRSPAY

3. Given the following READ and associated FORMAT statement, indicate what format codes go with which variables.

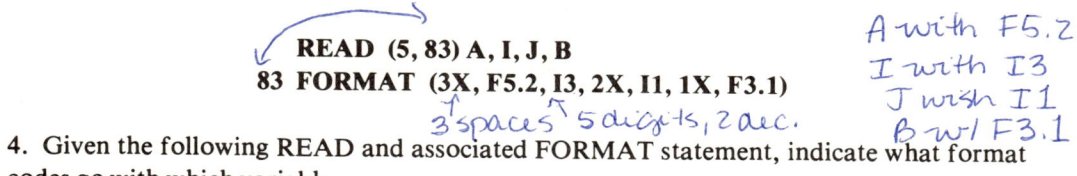

READ (5, 83) A, I, J, B
83 FORMAT (3X, F5.2, I3, 2X, I1, 1X, F3.1)

3 spaces, 5 digits, 2 dec.

A with F5.2
I with I3
J with I1
B w/ F3.1

4. Given the following READ and associated FORMAT statement, indicate what format codes go with which variables.

READ (5, 13) I, K, L, C
13 FORMAT (1X, I2, 3X, 2I4, 4X, F4.2)

5. Given the following READ and associated FORMAT statement, indicate the given error between any variable and listed format codes.

READ (5, 88) I, A, J
88 FORMAT (I4, 3X, 2I5)

A (real valued) should not be read w/ I5

6. Given the following READ and associated FORMAT statement, indicate the given error between any variable and listed format codes.

READ (5, 67) PRIN, RATE, TIME
67 FORMAT (F8.2, 3X, F4.3, 1X, I2)

T should not be read w/ I2

7. Denote the order of computations in the following Assignment statement.

A = B - C / (D + E) * S

*+, /, *, -*

8. Denote the order of computations in the following Assignment statement.

Z = X / Y * ((3.0 + T) / R)

w/in ( )
*+, /, /, **

9. Given the following WRITE and associated FORMAT statement, which format codes go with which variables?

WRITE (6, 18) A, B, C, I
18 FORMAT (1H0, F4.1, F5.1, 3X, F6.2, I3)

A with F4.1
B with F5.1
C with F6.2
I w/ I3

10. Given the following WRITE and associated FORMAT statement, which format codes go with which variables?

WRITE (6, 97) I, A, J, B, C
97 FORMAT (1H0, 2X, I4, 3X, F5.2, I3, 5X, 2F6.1)

I with I4
A with F5.2
J with I3
B,C with 2F6.1

3.12 PROGRAMMING PROBLEMS

1. Write a FORTRAN program to compute ones pay according to the equation:

pay = rate times hours worked

Your program should read the values of rate and hours worked as decimal numbers in the following card columns:

Column	Field	Value
1-4	Rate (Form = N.NN)	3.50
5-8	Hours worked (Form = NN.N)	85.0

Write the two input fields and computed pay on a single printer line. (Use a carriage control specification of 1H1.) The output positions are:

Field Description	Print Positions
Rate	1-4
Hours worked	7-10
Computed pay	15-20

2. Write a program to read three 2-digit integer numbers and solve the equation:

$$N = 2I + J/3K$$

Use I, J, and K as the integer variables assigned to adjacent input fields beginning in card column 1. Let I = 15, J = 90, and K = 5. Write the three input values and the computed value of N on a single-spaced printer line. Use print positions for the output fields as follows:

Field Description	Print Positions
I	1-2
J	6-7
K	11-12
N	17-20

3. Write a program to compute simple interest according to the formula i = p r t. Read the values of principal, rate, and time from a punched card as follows.

Card Columns	Field	Value
1-8	Principal (Form = NNNNN.NN)	20000.00
10-13	Rate (Form = .NNN)	.075
15-17	Time in years (Form = NN.)	05.

Write output results for the three input values and the computed value of interest on a double-spaced printer line. Use output field positions as follows:

Field Description	Print Positions
Principal	1-8
Rate	11-15
Time	18-20
Computed interest	25-32

4. Write a FORTRAN program to read the values of X and Y and solve the equation Z = 2X + 3Y. These values will be read from a card as described below:

Card Columns	Field	Value
1-6	X (Form = NNN.NN)	135.63
7-12	Y (Form = NNN.NN)	081.55

Write output results for X, Y, and Z on a double-spaced line with output field positions as follows:

Field Description	Print Positions
X	1-6
Y	10-15
Z	20-27 (Form = NNNNN.NN)

5. Write a program to read the values of length and width from a punched card and solve for the area of a rectangle according to the formula A = l w. The input fields are punched as follows:

Card Columns	Field	Value
1-5	Length (Form = NN.NN)	20.50
6-10	Width (Form = NN.NN)	08.75

Write the three output fields on a single-spaced line in the following print fields:

Field	Print Positions
Length	5-9
Width	15-19
Area	25-31 (Form = NNNN.NN)

6. Write a program to read the value of radius (15.62) off a punched card in the form NN.NN beginning in card column one. Compute the area of a circle according to the formula a = πr^2 where π = 3.14. Write output results on a double-spaced line with the value of radius and the computed area in print positions 2-6 and 10-15, respectively.

7. Write a program to read the values of rate and time from a punched card as follows:

Card Columns	Field Description	Value
2-5	Rate (Form = NN.N)	60.5
6-9	Time (Form = NN.N)	03.0

Solve for distance according to the formula d = r t. Write output values for rate, time, and distance beginning at the top of a new page. Each output value will be printed in the following locations:

Field	Print Positions
Rate	10-13
Time	20-23
Distance	30-35 (Form = NNN.NN)

8. Write a program to read an input value for X as the decimal number, 26.32, punched in columns 1-5 in the form NN.NN. Compute the square and square root of the value read for X. The square root of a real number may be obtained by raising the number to the one-half power. Print output results as a double-spaced print line with the following field positions:

Field	Print Positions
X	5-9
Square	13-19
Square root	23-27 (Form = NN.NN)

Processing Your FORTRAN Program

4.1 Introduction to Processing a FORTRAN Job

In chapter 3 we developed a basic FORTRAN program to compute the sum and average of three test scores. Now that we have a coded source program, how do we run it on the computer?

The purpose of this chapter is to take you through the remaining steps for the processing of your program. We will see how you convert your coded program to the input medium of punched cards so that it can be read by the computer. The additional computer control cards that are needed to tell the computer what to do with your program are discussed. This information is explained in section 4.2. The chances of your program producing the correct output results the first time it is run are very slim. Sections 4.3 and 4.4 discuss some of the types of errors you will probably encounter and how to deal with them.

Almost every program written for an application problem includes a program loop. The remaining sections of this chapter discuss the concept of looping in a program and the two basic control statements to provide this capability. Our basic program is expanded to compute the sum and average of the test scores for more than one student. Thus, a program loop is needed and requires only the addition of two control statements.

4.2 Preparing Your Program for the Computer and Running the Job

Once the FORTRAN source program has been coded, it must be transferred to some input medium to be input to the computer. FORTRAN statements are normally punched into the common 80-column punch card on the keypunch machine. Each coded line on the coding sheet is punched into one punch card. Specially designed FORTRAN program cards or any 80-column punch card may be used. The resulting program on punched cards is referred to as a *source deck*.

You may be required to keypunch your own program. However, many computer centers employ keypunch operators who will punch your program. You simply turn your coded forms in at a designated location and return the next day to pick up the punched deck. From then on, any corrections you make to your program become your responsibility. Thus, keypunch machines are usually available in the computer center for this need. You should read *Appendix B* on the operation of the keypunch if you are not familiar with its use.

After you pick up your punched program (if keypunching is provided by the computer center), you should carefully compare the punched cards with your coded forms. Since keypunch personnel normally have a heavy workload, they do not usually remove any cards they may have punched with an error. For the sake of time they usually leave the error card in the deck and punch the correction in a new card.

The process of checking your punched cards is known as *desk checking*. You should examine each card to remove any erroneously punched card left by the keypunch operator, to see if all statements start in column seven or later, to see if you have not left out a comma or other required punctuation, and to find any other error that may be easily detected. You should desk check your deck even if you punched your own cards. (Note: Desk checking is also known as **bench checking** by many programmers.)

In preparing a FORTRAN program for the computer, three different types of cards will normally be used. They are:

1. FORTRAN Source Statement cards
2. Job Control Language (JCL) cards
3. Data cards

FORTRAN Source Statements

FORTRAN source statements are the instructions that make up your program. These cards should be arranged in the proper order (as coded) to solve the problem.

Job Control Language (JCL) Cards

Job control cards, better known as JCL cards, are needed to communicate various resource needs to the computer. Modern computer systems can process many jobs (tasks) concurrently (called *multiprogramming*). A *"JOB" card* is used on most computers to identify the beginning of each new job. The JOB card normally contains the job name, the name of the programmer, and various job accounting information. Additional JCL cards are usually needed to request resources such as a specific compiler, to identify input and output devices, to indicate the beginning of the data cards, and to indicate the end of the data cards. JCL cards to process a FORTRAN job vary by computer and also by installation. Because of the wide variety of JCL cards, the subject of Job Control Language is beyond the scope of this text. The appropriate JCL cards will be explained to you by your instructor for the respective computer system used. *Appendix E* shows the JCL cards in common use on various computer systems.

Data Cards

Data cards are used to contain the data to be read and processed by the program. A data card is distinct from the FORTRAN source statement and is not subject to the same arrangement and rules as the FORTRAN statement. All eighty columns of a data card may be used to contain fields of data. The location of data fields on the data card is unique to each program.

Data cards may contain any of the characters acceptable to the computer system; however, many computers may not be able to print some of these characters because they are not provided on their printer. Data cards are placed after the end of the FORTRAN program and are preceded by a JCL card to indicate their beginning. The number of data cards and their order are, of course, unique to the execution of every program.

It is possible for a FORTRAN program to generate its own data for processing in the program. To provide much more flexibility and practicality in a program, however, the data to be processed is generally read from some input medium. Data may be read from any input medium, which includes not only punched cards, but also paper tape, magnetic tape, and magnetic disk. The most common input medium normally used in an academic environment is the punch card.

Job Deck

After the three categories of cards have been punched, they are arranged in the proper order to form a *job deck.* Most of the JCL cards will precede your source program. The data cards follow the FORTRAN source deck. The FORTRAN source deck and data cards are separated by a JCL card. A JCL card is normally required after the last data card to mark the end of the job deck. Figure 4.1 illustrates a constructed job deck setup which includes the JCL, FORTRAN source statements, and data cards.

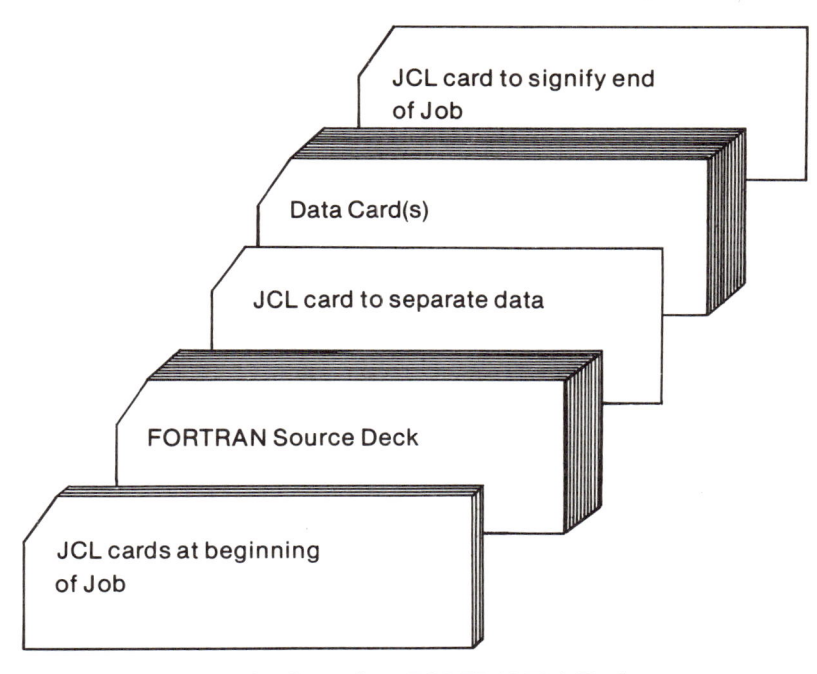

Figure 4.1. Setup for a FORTRAN Job Deck

The arrangement of the three different types of cards is also shown in **Appendix E.**

The job deck is now ready for the computer. The use of job decks to be read and processed by the computer is known as *batch processing.* The term batch processing comes from the method of collecting and processing jobs in a batch as the normal mode of operations. FORTRAN programs may also be entered into the computer for compilation and processing by interactive terminals (two-way communications between the user and computer). This mode of operation is called **timesharing.**

Timesharing means that you are sharing time on the computer simultaneously with other users. Terminals such as cathode ray tubes (CRTs) and typewriters allow the programmer to interact with the computer in a conversational mode. You can create data files on disk. You can type in your programs over the terminal. The programs may be compiled and executed according to timesharing commands from the terminal. Input data may even be entered at a terminal and output results may be displayed back to the terminal.

Timesharing operations are widely used and extremely beneficial in a production programming environment. The timesharing commands used to interact directly with a computer vary according to each computer system used. For this reason, the presentation of these commands is beyond the scope of this text. A user's manual or an experienced person is the best way to learn how to use a timesharing terminal and language if this capability is available at your school. A brief overview to the concepts of timesharing operations is given in **Appendix C.**

Timesharing terminals and commands may be covered in some advanced programming courses. But the students in a basic programming course will normally prepare their FORTRAN programs on punch cards. Thus, we will consider batch processing as our normal mode of operation in running FORTRAN jobs.

The FORTRAN job is now ready for submission to the computer facility for entry into the computer. The job deck is loaded into a card reader along with other jobs to be read and stored in the computer system. The start of each job is marked by a job card. You may load your own job deck or turn it in to an operator to load. The mode of operation varies among computer centers.

The jobs remain in a queue (holding area) until called for by the computer. When a new job is identified, the computer determines what is needed from the JCL cards. If a FORTRAN "compile and go" (compile and then execute) job is found, the system calls in the FORTRAN compiler to read the source program and generate the proper machine language instructions for execution. If an error is found in any of the source statements, a syntax error message is produced and printed out along with the source program; no program execution is normally attempted. Whenever the source program has no syntax errors, the compiler can generate all the necessary object code.

When the FORTRAN compiler encounters the END statement without any errors, all of the machine code has been generated. Next, the resulting object program is loaded into memory for execution. The computer then begins executing the object program at the first machine language instruction. The object program reads the input data, processes it, and produces the output results. Execution continues until it encounters the STOP instruction, which tells the computer to terminate processing. Occasionally, a "system error" (one the computer cannot handle) may be encountered during execution. This will terminate the program abnormally.

The output after execution will consist of a source program listing followed by the results (if printed). The results may be produced on any output medium; however, in a student environment they are usually printed.

The process of converting the FORTRAN source program to machine code and executing these instructions is illustrated in figure 4.2.

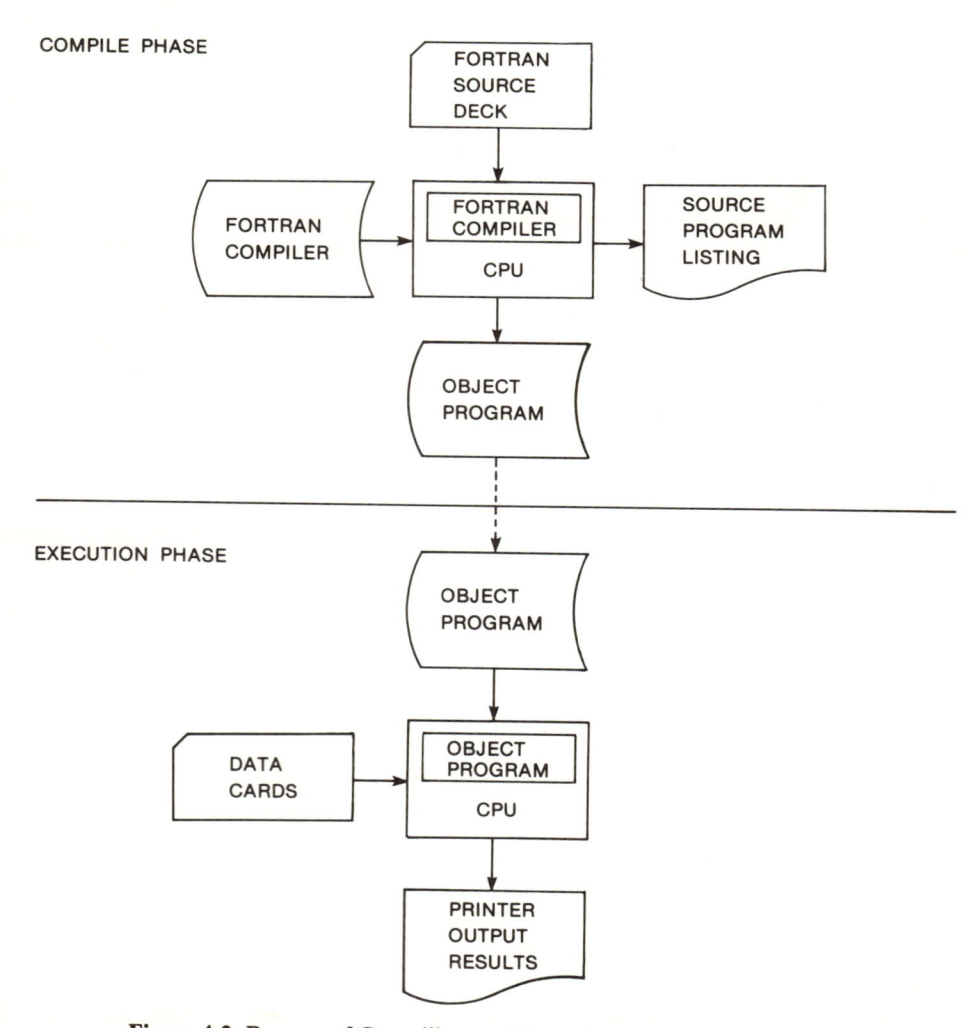

Figure 4.2. Process of Compiling and Executing a FORTRAN Program

Processing Your FORTRAN Program   61

In the compile phase the FORTRAN compiler (via JCL cards) is called into memory to read the FORTRAN program and generates the machine language instructions (object program). If the compile phase is successful, the execution phase is entered. The object program is processed and the output results produced.

If errors are found in the compile phase, the execution phase is not attempted. Even if the execution phase is entered, the output results may not be correct. Let us examine the errors that could occur in either of these two phases.

4.3 Program Compilation Errors

For the first run or so, it is very likely that the student will receive a listing that contains a number of error messages instead of the expected answers. These error messages are printed by the compiler during the program compilation. The FORTRAN compiler scans the program, one statement at a time, checking for the correct syntax. *Syntax* means the construction or grammatical arrangement according to the rules of the FORTRAN language. If it encounters a misspelled keyword, an extra comma, an incorrectly formed constant or variable, or any statement that is incorrectly formed, an error message is printed to flag the error.

These error messages are called *diagnostic or syntax messages* since their purpose is to help you "diagnose" the statement in error. The form of these diagnostic messages varies from compiler to compiler. Some compilers will print only an error message number which may be found in a manual with its corresponding description of the probable error. Other compilers, like the WATFOR and WATFIV compilers, print a highly descriptive error message which is much clearer and more understandable.

There are two types of diagnostic messages that may be produced in the program—error and warning. *Error messages* will cause termination of the program during the compiling phase, and no object code is generated. All errors must be corrected before the program can be processed. *Warning messages* tell us that a violation of some FORTRAN rule occurred, but was not severe enough to prevent the compilation of the program. The program may, in most cases, still be executed.

The error message is normally printed immediately below the statement in error. In some cases, however, the error message may actually appear below the statement just after the one in error. This happens because the compiler may not detect an error until the statement immediately following the one in error has been read. This is understandable, since statements may be continued onto additional cards.

Most compilers attempt to indicate the general area of the error by printing a special symbol such as a $ or * as near the error as possible. In most cases, the "flagging" symbol is printed below the character in error. For example, if we omitted a comma in the I/O list of a READ statement and thus exceeded the maximum length of a variable, the IBM G compiler would print diagnostic messages as follows.

READ (5,99) NTEST1, NTEST2 NTEST3
$

01) IEY003I NAME LENGTH

Note the $ printed beneath the N in the third variable. Since this is the seventh character in forming a variable name, the compiler knows that an error has been made. It doesn't know whether you omitted a comma or if you tried to use a variable name longer than six characters. So the error message number 01 (the first one found) that reads "IEY003I NAME LENGTH" is printed.

As mentioned before, the error message may be descriptive of the error or it may not be. But you know a syntax error has been found. A manual of error messages, that explains the IEY003I type, may be consulted for a better explanation of what may have caused the error. Some of the more common syntax messages found in IBM compilers are shown on the next page.

62   Processing Your FORTRAN Program

Message	Explanation
IEY001I ILLEGAL TYPE	The type of a constant, a variable, or an expression is not correct for its usage.
IEY002I LABEL	The statement in question is unlabeled.
IEY003I NAME LENGTH	The name of a variable exceeds six characters in length; or two variable names appear in an expression without a separating operation symbol.
IEY004I COMMA	The comma required in the statement has been omitted.
IEY005I ILLEGAL LABEL	Illegal use of a statement label. For example, an attempt is made to branch to the label of a FORMAT statement.
IEY013I SYNTAX	The statement or part of a statement does not conform to the FORTRAN IV syntax.
IEY015I NO END CARD	An END statement is missing.
IEY022I UNDEFINED LABELS	A reference is made to a statement label which does not appear in the problem.

The WATFOR and WATFIV compilers provide much better explanatory messages. Figure 4.3 illustrates some examples of syntax errors that one might make in our basic program to compute the total and average of three test scores.

The first message indicates that we have used too long a variable name, whereas, in reality, we omitted the needed comma. The second message tells us we forgot to include the closing parenthesis. The third message says we formed the format code for X incorrectly. There must be a width part even if we want to really say one space (1X). The fourth example tells us we have already used the statement number 98 and, thus, have a duplicate label. The fifth error is the type that appears when we misspell a keyword; the compiler doesn't know what we are trying to tell it. The sixth error is referring to a statement number which does not exist in our program. We punched the wrong number on the third FORMAT statement. It should have been labeled 97 instead of 98. The compiler doesn't know this, but it knows that we should have a statement with the label 97 since it is referenced in the second WRITE statement.

This last syntax message brings us to a discussion of the second location of syntax messages in our program. We may have formed all our source statements correctly according to the syntax rules. But if we forget to include a referenced statement in our program or make an error such as we did in punching the wrong label, the compiler will catch it. The statement may have been placed near the end of our program, so the compiler could not have detected a missing statement number until all statements had been scanned. Thus, any unresolved statement numbers, etc., are included in error messages that are printed after the end of our program. So make sure you also look for any syntax messages following your program as well as inside the program.

Most compilers provide a count of the number of syntax errors found in your program. This information is usually located after your program or on the next page. To make sure you have located all of your

Processing Your FORTRAN Program   63

                    READ (5,99) NTEST1, NTEST2 NTEST3
WARNING NAME NTEST2NTEST3 IS TOO LONG; TRUNCATED TO SIX CHARACTERS
                    99 FORMAT (I3, I3, I3
ERROR NO CLOSING PARENTHESIS
                    TOTAL = NTEST1 + NTEST2 + NTEST3
                    AVE = TOTAL / 3.0
                    WRITE (6,98) NTEST1, NTEST2, NTEST3
                    98 FORMAT (1H1, I3, X, I3, 2X, I3)
ERROR INVALID FIELD OR GROUP COUNT NEAR 3, X
                    WRITE (6,97) TOTAL, AVE
                    98 FORMAT (1H0, F6.2, 3X, F6.2)
WARNING STATEMENT NUMBER 98 HAS ALREADY BEEN DEFINED
                    SOTP
ERROR UNDECODEABLE STATEMENT
                    END
ERROR MISSING STATEMENT NUMBER 97

Figure 4.3. Sample Program with Syntax Error Messages

syntax errors, check the printed count with the number you have found. If your program does not contain any syntax errors, this count message will indicate 0 errors found.

Always try to locate and analyze the syntax errors yourself. Don't immediately rush off to find your instructor or someone else to point out the error for you. A trait of a good programmer is the ability to find his syntax errors himself. The compiler does the best it can to identify the error and its location in a statement. So use the diagnostic messages to help you find the error. Go to your manual and reread the rules (general notation) for the statements.

Some syntax errors are extremely hard to find. You may have punched the letter O instead of the digit 0 (zero) for a statement number. Both characters look very much alike on the printed page, but to the compiler, they are certainly different. The letter O is normally squared at the corners, where the digit 0 has rounded corners. You must concentrate on and examine each character very closely. You may also need to examine the preceding statements. The more practice you have in locating compilation errors, the easier it becomes for you to spot them. Skill in locating your errors comes only through experience. Being able to detect and correct syntax errors is a sign that you truly understand the FORTRAN language.

The best thing you can do is to correct all the errors you can find and resubmit your program for another run. Fixing one error will, quite often, cause other obscure syntax messages to disappear. But do not resubmit your program simply hoping all the errors will go away. If the compiler detected them during the first run and you did not make any corrections, you can rest assured they will be there in the next run.

Sometimes correcting one error may cause new errors to appear; however, this is highly infrequent. Don't panic if you can't find all the errors. Take a break and come back to the program later. You may have fixed your thoughts into a certain "rut," so looking at the problem later may give you new ideas. When you can't find the error after an intensive search, don't be afraid to ask your fellow students if they can spot the error. Everyone likes to be an expert at finding bugs, so they should be eager to assist you. When all else fails, see your instructor or another expert in the language.

Many types of compilation errors can occur from FORTRAN statements. Every error must be found and corrected in order for the computer to be able to execute your program. However, after the program is clean of all compilation errors, it may still contain logic errors.

4.4 Program Execution (Logic) Errors

After all compilation errors are removed, we now enter the cycle of testing and debugging a program. The output must be checked to verify the results. A program may produce answers, but they may be incorrect. We must, therefore, modify our program and rerun it for the new results. This cycle continues until the correct output results are obtained.

Execution errors are in the logic of the program, itself. They may be caused by performing a calculation incorrectly, reading the input data from the wrong card columns, using the wrong variable, or having the source statements in an incorrect sequence that performs the logic incorrectly. Sometimes the input data may be punched wrong; cards may also have been dropped from a source deck or the data deck.

The best way for the student to check as to whether the correct results are produced is to make hand calculations from some of the data used and determine what output results are expected. Hand calculations for one set of data required for each alternative logic route in the program will usually suffice. That is, if a program reads data from a card record, uses that data in various calculations, prints the results, and then returns to read additional records to do the same operations, it is necessary only to compute the results for the data from one record. If the hand calculations for the data on one record match the computer results, the rest of the data records will probably be correct.

When incorrect output results are produced, **play computer**. Follow the program through manually. One can normally pinpoint a logic error to a particular statement or group of statements. So examine each of these statements carefully and verbally describe what actions the statement is telling the computer to do—not what you want it to do. For example, you might be trying to read data on unit 6, the printer, or write data on unit 5, the card reader. These actions are impossible, but no error message may be issued as a result of these statements. Describing the specified actions orally often helps you detect your errors.

To ensure that the data has been read correctly, *echo check* the input values. That is, include a WRITE statement after the READ to make sure that the input values are read and stored correctly. If there are many calculations in your program, write out the intermediate results to determine whether the calculations are right. There are many techniques you can use to debug logic errors in your program. Chapter 14 describes the procedures and techniques you can use for complex logic problems. At this stage, you can follow your simple programs through manually and not have to go to the trouble of including extensive trace and debug routines. Again, experience in finding your own logic errors will benefit you in the long run.

Many logical errors can be avoided by spending a bit more time in the planning phase of programming; analyzing the problem, flowcharting the logic, writing the code, and desk-checking the program. The better job you do in these steps, the less time you will, more than likely, have to spend in the testing and debugging phase. The structured approach to programming emphasizes that one spends more time in the design phase, which generally results in a more error-free program.

We have gone through the steps necessary for developing and processing a FORTRAN program. A variety of procedures for submitting and running job decks are used by the different installations and universities. Your instructor will brief you on the procedures followed at your school. When in doubt, however, ask questions. Failure to observe your school's rules and procedures could mean a lost program deck or output results.

4.5 Expanding the Basic Program to Include a Loop

The calculation of the sum and average of three test scores for one student could certainly be done more quickly and easily by hand or with a pocket calculator. Solving this problem by means of a computer program, and getting the results will take hours, where by manually we could compute these values in a matter of minutes. So, why do we use the computer? The primary advantage in using a program manifests itself when a large volume of the same calculations must be made. Say we need to compute the sum and average of the test scores for all the students in six sections of a course. Then, the computer can be used for its primary purpose; it can calculate the values many times faster than we can, and without any mistakes. The program can be used for any class, semester after semester, with the only requirement being the punching of new data cards.

The only change we would need to make to our basic program would be to include a loop so that the same operations are repeated for new data values. To form a program loop all we need do is to add several control statements that will transfer execution back to a prior instruction. This technique or type of logic is referred to as a program loop or looping.

Looping is the process of repeating a series of instructions in the program. The group of repeated statements is known as the *loop*. The loop may consist of a section of a program or even the entire program. The technique of looping is used when one wishes to perform the same operations on more than one set of data. Thus, only one routine must be written within a loop that can be used to process any number of data sets.

The basic logic of a loop is illustrated as:

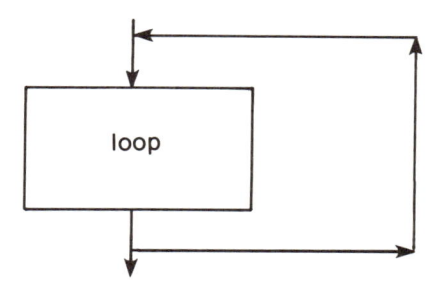

```
        ┌─────────┐
   ┌───▶│         │────┐
   │    │  loop   │    │
   │    │         │    │
   │    └─────────┘    │
   │         │         │
   └─────────┴─────────┘
```

A GOTO statement may be used in FORTRAN to transfer control back to the beginning of the loop. This statement consists of the keyword GOTO followed by a statement number. The statement number is one assigned to a FORTRAN command to which we wish to branch (transfer control of execution). Of course, the statement to which the branch is made must be an executable statement only. The first statement in the loop is assigned a statement number and is referenced by the GOTO to tell the computer where to transfer control. For example:

   1 READ (5,99) . . .
     . . .
     . . .
   GOTO 1

You may wonder how we get out of the loop. Another control statement is used to test for a special condition that tells us when to end our loop. The control statement most frequently used is the Logical IF. The Logical IF statement allows us to test various values to establish a condition. If the condition is true, we can branch out of the loop, stop the program, etc. If the condition is not met, we continue on in the loop. For example:

    IF (N .EQ. 0) GOTO 20

The condition is expressed within a set of parentheses. A relational operator .EQ., which means "is equal to," is used to make the test between the specified variables and/or constants. If N is equal to 0 then the statement GOTO 20 would be executed. Otherwise, the next statement after the IF is executed.

Let us expand our basic program to compute the sum and averages for, say, ten students. To identify the test scores for each student, we will include an extra field on our data cards to provide an identification number. We might use the last five digits of one's SSAN (Social Security Account Number) or some other number. Our new card format will be as shown at the top of page 66.

Card Columns	Field Description	Variable
1-5	Identification number	ID
10-12	Test score 1	NTEST1
13-15	Test score 2	NTEST2
16-18	Test score 3	NTEST3

Our output values for each student will be printed on a line as follows:

Field Description	Print Positions
Identification number	1-5
Test score 1	10-12
Test score 2	17-19
Test score 3	24-26
Total	31-36
Average	41-46

Thus, the calculated values of total and average will be printed on the same print line as the other values.

There are many types of loop logic. For example, we could count the number of students processed, or read a trailer card to indicate the end of our data. We will use the trailer card technique in our program. Various other techniques will be discussed in future chapters. The *trailer card* technique uses a special value in a selected field in a card following the last data card to signal the end of our data. We must use a number which is not a valid value for an input item to serve as our last card flag. We will use a value of 99999 in the ID field to indicate our trailer card. This trailer card technique is often referred to as *sentinal card* logic or *input-bound loops* by different people.

The flowchart of our logic is given in figure 4.4.

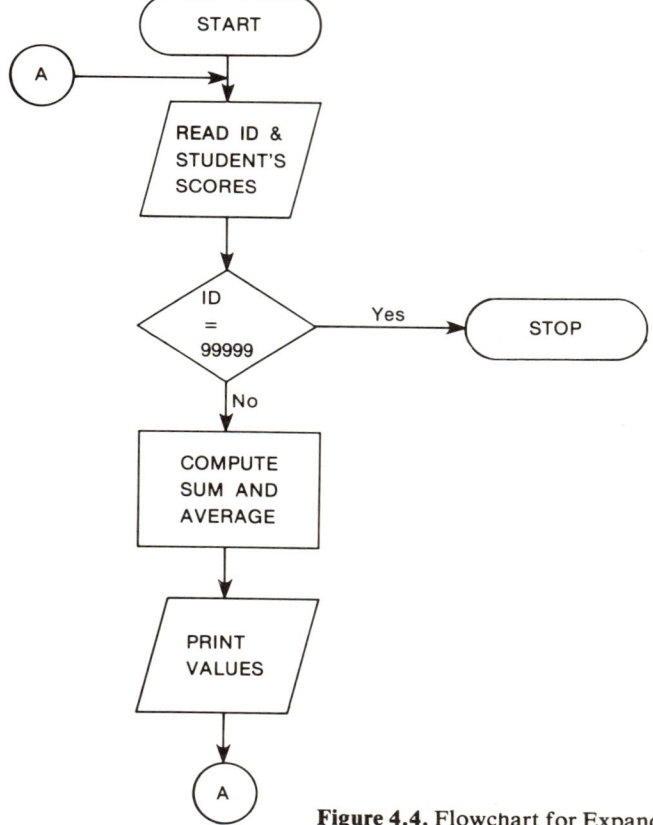

Figure 4.4. Flowchart for Expanded Program with a Loop

Figure 4.5 shows our new coded program.

```
99      FCRMAT (I5,4X,3I3)
98      FCRMAT (1H0,I5,4X,I3,4X,I3,4X,I3,4X,F6.2,4X,F6.2)
1       READ (5,99) ID, NTEST1, NTEST2, NTEST3
        IF (ID .EQ. 99999) STOP
        TOTAL = NTEST1 + NTEST2 + NTEST3
        AVE = TOTAL / 3.0
        WRITE (6,98) ID,NTEST1,NTEST2,NTEST3,TOTAL,AVE
        GOTO 1
        END
```

Figure 4.5. Expanded Program to Compute Ten Students' Grades

The Logical IF statement after our READ checks for an input value of ID equal to 99999. When this value is read, the condition will be true and the STOP statement on the Logical IF executed and the program terminated. If the input value of ID is not equal to 99999, the condition is false and processing of the test scores continues. The GOTO after the WRITE statement will cause control of execution to be transferred back up to the READ statement and the loop entered again.

The output results are shown in figure 4.6.

```
21234        100        83         76         259.00        86.33
22345        100        100        100        300.00        100.00
43456        90         70         80         240.00        80.00
34567        77         100        93         270.00        90.00
36789        68         79         74         221.00        73.67
67890        74         85         87         246.00        82.00
56901        84         55         76         215.00        71.67
59012        72         70         81         223.00        74.33
31234        82         90         89         261.00        87.00
22405        84         92         100        276.00        92.00
```

Figure 4.6. Output Results from Expanded Program with a Loop

How the computer selects which card is to be read by the READ statement is often a puzzle to the beginning programmer. One thing you must clearly understand about the execution of a READ statement. When the computer executes a READ statement, all the machine can do is to read the next data card in the card deck. It cannot recognize that card as being the correct one to read. It must depend on you to arrange the data cards in exactly the same order as required by the READ statements. The execution of the first READ statement reads the first data card. Execution of this same READ statement again (as in a loop), or another READ statement, simply reads the next data card. You must always keep this in mind and arrange your data cards to agree with the same order as the execution of the READ statements in your program.

The next section provides a more detailed explanation regarding the use of these two new control statements.

4.6 The Unconditional GOTO and Logical IF Statements

Unconditional GOTO Statement. The Unconditional GOTO statement causes program execution to be transferred to a specified FORTRAN statement instead of letting the processing of instructions continue in a sequential manner. As the name of the statement indicates, control of execution will always, under all conditions, be transferred to the FORTRAN statement with the referenced statement number assigned to it. The general form of the Unconditional GOTO statement is given in figure 4.7.

Unconditional GOTO Statement—General Form

        GOTO n

where:

    n   —   is a statement number assigned to an executable statement, and to which control is transferred.

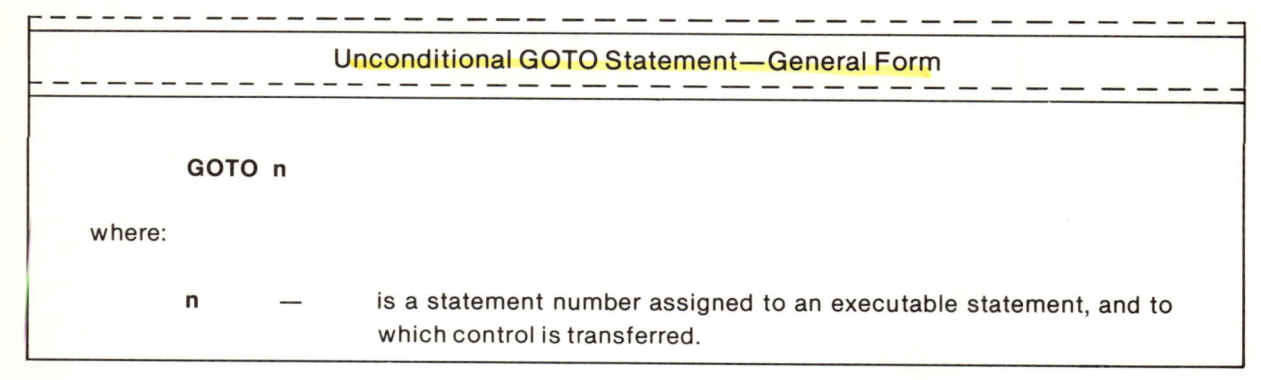

Figure 4.7. General Notation of the Unconditional GOTO Statement

The Unconditional GOTO can reference a FORTRAN statement which appears in the program prior to the GOTO or later in the program following the GOTO. If a branch is made to a previous statement in the program, a loop is formed. If the transfer is to a later statement after the GOTO, all the statements in between the GOTO and the later referenced statement are skipped in execution. Any statement referenced by the GOTO must be an executable FORTRAN statement; if it is not, a syntax error will occur. The most common usage of the Unconditional GOTO is to send control back "upstream" to a previous statement to form a loop.

The executable statement immediately following an Unconditional GOTO should have a statement number assigned to it. For example:

                GOTO 37
            23 C = A + B

Otherwise, the statement following the GOTO could never be executed. If this mistake is made, an error (warning) message will be produced.

The Logical IF Statement. The Logical IF statement is probably the most powerful and widely used decision-making statement in FORTRAN. It provides the programmer with the ability to evaluate variables and expressions and to choose one of two possible alternative paths of logic based on the evaluation.

The Logical IF statement is considered to be a conditional statement since it expresses a conditional command. If the condition that is evaluated is met (true), a certain action is taken. If the condition is not met (false), another action is taken. The condition to be evaluated is in the form of a logical expression. The expression is referred to as a logical one since only one of two conditions may result—true or false. Thus, the logical condition asks if the relationship between two values are true.

The logical expression is enclosed within a set of parentheses following the keyword IF. Following the parentheses is a FORTRAN statement which is executed if the expressed condition is true. The FORTRAN statement can be any executable statement except another Logical IF or a DO statement (another type of control statement). The general form of the Logical IF statement is presented in figure 4.8.

Logical IF Statement—General Form

        IF (logical exp) stmt

where:

    logical exp — represents any logical expression permitted in FORTRAN IV that is evaluated to determine a true or false result.

    stmt — is any executable statement except another Logical IF or a DO statement that will be executed if the result of the logical expression is true.

Figure 4.8. General Notation of the Logical IF Statement

The logical relationship between the values in the logical expression is represented by a *relational operator* such as .EQ. to mean "is equal to." We cannot use the equal sign (=) since it has a special meaning in FORTRAN to assign or replace. Also the relational symbols in math are not in the FORTRAN character set. Therefore a two-character relational operator is used to denote the respective math symbols for a logical expression.

The relational operators in FORTRAN equate to the relational symbols in math of $<, \leq, >, \geq, =$, and $\neq$ which are used to test the relationship between two values. The six FORTRAN relational operators with their equivalent math symbol and operation are:

Relational Operator	Math Symbol	Operation
.LT.	$<$	Less than
.LE.	$\leq$	Less than or equal to
.GT.	$>$	Greater than
.GE.	$\geq$	Greater than or equal to
.EQ.	$=$	Equal to
.NE.	$\neq$	Not equal to

We must precede and follow each relational operator with a period to distinguish it from a variable name which could be spelled in the same way.

The expressions before and after the relational operator may be a single constant, a single variable, or a complex arithmetic expression. All arithmetic computations are performed prior to the relational test. That is, the relational operator has a lower precedence of operation than any arithmetic operator and is performed last.

Examples of the Logical IF statement are:

```
      IF (RESULT .EQ. 0.0) STOP
      IF (COUNT .LT. 10.0) GOTO 1
   23 IF (X .GE. TEST) TEST = 0.0
      IF (R / S .GT. A * S – B) GOTO 74
      IF (TOTAL .NE. A + B) WRITE (6,91) A,B, TOTAL
```

The first example says if the value at the variable RESULT is equal to zero then stop execution. The second example tells the computer if the value of COUNT is less than 10.0, then transfer control to statement number 1. The third example (with the assigned statement number 23) says that if X is greater than or equal to the value at the variable TEST then to assign the value of zero to TEST. The fourth example tells the computer that if the quotient of R divided by S is greater than the result of A * S − B then to branch to statement number 74. The last example says that if the value of TOTAL is not equal to the sum of A + B then to write out these three variables.

When the evaluated logical expression is not met (false), the computer proceeds to the next executable statement in sequence. If the statement given on the logical IF is not a branch statement such as a GOTO, the next sequential instruction will be executed whether the evaluated condition is true or false. For example,

```
      ERRORA = 0.0
      ERRORB = 0.0
      IF (A .LE. 0.0) ERRORA = 1.0
      IF (B .GT. 100.0) ERRORB = 1.0
   87 FORMAT (1H0, 2F5.0)
      WRITE (6,87) ERRORA, ERRORB
```

First the two variables ERRORA and ERRORB are assigned a value of zero. Next, the variable A is tested to see if it is less than or equal to zero. If it is, the variable ERRORA is assigned the value 1.0. Whether this condition is true or false, control proceeds to the next IF statement. The variable B is tested to see if it is greater than 100.0. If it is, the variable ERRORB is assigned the value 1.0. Again control continues at the next executable statement no matter what the logical result is. The WRITE statement will be executed whether any of the conditions were true or false. Remember, the FORMAT statement is not an executable statement, so control proceeds to the WRITE statement after the execution of the second Logical IF.

The execution of the Logical IF is summed up in the following flowchart forms as to the way the Logical IF works.

Explanation   Flowchart Form

If the logical expression is true, the executable statement (S) is executed. If the logical expression is false, then the executable statement (S) is not executed. In either case, the next statement executed is the next one after the Logical IF.

If the executable statement is a GOTO, control is transferred to the statement number specified in the GOTO when the logical expression is true. Otherwise, the next statement after the Logical IF is executed.

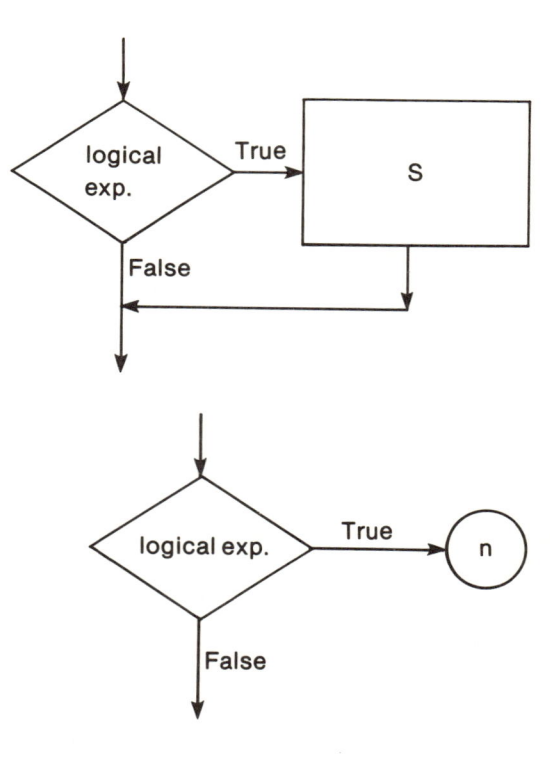

Processing Your FORTRAN Program   71

With the explanation of the Unconditional GOTO and the Logical IF control statements, the eight basic FORTRAN statements have now been discussed. You should be able to write a wide variety of application programs using only these basic statements. However, there are additional considerations and features for the advanced use of these statements which will be explained in the next three chapters.

4.7 REVIEW QUESTIONS

1. The process of carefully checking your source deck manually for errors is called *desk checking*
2. Name the three different types of cards normally associated with a FORTRAN job and give the purpose of each type. *Fortran statements, Data, JCL*
3. You normally place your data cards before your source program in a job deck. (True/False) *F*
4. What is the difference between batch processing and timesharing? — *must use cards to process prog, can use terminal to directly process your program*
5. What happens to your program during the compilation phase?
6. The source listing of the compiled program contains an exact image of the information punched in each source statement card. (True/False)
7. Syntax error messages are normally printed following the FORTRAN statement in error. (True/False)
8. What are the two locations where diagnostic messages may be printed by the compiler?
9. List several methods or procedures that can be used to help you in finding syntax errors.
10. What happens to your program in the execution phase?
11. How may one best determine if an execution (logical) error is made in the program?
12. What is probably the best technique you can use to find logic errors in short programs?
13. Name two other techniques that may be used to help you locate logic errors.
14. A loop is the reexecution of a series of statements in a program. (True/False)
15. Program loops allow one to perform the same operations on more than one set of data. (True/False)
16. How does one normally exit from a loop?
17. What is the purpose of the Unconditional GOTO statement?
18. What is the purpose of the Logical IF statement?
19. List the six relational operators and explain their function.
20. If the logical expression in the Logical IF statement is false, what is the location of the next statement to be executed?

4.8 EXERCISES

1. Indicate the syntax errors in the following statements.

   a) READ 5,99 X,Y   *no parenth.*
   b) FORMAT (2F5.2)   *no no. bef. FORMAT*
   c) WRITE (5,98) X,Y   *6 i.e. 5*
   d) XYZTOTAL = X + Y + Z   *XYZTOTAL is too long*
   e) 79 FORMAT (3X,F5.,I2)   *F5. does not specify no. of dec. places*
   f) GOTO 913576   *no. too long*
   g) IF(N .LT. 10)GOTO 30
   h) STOPTHEPROGRAM
   i) WRITE (6,97,) A,B

2. Indicate the syntax errors in the following statements.

   a) READ (5,95) A,,B,C,
   b) 93 FORMAT 2I3,7XF6.1)
   c) TOTAL = NEWTOTAL
   d) GOTO STOP
   e) READ (6,97) X,Y
   f) ENDOFPROGRAM
   g) IF (K EQ 0) READ (5,98) N
   h) WRITE (6 94) J,M
   i) 96 FORMT (1H0,I3,2X,F6.2)

3. Draw the flowchart to represent the following problem with looping logic. Read the values for X and Y from a punched card in the form 2F5.2. Sum up the two values and print the two values along with their sum. Transfer control back to the READ operation to do the same calculations for multiple data cards. When a value of 99.99 is read for X, terminate the program.

4. Draw a flowchart to represent the following problem with looping logic. Read the values for HOURS and RATE from a card in the form 2F6.2. Compute GRSPAY from the product of the two input values. Write the two input values along with the computed GRSPAY. Branch back to the READ statement to repeat the same operations for multiple employee payroll cards. When a value of −99.99 is read for HOURS, terminate the program.

4.9 PROGRAMMING PROGRAMS

1. Rewrite selected problems in section 3.12 of chapter 3 to read and process multiple data cards. That is, include a program loop to repeat the read, computations, and write operations for more than one set of data.

More on Arithmetic Expressions and the Assignment Statement 5

5.1 Introduction to the Use of the Assignment Statement

Arithmetic computations play an important role in programming. In putting together the required logic for programs, it is often necessary to compute net pay, grade averages, the area of a circle, or to make any other computations that the problem requires. It is the arithmetic expressions formed in the Assignment statement that allow us to perform calculations in FORTRAN. This statement also provides us the means of storing calculated values and retrieving these values as called for in program execution.

Arithmetic Expressions

An *arithmetic expression* is used to specify a computation involving two or more operands which may be all constants, all variables, or a combination of constants and variables. An expression may also consist of a single constant or variable.

Whenever two or more operands are used in an expression, each operand must be separated by an arithmetic operator. An *arithmetic operator* is a symbol that represents the arithmetic operation to be performed. The five arithmetic operators were detailed in chapter 3.

We shall look at the rules which must be followed when forming arithmetic expressions.

5.2 Rules for Constructing Arithmetic Expressions

Although arithmetic expressions generally follow the rules of algebra, certain other rules must be followed in FORTRAN. These rules are:

Rule 1

All computations must be explicitly specified by an arithmetic operator. That is, an arithmetic operator must separate all operands if more than one is used. For example:

$$(X) (Y) \text{ must be specified as } X * Y$$
$$2(K + 3) \text{ must be specified as } 2 * (K + 3)$$

Rule 2

No two arithmetic operators may be adjacent to each other in the same expression. For example:

$$A / - B \text{ must be specified as } A / (- B)$$
$$Z * - 3 \text{ must be specified as } Z * (- 3)$$

Rule 3

Computation of arithmetic expressions is performed from left to right according to the precedence or hierarchy of operations. A review of the hierarchy of operations with an additional unary minus sign precedence is as follows.

Hierarchy	Operations
1st	exponentiation (raising to a power)
2nd	evaluation of unary minus sign
3rd	multiplication and division
4th	addition and subtraction

Parentheses are also used to override the precedence of any operator and to indicate the order of computations as we do in algebra. Any expressions in parentheses are fully evaluated before continuing with further evaluation. Nested parentheses may also be used. When nested parentheses are present, calculations proceed from the innermost set to the outermost pair. Each left parenthesis must have a corresponding right parenthesis or a compilation error will occur.

Parentheses should be used to ensure that the calculations are correctly performed in the desired order. Only as many parentheses should be used as are needed, since excessive parentheses make an expression hard to follow. Within parentheses, calculations take place according to the hierarchy of operations.

The following are algebraic expressions with their equivalent FORTRAN expression to obtain the correct results:

$$\frac{A + B}{X - Y} \text{ would be coded } (A + B) / (X - Y)$$

$$(X - \frac{B + C}{D + E} - 5)^3 \text{ would be coded } (X - (B + C) / (D + E) - 5.) ** 3$$

In multiple consecutive exponentiation operations, calculation is performed from *right* to *left* as in algebra. The algebraic expression X^{2^C} would be coded in FORTRAN as X ** (2.** C) which would first raise 2 to the power of C and then raise X to the power of the resulting value of 2^C. Note that in ANSI FORTRAN parentheses must be used for multiple consecutive exponentiation. An exponent part of an expression may be a constant, a variable, or an arithmetic expression enclosed within parentheses.

A *unary minus sign* is simply the algebraic negation of a value. In the expression −X + Y, −X is a unary minus quantity. The value of X is complemented (sign reversed) by the unary minus sign which precedes it, before being added to Y. A unary minus sign may require a complementing operation before any of the four operations of *, /, +, or − is performed.

Consider the following FORTRAN expressions:

$$- A * B \text{ is treated as } (- A) * B$$
$$- A + B \text{ is treated as } (- A) + B$$
$$- A ** 2 + 1 \text{ is treated as } (- (A ** 2)) + 1$$

Whenever expressions contain operation symbols which are on the same level (such as addition and subtraction or multiplication and division), calculations are performed in a left to right order. The algebraic expression

$$A + X - C + \frac{Y}{(D)(S)}$$

would be coded in FORTRAN as:

$$A + X - C + Y / (D * S)$$

There are two ways to obtain the evaluated result of the arithmetic expression. First, one may simply perform the individual computations according to the given hierarchy. For those operations on the same level, do the computations from left to right.

Given the expression: 6 + 3 * 5 − 8 / 2 ** 2 one could obtain the correct result by

(1) obtaining the square of 2—4.
(2) multiplying 3 times 5—15.
(3) dividing 8 by the results of step (1)—2.
(4) adding 6 to the results of step (2)—21.
(5) finally, subtracting the results of step (3) from the results of step (4) to arrive at the final answer—19.

This order of computation can be illustrated as:

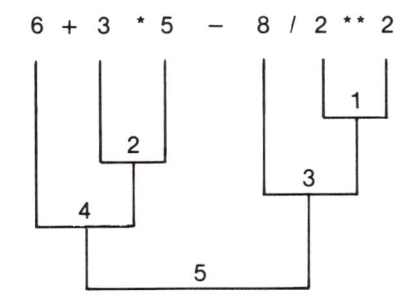

```
6 + 3 * 5  −  8 / 2 ** 2
        │          │  │
        │          │  1
        2          │
        │          3
        4          │
        │          │
        └────5─────┘
```

The second method in computing the result of an expression is to start at the beginning of the expression. If the first operator is equal to or higher in the hierarchy than the second operator, the calculation is performed. If not, the second operator is compared to the third, and so on through the expression. Whenever an operator is equal to or higher than the next operator, the operation is performed. When the end of the expression is reached, one returns to the beginning of the remaining expressions and repeats the same operations until one final value is obtained.

Given the same expression 6 + 3 * 5 − 8 / 2 ** 2, one would evaluate this expression by the second technique as follows: The first operator (+) is less than the second (*), so the first operation is skipped. The second operator (*) is compared to the third (−). Since the second is higher, the multiplication is performed (product of 15). The third operator (−) is less than the fourth (/), so the third operation is skipped. The fourth operator (/) is compared to the fifth (**). The fourth operation is skipped since it is less than the fifth. Finally, the last operator (**) is reached and performed. Two squared produces a result of 4. When the end of the expression is encountered, the remaining operations are: 6 + 15 − 8 / 4. If we start over and evaluate the remaining operations, we first add 15 to 6, giving 21. Next, we divide 8 by 4, giving us 2. Finally we subtract 2 from 21 to arrive at the same answer (19) that we computed by the first method.

Rule 4

The type and length of the result of an arithmetic operation depend upon the type and length of the two operands involved in the operation. *Type* refers to the *mode* (integer or real) of a value. *Length* refers to the size of the memory location used. Don't worry about the length attribute of a variable for now, since all variables we have been talking about occupy the same length storage location.

There are two basic modes of numeric values in FORTRAN—integer (fixed point) and real (floating point). Since we have two modes of variables and constants, there are two modes of arithmetic expressions. ANSI FORTRAN requires variables and constants in an expression to be in the same mode. That is, all operands in an expression must be either type integer or type real. If this rule is violated, a compile error will occur. To conform to this requirement, one often has to change the mode of a value by assigning the value to the proper type variable. If the value of the variable N is to be added to the constant 5.76, it might be accomplished by the following two Assignment statements:

$$XN = N$$
$$SUM = XN + 5.76$$

Thus, our first statement assigns the value of N to a real variable so that it can be added to the real constant in the second statement.

Or if the variable COUNT is to be subtracted from the variable M, we might perform this operation by:

$$K = COUNT$$
$$J = M - K$$

The use of the different modes of expressions are highly important in FORTRAN since the resulting value may be different than we expect.

An *integer mode expression* is one which consists entirely of integer constants and/or variables. *Integer mode arithmetic* is used to evaluate integer mode expressions. Integer mode arithmetic will always produce integer mode results. Any fractional part will always be truncated (chopped off), such as in integer division. Only the integer part of the result will be available. For example: 5 / 3 produces a result of 1. Consider the following examples of integer expressions:

```
M / 5 - 6 ** K
N + 4 * KT
4 / 3                      (result is 1)
3 - 5 / 2 + 4 ** 2         (result is 17)
3 ** 2 + 7 - 3 / 6         (result is 16)
2 / 6                      (result is 0)
```

Thus, we must be extremely careful when using integer mode expressions with a division operation.

A *real mode expression* consists of all real constants and/or variables. **Real mode arithmetic** is used to evaluate real mode expressions. Since real mode arithmetic is performed in floating point mode, the fractional part of the computation as well as the integer part is kept. Therefore, 5. / 2. would produce a result of 2.5; 3. / 4. would produce a result of .75. Consider the following examples of real expressions:

```
S / 6.3 + 4.5 ** D
A - 5.0 / X ** 3.5
4. / 3. - 6.                (result is - 4.666666)
3. - 5. / 2. + 7.           (result is 7.5)
2. ** 2.0 + 5. * 2. / 4.    (result is 6.5)
```

Exponentiation is the only arithmetic operation in which the modes may be mixed. Both real variables/constants and integer variables/constants can have integers as exponents. That is, both real and integer values may be raised to an integer power. But, a real exponent can be used only with a real variable or constant. That is, an integer variable/constant *cannot* be raised to a real power. Valid examples are:

$$XSQD = X ** 2$$
$$XSQD = X ** 2.0$$
$$NSQD = N ** 2$$
$$BCSQ = B ** C$$
$$BNSQ = B ** N$$

The following expressions are invalid.

$$NSQD = N ** 2.0$$
$$NBSQ = N ** B$$
$$SQ2B = 2 ** B$$

Rule 5

The type variable to which the resulting value of the expression is assigned in the Assignment statement determines the final mode of the assigned value in memory. No matter what the resulting mode of the value from the evaluation of an expression, it is the variable at which the value is stored that determines the final mode. For example, the expression 2. * 3. / 4.0 yields a real result of 1.5. If this value is assigned to an integer variable, however, the value stored is 1. Real values assigned to integer variables will have any fraction truncated and only the integer portion of the number stored in integer form. Integer values assigned to real variables will cause the number to be stored in floating-point form.

The use of integer mode expressions as opposed to real mode expressions, and the choice of integer variables as opposed to real variables, depend upon what we are doing. If we are counting or manipulating whole numbers, then integer expressions and variables should be used. If we are working with decimal values which contain a fraction, such as dollars and cents, grade point average, and so on, then real expressions and variables should be used. Arithmetic operations with integer expressions take far less time than with real expressions. So the rule is to use integer expressions and variables unless we need to manipulate real values.

Rule 6

Any variable used in an arithmetic expression must have been previously "defined" (had a value stored at the variable). To obtain a correct result from evaluating an expression, all variables used in the expression must have been given a value prior to the computation. Values may be stored at a variable either by an Assignment or READ statement during program execution. (Note: Some computers such as Honeywell and Burroughs initialize all variables to zero by clearing memory to zero before loading the program.)

Consider the Assignment statement:

$$N = N + 1$$

Only if the variable N had been previously given a value would the execution of the statement provide a valid result. Some compilers make no check to see if a variable has been given a value in the program before using it. The student should be extremely careful to ensure that all variables in an expression have a valid value. Failure to observe this rule may result in a very time-consuming debugging effort.

5.3 More on the Assignment Statement

The arithmetic Assignment statement is identified by the equal sign (=) following a variable at the start of a FORTRAN statement. This statement is a command to FORTRAN to compute the complete expression on the right of the equal sign and to assign the one resulting value to the variable on the left of the equal sign.

The named variable to the left of the equal sign may be a real or integer variable that the programmer has arbitrarily chosen to represent the storage location for holding the assigned value. The programmer should choose a variable name that conveys some meaning of the stored value and a type variable that stores the data in the proper form (i.e., integer or real). For example, if we wish to store computed interest in a variable, a likely name chosen might be XNTRST. Even though the variable INTRST is closer to the name "interest," it is undesirable since the computed value would be stored in integer mode and truncated of any cents amount calculated.

The equal sign is not used as it is in ordinary mathematical notation. We are not allowed to write statements such as A + X = B − C where X represents an unknown quantity and A, B, and C are known. Only one variable is permitted to the left of the equal sign. The equal sign implies that a value is to be *assigned* or *stored* in the left hand variable, and not mathematical equality.

The statement TOTAL = A + B + C means to compute the sum of the values stored at the variables A, B, and C and replace the value of the variable TOTAL with the result. Any previously computed value of TOTAL is lost. That is, a variable always contains the value most recently assigned to it in program execution. The value of any variable in the expression (in our example: A, B, and C) is unchanged.

To sum up the function of the arithmetic Assignment statement, we say it is used to provide four necessary operations in FORTRAN as follows:

(1) To calculate a value from a given formula which may be a simple or compound expression. For example:

        GRSPAY = HOURS * RATE
        AVERGE = SUM / 10.

(2) To initialize variables. Variables used to represent certain values must first be set to the desired initial value. For example:

        SUM = 0.
        KTR = 1
        PI = 3.1416

A common misunderstanding on the part of beginning students is their assumption that a variable must be assigned as many digits as needed in a maximum value. For example, if a counter may contain a maximum of three digits, the student assumes it necessary to assign that many digits to the variable; KTR = 000 is an example. This causes no harm, but is not needed. Simply, KTR = 0 will initialize the variable KTR to zero.

(3) To move values from one variable to another. Values may be moved to a different storage location by the Assignment statement. Examples are:

        HOLD = RESULT
        K2 = K1
        TEMP = XSQD

Often we need to retain the original value in a variable that is later changed in a calculation. We can assign the variable to a new variable in which either the new variable or the original variable can be used and modified in a calculation. But we always have access to the original value. When a variable is assigned to a new location, the value in the old variable remains unchanged.

(4) To change the mode of data. It is often necessary to change the mode of a stored value from integer to real or from real to integer. The arithmetic Assignment statement allows us to accomplish this easily, as illustrated by the following examples:

        REALN = N
        IRESLT = RESULT
        K = X

More on Arithmetic Expressions and the Assignment Statement 79

5.4 Truncation Considerations with Numeric Values

We have already discussed two types of truncation that occur in computer operations. The first type results from the division of two integer values in an expression. The second type of truncation results when we assign a real result to an integer variable. The entire decimal fraction is truncated and only the whole number portion is stored at the integer variable. The student must be constantly mindful of these two types of truncation operations when writing his program.

Another type of truncation occurs in the representation of real quantities that often causes the student much grief. This type of truncation, often called *round-off,* results from the form and limitation of real numbers used in computations or as stored in a variable. First, a storage location can hold a maximum of approximately seven digits. Second, some quantities cannot be exactly represented by the computer. For example, the value of one-third is stored as .3333333. Since the fraction contains a nonending repetition of 3's and is chopped off after seven digits, the exact value of 1./3. can never be represented in a memory location. The same is true for other fractions such as 2./3., 1./9., 4./3., etc., which have a repeating fractional digit.

The comparison of these values may, therefore, never be equal to what we might expect if we performed the calculations by hand. For example:

$$A = 1./3.$$
$$\text{IF } (A * 3.0 \text{ .EQ. } 1.0) \text{ GOTO } 33$$

we would never get an equal comparison since A * 3.0 would be .9999999. Thus, the comparison would never provide a true condition and we would never branch to the statement numbered 33.

To avoid such errors with the round-off of real quantities, we must use relational operators which include both the equal condition and either a "less than" or "greater than" condition. If we wish to branch to the statement numbered 33 whenever A * 3.0 is less than or equal to 1.0, we must code:

$$\text{IF } (A * 3.0 \text{ .LE. } 1.0) \text{ GOTO } 33$$

Now, we have taken into consideration the possibility of fractional truncation and can take the proper course of action. This type of truncation may cause such subtle errors that the student thinks the computer is playing tricks with him, or is wrong.

5.5 Arithmetic Assignment Statements in Sequence

The computer must be informed of the values contained in an expression before the expression is evaluated. A statement such as: X = A + 5.75 would be valid only if the variable A had been assigned a previous value (by an Assignment statement or READ statement).

Given the statements:

$$A = 2.3$$
$$X = A + 5.75$$

the computer executes the first statement and assigns a value of 2.3 to the variable A. When the second instruction is executed, the computer supplies 2.3 as the value of A. Therefore, 2.3 is added to 5.75 and a value of 8.05 is stored as the value of X.

We may have many arithmetic Assignment statements in sequence. For example:

$$T = 3.2$$
$$U = 5.5 - T$$
$$V = 2.0 * U$$
$$X = T + U + V$$

when all five of these statements have been executed, the value of X is 10.1.

If the same variable appears the second time in a sequence of arithmetic Assignment statements, the latest value is assigned to the variable, and the previous value is lost. For example:

$$K = 4$$
$$\ldots$$
$$\ldots$$
$$K = K + 1$$

First, the value 4 is assigned to K. In the last statement, the value of 4 is supplied for K to the right of the equal sign and 1 is added to 4. The result, 5, is then assigned to K; and the previous value of K is lost. Statements in sequence, such as the above example, occur very frequently in computer programs in accumulating totals and counter logic.

5.6 Counting and Accumulating Totals with the Assignment Statement

A common requirement in many computer programs is to keep a count of a number of items and to accumulate a "running" total of values. These types of operations are often referred to as *counter* and *accumulated totals logic,* respectively.

Counter and totals operations are normally achieved by the Assignment statement in a program loop. For each pass through the loop the counter is incremented and the value of an item is added to the variable used to accumulate the total. The counter and total variable must, of course, be initialized to some value, usually zero, before entering the loop. Counter logic may be used independent of total logic or they can both be included in the same loop.

Let us illustrate these operations in a program which reads an unknown number of test scores and computes their average. Each test score is read from a separate card under the format specification I3. The last data card (trailer card) contains a value of 999 to indicate the end of the input data. The program is:

```
99 FORMAT (I3)
   K = 0
   NTOTAL = 0
10 READ (5,99) NTEST
   IF (NTEST .EQ. 999) GOTO 20
   K = K + 1
   NTOTAL = NTOTAL + NTEST
   GOTO 10
20 NAVE = NTOTAL / K
   . . .
   . . .
```

Each time a valid test score is read in the loop the counter K is incremented by 1 and the test score accumulated in the variable NTOTAL. When the trailer card of 999 is read, we know that all the test scores have been read and we branch to statement number 20. Then we can compute the average of the test scores by dividing the accumulated total by the counter.

Assignment statements are very powerful statements. They are the heart of most FORTRAN programs because of their frequent usage. For this reason, one is encouraged to become very familiar with the many capabilities of the arithmetic Assignment statement.

5.7 Nonstandard Language Extensions (Mixed Mode Expressions)

Many FORTRAN compilers allow mixed mode expressions. That is, both integer and real constants and/or variables may be included in an expression. Thus, a *mixed mode expression* is defined as one in which both integer and real constants and/or variables are used in an arithmetic expression. All internal calculations performed by the computer, however, must be on values in the same mode. Therefore, compilers which allow mixed mode expressions must convert the integer mode operand which is used with a real operand to real mode before doing the calculation. You may wonder then, why have mixed mode expressions at all? They are allowed primarily for the convenience of the programmer—to avoid his having to write additional statements to establish a new mode for a variable. Mixed mode expressions are not recommended, but are explained here since they are allowed on so many compilers (but not in ANSI FORTRAN).

In a mixed mode expression consisting of multiple operations, the arithmetic operation with two integer operands will be performed in integer mode. All other combinations of operands are performed in real mode. It is the type of the two operands involved in each operation that determines the arithmetic mode. Consider the following mixed mode expressions:

$$5 / (A + 3) - X ** 7$$
$$86.3 * K + R / 16.5 ** J$$
$$7.5 - 3 / 2 \quad \text{(result is 6.5)}$$
$$4 * 3.5 - 16 ** .5 \quad \text{(result is 10.0)}$$
$$10 / 3. - 1 / 3 + .6 \quad \text{(result is 3.933333)}$$

Remember, in a single mixed mode expression between two operands, the computer will convert the integer value to real before doing the calculation. The result is, therefore, a real value. The type of the variable to which the calculated value is assigned determines the mode of the stored result.

Multiple Assignment Statements

Some compilers such as WATFIV and CDC 6000 allow the assignment of a value to multiple variables in the same Assignment statement. This form of the Assignment statement is often referred to as multiple replacement. The general form for this type of Assignment statement is:

variable$_1$ = variable$_2$ = variable$_n$ = expression

where any number of variables may be assigned the value obtained from the expression. For example:

$$X = Y = Z = 0.0$$
$$A = I = V = W = 3.2 * R + X$$

The assignment of the values is from *right* to *left*. That is, the value from the expression is assigned to the rightmost variable, then the value in this variable is assigned to the immediate next left variable, and so on. One must be careful when using this type of Assignment statement since different types of variables may result in a truncation of fractional digits. Thus, in the latter example, the variable A would contain only the whole number portion of the value since the value in I is assigned to A.

5.8 REVIEW QUESTIONS

1. Why is the Assignment statement so important in FORTRAN programs?
2. Name the possible elements of the FORTRAN language that may be used in forming an arithmetic expression.
3. List the five arithmetic operators used in FORTRAN. Indicate their order or hierarchy in computations.
4. What is the importance of using parentheses in arithmetic expressions?
5. Why are the following expressions invalid?
    a) Z / − 3
    b) 5X + 7Y
    c) (A + B * (C − D)
    d) 6 − X **

6. The type and length of the result from an arithmetic operation depend upon what?
7. The two types of arithmetic expressions are _____ and _____ mode.
8. Describe the general form of the Assignment statement. Give an example.
9. The " = " in the Assignment statement means to S _____ or to R _____ and does not mean algebraic equality.
10. In an Assignment statement, how many values may be assigned to the variable to the left of the equal sign?

5.9 EXERCISES

1. Evaluate the following arithmetic Assignment statements. Determine the resulting value from each and show whether the result assigned to the variable is in integer or real form.

   a) A = 5 / 3 + 8 * (4 − 2)
   b) K = 9 − 3 * 2
   c) I = 9 − 2 ** 3 + 7
   d) R = (6.0 / 3.0 * 2.) ** 3.0 − 15.6
   e) J = 8.2 + .8 − 6. ** 2.0

2. Evaluate the following arithmetic Assignment statements. Determine the resulting value from each and show whether the result assigned to the variable is in integer or real form.

   a) X = 5. / 2.5 * 6. − 8. ** 2.0
   b) L = (3 − 5) * (8 + 3)
   c) N = 5.68 − 3.4 + 7.0 / 2.0
   d) S = 5 / 2 + 3 ** (2 ** 2)
   e) M = (3 + 2) * (6 − 2) / 4

3. Consider the following arithmetic Assignment statements. If the statement is valid, check the valid column. If invalid, write a correct statement. Mixed mode expressions are not allowed.

	Statement	Valid	Rewritten Statement
a.	Q = R − .36 + 7.0		
b.	DISTANCE = RATE*TIME		
c.	SQROOT = X ** .5		
d.	RATE = DIST / TIME		
e.	C**2 = A*A + B*B		
f.	GRSPAY = $210.75		
g.	C = A * −B		
h.	TOTAL = 0.0		
i.	PAGECT = 0001		
j.	A * B / Z = T		

4. Evaluate the following arithmetic Assignment statements as to validity in the same manner as required in exercise 3.

	Statement	Valid	Rewritten Statement
a.	RESULT = 8.4 + R / 5.		
b.	AMT = DOL + CENTS		
c.	ADD A TO B GIVING C		
d.	SUM=SUM+CTR.5		
e.	E = 5X**2 + 2Y + Z		
f.	HOLD = 5./3.*(R+S)−B*4.0		
g.	L = (Z+5.)*((R−3.2)+V)		
h.	CUBERT = XNUM ** (1./3.)		
i.	AREAT = 1./2. B (H)		
j.	SUM = SUM + VALUEX		

5.10 PROGRAMMING PROBLEMS

1. The XYZ Company pays its night watchman $3.00 an hour for all time up to 40 hours. All hours over 40, he is paid time and a half of the normal rate. Last week, the night watchman worked 47 hours. Write a program to compute his gross pay and print his regular time pay, overtime pay, and total gross pay. Print the output results as follows:

Field Description	Print Positions
Regular time pay	5-10
Overtime pay	20-25
Gross pay	35-41

2. Write a program to: (1) read a temperature recorded in Fahrenheit off a punched card in card columns 1 through 5 in the form NNN.N, (2) convert the temperature given in Fahrenheit to its corresponding Centigrade temperature according to the formula $C = \frac{5}{9}(F-32)$, (3) print the Fahrenheit temperature and the converted Centigrade temperature under the format specifications 2F10.1.

3. Write a program to compute the sum and average of four integer values read from a punched card. The four fields are to be referred to as variables N1, N2, N3, and N4, respectively. They are punched in the following card columns.

Card Columns	Field	Value
1-3	N1	321
4-6	N2	196
7-9	N3	38
10-12	N4	639

Print the input fields and computed results as follows:

Field Description	Print Positions
N1	3-5
N2	10-12
N3	17-19
N4	24-26
SUM	31-36 (Form NNNN.N)
AVER	41-45 (Form NNN.N)

4. Write a program to find the volume of a cylinder; the radius of its base is 1.75 inches; and its height is 4 inches. The formula is $V = \pi r^2 h$ (use 3.14 as π). Print the radius, height, and computed volume under the format specifications 3F10.2.

5. Write a program to perform the same function as problem three but with five sets of data values. Print the results of each set of values on a double-spaced line.

6. A salesman earns an 8 percent commission on each item he sells. Read a card file with the sales price of each item sold. The sales price is punched in card columns 1-7 in the form NNNN.NN. Compute the commission earned for each sales price in the card file. Print each sales price with its respective earned commission on a new line. Also compute the total of all items sold and the total commission earned. Output each input value and earned commission as double-spaced lines in the following format:

Field Description	Print Positions
Sales price	3-9
Commission	15-20

Output the total sales and total earned commissions on a separate line beneath their respective column values.

More on Reading/Writing Numeric Data

6

6.1 Introduction to the Use and Importance of Input/Output Statements

If a program consists of a number of calculations with only one set of quantities, the data values can easily be established in the form of constants. This is seldom the case, however, since one could probably obtain the result more quickly on a calculator than by having to go through the steps needed to obtain the results by computer. Most programs are set up to read the quantities as input data inasmuch as multiple sets of data are normally entered. Constants are coded for only those values which remain fixed throughout the program. The program reads a set of quantities into the named variables, manipulates them according to the proper formulas, outputs the results, and loops back to repeat the same processes for another set of input fields.

Thus, the use of input/output statements provides the programmer with much power and flexibility. After calculating and outputting the results for one set of variables, new values can be read into the variables and these same procedures used for the new set of values. Therefore, any time a new value is read for a variable, the new input value replaces the old one at that storage location. The READ and WRITE statements are widely used in programs to input and output new sets of values for the variables used in calculations and decision making.

The input/output (I/O) statements in FORTRAN allow the programmer to control the peripheral input/output devices connected to the computer system and to transfer data between memory and the various I/O units. The term *I/O* is an abbreviation for input/output, and it is used by programmers to talk about input/output operations or storage media. I/O may be used to refer only to an input operation, only to an output operation, or to both input and output operations.

Let us review the terms used in connection with the organization of data. Some terms used regarding the organization of input and output data are field, record, and file. A *field* is the smallest division of data and identifies a single item of information. Examples of a field are name, sex, unit cost, radius of a circle, test score, etc. A *record* is a complete unit of information, usually about a person or transaction. A record consists of a collection of related data fields; examples: a personnel pay record, an accounts receivable record, an inventory control record, and a student's registration record. This type of record is also referred to as a *logical record*. A record is usually one punched card or a single print line. A *file* contains all the related records for one data processing application. For example, a payroll file contains all the employee pay records for persons employed by a company; an inventory file contains all the stock records maintained by a firm; a student grade file contains grade information on a group of students.

A card file is a collection of records on punched cards. Card files are known as sequential data files. In a *sequential* file the records are read in a serial manner. That is, record one is read, then record two, then record three, and so on. Before the tenth record can be read, the first nine must have been read previously. Therefore, data cards are read in the order in which they are arranged in the data deck. The first READ statement executed in a program reads the first data card, the next READ executed reads the second data card, and so on.

Multiple cards can be read with a single READ statement depending upon how you establish the list of

86   More on Reading/Writing Numeric Data

variables on the READ, and how you organize the format specifications in the FORMAT statement. The same is true with the WRITE statement and its output operations. This chapter will cover the finer points of the READ, WRITE, and FORMAT statements with the input and output of numeric data. A large percentage of the errors (roughly 40 percent) made by students are in the area of I/O specifications. It is very important that you try to understand the material covered in this chapter on I/O operations. You will use the information and knowledge learned from these subjects in every program you write. The better understanding that you acquire now of the I/O operations, the less mistakes you will make in writing programs.

Sometimes you might make a mistake with the I/O statements and read or write more records than you intended. An in-depth discussion of the rules which govern the use of the FORMAT statement and I/O lists with READ/WRITE statements are provided to explain their operation. These rules explain what will happen if you include more variables in an I/O list than you have format codes in the FORMAT statement, etc. Study these rules well and you will be able to answer many of the curious questions that students often have, like what will happen if I do this, that, or the other.

6.2 Rules for Using the FORMAT Statement and the I/O List of Variables

The FORMAT statement is used in conjunction with the I/O list of variables on the READ and WRITE statements to specify the location and form of the data fields within the records. A FORMAT statement must be provided for each READ and WRITE statement that uses format specifications with input and output fields.

A *format code* is used to describe the form and size of each data field transferred to or from memory. The format codes are coded in the FORMAT statement and are enclosed in a pair of parentheses following the keyword, FORMAT. A comma must be used to separate each format code.

Some examples of FORMAT statements associated with card input operation are:

```
 25 FORMAT (I2, I3, F7.2)
 98 FORMAT (5X, F5.2, I3, 2X, I4)
359 FORMAT (F6.1, 3X, 2I5, 1X, F4.2)
```

Some examples of FORMAT statements associated with printer output operations are:

```
 97 FORMAT (1H1, I4, 5X, F6.2)
186 FORMAT (1H0, F5.1, 3X, I6)
 73 FORMAT (1X, I3, 2X, I3, 2X, F4.1)
```

Remember, a carriage control specification must always be given as the first format code in the FORMAT statement for line printer output in order to control the movement of the paper.

The following rules are presented to explain the function of the FORMAT statement and to help you better understand the relationship that exists between the FORMAT statement and its contained format specifications.

Rule 1

FORMAT statements are not executed; their function is to supply information concerning I/O data to the program. FORMAT statements, therefore, should only be referred to by an I/O statement (READ or WRITE).

Rule 2

FORMAT statements may be placed anywhere in a source program before the END statement except between continuation statements. Some programmers may place all FORMAT statements at the beginning or end of a program for easy reference. Or they are often placed before or after their related I/O statement.

Rule 3

When defining a record by a FORMAT statement, the programmer must consider the maximum size of the record allowed for that I/O medium. For example: the maximum size of a card record is 80 characters; the maximum size of a printer output line is 132 characters.

Rule 4

It is not necessary to specify blanks (X format code) to fill out the remaining positions of a record. For example, to read five items of 4 positions each off a punch card (columns 1-20), the format specifications need only be:

9 FORMAT (5I4)

and not:

9 FORMAT (5I4, 60X)

Or to print a double-spaced line using only the first 70 print positions for seven fields of ten position integer numbers, we need only to code:

10 FORMAT (1H0,7I10)

and not:

10 FORMAT (1H0,7I10,62X)

Rule 5

When formatted records are prepared for printing or output to a terminal, the first character of the format specifications is taken and used as a carriage control character. A carriage control character controls movement of the paper in the printer and specifies how the device is to mechanically move the paper (i.e., to the top of the next page, one line, two lines, or no movement). The carriage control character is specified as the first format character of each printer output line.

For output media other than the line printer or remote terminals, the first character of the record is treated as data, and no carriage control character is used. If a carriage control specification is not provided for a line printer or terminal output line, the first character in the first data field will be used by the system for carriage control operations. If an invalid carriage control character is generated, one of two things may occur. On some computers a default character of a single space control is provided; on others a "runaway" printer may occur (i.e., continuous paper feed).

Additional rules in using the FORMAT statement relate to the variables in the I/O list.

Input/Output List Rules

The following rules are presented to help the student understand the relationship between the I/O list variables and the FORMAT statement specifications.

Rule 1

The I/O list can contain any type of variable names.

Rule 2

The I/O list must not include any constants or arithmetic expressions.

Rule 3

Each variable name in the I/O list must be separated by a comma.

Rule 4

For each variable that appears in the I/O list, one item is transmitted to or from memory.

Rule 5

A one-to-one relationship exists between the names in the I/O list and the format codes (except the H or X) in the associated FORMAT statement.

Rule 6

The order of the names in the I/O list must be consistent with the left-to-right order of the data fields in the I/O record and also with the order of the format codes specified in the FORMAT statement. For example:

READ (5, 98) A, I, Z
98 FORMAT (2X, F8.2, I6, 5X, F5.1)

The variable A corresponds to the F8.2 format code, the variable I corresponds to the I6 format code, and the variable Z corresponds to the F5.1 format code. (Remember that the X format codes do not relate to a variable or data field, but are used to specify how many positions are to be skipped.)

Rule 7

The types of names assigned, format codes used, and data fields used must all agree with the type of I/O data. For example, an integer (I) format code must be used with an integer variable to read integer data fields. A real variable must be assigned to a numeric field with a decimal point and be transmitted under a F format specification.

Rule 8

If there are fewer names in the I/O list than there are format codes, the remaining format codes are ignored. The system first checks to see if additional names are present in the I/O list before selecting the next format code. For example:

READ (5,87) X, Y, J
87 FORMAT (F3.1, F4.0, I5, F8.2, I7)

The use of format codes and transmission of data terminates with the code I5 and variable J. The remaining format codes F8.2 and I7 are not used.

Rule 9

If there are more names in the I/O list than there are format codes in the FORMAT statement, the system will automatically supply additional format codes by reverting back to the format code after the first left parenthesis in a right-to-left scan (i.e., rightmost left parenthesis). The system will also cause a new data record to be read at that time. That is, each time the last (rightmost) parenthesis is reached and additional items are to be transferred in the I/O list, a new record is automatically read. The same principle also holds true for output records. For example:

READ (5,40) A, B, C, D, E
40 FORMAT (F4.1, F5.2)

One record will be read with the fields A and B transferred as F4.1 and F5.2 specifications. Since the end of the FORMAT statement is reached but more items need to be transmitted, the system will read a new record and transfer the field C as an F4.1 (positions 1-4 of new record) specification and field D as an F5.2 (positions 5-9) specification. The end of the FORMAT statement is once more encountered with still a remaining item that needs to be read. A third record will be read and the field E transferred as an F4.1 (positions 1-4 of new record) specification. A total of three records would be read by the illustrated READ and FORMAT statements.

Rule 10

If the I/O list is omitted from the READ statement, at least one record is still read.

Rule 11

The same FORMAT statement may be referenced by different I/O statements in the program as long as the format specifications are the same for the I/O list of variables on the different I/O statements.

6.3 More on the Integer (I) and Floating-point (F) Format Codes

Integer (I) Format Code

The *integer (I) format code* is used in transmitting integer data fields only. This format code consists of the capital letter I followed by an integer constant which represents the width (size) of the data field. The general form of the integer format code is:

$$rIw$$

where:

r represents an optional repetition factor that specifies the number of times the format code is to be used. The value of r, when used, must be an unsigned positive integer constant whose value is less than or equal to 255. w specifies the number of record positions in the field. w must be an unsigned positive integer constant with a value less than or equal to 255. The specified width includes the sign, if present.

Examples are:

   18 FORMAT (I3, I7, I2)
   47 FORMAT (I6, 2I4, I9)

Input Considerations

On input, any leading, embedded, and/or trailing blanks in the data field are interpreted as zeroes. The sign, if present, must appear in the leftmost position of the field or prior to the first digit. The input value must be right-justified in its field positions.

Examples to illustrate the input rules when using the I format code for input fields are as follows (b́s represent blank positions in the field):

Field Size and Value	I Format Code
123	I3
+123	I4
+b́123	I5
b́−123	I5
−123	I4
b́b́123	I5
b́+123	I5

Output Considerations

On output, the **w** represents the total positions in the output field. A sign is printed to the immediate left of the first digit only for negative values. However, students should leave an extra space in **w** for the sign, even if they expect it to be positive (because it might not be), i.e., **w** should be number of output digits + 1. If the number of digits (and minus sign whenever the quantity is negative) that represents the quantity in memory is less than **w**, the leftmost output positions are filled with blanks. If the output quantity plus the minus sign (when negative) is greater than **w**, asterisks are generally printed on most computers in place of the output value. As many asterisks will be printed as the width specification. All output values will be right-justified in the output field positions.

Examples to illustrate the output rules when using the I format code for output fields are as follows (bs represent blank positions):

Value in Memory	I Format Code	Output Field
123	I3	123
123	I5	bb123
123	I2	**
−123	I4	−123
−123	I5	b−123
−123	I3	***

Floating-point (F) Format Code

The *floating-point (F) format code* specifies the number of record positions in a field to be transmitted as real data. Whole numbers or decimal fields may be read as input. The general form of the F format code is:

rFw.d

where:

r represents an optional repetition factor that specifies the number of times that the format code is to be used. The value of **r**, when used, must be an unsigned positive integer constant whose value is less than or equal to 255. **w** specifies the number of record positions in the input or output field. **w** must be an unsigned positive integer constant with a value less than or equal to 255. The decimal point (.) is a necessary separator between the **w** and **d** parameters. **d** represents the number of digits in the fractional part of the number (i.e., number of digits following the decimal point). **d** must be an unsigned integer constant between 1 and 255, inclusively.

Examples are:

107 FORMAT (F5.2, I3, F7.1)
96 FORMAT (2F4.1, F8.3, 4F3.0, F4.3)

Input Considerations

On input, any leading, embedded, or trailing blanks in the field are interpreted as zeroes. The sign if present, must appear prior to the first digit. Additional input rules include:

1. The width of the field must include a position for the decimal point and/or sign, if present.
2. If the position of the decimal point in the data field is different from the position specified by the **d** parameter in the F format code, the actual position of the decimal point in the input data overrides the **d** specification and determines the value of the item. For example:

```
         READ (5,99) A
      99 FORMAT (F5.1)
         23.45
```

The value is stored as 23.45 and not 234.5.

More on Reading/Writing Numeric Data   91

3. Since the position of the actual decimal point in the data overrides the **d** specifications, the quantity does not have to be right-justified in the field, but normally is and should be right-justified.
4. If the decimal point is omitted in the field, the **d** specification determines the position of the decimal point in the quantity and, thus, the true value of the number.
5. Any permissible form of constructing an integer or real constant may be read by the F format code. No commas must be included in a data field. For example, a data field like 43,276.5 is illegal.

Examples to illustrate the input rules when using the F format code for input fields are as follows (b̸s represent blank positions):

Field Size and Value	F Format Code	
12.34	F5.2	
−12.34	F6.2	
+12.34	F6.2	
b̸b̸12.3	F6.1	
b̸−12.3	F6.1	
+b̸12.3	F6.1	
12.34	F6.0	(Value of 12.34)
1.234	F6.2	(Value of 1.234)
1234	F4.2	(Value of 12.34)
123	F3.0	(Value of 123.)

Output Considerations

On output fields, the following rules apply:

1. A decimal point will always appear in the field followed by as many fractional digits as given in the **d** parameter.
2. The output quantity will always be right-justified in the output field positions.
3. The width specifications must provide sufficient spaces for the integer digits of the value, the decimal point, the number of fractional digits specified by **d**, and a sign if the output quantity is negative. Like the output specifications for the I format code, students should always allow room for the sign.
4. If insufficient output field positions are provided by **w**, asterisks instead of data are written by most computers. The number of asterisks written is equal to the **w** specification.
5. If excessive output field positions are provided (quantity does not fill the total output positions), the leftmost excessive positions are filled with blanks.
6. Other important considerations such as rounding are discussed in the following paragraphs.

If an output quantity is a total fraction (i.e., contains no whole number or no significant digit to left of decimal point) and excessive output positions are provided, a single zero will be included before the decimal point. Examples are:

1. Given the value .934 in memory to be output under an F5.3 format code, the output field would appear as 0.934.
2. Given the same value to be output under an F7.3 format code, the output field would be b̸b̸0.934 (where the b̸s represent blank positions).
3. Given the same value to be output under an F4.3 format code, the output field would appear as .934.

In some compilers the output field specifications for a real quantity must always allow room for at least one integer position. Otherwise, the output field size is not considered wide enough and asterisks will be output. For example, a field specification of F4.3 would not allow room for an integer position even though the quantity in memory is a total fraction. A field specification of F5.3 would have to be coded to allow the output of three fractional digits. This is true with the WATFIV and Honeywell 6000 compilers.

If the output requirements specified by the **d** parameter truncate any fractional digits of the quantity, FORTRAN provides automatic rounding of the output results. When the first digit to be dropped in the fraction is 5 or greater, the system will automatically round the next left position up (increase) by one. Rounding could continue into the integer positions if the fraction digits are 9's. If the beginning digit of the fractional part that is truncated is less than 5, no rounding up occurs. That is, the output digits remain as they are. Examples to illustrate the *rules of rounding* are:

1. Given the value in memory of 3.8916 to be output under an F5.3 format code, the output value would be 3.892.
2. Given the same value in memory to be output under an F4.2 format code, the output field would appear as 3.89.
3. Given the value 3.98 in memory to be output under an F3.1 format code, the output field would be 4.0.
4. Given the value 29.99 in memory to be output under an F3.0 format code, the output field would appear as 30.; thus, rounding could begin with integer position digits.

Output examples of the F format specifications to illustrate the output rules are as follows (ɓs represent blanks):

Value in Memory	F Format Code	Output Field
87.153	F7.2	ɓɓ87.15
87.153	F7.4	87.1530
87.153	F7.6	*******
87.153	F7.1	ɓɓɓ87.2
−15.483	F6.3	******
−15.483	F6.0	ɓɓ−15.
−15.483	F6.1	ɓ−15.5
−3.967	F5.2	−3.97
3.967	F8.5	ɓ3.96700
.9348	F7.4	ɓ0.9348
.9348	F4.3	.935

Now let us illustrate the use of the rules and other information we have covered by some examples of reading and writing numeric data.

6.4 Reading Numeric Data

Each variable in the I/O list must have a corresponding format code to transmit data fields to memory. The type format code must also match the type variable established. The variables are matched up with the data transferring format codes in a one-to-one, left-to-right order.

To begin reading input data fields, the programmer must know four things: (1) the location (record position) of the field, (2) the type of the data field, (3) the size (width) of the field, and (4) the location of the decimal point for numeric fields. With this knowledge, the proper type variables can be set up, the proper format codes established, and the correct FORMAT statement constructed.

To begin our discussion on reading numeric fields, we shall establish a record layout with field positions as given at the top of page 93.

The first field consists of an integer item of seven positions. The proper format code would be I7. An integer variable name must be chosen to represent the storage location into which the value can be read. Let us choose the name NRSREG. Card column eight is blank, so we select a format code of 1X to skip that position. No variable is assigned to the X format code since no data is transferred. The second data field is the student's classification in coded form (i.e., 1 = freshman, 2 = sophomore, 3 = junior, and 4 = senior). Since only one integer digit is coded as the input field, we will establish the format code of I1 to read the field and assign an integer variable of LVLCLS. Columns 10-14 are blank, so a format code of 5X is needed to skip those positions.

More on Reading/Writing Numeric Data   93

Card Columns	Field
1-7	Student registration number (in the form NNNNNNN)
8	Blank
9	Classification of student (form = N)
10-14	Blank
15-21	Tuition fee (form = NNNN.NN)
22-23	Semester hours enrolled (form = NN)
24-80	Blank

The third data field is a real quantity occupying seven positions with two fractional digits. A field specification of F7.2 is established to read that field. A real variable of TFEE is suggested for the item of tuition fee. The fourth data field, "semester hours enrolled," consists of two integer positions. A format code of I2 is established to read this field and an integer variable of NRSHRS assigned to the field. Since the remainder of the card is blank, no further field specifications are needed.

These fields would be punched in a card as illustrated in figure 6.1.

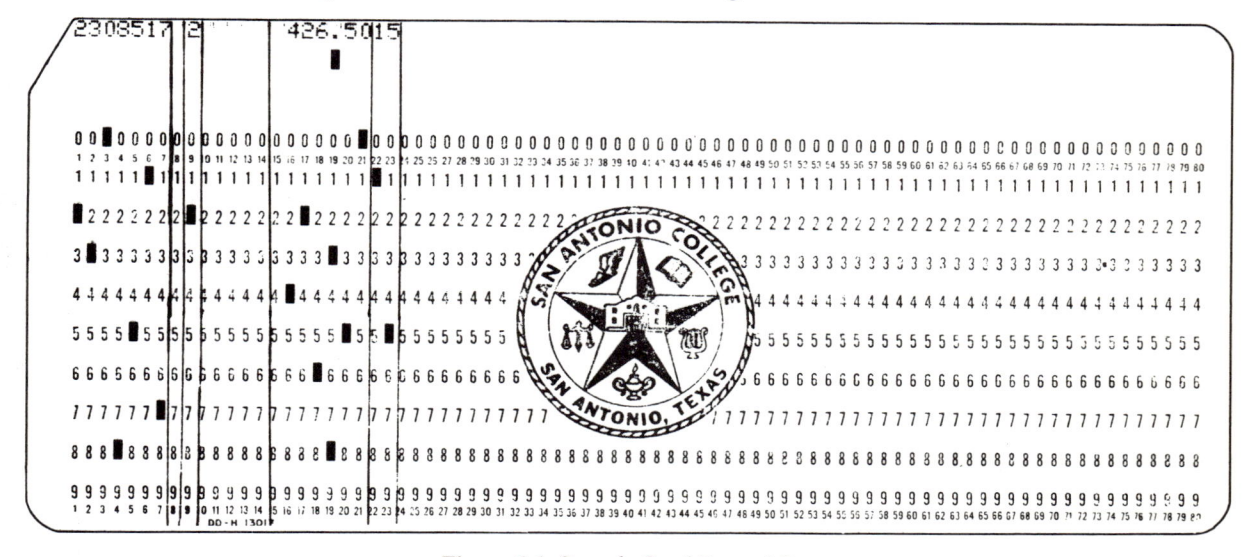

Figure 6.1. Sample Card Record One

Using the assigned variables and a FORMAT statement (number 99) containing the established format codes, our READ statement and related FORMAT statement appear as follows:

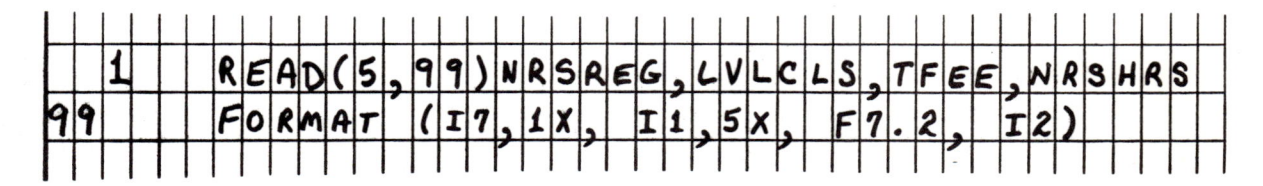

```
     1     READ(5,99)NRSREG,LVLCLS,TFEE,NRSHRS
  99       FORMAT (I7,1X,I1,5X,F7.2,I2)
```

This READ statement would normally be executed repeatedly in a program loop to read many data records. The values in the input fields would not, of course, all be the same. But the arrangement of the fields on the records would have to be the same. This concept of having the data fields located in the same positions as those of the same type record is known as a ***record layout design.***

To study more flexibility in reading data fields, let us use the data card in figure 6.2. Its field layout specifications are represented by vertical lines on the card. These vertical lines (along with the assigned variables written on the card) are solely for our benefit. The computer interprets the data by the holes and the field specifications that the programmer provides.

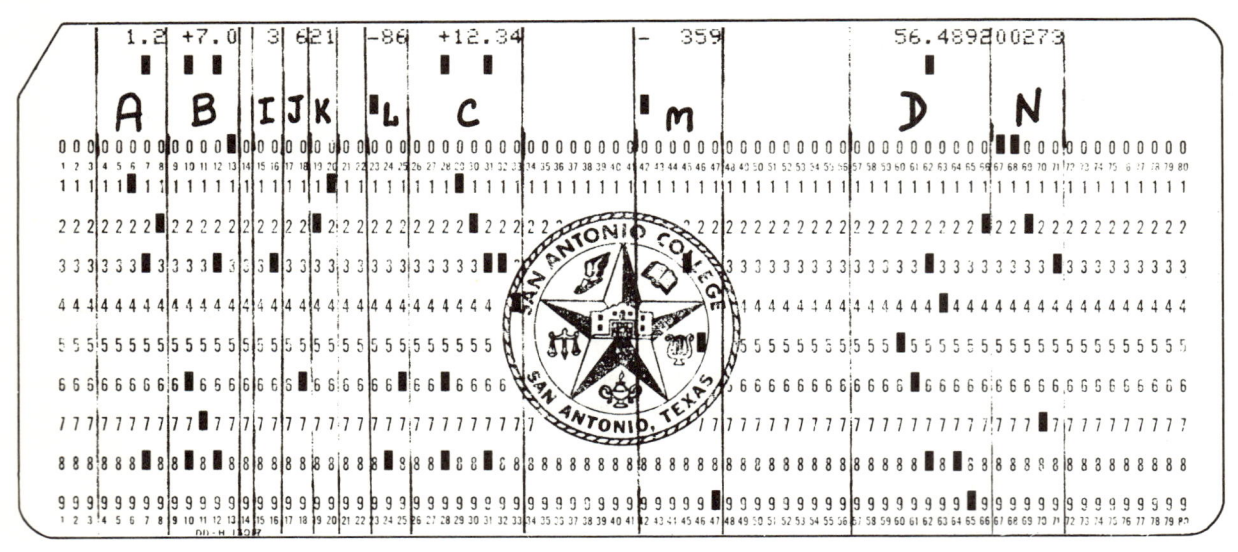

Figure 6.2. Sample Card Record Two

If we wished to read and store all the numbers on the card, we could do so with the following READ statement and its associated FORMAT statement.

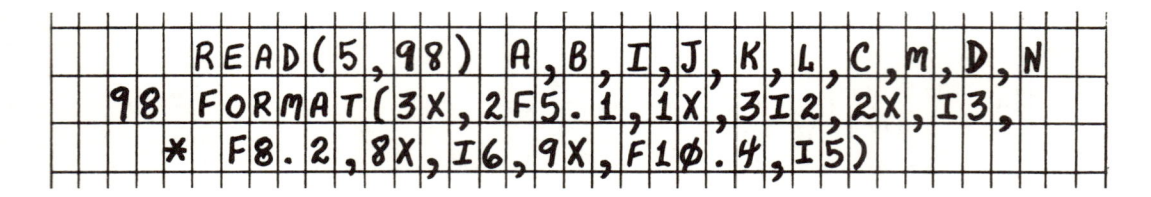

```
      READ(5,98) A,B,I,J,K,L,C,M,D,N
   98 FORMAT(3X,2F5.1,1X,3I2,2X,I3,
  *    F8.2,8X,I6,9X,F10.4,I5)
```

If we wished only to transfer the second field (B), third field (I), and the sixth field (L) to memory, we could do so with the following READ statement and its associated FORMAT statement.

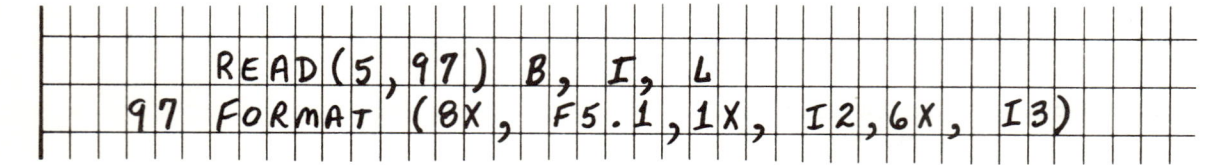

```
      READ(5,97) B,I,L
   97 FORMAT(8X,F5.1,1X,I2,6X,I3)
```

If we wished to transfer the third field (I), fourth field (J), seventh field (C), and the tenth field (N), we could do so with the following READ statement and its associated FORMAT statement.

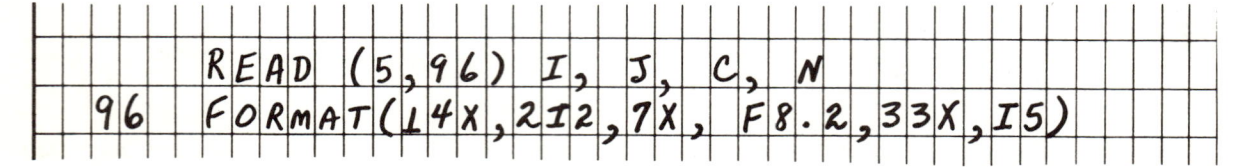

```
      READ(5,96) I,J,C,N
   96 FORMAT(14X,2I2,7X,F8.2,33X,I5)
```

There may be situations in which we wish to cause the system to automatically read a new record for us without specifying multiple READ statements. That is, when the end of the FORMAT statement is reached and there are still additional items in the I/O list, the system will cause a new record to be read.

The system will also return to the first encountered left parenthesis in a right-to-left scan (i.e., rightmost left parenthesis) to find new format codes for the remaining variables to be transferred. For example:

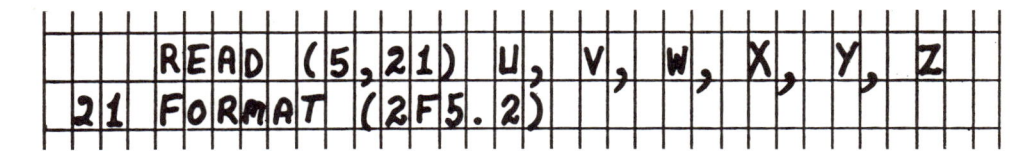

```
      READ (5,21) U, V, W, X, Y, Z
   21 FORMAT (2F5.2)
```

This READ statement and associated FORMAT statement cause three records to be read. First, the items U and V are transferred off the first record under the field specifications of 2F5.2. Since no more format codes exist in the FORMAT statement for the remaining variables, the system will cause the next record to be read and will return to the beginning of the FORMAT statement for additional format codes.

The items W and X are read off the second record under the field specifications of 2F5.2. Once again, the end of the FORMAT statement is reached and additional items exist in the I/O list. The system again causes a new record to be read and returns to the first format code to use it.

The items Y and Z are transferred off the third record under the field specifications of 2F5.2. At this point the end of the FORMAT statement is reached, but there are no remaining items to be transferred. Execution of the READ statement is therefore terminated, and control of execution continues with the next instruction in sequence.

6.5 Writing Numeric Data

The output statements are widely used statements inasmuch as they provide the key for the programmer to see the results of all his hard labor. Numeric data may be output in the form of printed results: punched cards, magnetic tape, magnetic disk, and other output media. This section will discuss the output of numeric data in the form of the printed page since it is the most common form of output results, especially to the student programmer.

If many data fields are to be printed, the programmer should first construct the output field locations on a printer layout form. It is from this sheet that the output field specifications can begin to take place. Instead of using a printer layout form here, we will describe the values in memory with their assigned variable and the desired output field locations. We will print the values read from the student registration card record given in figure 6.1.

Desired Print Line Position	*Value in Memory*	*Assigned Variable*
5-11	2308517	NRSREG
17	2	LVLCLS
25-26	15	NRSHRS
30-36	426.50	TFEE

If we wish to print these values on a double-spaced line, we could print them using the WRITE statement as follows:

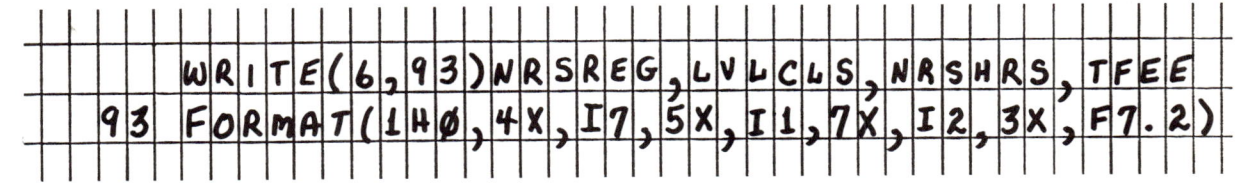

```
      WRITE(6,93)NRSREG,LVLCLS,NRSHRS,TFEE
   93 FORMAT(1H0,4X,I7,5X,I1,7X,I2,3X,F7.2)
```

The first character of data in the output record specifications is a carriage control specification of 1H0 to provide a double-spaced line feed. Next, four blank positions are specified as 4X to start the first put field at print position 5. The data item, NRSREG, is output under an I7 field specification. The data item will be right-justified in the field positions 5-11. We wish to print LVLCLS starting at print position 17, so we must specify five blank positions (5X) to align us at position 17. The item LVLCLS is transferred under an I1 field specification.

We want the next field to begin at position 25, so we must again specify seven blank positions (7X) so that we are located at this position when the data field is printed. The item NRSHRS is transferred under an I2 field specification. The last field is to start at column 30, so we specify three blank positions (3X) to align us at this print column. The item TFEE is transferred under an F7.2 field specification. The item will be right-justified in the output field. The output field will consist of any integer digits present, a decimal point, and two fractional digits.

The printed results for the given output statement are:

```
Print                 1         2         3         4
Positions:   1234567890123456789012345678901234567890
             2308517   2      15      426.50
```

We can print any item in memory at any print position we wish and in any sequence we desire. Just because data fields are input in a certain order does not mean that we must output them in that order.

The type format code used to output a data field must agree with the type variable with which it is associated (the same type as input requirements). The size need not agree, but the output format codes must agree with the output list in a one-to-one, left-to-right order.

To further illustrate the printing of numeric data, we shall print some of the values read from the data card given in figure 6.2. Our desired output specifications are:

Desired Print Line Position	*Value in Memory*	*Assigned Variable*
1-5	1.2	A
9-10	6	J
14-15	21	K
20-22	−86	L
26-35	56.4892	D

If we wish to print these items on a single-spaced line, we would code our WRITE and FORMAT statements as follows:

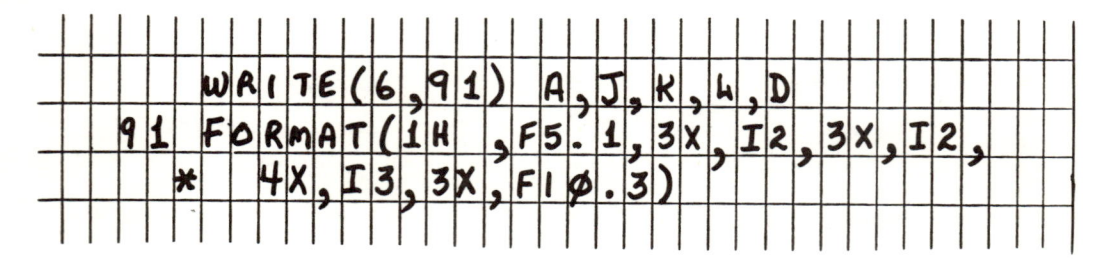

```
      WRITE(6,91) A,J,K,L,D
   91 FORMAT(1H ,F5.1,3X,I2,3X,I2,
     * 4X,I3,3X,F10.3)
```

The first character of data in the output record specifications is a carriage control specification of 1H␢ to provide a single-spaced paper feed. The first item is to be printed at position one, so we do not have to provide any blank positions before the first field. The item A is transferred under an F5.1 format code. Next, we provide three blank positions so that the item J is transferred under an I2 field specification starting at print column 9. Three blank positions are provided with the format code 3X to align the next field at position 14. The field K is transferred as an I2 field specification. Four blank positions are specified to start the field L at position 20. The item L is transferred under an I3 format code. Three blank positions (3X) are specified to start the last field at position 26. The field D is output under an F10.3 field specification. No more items are to be output, so the FORMAT statement is ended with the close parenthesis.

The printed results for this output statement are:

```
Print              1         2         3
Positions:  123456789012345678901234567890123456789
              1.2   6   21    -86     56.489
```

The same output results could be obtained by eliminating the X format codes and expanding the width of the other format codes. Remember that when the width of the output field specifications exceeds the number of output characters, the excessive high-order positions are filled with blanks. It is considered a better practice to use the X format code to separate data fields, since more accurate internal documentation is maintained thereby.

6.6 Using the Slash to Select Records with Reading and Writing Operations

Normally, one card record is read with the execution of a READ statement and one print line is written with the execution of the WRITE statement. There are occasions when a programmer may wish to read multiple input records or to skip various records in a file. The FORTRAN programmer also finds it highly desirable to be able to print multiple lines of output with only one WRITE and its associated FORMAT statement. Data fields can be read from more than one input record, input records can be skipped, and multiple output records can be produced by the same FORMAT statement with the use of the slash (/) specifications.

A slash in the FORMAT statement indicates the end of a FORTRAN record and causes a new record to be read or written. Slashes at the beginning or end of a FORMAT statement and/or consecutive slashes in the middle of a FORMAT statement are used to skip input or output records. Rules governing the use of slashes in the FORMAT statement are:

Rule 1

Slashes may be used instead of (or with) commas to separate format codes.

Rule 2

Whereas a comma used to separate two format codes means that the next field is contained on the same record, a *slash* used to separate two format codes means that the next field is contained on a new record.

Rule 3

If there are **n** (where **n** is the count) consecutive slashes at the beginning or end of the FORMAT statement associated with a READ statement, then **n** input records are skipped and no data fields are transferred from them.

Rule 4

If there are **n** consecutive slashes at the beginning or end of the FORMAT statement associated with an output statement, then **n** blank records are produced as output.

Rule 5

If **n** consecutive slashes appear other than at the beginning or end of a FORMAT statement associated with a READ statement, then **n − 1** (n minus one) input records are skipped.

Rule 6

If **n** consecutive slashes appear elsewhere (other than the beginning or end) in a FORMAT statement associated with an output statement, then the number of blank output records is **n – 1**.

Input Considerations

Examples in using slashes to skip input records are as follows:

Example one:

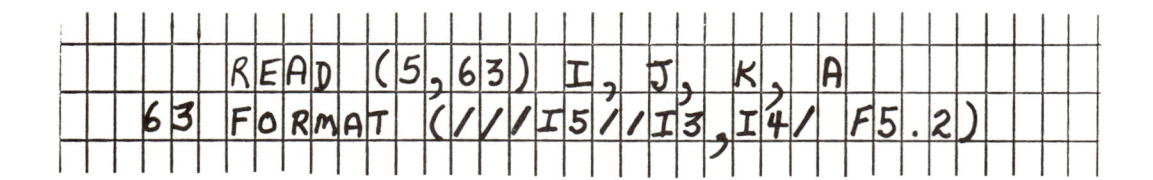

```
      READ (5,63) I, J, K, A
   63 FORMAT (///I5//I3,I4/ F5.2)
```

The first three slashes at the beginning of the FORMAT statement indicate that the first three card records are to be skipped (read without transferring any data fields). The item I is transferred off the fourth record under the I5 field specification. The next two consecutive slashes indicate that one (**n – 1**) record is to be skipped, so the fifth record is bypassed. The items J and K are transferred off the sixth record under the format codes of I3 and I4, respectively. The slash after the I4 format code indicates that the next field is to be taken off a new record. Consequently, the item A is read off the seventh record.

A total of seven records is, therefore, read in storing the four data fields in memory. Whenever the number of variables in the I/O list agrees in count with the number of format codes in the FORMAT statement, one may calculate the total records read by counting the number of slashes encountered and adding one. More difficulty occurs when more or less variables are given in the I/O list than there are format codes in the FORMAT statement.

Example two:

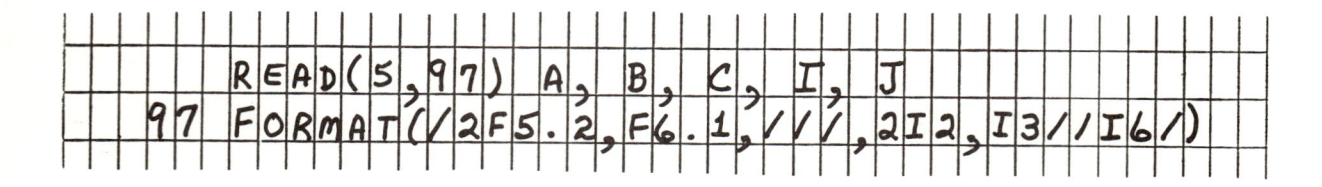

```
      READ(5,97) A, B, C, I, J
   97 FORMAT(/2F5.2,F6.1,///,2I2,I3//I6/)
```

This is an example in which there are less variables than format codes, so we know that some of the format codes will be ignored. The one slash at the beginning of the format specifications indicates that the first record is skipped. The three items A, B, and C are transferred off the second record under the format codes 2F5.2 and F6.1, respectively. The commas before and after the next three slashes have no significance and are ignored. The three slashes cause the next two (**n – 1**) records, numbers three and four, to be skipped. The items I and J are transferred from the fifth record under the two integer field specifications of the format code 2I2. No more items exist in the I/O list, so the remaining part of the FORMAT statement is ignored.

A total of five records were read and five data fields transferred to memory. Note that four slashes had been encountered—four plus one equals five records read.

Example three:

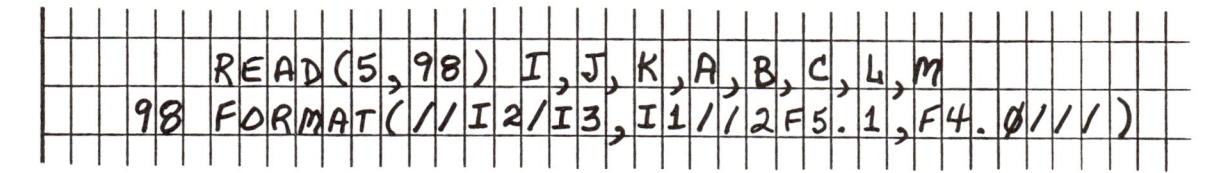

```
        READ(5,98) I,J,K,A,B,C,L,M
98 FORMAT(//I2/I3,I1//2F5.1,F4.0///)
```

This is an example of having more variables in the I/O list than format codes. The first two slashes indicate that records one and two are skipped. The item I is transferred off the third record under the format code I2. The next slash following the I2 format code implies that the next field is taken off a new record. The items J and K are transferred off record four under the field specifications of I3 and I1, respectively. The next two slashes imply that one record (**n** − 1) is skipped (record five). The items A, B, and C are transferred off record six under the format codes 2F5.1 and F4.0, respectively. The three slashes at the end of the FORMAT statement imply that records seven, eight, and nine are skipped.

Since more items exist in the I/O list, the system reverts to the first left-hand parenthesis in the right-to-left scan to find additional format codes. The two slashes at the beginning of the FORMAT statement imply that the two records, numbers ten and eleven, are skipped. The item L is transferred off record twelve under the I2 format code. The next slash implies that a new record is to be read. The item M is transferred off record thirteen. Since the I/O list has been exhausted, data transfer ends. A total of thirteen records is read and eight data fields are stored in memory.

Output Considerations

Now let's examine the use of slashes in output records. Suppose we want to print the values of A, B, and C on one line and the values of I and J on a new line. The slash could be used in the format specifications to produce these two lines of output with only one WRITE statement. The coded statements would be

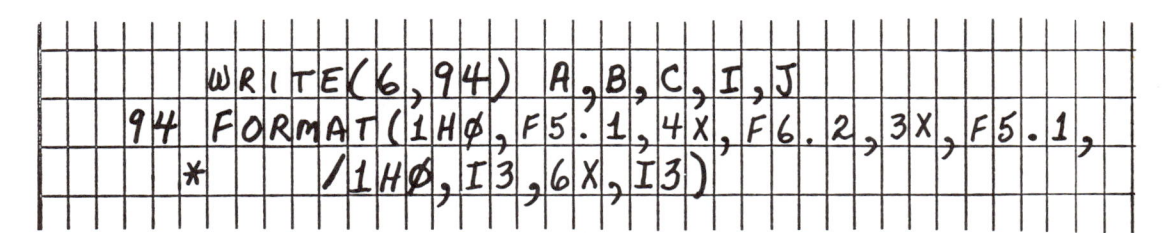

```
        WRITE(6,94) A,B,C,I,J
94 FORMAT(1H0,F5.1,4X,F6.2,3X,F5.1,
*       /1H0,I3,6X,I3)
```

The values of A, B, and C are printed beginning in positions 1, 10, and 19, respectively on the first line. The slash says start the next values on a new line, so the values of I and J are printed on a new line beginning in positions 1 and 10, respectively. Note the carriage control specification given after the slash for the second line of output. (Remember, you must include a carriage control specification for each line of printer output. Thus, a carriage control character must be provided for in the format specifications for any set of variables to be printed on a new line.)

The slash specification provides a simple way of moving the printer paper to include blank lines. Each slash moves the paper one line. Suppose we want to triple-space our output lines. We could easily specify this by including the slash as follows:

WRITE (6,98) A, B, C
98 FORMAT (/1H0, 3F8.2)

The slash moves the paper one line and then we have a double-space carriage control. Thus, the values of A, B, and C will be written on the third line below the last one.

It is often more convenient to place the slash at the end of the format specifications. We could also provide triple-spacing between output records as follows.

WRITE (6,97) A, B, C
97 FORMAT (1H0,3F8.2/)

We double-space to write our three values. Then the slash causes the paper to be moved one more line. If the next output record is written as a double-spaced line, it will be printed on the third line below our last one.

6.7 A Sample Program to Prepare a Student Tuition Register

A listing of the records in a file is often used for quick reference by clerical personnel. This listing is often referred to as a register or a roster. It is a handy reference when we want to check on the status of a person or an item of information. We have payroll registers, leave registers, grade registers, registration registers, inventory registers, etc. To further illustrate the techniques of reading and writing numeric data, let us write a program to prepare a tuition register.

We will read the input records as described in figure 6.1. The items will be printed as follows.

Field Description	*Print Positions*
Student registration number	1-7
Classification	12
Semester hours enrolled	20-21
Tuition fee	30-36

Each record will be printed double-spaced.

Let us add the logic to count the number of records printed on a page and print a maximum of only twenty-five per page. The logic to control the number of lines we print per page is called *line counter logic*. We will use a counter (LINECT) and increment it by one each time a new record is printed. When the counter gets to twenty-five we will perform a page eject and write the next record beginning at the top of the new page. We must, of course, reset our counter to zero when we perform the page eject operation. A trailer card with a value of 9999999 for student registration number will be used to terminate our processing.

A flowchart of the logic and our coded program are given in figures 6.3 and 6.4.

First, we initialize our line counter to 25 to force a page eject before printing the first record. After each student record is read we test for our trailer card to indicate an end of file. If our line counter is less than 25, we will print our detail record, bump our line counter by one, and return to process another record. When our line counter is equal to 25, we will perform a page eject and reset our line counter back to zero. We then print the next record, bump our counter, and return to our READ. You may wonder why we didn't have a WRITE with the page eject at the beginning of our program and arrange our output operation with line counter logic as follows:

```
 5  WRITE (6,99)
    LINECT = 0
10  READ (5,98) NRSREG,LVLCLS,TFEE,NRSHRS
    IF (NRSREG .EQ. 9999999) STOP
    WRITE (6,97) NRSREG,LVLCLS,NRSHRS,TFEE
    LINECT = LINECT + 1
    IF (LINECT .LT. 25) GOTO 10
    GOTO 5
```

More on Reading/Writing Numeric Data   101

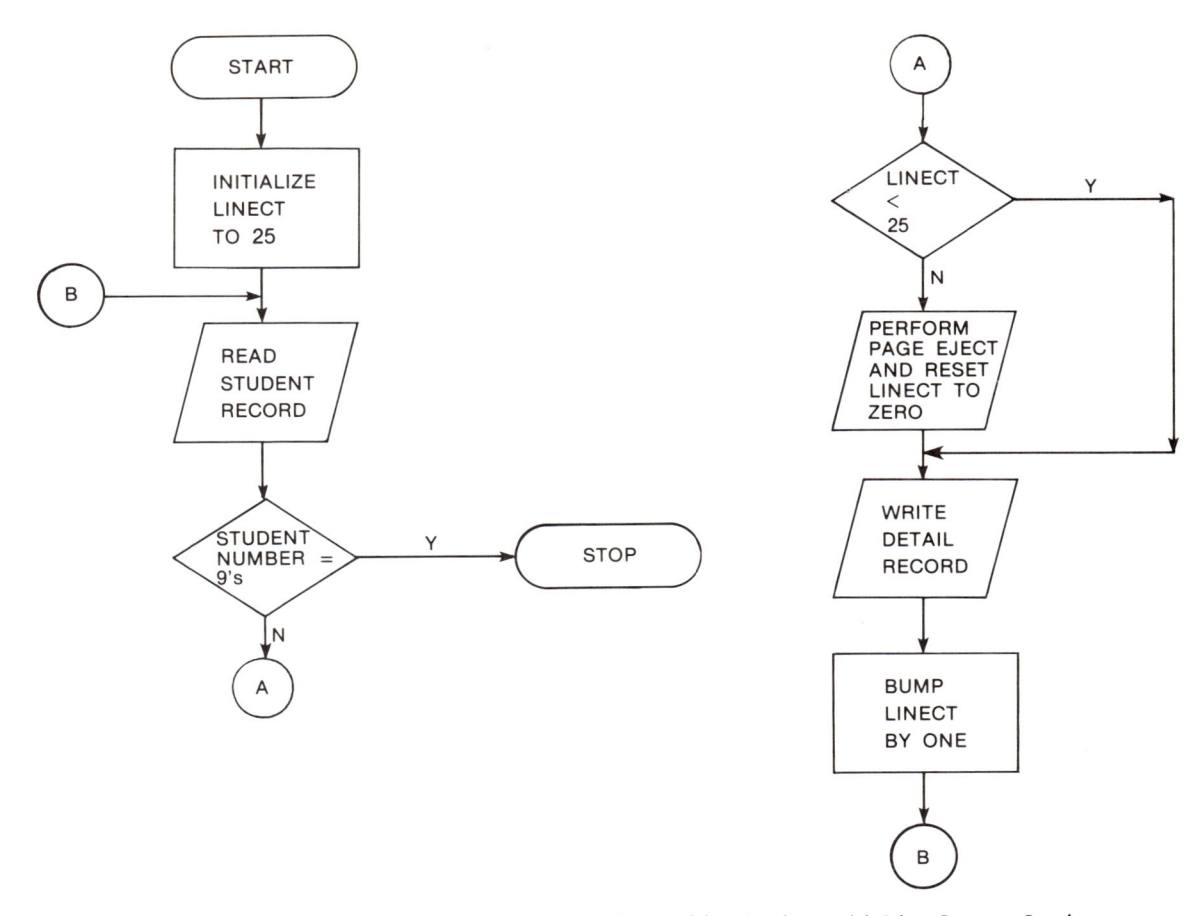

Figure 6.3. Flowchart of Logic to Prepare Student Tuition Register with Line Counter Logic

```
C *** PROGRAM TO PREPARE STUDENT TUITION REGISTER
C *** WITH LINE COUNTER LOGIC.  MAX 25 LINES/PAGE.
C
   99   FORMAT (1H1)
   98   FORMAT (I7,1X,I1,5X,F7.2,I2)
   97   FORMAT (1H0,I7,4X,I1,7X,I2,8X,F7.2)
        LINECT = 25
   10   READ (5,98) NRSREG, LVLCLS, TFEE, NRSHRS
        IF (NRSREG .EQ. 9999999) STOP
C *** CHECK LINE COUNTER FOR NEW PAGE EJECT
        IF (LINECT .LT. 25) GOTO 20
C *** WHEN LINE COUNTER IS EQUAL TO 25, WRITE PAGE EJECT AND
C *** RESET LINE COUNTER BACK TO ZERO.
        WRITE (6,99)
        LINECT = 0
   20   WRITE (6,97) NRSREG, LVLCLS, NRSHRS, TFEE
        LINECT = LINECT + 1
        GOTO 10
        END
```

Figure 6.4. Sample Program to Prepare Student Tuition Register

We don't want to set up our logic this way because we wish to avoid having a blank page in case we have an even multiple of 25 input records. If we used the logic given on page 100 and had an even multiple of 25 input records, we would perform a page eject in anticipation of printing a new page of output. When we branched back to the READ statement we would get an end-of-file condition. The blank page may not be so bad, but what would a clerk think when we write headings for each new page? A last page with only headings would be undesirable since a user wouldn't know whether we had missed someone or what had happened. This is an example of some of the subtle points we must remember when developing our program logic.

6.8 Nonstandard Language Extensions (END = option with the READ)

A very useful option included with the READ statement on most compilers is the **END** = parameter. This parameter may be included in the READ to check for the end of a data file. The general form of the READ with the **END** = option is:

READ (unit, n, END = a) list

The **END** = parameter is included after the format statement number and separated by a comma. The "a" represents a statement number assigned to an executable statement to which control is transferred whenever an end of file is encountered on the given file unit by the computer.

A JCL card marking the end of a data file is normally used to invoke the **END** = option. That is, when the computer attempts to read another data card, but reads the end-of-file JCL card, the **END** = option is then taken.

The use of the **END** = option in a READ statement eliminates the need for the programmer to include a trailer card at the end of his data deck and to check for this card. Thus, both the trailer card and Logical IF statement can be omitted in the program. For example:

```
   99 FORMAT (I5,1X,3I3)
   10 READ (5,99,END=20) ID, NTEST1, NTEST2, NTEST3
      . . .
      . . .
      GOTO 10
   20 . . .
```

A loop is formed to read and process multiple sets of data until an end-of-file (EOF) condition is encountered on unit 5. When an EOF condition is found, control will be transferred to the statement labeled 20.

An abnormal program termination occurs when the programmer tries to read past the end-of-file of a data deck and no provision has been made to detect this condition. That is, if the trailer card technique is not used or the **END** = option is not included in the READ and one tries to read a data card when no more are present, the computer will abort the program. The error termination message given on the IBM G compiler is:

217 END OF DATA SET ON UNIT x

where **x** is the file unit number. This means the programmer tried to read another data record when none was available and no provisions were included to handle the end-of-file condition.

CDC compilers allow the testing for an end-of-file condition by the following statement:

IF (EOF(unit) .NE. 0) s

where **unit** is the file unit number and s is an executable statement. **EOF** (unit) is a call to a system routine to see if an end-of-file has been encountered on the specified unit. If an end-of-file has *not* been reached, a value of zero is returned. Thus, the IF statement tests for a value other than zero to detect an end-of-file condition and executes the given statement when the condition is true. Examples are:

IF (EOF (5) .NE. 0) STOP
IF (EOF (5) .NE. 0) GOTO 100

The **END** = option is available on most compilers and is widely used by programmers. It would be nice to have had the option included in ANSI FORTRAN, but it isn't. Therefore, programs in this text will use the trailer card technique whenever an undetermined number of data records are to be read and processed.

Unexpected Output Results

When an incorrect numeric format code is specified to output a certain type variable (for example: an integer variable under an F format code or a real variable under an I format code), the value normally written with the IBM "G" level compiler is zeroes for the field. If an improper format code is specified with a type variable in *WATFIV*, an execution error will occur with the diagnostic message of:

"FORMAT SPECIFICATIONS AND DATA TYPE DO NOT MATCH"

If the programmer has not stored a value at a variable or misspells the name in the output I/O list, unpredictable output results will be printed with the IBM "G" level compiler. WATFIV, when this situation occurs, will print U's for the output field. As many U's as specified in the field width are printed.

Arithmetic Expressions in an Output I/O List

Some compilers like WATFIV and the B6700 allow arithmetic expressions to be included in the I/O list for the WRITE. For example:

WRITE (6,99) A, B, 10.0, A + B + 10.0
99 FORMAT (1H0, 4F10.2)

This capability comes in very handy for use when one wishes to quickly obtain a required answer, or in debugging.

The J Format Code

The **J** format code is unique to the Burroughs 6700 and 7700 compilers. On input fields, the **Jw** specifications function identically to the Iw specifications. On output, the **Jw** specifications cause the integer value to be printed in the minimum field necessary to contain the value without exceeding the width specification. That is, any high-order blanks are eliminated. If the output value requires more than **w** positions, then **w** asterisks are printed.

6.9 REVIEW QUESTIONS

1. What capability in the FORTRAN language do the various I/O statements provide?
2. A collection of related data fields is called a _____ .
3. A collection of related characters to represent one item of information is a _____ .
4. A collection of related records for a specific application is called a _____ .

5. The FORTRAN I/O statement to read data into the computer is the _____ statement.

6. The FORTRAN I/O statement which is used to write output data onto many different types of media is the _____ statement.

7. What purpose does the FORMAT statement serve in FORTRAN?

8. The format code to read numeric data in integer form is _____ .

9. The format code used to read numeric data in decimal form with a decimal point is _____ .

10. The format code to skip positions on input or to insert blanks on output is the _____ format code.

11. The integer constant which follows the format code letter is used to specify the field's W_____ .

12. Where are carriage control specifications used?

13. FORMAT statements must always have a unique statement number assigned to them. (True/False)

14. FORMAT statements may or may not be executed by the program. (True/False)

15. FORMAT statements may generally be placed anywhere in a program as long as they appear before the END statement. (True/False)

16. Constants and/or expressions may appear in the I/O list of a READ or WRITE statement. (True/False)

17. There is a one-to-one relationship between the names in the I/O list, data fields in the record, and format codes in the FORMAT statement. (True/False)

18. The type format code used must always match the type variable name assigned to an item. (True/False)

19. Numeric data fields are constructed in I/O records in much the same way as numeric constants in FORTRAN statements. (True/False)

20. Can decimal points and/or commas be included in integer data fields which are read by the I format code?

21. Leading, imbedded, and/or trailing blanks in an integer input data field are ignored. (True/False)

22. If an integer quantity in memory is too large to be output under the width specifications for the output field, what values will normally be printed by most computers?

23. If an integer quantity in memory requires less positions as specified in the **w** parameter, what values will be printed in the excessive positions?

24. The width specifications for the F format codes must include a position for the decimal point, if present, in an input field. (True/False)

25. On output fields, as many digits as specified in the **d** parameter will be output behind the decimal point. (True/False)

26. Any permissible form of an integer or real numeric field may be read by a F format code. (True/False)

27. All numeric output quantities are left-justified in their output field. (True/False)

28. FORTRAN provides automatic rounding by the system if the first digit of the truncated fractional part is equal to or greater than _____ .

29. What four things must a programmer know about a field before the proper type variable and field specifications can be constructed?

30. Where must the carriage control specification be given in the FORMAT statement for printer output?

31. The use of commas to separate format codes indicates that the next field comes from the _____ record.

32. The use of slashes to separate format codes for input specifications indicates that the next field comes from a _____ record.

6.10 EXERCISES

1. Construct a READ and its associated FORMAT statement to read the input values of J, K, and L from a punched card in card columns 1-3, 6-8, and 10-12, respectively. Use a statement number of 95 for the FORMAT statement.

2. Construct a READ and associated FORMAT statement to read the two variables N and X from a punched card. The variable N is punched in columns 2-4; the variable X is punched in columns 10-13 in the form NN.N. Use a statement number of 78 for the FORMAT statement.

3. Construct the WRITE and its associated FORMAT statement to print the input values read in exercise one. Use statement number 89 for the FORMAT statement. Output the variables as a double-spaced line with the following field positions.

Field Description	Print Positions
J	3-5
K	10-12
L	15-17

4. Construct the WRITE and its associated FORMAT statement to print the input values of N and X read in exercise two. Use statement number 17 for the FORMAT statement. Output the variables on a single-spaced line with N in print positions 5-7 and X in positions 10-13.

5. Give the required format codes (I or F) to read the following numeric input fields into memory. A ƀ indicates a blank position.

Data Field Value	Required Format Code
12	
ƀ123	
−12	
+12.34	
12.345	
−ƀ1.23	
ƀƀ123.	

6. Give the required format codes (I or F) to read the following numeric input fields into memory. (ƀ = a blank.)

Data Field Value	Required Format Code
1234	
ƀ12	
+123	
ƀƀ12.	
ƀ+1.2	
+012.3	
−12.3	

7. Show the output values printed for the following values in memory with their associated format code. Include a ƀ to represent a blank position.

Value in Memory	Format Code
−567	I4
−1819	I4
0	I5
21	I3
0.	F3.2
0.	F3.1
49.925	F7.2
−49.925	F7.1
49.925	F7.0
49.925	F7.5
49.925	F8.4
.6347	F6.4
.6347	F4.3
.6347	F4.1
.6347	F3.0
−.6347	F7.4
−.6347	F8.4
−.6347	F6.4
−.6347	F10.6

8. Show the output values printed for the following values in memory with their associated format code. Include a ∅ to represent a blank position.

Value in Memory	Format Code
−1234	I3
−1234	I6
1234	I4
0	I3
0.	F3.0
12.563	F5.1
12.563	F5.3
12.563	F5.0
12.563	F7.2
−12.563	F7.2
.1256	F6.2
.1256	F5.1
.1256	F2.1
.1256	F7.5

9. How many input records will be read with the following READ statements? Count all records read whether data fields are transferred off them or not.
   a) **READ (5,81) M, N, O, P**
      81 FORMAT (I4//I2///F6.2/F4.2////)
   b) **READ (5, 82) I, J, K, A, B**
      82 FORMAT (//I5, I7///I3, F4.2/2F5.0///)

10. How many input records will be read with the following READ statements? Count all records read whether data fields are transferred off them or not.
   a) **READ (5, 83) X, Y, Z, L, M, N**
      83 FORMAT (///F6.0, 2F5.1, I3////, 2I5/)
   b) **READ (5, 84) I, J, A, K, L, B, M**
      84 FORMAT (/I6, I7//F6.2)

11. How many lines of output values will be printed with the following WRITE statement?

WRITE (6, 93) A,B,J,X,K,Y
93 FORMAT (1H0,2F5.2,I3/1X,F4.1/1X,I2/1X,F6.2)

12. How many lines of output values will be printed with the following WRITE statement?

WRITE (6,92) X,Y,Z,I,J,K,G,H
92 FORMAT (1H0,2F6.2/1X,F5.1,I2/1X,I2/1X,I3/1X,F5.2)

6.11 PROGRAMMING PROBLEMS

1. Write a program to read a data card with values of X and Y punched in columns 1-5 and 6-10, respectively. The form of each number is NNN.N. Sum the two numbers and print the following output results beginning on a new page.

Field Description	Print Positions
X	1-5
Y	10-14
SUM	20-25 (Form = NNNN.N)

2. Write a program to read a card file of employee labor records with the following values.

Card Columns	Field Description
1-5	Employee number (Form = NNNNN)
6-9	Blank
10-14	Hours worked (Form = NNN.N)
15-19	Hourly pay rate (Form = NN.NN)

Compute the gross pay of each employee. Print each employee record on a double-spaced line with the following specifications. Use a trailer card with 99999 for employee number to indicate the end of the data file.

Field Description	Print Positions
Employee number	5-9
Hours worked	15-19
Hourly pay rate	25-29
Gross pay	40-47

3. Write a program to perform the same function as problem 1, but with multiple data records. Double-space each output line. Use a trailer card with a value −99.9 for X to indicate the end of the data file.

4. Write a FORTRAN program to read a weekly payroll transaction file on punched cards and list the data in each record. Card layout is as follows:

Card Columns	Field
1-6	Employee number (form = NNNNNN)
7-19	Blank
20-24	Hourly wage rate (Form = NN.NN)
25-30	Hours worked (Form = NNN.NN)

Write the output values in the following print positions.

Field Description	Print Positions
Employee number	6-11
Hourly wage rate	20-24
Hours worked	33-38

Use a trailer card with a value of 999999 for employee number to indicate the end of the data file.

5. Write a FORTRAN program to read an inventory file and extend (compute) the total value of each item in stock. The input values are to be read from a card in the following form:

Card Columns	Field Description
1-4	Item number (Form = NNNN)
5-7	Quantity on hand (Form = NNN)
8-13	Unit price (Form = NNN.NN)

As each card is read, compute the total value oy each stock item by multiplying quantity on hand by unit price. Print the input values from each card and computed item worth on a double-spaced line according to the following print positions.

Field Description	Print Positions
Item number	5-8
Quantity on hand	15-17
Unit price	24-29
Item worth	36-44

Use a trailer card with a value of 0000 for item number to indicate an end of the data file.

Writing Literal Data and Additional Format Specifications 7

7.1 Introduction to Literal Constants and the Hollerith (H) Format Code

When only numeric values are written for output, determining which value belongs to a variable is often confusing. One of the most important requirements for output results is clarity, otherwise the values serve little use. So most reports and printed output include column headings or some textual information to identify the output values. This textual output is known as literal data. The **literal data** is simply a string of characters used to make our output more readable. Literal data is used to:

1. provide report headings which appear at the top of the first output page to identify the report.
2. provide column headings which are printed at the top, above each column of output values, to identify the columnal values.
3. provide any type of titles or textual information to identify output results.

Literal data is usually not read and stored into memory. Since we know what report heading, column headings, and other textual information we want to have on our output, we can write the values as *literal constants.* This is not to say that we cannot read alphabetic and other characters and use them in our program. We do this with the alphanumeric (A) format code and can read any character string we wish. This type of format capability will be discussed later. We are now only interested in including literal data in our output results.

Literal data is specified by the Hollerith (H) format code. This format code is named after Dr. Herman Hollerith who first devised the coding scheme for representing characters in the punched card. Thus, literal data is often referred to as *Hollerith strings.* The H format code identifies what we call a literal constant. A *literal constant* is a string of literal data in character form that is preceded by the H format code. The letter H used to identify a literal constant must be preceded by a count of the number of characters in the string. The general form of the H format code is:

$$wH$$

where:

w is an integer constant that represents the width or number of characters in the literal constant. w may range from 1 to 255.

Examples of valid literal constants are:

```
32HINVENTORY REPORT FOR XYZ COMPANY
16HPAYROLL REGISTER
 5HTEST1
 9HX-SQUARED
 6HSUM =                              (one blank follows the  =)
11HJOHN'S DATA
```

All characters immediately following the H and through the last character in the string must be counted. Each space is also counted as a character. One must be extremely careful in providing the *exact* count of the characters since one less or one greater than the actual length will result in a syntax error. For example:

   4HTOTAL   (w is too short)

   8HAVERAGE   (w is too long)

What do we do when we have a long literal constant and we have to use continuation cards? We could break up the literal constant into several parts and code each part on a different card. This is probably the easiest way since the beginning of a second literal constant is placed immediately after a prior literal. If the literal is not broken up, but continued into a new card, the character in card column 7 is included in the literal string immediately after the character in card column 72 of the previous card.

Now let's see how we use these literal constants to include them in our output.

7.2 Writing Literal Data

The literal constant is included in the FORMAT statement specifications at the location where we want the literal to be printed. Since the literal is a constant that we form and not stored at a variable in memory, there is no variable name associated with the literal constant in the I/O list of the WRITE. A variable name *must never* be associated with a literal constant or the H format code.

We could print a report heading beginning on a new page starting in print position 14 as follows:

   WRITE (6,99)
   99 FORMAT (1H1, 13X, 20HSTUDENT GRADE REPORT)

The 1H1 tells the computer to do a page eject to begin our output on a new page. The carriage control operations such as 1H1, 1H0, 1Hb, and 1H+, that you have been using all along, have been specified as literal constants with the H format code. The 13X tells us to skip the first 13 positions. Then the literal constant of STUDENT GRADE REPORT, which is 20 characters in length, is printed starting at print position 14. Note that we do not have any variables in the I/O list of the WRITE since only the literal constant is printed.

Column headings are printed the same way. Suppose we wish to have column headings that identify our columns of output values as follows.

Print Positions	*Literal Constant*
1-7	STUD-ID
10-14	TEST1
17-21	TEST2
24-28	TEST3
32-36	TOTAL
40-46	AVERAGE

We could print these column headings to identify the columns of output values derived from our expanded student grade problem in chapter 4, as follows.

   WRITE (6,98)
   98 FORMAT (1H0,7HSTUD-ID,2X,5HTEST1,
    * 2X,5HTEST2,2X,5HTEST3,3X,5HTOTAL,
    * 3X,7HAVERAGE)

These literal constants would be printed over their corresponding columns of output values to clearly identify the output results. Report and column headings, as we have described them above, are normally

Writing Literal Data and Additional Format Specifications 111

printed on every output report. Otherwise, the report would have little value to our users unless they were very familiar with the locations of the output values.

Now let us look at how we can intersperse literal data and numeric output fields. That is, we can output a literal constant followed by a numeric value, or vice versa. Suppose we want to print the values of X, Y, and SUM with a literal constant before each value to identify which value goes with which variable. Our output specifications are:

```
      WRITE (6,96) X, Y, SUM
   96 FORMAT (1H0,4HX  =  ,F4.1,3X,
  *           4HY  =  ,F4.1,4X,6HSUM  =  ,F5.1)
```

We write the first literal constant followed by the value of X, the second literal constant followed by the value of Y, and the third literal constant followed by the value of SUM. Assuming the values of 7.2, 11.4, and 18.6 for X, Y, and SUM, respectively, our output results would appear as follows:

```
Print               1         2         3         4
Positions:  1234567890123456789012345678901234567890
            X  =  7.2   Y  =  11.4    SUM  =  18.6
```

When a literal constant is written as the first output value on a line, many programmers will include the carriage control character as the first character in the literal. For example,

```
      WRITE (6,91) I, N
   91 FORMAT (5H0I  =  ,I2,3X,4HN  =  ,I4)
```

This provides a double-spaced line with the literal constant, I = , beginning in position one. It is highly recommended that you code the carriage control characters as a separate literal to avoid forgetting to include them, and for better documentation.

Literal data may be written to punched cards, magnetic tape, and magnetic disk, as well as to printer output. Now let's see how we can use the slash specification to help us in writing various textual headings on our printed products.

7.3 Using the Slash to Write Multiple Headings

The slash specification is frequently used to produce multiple lines of literal data. That is, we can write more than one heading with only one WRITE statement by separating the literal headings with a slash. For example, we can write a report heading and column headings all in the same WRITE statement as follows.

```
      WRITE (6,92)
   92 FORMAT(1H1,10X,24HSTUDENT GRADES FOR DP101,
  *          //1H0,7HSTUDENT,5X,5HTEST1,5X,5HTEST2,
  *          5X,5HTEST3,5X,5HTOTAL,5X,7HAVERAGE)
```

The literal constant, STUDENT GRADES FOR DP101, is printed as a report heading beginning in position 11. The first slash says that is all for that line of output and any other data will appear on a new line. The second slash moves the paper one line. The carriage control of 1H0 moves the paper two more lines so the column headings are printed triple-spaced below the report heading.

The slash is frequently used to place new literals and their respective values on a new line. Assume the values of 36.87, 12.34, and −2.59 for the variables A, B, and C, respectively. We can print each literal title followed by its value on a new line as follows.

```
              WRITE (6,93) A, B, C
           93 FORMAT (1H1,4HA  =  ,F6.2/
          *          1H0,4HB  =  ,F6.2/
          *          1H0,4HC  =  ,F6.2)
```

The values would be printed as follows:

```
         Print                  1
         Positions:    12345678901234567
                      ┌─────────────────
                      │ A =   36.87
                      │
                      │ B =   12.34
                      │
                      │ C =  -2.59
```

Writing multiple lines of headings and literal values with the slash specification proves a shorter and more efficient means of outputting multiple print lines. The longer the format specifications, however, the more chances for an error. Thus, one should use discretion when using the slash for multiple output lines and not attempt to write dozens of lines this way. The simpler you keep your format specifications, the less errors you will encounter.

7.4 Repetition of Group Format Specifications

We often wish to print a group of values that require a set of repeating format specifications such as I3, 2X, F5.2, I3, 2X, F5.2, I3, 2X, F5.2. As you can see there is a group pattern to the format specifications of I3, 2X, F5.2 which is repeated three times. The **group format repetition** is a short-cut notation for repeating a *group* of format codes. The group of format codes to be repeated is enclosed within a set of parentheses. A repetition factor is specified before the beginning parenthesis to denote the number of times they are to be repeated. The general form of the group format repetition is:

$$r(fc_1, \ldots, fc_n)$$

where:

r specifies the number of times the group of format codes within the pair of parentheses is to be used.
(fc₁, . . . , fc_n) represent the group of format codes to be repeated. Any combination of FORTRAN format codes may be used in the group.

Examples of the group format repetition are as follows:

```
  31   FORMAT (I6,2X,3(I4,1X))                  (input specifications)
  86   FORMAT (4(I2,2X,F6.2),4X,2(1X,I7))       (input specifications)
  95   FORMAT (1H0,5I4,4(5X,I3),5X,F7.2)        (output specifications)
  91   FORMAT (1H0,16HINPUT VALUES ARE/         (output specifications)
      *        10(1H0,3(I5,5X)/)
```

Example one inputs a six-digit integer field in columns 1-6; columns 7 and 8 are skipped. Then three groups of four-digit integers, each followed by a blank column, are read. The group format specification of 3(I4,1X) is another way of saying I4,1X,I4,1X,I4,1X. Example two reads four groups of two fields. Each group consists of a two-digit integer, followed by two blank columns, and then by a floating-point field of six positions. The two-digit integer of the first group beginning in card column 1; the two-digit integer of the second group begins in column 11; the two-digit integer of the third group begins in column 21; and the two-digit integer of the fourth group begins in column 31. Then four columns are skipped and two groups of a blank column followed by a seven-digit integer are read.

The third example writes five four-digit integer fields. Beginning in print position 21, four groups each consisting of five spaces followed by a three-digit integer are written. Then five blank positions are skipped and a seven-position floating-point field written in positions 58-64. The fourth example writes a literal heading on one printer line. The group format specification writes ten double-spaced print lines, each containing three groups of five-digit integers, followed by five spaces on the same line. Thus, a total of 30 integer values are printed, three per line.

Group format repetition may be nested to a depth of two deep. That is, we can have a group format repetition within one other group format specification as shown in the fourth example. Consider another example:

96 FORMAT (I6,2(3(I2,F5.2),3X))

This format specification tells us to read a six-digit integer followed by two groups each containing three groups of a two-digit integer followed by a five-position floating-point field. Three columns are to be skipped at the end of each of the two groups. The equivalent format specification in expanded form would be:

96 FORMAT (I6,I2,F5.2,I2,F5.2,I2,F5.2,3X,I2,F5.2,I2,F5.2,I2,F5.2,3X)

In order to use the group format specifications, the format codes must have a consistent pattern. Then the format codes can be grouped according to this pattern and the group format repetition used. All too many students believe that each format code is repeated the specified number of times before the next format code is used. Remember the entire group of format codes must be used before they are repeated. Thus, 2(A1,I3) means A1,I3,A1,I3 and *not* A1,A1,I3,I3.

An abbreviated form of the group format repetition is handy for repeating various literal constants. The specification 100 (1H*) would cause 100 asterisks to be formatted. For example:

WRITE (6,98)
98 FORMAT (1H0,100 (1H*))

The specifications 40 (3H – * –) would cause 40 groups of dash, asterisk, dash to be printed. For example:

WRITE (6,97)
97 FORMAT (1H0,40(3H – * –))

The group format repetition is permitted with single format codes, but does not provide anything that the normal repetition factor does not do. For example, 3(I4) does the same time as 3I4, but the 3I4 is much simplier to code. For sake of clarity, always use the simple repetition factor for single format codes.

7.5 A Sample Program to Illustrate the Use of Headings

To illustrate the use of literal data and output headings let us develop a FORTRAN program to read an inventory file on cards and prepare a listing of the items in stock. Our input cards will contain the following items.

Card Columns	*Field Description*
1-7	Item number (Form = NNNNNNN)
8-9	Blank
10-13	Quantity on hand (Form = NNNN)
14	Blank
15-21	Unit cost (Form = NNNN.NN)
22-28	Unit price (Form = NNNN.NN)

Writing Literal Data and Additional Format Specifications

Let's extend the unit cost and unit price of each item in inventory by multiplying these values by their quantity on hand. We will also accumulate the total of the extended unit cost and unit price for all items in inventory. These accumulated totals will be printed at the end of the report with a literal that identifies these values.

The output will contain a report heading to identify the inventory of items in the ABC WAREHOUSE COMPANY. Report headings are normally centered over the other output values. Column headings will be included to identify the output values of each record. Each record will be printed as follows.

Field Description	Print Positions
Item number	1-7
Quantity on hand	14-17
Unit cost	26-32
Unit price	38-44
Extended unit cost	51-58
Extended unit price	65-72

The printer layout form should be used to identify the location of these output requirements in order to supply the correct format specifications. Our printer layout form is illustrated in figure 7.1.

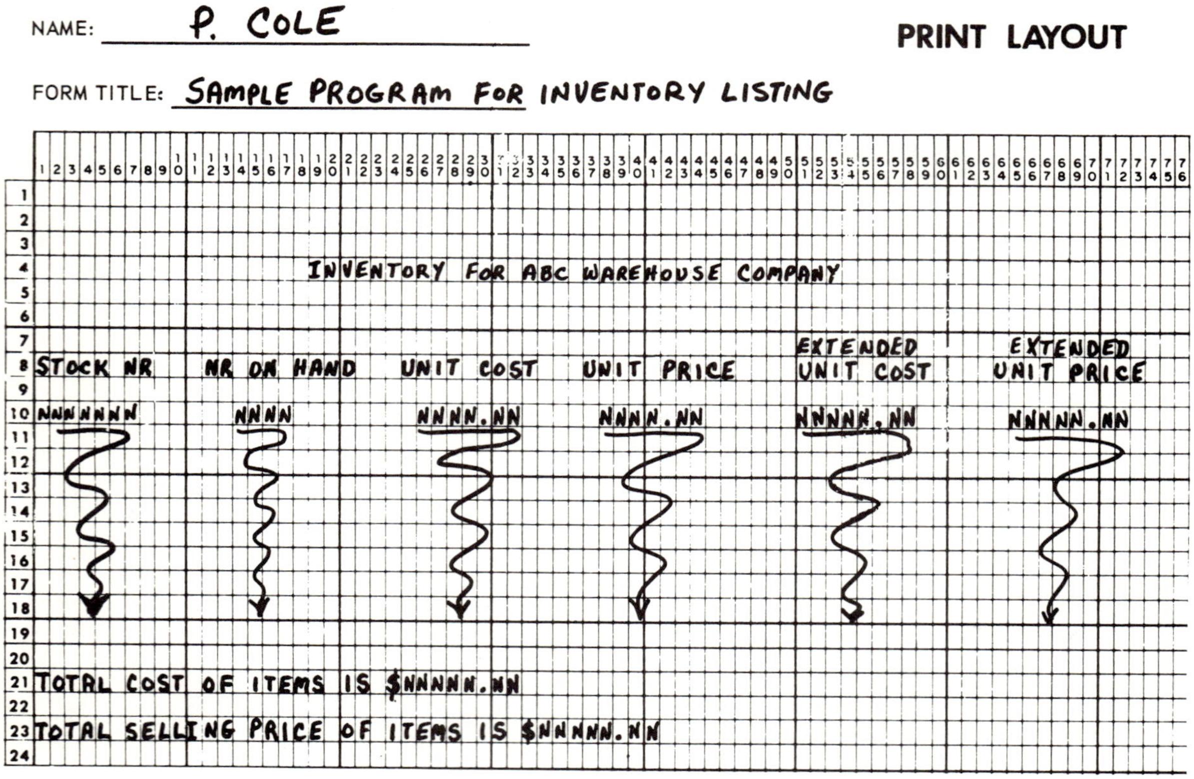

NAME: P. COLE   PRINT LAYOUT

FORM TITLE: SAMPLE PROGRAM FOR INVENTORY LISTING

```
INVENTORY FOR ABC WAREHOUSE COMPANY

STOCK NR    NR ON HAND    UNIT COST    UNIT PRICE    EXTENDED       EXTENDED
                                                     UNIT COST      UNIT PRICE
NNNNNNN     NNNN          NNNN.NN      NNNN.NN       NNNNN.NN       NNNNN.NN





TOTAL COST OF ITEMS IS $NNNNN.NN
TOTAL SELLING PRICE OF ITEMS IS $NNNNN.NN
```

Figure 7.1. Printer Layout for Sample Program

Our developed logic would be to first print our headings. Then we wish to create a loop that will read an inventory record, do our calculations, and print our output results until all card records have been processed. Remember, we must be careful not to include the printing of our headings inside our loop. Our output results would hardly be readable if we had a line of headings in between the output line of each stock item. Yet, this mistake is often made by beginning students. Our accumulated totals are printed at the end of the report. The developed flowchart is presented in figure 7.2.

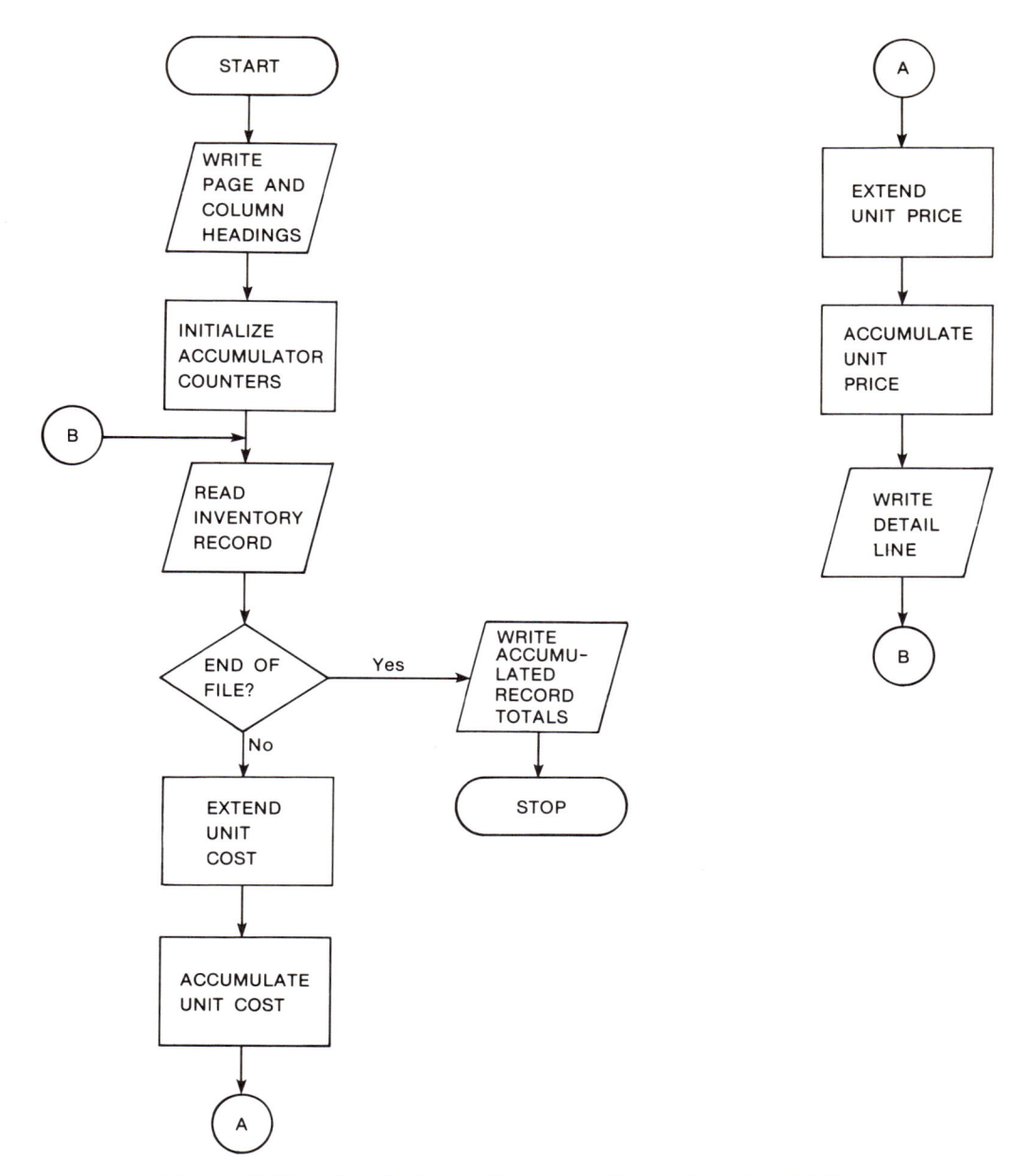

Figure 7.2. Flowchart for Sample Program to Prepare Inventory Listing

Our coded program is given in figure 7.3.

```
C *** FORTRAN PROGRAM TO LIST THE ITEMS IN INVENTORY
C *** FOR THE ABC WAREHOUSE COMPANY.  IN ADDITION,
C *** THE ACCUMULATED COST AND PRICE OF ALL ITEMS
C *** ON HAND ARE COMPUTED.
C
99      FORMAT(1H1,18X,35HINVENTORY FOR ABC WAREHOUSE COMPANY//
       *        1H0,50X,8HEXTENDED,6X,8HEXTENDED/
       *        1H ,8HSTOCK NR,3X,10HNR ON HAND,3X,9HUNIT COST,
       *        3X,10HUNIT PRICE,4X,9HUNIT COST,4X,10HUNIT PRICE)
98      FORMAT (I7,2X,I4,1X,2F7.2)
97      FORMAT (1H0,I7,6X,I4,8X,F7.2,5X,F7.2,6X,F8.2,6X,F8.2)
96      FORMAT (/1H0,24HTOTAL COST OF ITEMS IS $,F8.2/
       *        1H0,33HTOTAL SELLING PRICE OF ITEMS IS $,F8.2)
        WRITE (6,99)
        TOTALC = 0.0
        TOTALP = 0.0
10      READ(5,98) ITEM, NOHND, UNITCT, UNITPR
        IF (ITEM .EQ. 9999999) GOTO 20
        RNOHND = NOHND
        EUNITC = UNITCT * RNOHND
        TOTALC = TOTALC + EUNITC
        EUNITP = UNITPR * RNOHND
        TOTALP = TOTALP + EUNITP
        WRITE (6,97) ITEM,NOHND,UNITCT,UNITPR,EUNITC,EUNITP
        GOTO 10
20      WRITE (6,96) TOTALC, TOTALP
        STOP
        END
```

Figure 7.3. Sample Program to Prepare Inventory Listing

Assuming five inventory records in our input file, our printed results would appear as illustrated in figure 7.4.

INVENTORY FOR ABC WAREHOUSE COMPANY

STOCK NR	NR ON HAND	UNIT COST	UNIT PRICE	EXTENDED UNIT COST	EXTENDED UNIT PRICE
123456	10	20.00	30.00	200.00	300.00
235678	5	156.75	199.95	783.75	999.75
238796	100	7.60	19.50	760.00	1950.00
239032	250	3.25	9.99	812.50	2497.50
316738	51	10.33	25.50	526.83	1300.50

TOTAL COST OF ITEMS IS $ 3083.08

TOTAL SELLING PRICE OF ITEMS IS $ 7047.75

Figure 7.4. Printer Output Results from Sample Program

The use of literal constants in our output products make them more readable and fancy. That is, literal constants enable us to make our reports very attractive and pleasing to the eye of the user. Some people refer to this as **"bells and whistles."** You can get "carried away." Some programmers go overboard and place a border of asterisks around certain values or output. Keep your output simple and use literal constants only when it helps the user to understand the report.

7.6 Nonstandard Language Extensions (T and ' Format Codes)

Tabulation (T) Format Code

It is often very time-consuming to count the number of blanks between input and output fields. We have a special format code in most compilers that eliminates the need for this operation. The Tabulation (T) format code is used to indicate the starting position of input and output fields. That is, we can specify the input position on the punched card of the various fields to be read. Or we can specify various print positions for the beginning of our output fields. This format code is very useful when many blank columns or positions are to be skipped on our input and output records.

The general form of the T format code is:

Tw

where:

w specifies the position in the input or output record where the transfer of data is to begin. **w** must be an unsigned integer constant with a value less than or equal to 255.

The T format code does not correspond to any variable in an I/O list since it does not cause any data to be transferred. The T format code simply specifies the beginning position of a field in the input or output record. Thus, the T format code is always followed by the format code of the data to be transferred.

Examples in using the T format code are:

99	FORMAT (T10, I2, T25, I5)	(input specifications)
98	FORMAT (I3, T11, I4, 2X, F6.2)	(input specifications)
97	FORMAT (IH0, T4, I4, T31, I3)	(output specifications)

The first example tells us that the first input field begins in card column 10 and is transferred as a two-digit integer field. The next field begins in column 25 and a five-digit integer field is read. The second example illustrates an input field of three integer digits beginning in implied card column one, in the same way as that by which we have always expressed a field position. The second field of four integer digits begins in card column 11, followed by two skipped columns and a floating-point field of six positions beginning in column 17. The third example tells us to double-space the print line and to output a four-digit integer field beginning in print position 3. A three-digit integer field is also printed beginning at print position 30. The output positions are confusing. Why did we start at a position one less than was given? The starting position for T specifications on printer output is always one less than what is given.

On input the specifications given in the T format code indicates the high-order (beginning) position of the field on the input medium. The T format code does not have to precede each of the data transferring format codes. Any format code not preceded with the T specifications begins at the next card column where the last data transferring format code left off. For example:

98 FORMAT (I2,3X,I3,T20,I5,F5.1)

The first field begins in card column one and the second field begins in card column six, just as we have always indicated the start of each field. The T20 indicates the third field begins in column 20 and goes through column 24. The fourth field then takes up where the third field left off and begins in column 25.

The same specifications apply for output results written to cards, tape, or disk. The T specifications indicate the beginning position of the next field to be written. But for printer output, the T specifications must be *one greater* than the desired position of the printed field. If we wish to begin printing at print position 50, we must specify a T51 format code. The reason for this different specification is that the computer will always strip off (remove) one position for carriage control. Thus we do not have to supply a carriage control character when using the T specification for the printed line. One position is stripped off and used for single-spacing control. If a carriage control character is supplied for the print line, it is stripped off and provides the normal control for the movement of the paper. But one position is still stripped off for all T format code specifications on printer output. Consider the following examples:

```
      WRITE (6,99) N1,N2
99 FORMAT (T11,I2,T21,I2)
      WRITE (6,98) N3,N4
98 FORMAT (1H0,T11,I2,T21,I2)
```

The first WRITE and FORMAT statements will cause the variable, N1, to be printed at position 10; a carriage control operation of single-spacing is obtained from the T11 format code. The second variable, N2, is printed at position 20. The second WRITE and FORMAT statements cause the variable, N3, to be printed at position 10 and the variable, N4, to be printed at position 20. Since the carriage control character of 1H0 was found, the line will first be double-spaced.

The T format code does not have to indicate field positions in ascending order on the input record. For example, we could specify

```
      READ (5,99) N1, N2, N3, N3DIG
99 FORMAT (T1,I1, I1, I1, T1,I3)
```

Thus, the T format code can be used to give us a reread capability of a record, and to reformat the same record under different format codes. The example allows us to read three integer digits as separate fields and then to reread the same three digits as one three-digit integer field. The same capability is permissible with output values; but this operation provides no advantages, inasmuch as the last character formatted at a print position is the character printed.

The T format code is a simple way of telling us to skip certain positions on an input or output record. It is really accomplishing the same purpose as the Blank (X) format code. But, if we have a long gap between fields, we may make a mistake by counting the columns and using the X format code. The T format code saves us time and permits less opportunity for error by allowing us to specify the exact position in which our field starts. The T and X format codes, however, may both be used in the same FORMAT statement. IBM, Honeywell, Burroughs, and WATFIV compilers all allow the T format code.

Single Quote to Express Literal Constants

Most compilers (IBM, Honeywell, Burroughs, and WATFIV) also provide the single quote (') format code to represent literal constants. The single quote is the 5/8 multipunch character. When the single quote is used to represent a literal constant, the character string is enclosed within quotes. That is, a ' is used to designate the beginning of the literal and a ' is used at the end of the string. (Note: the ending single quote is the same character as the beginning single quote.) If we wish to include a quote within the literal constant, two consecutive quotes must be used.

For example:

```
'SUM = '
'LEAVE REGISTER'
'X = '
'JOHN''S DATA'
```

The single quote is usually preferred by most programmers since they do not have to count the number of characters in the literal. Thus, the mistake of specifying too short or too long a literal, especially in long strings, is avoided. For example:

WRITE (6,99)
99 FORMAT (1H1, 20X, 'GRADE REPORT FOR DP101')

The use of the H format code, however, is usually more documentational, and prevents you from exceeding the maximum characters allowed on a print line. It is usually easier to modify programs which use the H format code to represent literals since the number of characters in each string will not have to be recounted. The main advantage of using the H format code over the single quote is that the expression of literals with the H format code is standard on all compilers.

Not all compilers allow the single quote. CDC uses an asterisk (*) to represent literal constants in addition to the H format code. Honeywell and Burroughs compilers allow both the single quote and the double quote (") to be used to represent literal constants in addition to the H format code.

7.7 REVIEW QUESTIONS

1. How is a literal constant expressed in FORTRAN?
2. Name three uses of literal constants.
3. A space is not counted in a literal constant. (True/False)
4. What normally happens when the width count does not match the length of a literal constant?
5. Literal constants may only be written on printer output files. (True/False)
6. What is the advantage of using the group format repetition?
7. The group of format codes to be repeated must be enclosed within a pair of _____ .
8. How many levels deep can the group format repetition be used?
9. To use the group format repetition the groups of format codes must have the same pattern. (True/False)
10. A valuable tool used to display the location of literal constants on output is the _____ _____ form.

7.8 EXERCISES

1. Construct the FORMAT statement for the following output specifications.
   a) Write a report heading of STUDENT REGISTRATION ROSTER to begin on a new page in print position 50.
   b) Write column headings as follows (double-spaced):

EMPLOYEE NUMBER	beginning in position 1.
HOURS	beginning in position 20.
RATE	beginning in position 28.

   c) Write output values that produces the following line of print (single-spaced):
   the literal X = ⌿ in positions 1-4.
   the value of X as NN.N in positions 7-10.
   the literal X – SQD = ⌿ in positions 15-22.
   the value of XSQD as NNN.N in positions 25-29.

2. Construct the FORMAT statement for the following output specifications.
   a) Write a report heading of FOOTBALL TEAM ROSTER to begin on a new page in print position 45.
   b) Write column headings as follows (double-spaced):

   BOOK NUMBER   beginning in position 1.

   QTY IN STOCK   beginning in position 15.

   COST   beginning in position 30.

   c) Write output values that produces the following lines of print:
      the literal A = ∅ in positions 1-4
      the value of A as NN.NN in positions 8-12.
      the literal J = ∅ in positions 1-4 (double-spaced on a new line).
      the value of J as NNN in positions 7-9.
3. Use the group format repetition to shorten the following format specifications. Also use the simple repetition factor where possible.
   a) 37 FORMAT (I2,I2,3X,I3,3X,I3)
   b) 86 FORMAT (F6.2,I3,F6.2,I3,F6.2,I3)
   c) 97 FORMAT (5X,I6,I6,F5.2,2X,F5.2,2X)
   d) 94 FORMAT (I4,I3,2X,I3,2X,5X,I4,I3,2X,I3,2X,5X)
   e) 89 FORMAT (1H0,I3,3X,I3,3X,I3,3X,F8.2)
4. Use the group format repetition to shorten the following format specifications. Also use the simple repetition factor where possible.
   a) 56 FORMAT (I7,3X,F6.2,I7,3X,F6.2)
   b) 58 FORMAT (I4,I4,I4,3X,F6.2,2X,F6.2,2X)
   c) 64 FORMAT (5X,I3,I1,1X,I1,1X,I3,I1,1X,I1,1X)
   d) 76 FORMAT (F6.2,F6.2,5X,F5.2,I3,I1,I3,I1)
   e) 84 FORMAT (1H0,5X,F6.2,4X,F6.2,4X,F6.2,4X)

7.9 PROGRAMMING PROBLEMS

1. Write a program to write one's initials in block form as presented below. For example, my initials of PC would appear as:

```
          XXXXXX          XXXXX
          XX    XX        XX      X
          XX    XX        XX
          XXXXXX          XX
          XX              XX
          XX              XX      X
          XX              XXXXX
```

The first initial should begin in print position 20. The second initial should begin in print position 35. Each character should be seven units (lines) high.
2. Rewrite programming problem 2 in chapter 5 to include identifying literals as follows.
   the literal FAHRENHEIT TEMPERATURE = ∅ beginning in position 1.
   the value of F (fahrenheit) as NNN.N in positions 30-34.
   the literal CENTIGRADE TEMPERATURE = ∅ beginning in position 1.
   the value of C (centigrade) as NNN.N in positions 30-34.
3. Rewrite programming problem 5 in chapter 5 to include appropriate column headings.

4. Rewrite programming problem 3 in chapter 6 to include appropriate column headings.
5. Rewrite programming problem 4 in chapter 6 to include a report heading of PAYROLL TRANSACTION FILE FOR WEEK ENDING OCT. 8, 1976 and appropriate column headings.

Control and Decision-making Statements 8

8.1 Introduction to the Use of Control Statements and Types of Program Loops

Control statements provide the programmer with a decision-making ability in the execution of a program. These statements may allow one to reexecute a series of statements a desired number of times, to execute a statement only if some condition is met, to bypass statements that must not be executed in certain situations, and to terminate the program.

The function of the control and decision-making statements, therefore, is such as to allow the programmer to control the sequence in which the instructions will be executed by the computer. Statements are executed in a sequential manner until a control statement is encountered that normally directs the computer to resume execution at a specific location in the program. The use of control statements to repeat a group of instructions greatly reduces the size of programs. Decision-making statements provide flexibility in a program and make the computer appear to be almost "human" in its processing capability.

The functions of the control statements may be summed up by the three operations as follows: (1) Terminating the program, (2) Decision making, and (3) Branching/Looping. A control statement may perform any of the separate operations or it may be used to combine several of these operations. For example, a Logical IF statement may be used in decision making to execute or not to execute a certain statement. Or an IF statement may be used to make a decision as to whether to terminate the program or branch to another location in the program.

A program automatically starts execution at the first executable statement. We must instruct the computer to *terminate execution,* when we have finished processing, to obtain a *normal end of job.* The STOP statement (or CALL EXIT) that you have been using all along in your programs performs this termination.

Decision making is the process of evaluating a certain condition and determining what action is to be performed based on the result. The decision-making capability is achieved through the use of the IF statements. A branch may be made in some situations. Or a specific statement (other than a branch) may be executed when a certain condition is met. The Logical IF statement is used to perform the evaluation of a logical condition with the use of relational operators. The Arithmetic IF (another type of IF statement to be covered in this chapter) may be used to perform an evaluation of an arithmetic expression to decide what course of action to take.

Branching is the process of causing the computer to resume execution at a statement either preceding or following the statement that caused the branch. Thus, the computer can transfer control of or jump to a statement out of the order in which it appears in the program. The programmer can control the order in which the statements are executed; they do not have to be processed in a single sequential order as they are given in the program.

Branching may be one of two types—unconditional and conditional. An *unconditional branch* always causes a transfer to the indicated statement. A *conditional branch* allows the computer to choose between different courses of transfer. The choice of where to branch and resume execution is usually determined by an evaluated variable or condition.

The Unconditional GOTO always causes a branch to the specified statement. The Logical IF may contain an Unconditional GOTO to cause a branch to a location in the program when the evaluated condition is true. The Arithmetic IF statement always causes an uncondition branch to one of three provided locations based on the resulting evaluation of the contained arithmetic expression. A Computed GOTO statement can be used to transfer control to one of several alternative locations based on the value contained in an integer variable.

Looping, as we learned in chapter 4, is the process of repeating a series of instructions in the program. The technique of looping always causes a branch back to a prior program statement. A program loop may be established by using any form of the GOTO statements, Logical IF or Arithmetic IF statements.

These new control statements which you have not yet learned are explained in this chapter. Before covering these statements, let us talk more about the different types of loops that one can establish in a program.

There are basically three types of program loops. These are: (1) Input-bound Loops, (2) Counter Loops, and (3) Iterative Loops.

Input-bound loops terminate normally when some end-of-file card is read. The end-of-file card may be a system JCL card or a trailer card such as we have been using so far. The illustrated logic for an input-bound loop is:

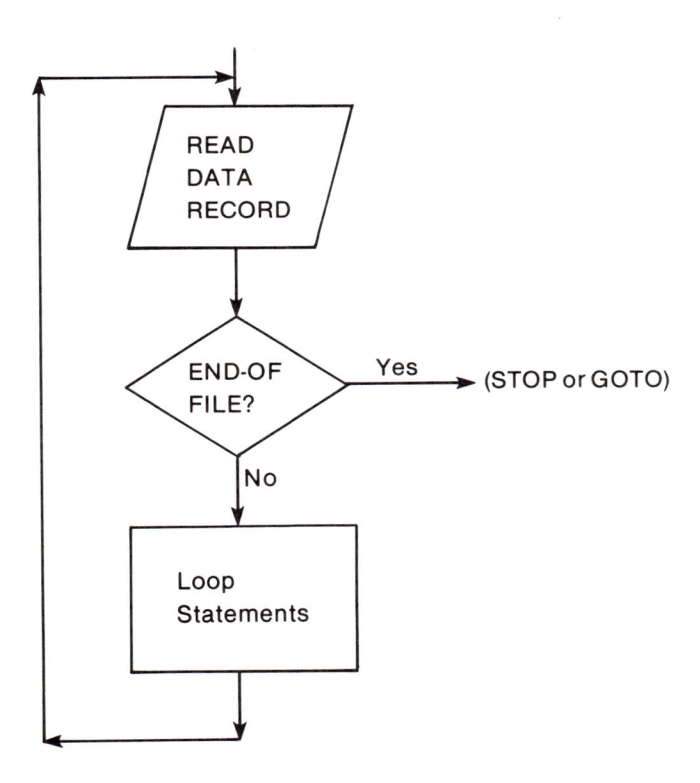

For example, application files such as payroll, inventory control, accounts receivable, etc., are usually programmed as an input-bound loop. The contents of the loop involve reading an input record, processing the data, and writing the results. The loop is terminated when there are no more records to process.

Counter loops are used to perform various functions when the programmer knows beforehand how many times the loop will be executed. An *index variable* is used to keep track of how many times the loop is executed. This index variable must be *initialized* before the loop is started, *incremented* by some value each time through the loop, and *tested* to see if the loop has been executed the proper number of times. The illustrated logic for a counter loop is given at the top of page 124.

124 Control and Decision-making Statements

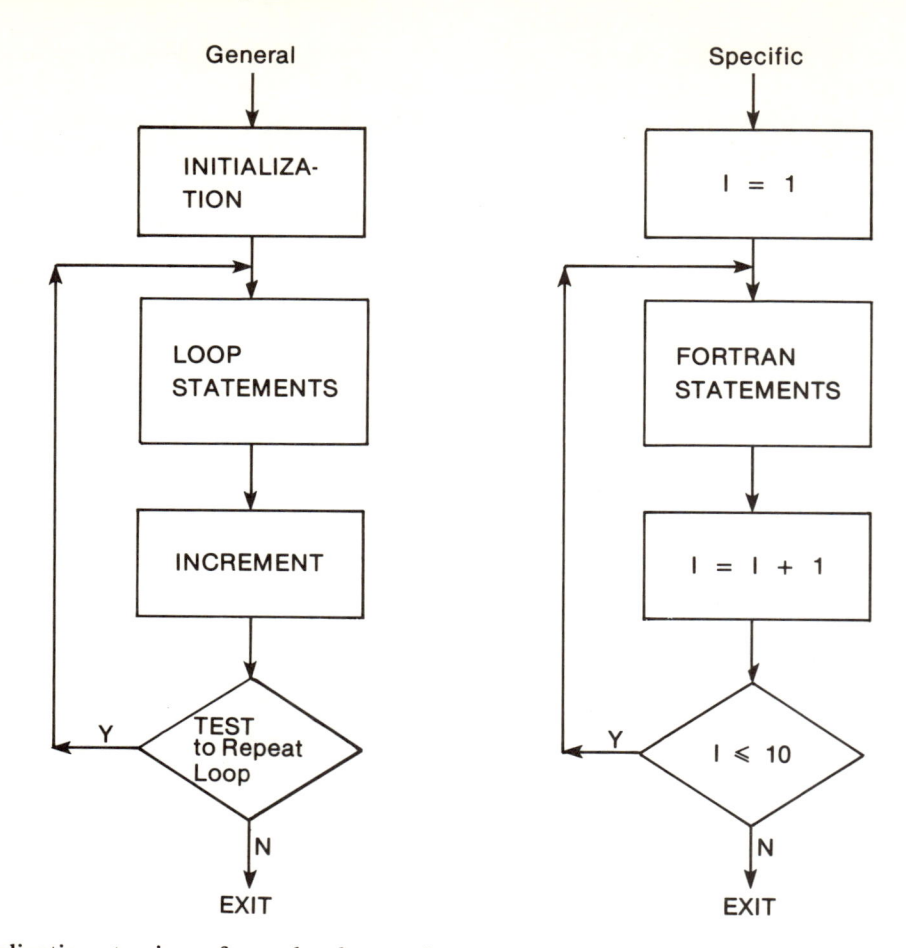

General

- INITIALIZATION
- LOOP STATEMENTS
- INCREMENT
- TEST to Repeat Loop → Y (loop back) / N → EXIT

Specific

- I = 1
- FORTRAN STATEMENTS
- I = I + 1
- I ≤ 10 → Y (loop back) / N → EXIT

The initialization step is performed only one time to set the index variable to a specific value. The index variable is usually incremented after the other instructions in the loop. Finally, the index variable is evaluated to determine whether to repeat the loop or exit from it.

For example, a loan payment schedule is used to show the payment amounts, interest amounts, and balances during the term of the loan. The payment number is the index variable that is used to determine how many payments are made in the loop. That is, if we take out a loan for two years and make a payment every month, then we want to execute the loop twenty-four times. The loop which computes the interest and balance, and prints the appropriate information is executed once for each payment.

Let us illustrate the technique of counter loops in a FORTRAN program since they are so widely used. The problem logic is to compute the square and square root of the integers from 1 to 10. The program is:

```
        INTEGR = 1
    7   ISQ = INTEGR * INTEGR
        RINTGR = INTEGR
        SQROOT = RINTGR ** .5
        WRITE (6,99) INTEGR, ISQ, SQROOT
   99   FORMAT (1H0,I2,5X,I3,5X,F6.2)
        INTEGR = INTEGR + 1
        IF (INTEGR .LE. 10) GOTO 7
        STOP
        END
```

The variable INTEGR is used as the index variable to control the number of times the loop is executed. Each time the loop is processed, the square and square root are computed. The index variable, representing the integer, and the two computed values are printed. The index variable is incremented by one on each pass of the loop. When the index variable is less than or equal to 10, the loop is repeated. When the index variable reaches 11, the loop is exited. Thus, we have produced a table of squares and square roots of the integers from 1 to 10.

Counter loops could be used to process the records in an application file when we know the exact number of records. But, the number of records usually varies from one processing to another, for people leave a company, new people are hired, new accounts are added, old accounts are closed out, etc. Thus the variation in the number of records processed prevents us from effectively using counter loops with application programs unless we modify our program each time, which is prohibitive.

Perhaps we do want to use counter loop logic in our program, but the number of times the loop is executed varies from time to time, such as for the number of students in a class. A **header card** is sometimes used with counter loops to supply the count for the number of times the loop is to be executed. This is often referred to as header-card logic. *Header-card logic* is simply the reverse of trailer card logic; an extra data card is included ahead of the data, before the detail data cards. This header card is often referred to as a *control card,* since it is used to control the looping process. A value is read from the header card and tested against our counter to see if we have executed our loop the proper number of times.

Following is a program to illustrate header-card logic. The problem is to read the number of data cards as indicated in the header card and compute their sum.

```
99      FORMAT (I3)
98      FORMAT (F5.2)
97      FORMAT (1H0,F5.2)
96      FORMAT (1H0,I5,5X,F7.2)
        KTR = 1
        SUM = 0.0
C *** READ HEADER CARD VALUE
        READ (5,99) NRCRDS
C *** READ DATA CARDS
10      READ (5,98) X
        WRITE (6,97) X
        SUM = SUM + X
        KTR = KTR + 1
        IF (KTR .LE. NRCDRS) GOTO 10
        WRITE (6,96) KTR, SUM
        STOP
        END
```

The first READ statement reads our value, NRCRDS, from the header card, which tells us how many detail cards to read. The Logical IF statement tests our counter, KTR, to see if it is still less than or equal to the variable NRCRDS. If the condition is true, the loop is repeated to read and process another data card.

Iterative loops simply execute a loop until a certain condition is met. Thus, iterative loops are sometimes called Repeat UNTIL loops or WHILE loops depending upon the placement of the test statement in the loop. If the loop is executed first and then the condition tested, we have a *Repeat UNTIL* loop. If the condition is tested before the loop is executed, we have a *WHILE* loop.

The illustrated logic for the two kinds of iterative loops is shown at the top of the next page.
For example, a machinery depreciation program computes the depreciation on a piece of machinery by applying the depreciation formula to the book value of the machine repeatedly until the book value is less than or equal to the scrap value. The Logical and Arithmetic IF statements are normally used to set up this type of loop.

Sometimes the various types of loops will be mixed in a program to accomplish the desired output. For

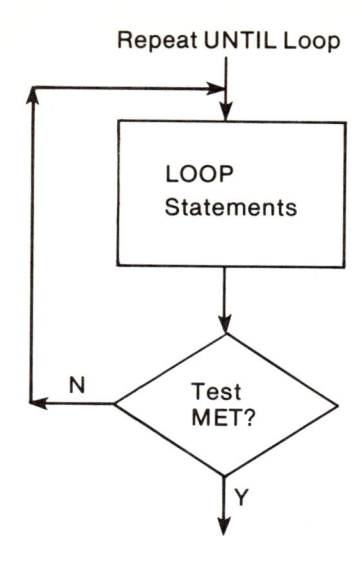

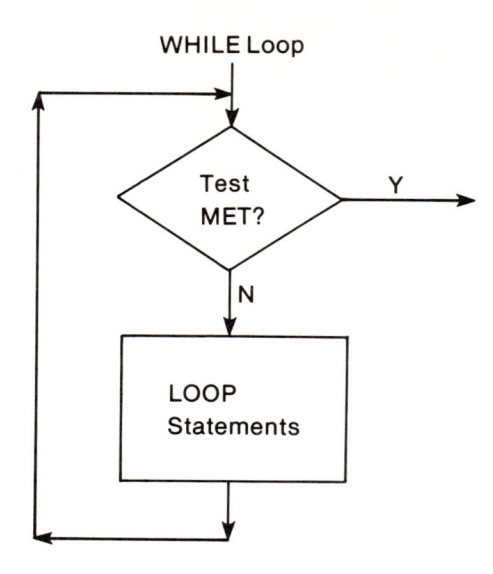

```
         Repeat UNTIL Loop                          WHILE Loop

              ┌──────────┐                         ┌──────────┐
              │   LOOP   │                         │   Test   │──Y─→
              │Statements│                         │   MET?   │
              └──────────┘                         └──────────┘
                   │                                    │ N
                   ▼                                    ▼
              ┌──────────┐                         ┌──────────┐
           N──│   Test   │                         │   LOOP   │
              │   MET?   │                         │Statements│
              └──────────┘                         └──────────┘
                   │ Y
                   ▼
```

instance, an input-bound loop may be used to read the data cards. Inside this loop, a counter or iterative loop might be used to compute various values.

We need to look at some of the most frequent errors encountered with loops, and some of the precautions you should take before we proceed to discuss the addition control statements. When a loop is established within our program, we naturally expect to exit from the loop after so many iterations. If we fail to provide an exit from the loop, then we have an infinite or endless loop. An *infinite loop* is defined as a set of instructions which have no exit from the loop, but are always repeated. In other words, we have an *endless* loop in our program. For example,

$$\begin{aligned} X &= 0.0 \\ 12 \ SUM &= SUM + X \\ X &= X + 1.0 \\ GOTO\ &12 \end{aligned}$$

We have not provided any means of exiting from our loop. The computer repeats the execution of these statements until the alloted amount of computer time has been exhausted. Infinite loops are bad, as you can see, and should be carefully avoided. Student jobs are allocated only a small amount of time, such as 30 seconds, to prevent the computer from running endlessly in case an infinite program loop is encountered.

There are several ways in which we accidentally cause infinite loops. The most frequently encountered causes are:

1. Using a poor loop structure where one can never get out of the loop. In our previous example of using an Unconditional GOTO, we did not provide any test or way to exit the loop. So always include a test or a way to exit from any loop.
2. Not initializing an index variable. We must always set our index variable to some initial value in order to produce the correct results in our loop. The most common mistake is in initializing our index variable in our first loop and completing it correctly, then failing to *reinitialize* our index variable in a later second loop. This nearly always produces a logic error. Say we want two loops—the first one to be for 20 iterations and the second one for 10 iterations. For example:

```
            I = 1
     10  . . .
            . . .
            I = I + 1
            IF (I .EQ. 21) GOTO 20
            GOTO 10
     20  . . .
     30  . . .
            I = I + 1
            IF (I .EQ. 11) GOTO 40
            GOTO 30
     40  . . .
```

Our first counter loop in which I should go from 1-20 is executed properly. But, our second loop in which I should go from 1-10 results in an endless loop, because we did not reinitialize our index variable I back to 1. Another cause of the endless loop is that we used an .EQ. relational operator. Always use a .LE. or .GE. operator to cover yourself (in most cases) for error. Thus, many endless loop situations can usually be avoided if you always make sure you properly initialize the index variable before entering each loop and/or using the less than or greater than conditions with the equal condition in the test.

3. Using two different control variables—incrementing one and testing the other. That is, we may often use two variables in our loop and forget which one is intended to be the index variable. For example,

```
            I1 = 1
            I2 = 1
            X = 0.0
     20  SUM = SUM + X
            X = X + 1.0
            I1 = I1 + 1
            IF (I2 .LE. 20) GOTO 20
```

We should have tested I1 in the IF statement, since that is our index variable. I2 is erroneously used, which results in an endless loop.

Any of the above errors can cause an infinite loop which is repeated until the allocation execution time is exhausted. What we really want to establish is a finite loop. A *finite loop* is one in which we always provide a way to properly exit from the loop, and execute the loop a only finite (specific) number of times. The Logical IF statement is normally used in most loops, either at the beginning or at the end of the loop, to test a certain input or index variable to properly terminate the loop. Now let us proceed to discuss some additional capabilities of the Logical IF statement.

8.2 More on the Logical IF Statement

Sometimes we need to test multiple conditions to decide whether to perform certain actions. That is, we may execute a certain routine only if several conditions are true or if any one of several conditions is true. For example, we may execute a routine to calculate electricity consumption for kilowatt hours in the range of 1 to 200. Thus, our conditions would be KWTHRS (kilowatt hours) greater than 0 **and** less than 201. We may want to compute a certain value for a variable only when we have a certain value for one of several variables. For example, we may want to add one to a counter only when X is not equal to 1 **or** Y is less than 0.

We could use multiple Logical IF statements in succession to test these multiple conditions. However,

FORTRAN permits the use of *logical operators* to allow us to include multiple conditions in the same Logical IF statement. For example:

IF (KWTHRS .GT. 0 .AND. KWTHRS .LT. 201) GOTO 20
IF (X .NE. 1 .OR. Y .LT. 0) KNT = KNT + 1

There are three logical operators in FORTRAN that correspond to these operations in math. Each logical operator is preceded and followed by a period. The *three logical operators,* their order of precedence, and their math function are:

Logical Operator	Order of Precedence	Meaning	Mathematical Notation
.NOT.	1st	Not (negation)	⌐
.AND.	2nd	And	∧
.OR.	3rd	Or (inclusive or)	∨

The .NOT. operation is performed before any .AND. or .OR. operation. The .AND. operation is performed before the .OR. operation.

The .AND. and .OR. logical operators are known as binary logical operators since two operands (relational expressions) are required in their use. They may be compared to an electrical circuit which connects a power source to a light bulb as follows.

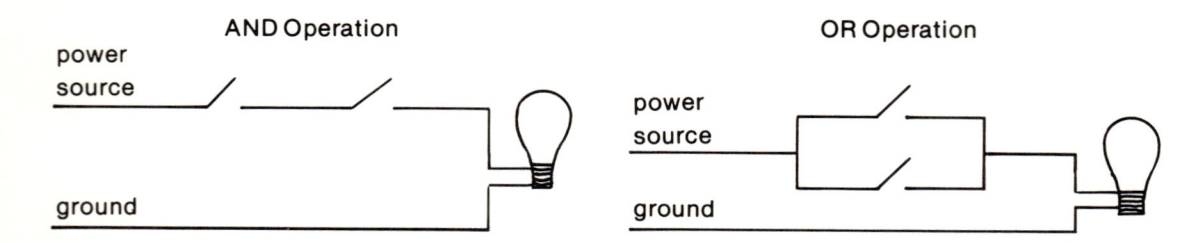

The raised (slanted) lines represent a switch. In order for the light bulb to be lit, both switches must be closed in an "AND" operation. To light the bulb in an "OR" operation, either one of the two switches (or both) may be closed. Our .AND. and .OR. logical operators work the same way. Both conditions must be met in an .AND. operation for the result to be true. However, only one of the two conditions need be met in an .OR. operation for the result to be true.

The .NOT. logical operator is a unary operator since only one operand is used with it. The .NOT. operator simply reverses the result of the condition. For example:

IF (.NOT. K .EQ. 1) GOTO 40

If K is equal to 1, the condition would be true, but the .NOT. operator changes it to false. If K is not equal to 1, the condition would be false, but the .NOT. operator changes it to true. You may ask why we didn't just say

IF (K .NE. 1) GOTO 40

This provides the same result. It is not apparent now, but the .NOT. operator can be highly useful in programming logic, especially in negating compound .AND. or .OR. conditions. For example,

IF (.NOT. (K .EQ. 1 .AND. J .NE. 0)) GOTO 50

The result of the "AND" operation is reversed by the .NOT. operator. The .NOT. operator, however, is generally not used as frequently as the .AND. and .OR. operators.

The .NOT. operator must *never* immediately *precede* the .AND. or .OR. operators. It may immediately *follow* an .AND. or .OR. operator and must precede an operand. For example,

IF (K .EQ. 1 .AND. .NOT. J .EQ. 0) GOTO 75
IF (L .EQ. 1 .OR. .NOT. M .EQ. 0) GOTO 77

We have now learned three different types of operators that can be used in a conditional expression. Arithmetic operations are always performed first, then the relational operators. The logical operators are lower in the precedence of all operators and are performed last. For example,

IF (K .EQ. 1 .AND. J .GT. 3*L + 2) VAL2 = VAL1

The arithmetic expression (3*L + 2) is evaluated first. Then each relational expression is evaluated next. And the logical .AND. operation is evaluated last. However, we must remember that there are different levels of precedence for the various arithmetic and logical operators within these types. Parentheses must be used to override the precedence of the logical operators when needed, just as we use them with arithmetic operators.

If we want our .OR. operation to be performed before our .AND. operation (since .AND. takes precedence over .OR.) we must enclose our .OR. operation within parentheses. Suppose we want to branch to a routine when either A or B is equal to 0, and C is equal to 1. We must code our statement as follows:

IF ((A .EQ. 0.0 .OR. B .EQ. 0.0) .AND. C .EQ. 1.0) GOTO 60

If we had not enclosed our .OR. condition with parentheses, the computer would have done the AND operation first between the second and third operands and then ORed the results with the first operand.

We are not limited to only one .AND. or .OR. operation within our condition. There can be any number of each type. Examples are:

IF (J .EQ. 0 .AND. K .EQ. 1 .AND. L .NE. 0) GOTO 80
IF (J .EQ. 0 .OR. K .EQ. 0 .OR. L .EQ. 0) GOTO 85
IF (J .EQ. 1 .AND. .NOT. (K .EQ. 0 .AND. L .EQ. 0)) KOUNT = KOUNT + 1

The condition in the first example would be true only if all the relational expressions are true. The condition in the second example would be true if any of the three relational expressions is true. The condition in the third example would be true when the first relational expression is true and either of the remaining two relational expressions is false.

Whenever the same variable is to be tested for different values, the variable must be repeated in the operands. Suppose we want to test N for a value greater than 0 and less than or equal to 100. We *cannot* code,

IF (N .GT. 0 .AND. .LE. 100) GOTO 90

We must repeat the variable in each operand and, thus code,

IF (N .GT. 0 .AND. N .LE. 100) GOTO 90

to provide the correct syntax.

The use of logical operators in the Logical IF statement is a complex operation. They can, however, provide flexibility and a short cut in programming.

8.3 The Arithmetic IF Statement

The Arithmetic IF statement is used to test an arithmetic expression and to provide branching to one of three possible program statements. The Arithmetic IF has an arithmetic expression within parentheses following the keyword IF. Three statement numbers follow the closing parenthesis to provide the statements that control can be transferred based on the evaluation of the arithmetic expression. If the resulting value of the expression is negative (less than zero) a branch is made to the first statement number. If the value of the expression is equal to zero, the branch is made to the second statement number. And if the value of the evaluated expression is positive (greater than zero), the branch is made to the third statement number. The general form of the Arithmetic IF statement is illustrated in figure 8.1.

Arithmetic IF Statement—General Form

IF (arith exp) n_1, n_2, n_3

where:

    arith exp — represents any arithmetic expression permitted in FORTRAN IV.

    n_1 — is a statement number assigned to an executable statement to which control will be transferred if the value of the arithmetic expression is less than zero (negative).

    n_2 — is a statement number assigned to an executable statement to which control will be transferred if the value of the arithmetic expression is equal to zero.

    n_3 — is a statement number assigned to an executable statement to which control will be transferred if the value of the arithmetic expression is greater than zero (positive).

Figure 8.1. General Notation of the Arithmetic IF Statement

Examples of the Arithmetic IF are:

```
        IF (R) 100, 200, 300
        IF (KOUNT  -  10) 1,1,2
     28 IF (A / T  -  S) 3,5,3
        IF ((N / 2) * (J  -  6)) 7,34,34
```

From these examples we see that the statement numbers may all be different ones or we can have any two the same. The logic required determines which form we use. For example, if we wish to execute a loop ten times we would probably construct our Arithmetic IF statement according to the second example. A counter would be kept and 1 added to it each pass through the loop. If the counter was less than or equal to 10, we would get a negative or zero result and branch back to statement number 1. When the counter is greater than ten we would execute statement number 2 next.

You may wonder why we have the Arithmetic IF in addition to the Logical IF statement. The Arithmetic IF can provide branching for up to three different statements, is sometimes easier to use, and is often more efficient. However, it is not usually accepted in structured programming. Thus, the Logical IF is recommended over the Arithmetic IF unless a need exists for a three-way branch.

The Arithmetic IF is flowcharted as follows when a three-way branch is used.

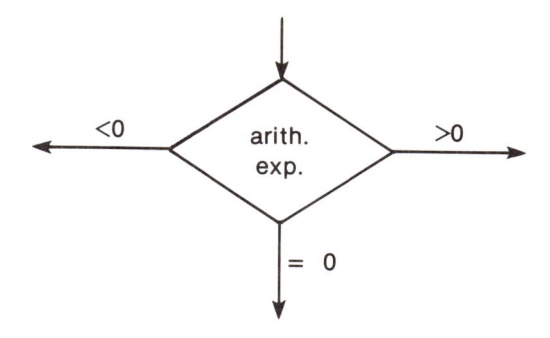

8.4 The Computed GOTO Statement

The Computed GOTO statement permits the transfer of control to one of several possible statements based on the current value of an integer variable. Thus, the computer can choose the statement that it will execute next from many paths, according to an evaluated variable included in the statement. The general form of the Computed GOTO statement is illustrated in figure 8.2.

Computed GOTO Statement—General Form

GOTO ($n_1, n_2, \ldots, n_n$), ivar

where:

$n_1, n_2, \ldots, n_n$ — are statement numbers of various statements to which transfer of control can be made.

ivar — is an integer variable containing an integer value which determines the specific statement to which control is transferred.

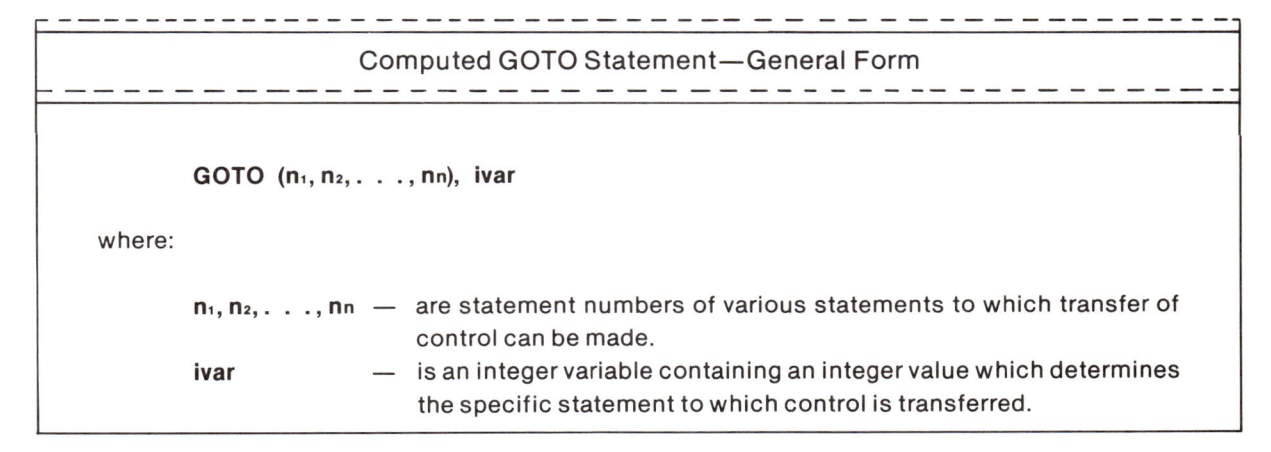

Figure 8.2. General Notation of the Computed GOTO Statement

Examples are:

      GOTO (47, 12), J
      GOTO (10, 20, 30, 40, 50, 60), KOUNT
129 GOTO (1804, 37, 456), N

Control is transferred to the first, second, third, etc., statement number depending upon whether the value of the integer variable is 1, 2, 3, etc., respectively. That is, if the value of the variable is 1, control is transferred to the first statement number inside the parenthesis. If the value of the variable is 2, control is transferred to the second statement number, and so on.

If the value of the integer variable is less than 1 or larger than the number of given statement numbers, different things happen on different computers. Some compilers such as IBM, WATFIV, and Burroughs, will fall through, i.e., continue at the next statement after the Computed GOTO. Other compilers, such as Honeywell, will abort the program and give an error message. Thus, good programming practice would call for an IF statement to check the value of the integer variable to see if it is within the permitted range before executing the Computed GOTO. If it is not, an error message should be printed, and the Computed GOTO should not be performed. Many compilers place a limit on the quantity of statement numbers within the Computed GOTO. The student need not worry about this limit, since it is extremely high, such as 63 on one specific compiler.

Note the use of the comma to separate each statement number and also to separate the closing parenthesis from the integer variable outside. You should take special care to remember the comma separating the right parenthesis and the integer variable. The omission of this comma is a mistake made by programmers very frequently.

The flowcharting symbol to represent the Computed GOTO statement is a variation of the decision symbol. The Computed GOTO statement is normally represented in a flowchart in the manner following:

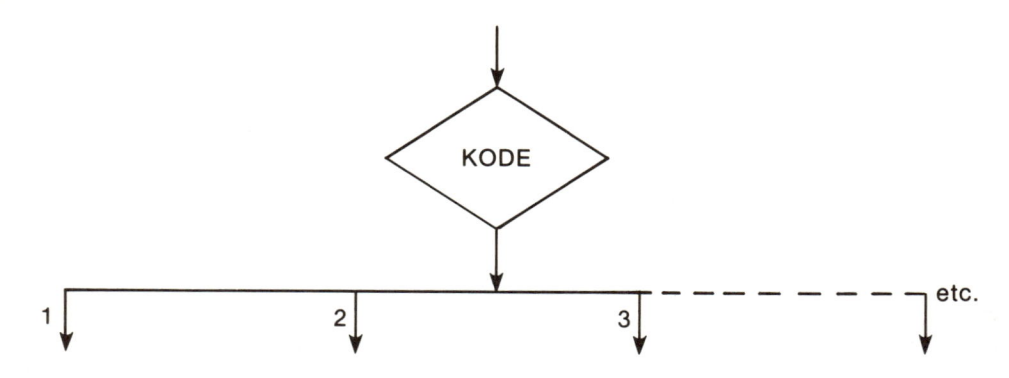

KODE is used in the above example for the integer variable required in the Computed GOTO. Now let's see how we might use these control statements in a FORTRAN program.

8.5 A Sample FORTRAN Program to Illustrate Control Statements

Let us develop a FORTRAN program to illustrate the use of the control statements in the solution of more complex problems. Our problem statement is to produce a company gross pay register. Detailed problem specifications are:

1. A three-shift operation (day, swing, and graveyard) is run in the XYZ Company.
2. Regular pay rate is paid for all hours worked up to and including 40 hours.
3. Overtime pay rate of one and one-half times regular pay rate (regular pay plus 50 percent) is paid for all hours over 40.
4. Personnel on the day shift (8:00 A.M.-4:00 P.M.) earn normal regular and overtime pay rates.
5. Personnel on the swing shift (4:00 P.M.-12:00 A.M.) earn an extra 10 percent shift differential pay for all hours worked.
6. Personnel on the graveyard or midnight shift (12:00 A.M.-8:00 A.M.) earn an extra 15 percent shift differential pay for all hours worked.

The employee payroll file is on punched cards with a record layout as follows:

Card Columns	Field Description
1-9	SSAN (Integer form)
10-14	Employee number (Integer form)
15-19	Blank
20	Shift code (1 = day, 2 = swing, 3 = graveyard)
21-26	Hours worked (Form = NNN.NN)
27-31	Hourly pay rate (Form = NN.NN)
32-80	Blank

Control and Decision-making Statements 133

The desired processing is as follows:

a) Read employee's data card.
b) Edit input shift codes to be 1, 2, or 3.
c) Edit hours worked to have a range of 0.00 to 168.00 hours.
d) Edit hourly pay rate to have a range of 2.25 to 25.00.
e) Print error messages for all data records failing their edits.
f) Compute employee's regular, overtime, shift differential, and gross pay.
g) Print input values, computed regular pay, overtime pay, shift differential pay, and total gross pay for all employee records.
h) Compute the total company gross payroll.
i) Print total computed company gross pay after all records have been processed.

Edit processing is simply checking the input data to verify its validity. We may not be able to insure that all input values are correct, but we can check for a range of values if such a variation in data is possible. That is, we can make sure that our shift codes, hours worked, and hourly pay rate are not negative or beyond the maximum value allowed. To process erroneous data would vastly distort our output results. So when an error is detected, we do not want to process the data as we normally would. We would write a message denoting whatever errors were found in the data so that they could be corrected later. Of course, the input values should be printed along with the error message to identify the record that is in error.

The output requirements are illustrated in the print layout form shown in figure 8.3.

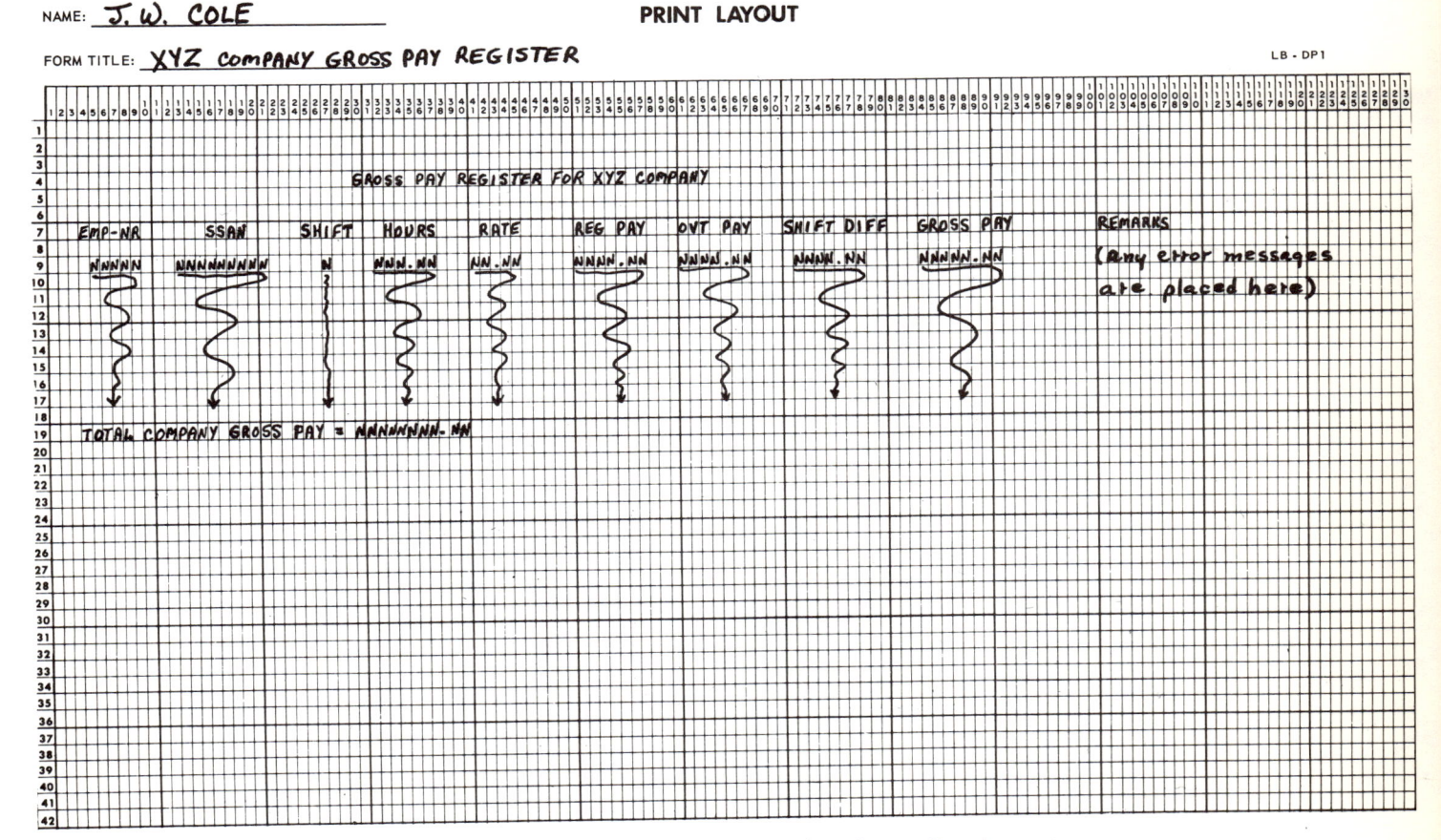

Figure 8.3. Print Layout Form for Gross Pay Register Output Requirements

134   Control and Decision-making Statements

Let us illustrate the problem and the control logic requirements in a structured program form. Figure 8.4 illustrates the structured program definitions down to three levels.

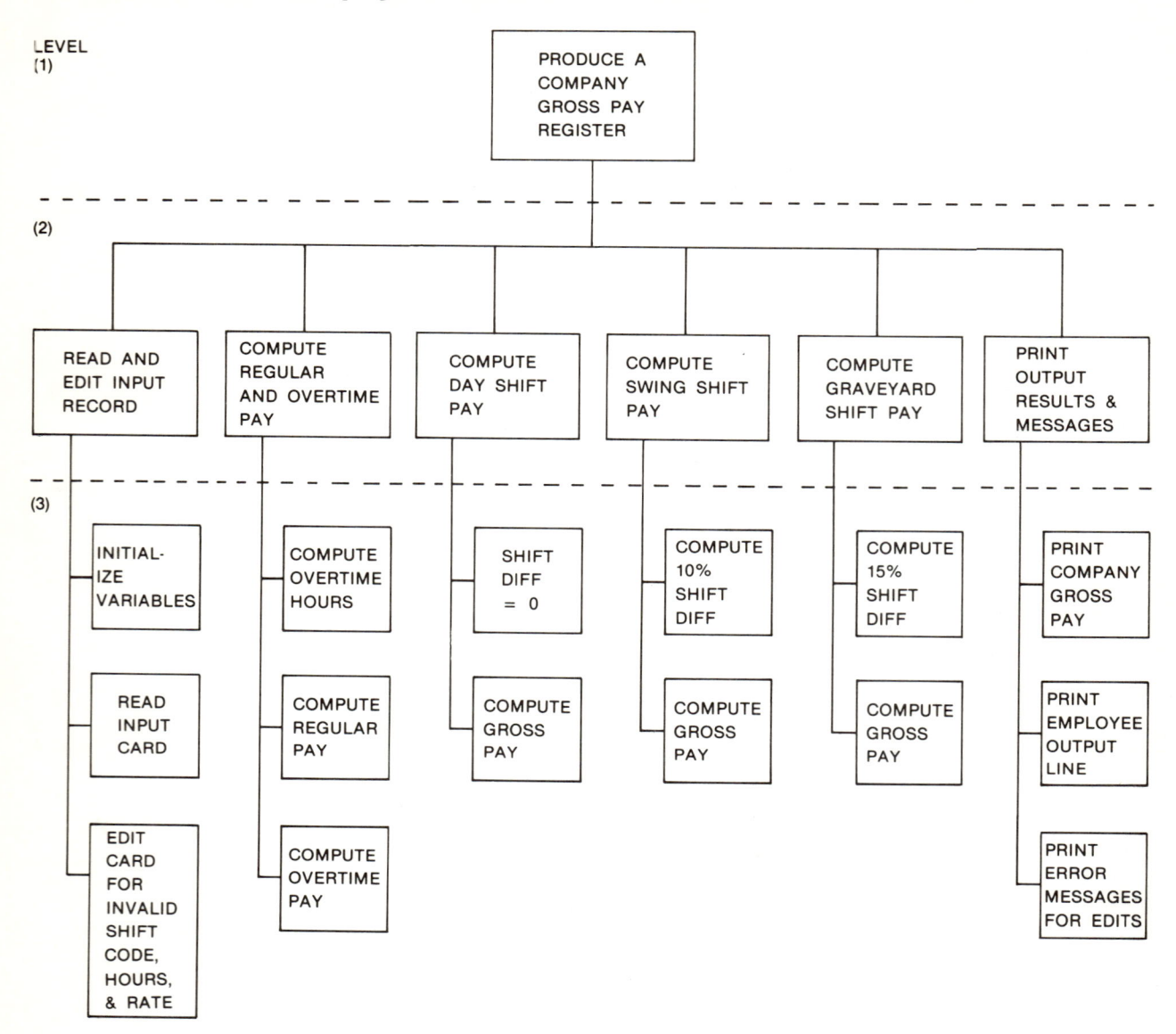

Figure 8.4. Structured Program Approach to Company Gross Pay Register (down to 3 levels)

The control logic is then converted to flowchart form. See figure 8.5 for our developed flowchart.

Control and Decision-making Statements   135

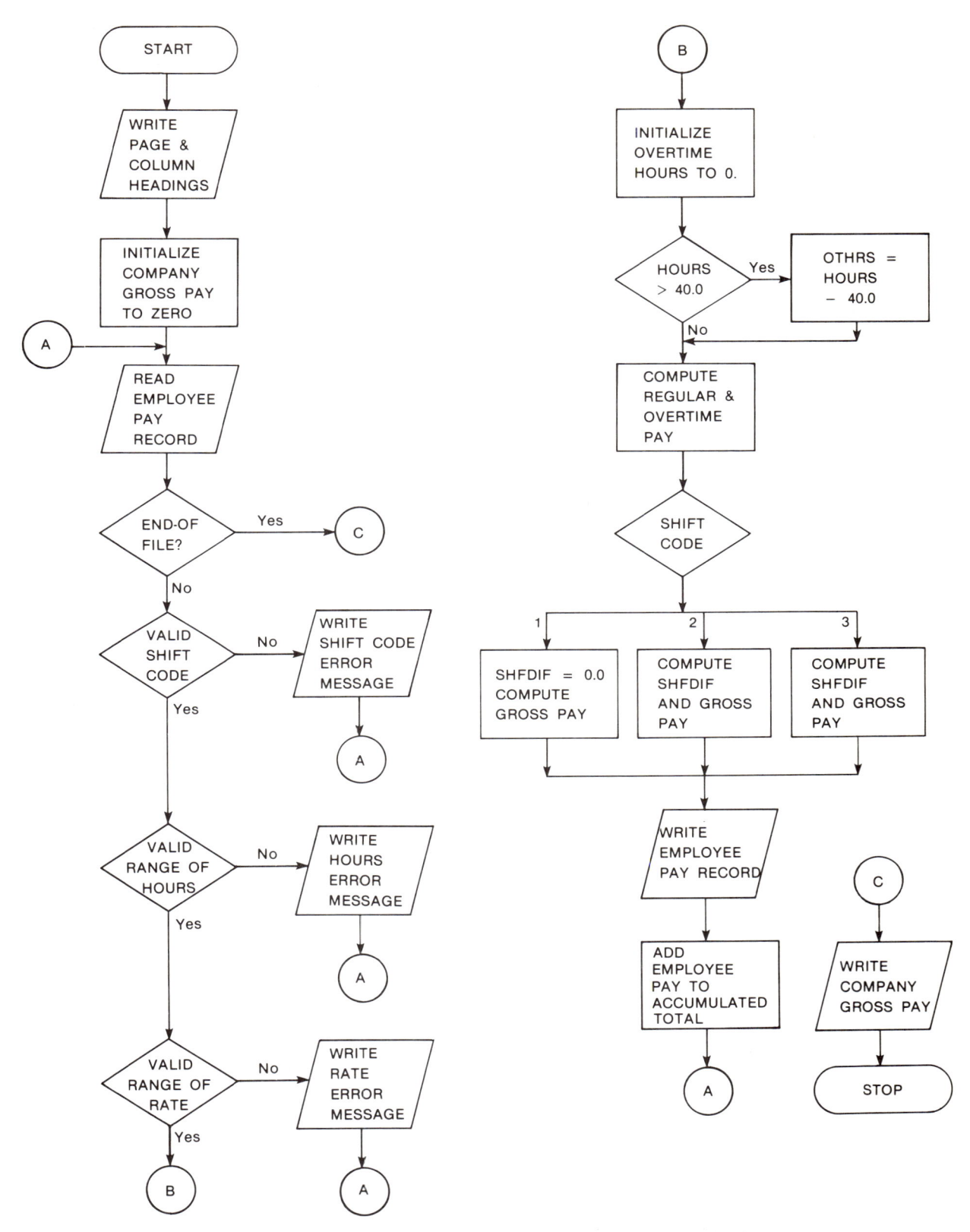

Figure 8.5. Flowchart of Logic to Produce Company Gross Pay Register

Now to the coding of our program. The resulting program is presented in figure 8.6.

```fortran
C *** FORTRAN PROGRAM TO PRODUCE COMPANY
C *** GROSS PAY REGISTER FOR WEEKLY PAYROLL
C
99      FORMAT (1H1,29X,23HGROSS PAY REGISTER FOR ,
       1    11HXYZ COMPANY//1H0,3X,6HEMP NR,6X,4HSSAN,5X,5HSHIFT,
       2    3X,5HHOURS,4X,4HRATE,5X,7HREG PAY,3X,7HOVT PAY,3X,
       3    10HSHIFT DIFF,3X,9HGROSS PAY,8X,7HREMARKS)
98      FORMAT (I9,I5,5X,I1,F6.2,F5.2)
97      FORMAT(1H0,4X,I5,3X,I9,5X,I1,4X,F6.2,3X,F5.2,
       *    55X,25H***ERROR IN SHIFT CODE***)
96      FORMAT(1H0,4X,I5,3X,I9,5X,I1,4X,F6.2,3X,
       *    F5.2,5X,F7.2,3X,F7.2,4X,F7.2,5X,F8.2)
95      FORMAT(1H0,3X,26HTOTAL COMPANY GROSS PAY = ,F11.2)
94      FORMAT (1H0,4X,I5,3X,I9,5X,I1,4X,F6.2,3X,F5.2,
       *    55X,26H***ERROR IN HOURS FIELD***)
93      FORMAT (1H0,4X,I5,3X,I9,5X,I1,4X,F6.2,3X,F5.2,
       *    55X,25H***ERROR IN RATE FIELD***)
        WRITE (6,99)
        SUMPAY = 0.
5       READ(5,98) NOSSAN,NOEMPE,KODE,HOURS,RATE
        IF (NOSSAN .EQ. 999999999) GOTO 60
C *** EDIT SHIFT CODE TO BE IN RANGE OF 1 - 3.
        IF (KODE .GT. 0 .AND. KODE .LT. 4) GOTO 7
        WRITE (6,97) NOEMPE,NOSSAN,KODE,HOURS,RATE
        GOTO 5
C *** EDIT HOURS WORKED TO BE IN RANGE OF 0.00 - 336.00
7       IF (HOURS .GE. 0.00 .AND. HOURS .LE. 336.00) GOTO 8
        WRITE (6,94) NOEMPE,NOSSAN,KODE,HOURS,RATE
        GOTO 5
C *** EDIT HOURLY RATE TO BE IN RANGE OF 2.25 - 25.00
8       IF (RATE .GE. 2.25 .AND. RATE .LE. 25.00) GOTO 9
        WRITE (6,93) NOEMPE,NOSSAN,KODE,HOURS,RATE
        GOTO 5
9       OTHRS = 0.
        IF ( HOURS .GT. 40.0)  OTHRS = HOURS - 40.0
        REGPAY = HOURS * RATE
        OTPAY = OTHRS * (RATE * .50)
        GOTO ( 10, 20, 30), KODE
C *** COMPUTE DAY SHIFT EMPLOYEE PAY
10      SHFDIF = 0.
        EMPAY = REGPAY + OTPAY
        GOTO 50
C *** COMPUTE SWING SHIFT EMPLOYEE PAY
20      SHFDIF = (REGPAY + OTPAY) * .10
        EMPAY = REGPAY + OTPAY + SHFDIF
        GOTO 50
C *** COMPUTE GRAVEYARD SHIFT EMPLOYEE PAY
30      SHFDIF = (REGPAY + OTPAY) * .15
        EMPAY = REGPAY + OTPAY + SHFDIF
50      WRITE(6,96) NOEMPE, NOSSAN, KODE, HOURS,
       1    RATE, REGPAY, OTPAY, SHFDIF, EMPAY
        SUMPAY = SUMPAY + EMPAY
        GOTO 5
60      WRITE (6,95) SUMPAY
        STOP
        END
```

Figure 8.6. Sample Program to Compute Company Gross Pay Register

Assume that we have constructed fifteen data records to test our program. The output results are displayed in figure 8.7.

```
                    GROSS PAY REGISTER FOR XYZ COMPANY

EMP NR    SSAN      SHIFT   HOURS    RATE    REG PAY   OVT PAY   SHIFT DIFF   GROSS PAY    REMARKS

12207    456231897    1    40.00    3.50    140.00     0.00       0.00        140.00
12327    503264178    2    40.00    3.00    120.00     0.00      12.00        132.00
12372    482370916    3    40.00    5.00    200.00     0.00      30.00        230.00
12468    123456789    1    30.00    4.00    120.00     0.00       0.00        120.00
12345    245781390    2    30.00    3.00     90.00     0.00       9.00         99.00
12370    341260872    3    35.00    4.00    140.00     0.00      21.00        161.00
12239    464209135    1    60.00    4.00    240.00    40.00       0.00        280.00
12496    522839014    2    45.00    3.00    135.00     7.50      14.25        156.75
12355    582134607    3    70.00    4.00    280.00    60.00      51.00        391.00
12367    502446712    0    40.00    3.00                                                   ***ERROR IN SHIFT CODE***
12401    532189264    5    40.00    3.00                                                   ***ERROR IN SHIFT CODE***
12357    301258674    1   -40.00    3.00                                                   ***ERROR IN HOURS FIELD***
12385    420485261    2   540.00    3.00                                                   ***ERROR IN HOURS FIELD***
12265    435126007    3    55.00    0.25                                                   ***ERROR IN RATE FIELD***
12461    466320118    3    50.00   33.00                                                   ***ERROR IN RATE FIELD***

TOTAL COMPANY GROSS PAY =    1709.75
```

Figure 8.7. Output Results of Company Gross Pay Register Using Test Data

Note that the test records were constructed to check out every logic path in our program. A data record was included for all three shifts, and for overtime hours as well as regular hours. Data records were also included to verify our edit checking in the program. With our program thoroughly checked out with detailed test data, we can be confident of its validity with desired "live" data files.

The control statements presented in this chapter will allow the programmer to create loops for repetitive execution of statements. The DO statement covered in the next chapter, however, provides a more powerful and flexible capability in program loops. The DO statement also produces a more orderly approach to program logic involving loops, since it eliminates many of the GOTO statements.

8.6 Nonstandard Language Extensions

The standard control statements discussed in this chapter are found on nearly all FORTRAN compilers. The CDC 6000-7000 compilers, however, also include a two-way branch for their Logical and Arithmetic IF statements, in addition to the standard forms.

The general form of the CDC two-way branch on the Logical IF statement is:

IF (logical expression) n_1, n_2

where n_1 and n_2 are statement labels. If the logical expression is true, a branch is made to statement label n_1. If the logical expression is false, the branch is made to statement label n_2. For example,

IF (K .EQ. 10) 50, 60

Thus, if K is equal to 10, control is transferred to statement label 50. If K is not equal to 10, control is transferred to statement label 60.

The general form of the CDC two-way branch on the Arithmetic IF statement is:

IF (arithmetic expression) n_1, n_2

where n_1 and n_2 are statement labels. If the arithmetic expression is not equal to zero, a branch is made to statement label n_1. If the arithmetic expression is equal to zero, a branch is made to statement label n_2. For example,

$$\text{IF } (M - 100) \ 30, 40$$

If M is not equal to 100 and the resulting expression is unequal to zero, control is transferred to statement label 30. If M is equal to 100 and the resulting expression is equal to zero, control is transferred to statement label 40.

8.7 REVIEW QUESTIONS

1. List the three functions performed by the control statements in a program.
2. What is the difference between an unconditional and a conditional branch?
3. Name the three types of program loops.
4. How is an input-bound loop terminated?
5. What are the three basic operations that must be supplied in a counter loop?
6. When is the loop terminated (exited) in a counter loop?
7. A control card that is read before the detail data cards and used in counter loop logic to specify the number of times the loop is to be executed is a _____ _____ .
8. What are the two kinds of iterative loops?
9. How is an exit made from an iterative loop?
10. Name the three logical operators available for forming compound conditions.
11. The logical operator that provides a true condition only if all expressions are true is the _____ .
12. The logical operator that provides a true condition if any of the expressions is true is the _____ .
13. Give the precedence order of the logical operators.
14. The Arithmetic IF can only be used to evaluate an arithmetic expression. (True/False)
15. The Arithmetic IF provides a branching capability to one of _____ possible statement labels.
16. The Computed GOTO provides a branching operation to one of several statement labels depending upon the evaluated value in an _____ _____ .
17. The process of checking input values to ensure their validity, such as an acceptable range of values, is known as _____ .

8.8 EXERCISES

1. Write an Unconditional GOTO statement to branch to the FORTRAN command with statement number 11 assigned to it.
2. Write an Unconditional GOTO statement to branch to statement number 37.
3. Write Logical IF statements to accomplish the following actions:
    a) If C1 is greater than C2, C3, or C4, then branch to statement label 76.
    b) If R is less than or equal to SUM / 2.0, then read another card record with a value of R.
    c) If ROOT is equal to zero, then stop execution of the program.
    d) If A is equal to 10.0 and B is greater than 0.0, then branch to statement label 41.
    e) If X or Y is equal to 0.0 and Z is unequal to 0.0, then add one to KTR.
4. Write Logical IF statements to accomplish the following actions:
    a) If T is not equal to A * B + C, then transfer control to statement number 85.

Control and Decision-making Statements   139

  b) If LINECT is greater than or equal to 50, then assign to LINECT a value of zero.
  c) If A is less than B or C, then branch to statement label 26.
  d) If 2.0 * A is less than B + C, then assign C a value of zero.
  e) If X and Y are equal to 0.0 or S and T are unequal to 0.0, then branch to statement label 37.
5. Write the Arithmetic IF statements to accomplish the following actions.
  a) If the expression 3.5 * A + C is negative, zero, or positive, then branch to statement labels 10, 20, and 30, respectively.
  b) If the expression N / 2 * 2 − N is zero, then branch to statement label 33; otherwise, branch to statement label 12.
  c) If the value of SUM is negative or zero, then branch to statement label 46; otherwise, branch to statement label 18.
6. Write the Arithmetic IF statements to accomplish the following actions.
  a) If the value in K is zero, then branch to statement label 39; otherwise, branch to statement label 66.
  b) If the expression X + Y / (3.7 − Z) is positive, then branch to statement label 46; otherwise, branch to statement label 71.
  c) If the expression 100 − KTR is negative, zero, or positive, then branch to statement labels 13, 86, and 53, respectively.
7. Write a Computed GOTO statement to branch to statement numbers 32, 47, 15, 53, and 27 if the value of M is 1, 2, 3, 4, or 5, respectively.
8. Write a Computed GOTO statement to branch to statement numbers 103, 17, 68, and 3 if the value of K is 1, 2, 3, or 4, respectively.
9. Indicate the errors in each of the following statements.
  a) **GOTO (2, 18, 5, 162) KNT**
  b) **IF (J .LT. 2M − 3) GOTO 42**
  c) **IF (SUM .LT. 10.0) 13, 14, 15**
  d) **GOTO (3, 1, 16), CODE**
  e) **IF (A EQ B) C = A**
  f) **IF KNT 10, 11, 10**
  g) **GOTO 199999**
  h) **IF (N .EQ. 1 AND L .GT. 5) STOP**
10. Indicate the errors in each of the following statements.
  a) **IF (XRESULT .GT. 25) STOP**
  b) **IF (N − 10) STOP**
  c) **IF (A .GE. C/D) END**
  d) **GOTO 20.**
  e) **GOTO (3,81,5,), L**
  f) **GOTO (1,2) I**
  g) **IF (SUM − 100) 9,8,8**
  h) **IF ((X .LT. 0.0 OR Y .EQ. 0.0) AND Z .NE. 0.0) NERR = NERR + 1**

8.9 PROGRAMMING PROBLEMS

1. Develop a FORTRAN program to produce a table of simple interest according to the formula i = prt. Use a constant of 10000.00 for principal and a constant of .085 for rate. Vary the index variable, ITIME, from values of 1 to 10 to produce a table of simple interest for one to ten years. Write output values for time, principal, rate, and computed interest according to the following output positions.

Field Name	Print Positions
Time	3-4 (Form = NN)
Principal	10-17 (Form = NNNNN.NN)
Rate	23-27 (Form = N.NNN)
Computed interest	33-40 (Form = NNNNN.NN)

Print appropriate column headings for each output field.

2. Write a program to produce a table of squares, square roots, cubes, and cube roots for the integer numbers 10 through 25. The square root of a number may be obtained by raising the number to the one-half power. The cube root of a number may be obtained by raising the number to the one-third power. Output each line of computed values as follows:

Field Name	Print Positions
Number	5-6 (I2 format code)
Square	11-13 (I3 format code)
Square root	18-23 (F6.3 format code)
Cube	28-32 (I5 format code)
Cube root	37-42 (F6.3 format code)

Print appropriate column headings for each of the five output fields. Also print an appropriate page heading.

3. Write a FORTRAN program to read a student registration card file with records containing the following information.

Card Column	Field Description
1-7	Student number (I7 format code)
8-9	Blank
10	Level of classification (1 = Freshman, 2 = Sophomore, 3 = Junior, 4 = Senior, 5 = Postgraduate)
11-12	Number of courses in which enrolled

Count the number of students within each level of classification. Also accumulate the number of enrolled courses within each classification. Print out this computed information on separate lines with proper literals to identify each output item. A trailer card with a student number of 9999999 is used to signal the end of the input file.

4. Write a FORTRAN program to read a student registration card file with records containing the following information.

Card Columns	Field Description
1-7	Student number (I7 format code)
8-9	Blank
10	Tuition rate code (1 = resident undergraduate 2 = nonresident undergraduate 3 = resident graduate 4 = nonresident graduate)
11	Blank
12-15	Semester hours enrolled (F4.1 format code)

A trailer card with a student number of 9999999 is used to signal the end of the input file.

Compute the required tuition fee for each student according to the following rate code:

Tuition Rate Code	Tuition Fee per Semester Hour
1	20.00
2	50.00
3	30.00
4	75.00

(The required tuition fee is calculated by multiplying the semester hours enrolled by the semester-hour fee.)

Edit semester hours enrolled to be in the range of 1.00 through 25.00. Print an appropriate error message if the record fails the edit. Count the number of students registered within each tuition rate code. Also accumulate the total tuition fee for each rate code and the overall tuition fee for all students.

Print a double-spaced detail line of the input values and computed tuition fee for each student which passes the edit check. Suggested output print positions are:

Field Name	Print Positions
Student number	2-8
Tuition rate code	15
Semester hours enrolled	20-23
Tuition fee	30-36

Print appropriate column headings for the above output fields. Print the accumulated count of students and total tuition fee within each rate code after the last student output line. Use appropriate literals to identify each accumulated item.

5. Write a program to read three values from a punched card and determine the largest value. The input values are each two-digit integer numbers punched in adjacent card columns beginning in card column two. For output results, print the three input integers and the selected largest value. Identify each output field by either a column heading or a literal title before the output item.

6. Write a program to perform the same operations as in problem 5, except this time find the smallest of the three integer values.

7. Write a program to read a value for the variable X from an input card in columns 1-4 (Format code of F4.1). Evaluate the input value according to the following conditions and compute the respective value of Y. If the value of X is less than zero, then compute $Y = 2X^2$. If the value of X is equal to zero, compute $Y = 1$. If the value of X is greater than zero, compute $Y = 4X - 1$. For output results write the value of X and Y, each preceded by proper literal titles.

8. Write a program to perform the same function as problem 7. Include a loop to read and process multiple sets of data until the input file is exhausted. Terminate the loop with a data card containing a value of 99.9 for X.

9. Write a program which includes counter loop logic to compute the sum and average of five real numbers. Each number is punched in columns 1-5 (form = NN.NN) on a separate card. Print each input value on a new double-spaced line. Write the computed sum and average on separate output lines with appropriate literal titles.

10. Write a program to perform the same operations as in problem 9. Include three sets of five real numbers (total of 15 data cards) to be processed. Print the output results from each set of five numbers on a new page. This problem has two nested counter loops (i.e., a loop within a loop).

Program Looping and the DO Statement   9

9.1 Introduction to Counter Loop Logic with the DO Statement

Program looping (as previously explained) consists of a branch to a previous statement which, thereby, causes a series of instructions to be reexecuted. This looping concept is one of the most powerful and frequently used techniques in programming. To read 1,000 data records requires only one READ statement executed within an input-bound loop 1,000 times. To compute the sum of the integers from 1 to 100 could be accomplished with an Assignment statement executed in a counter loop 100 times. To build a table of compound interest for so many periods of time requires us only to compute the value for one period and then to reexecute these statements for the desired number of periods. Thus, the technique of looping greatly reduces the size of a program which involves repetitive operations. For these reasons most programming languages include a control statement that simplifies loop control and implementation.

Counter-type loops are frequently used in most programs and result in a large number of errors. The biggest problem is in cycling through the loop the correct number of times. Usually the last pass through the loop causes the most problems. The index variable is incorrectly tested, which either causes the loop to be executed an extra time or lacks one pass of completing the loop the correct number of times.

It would be convenient if we had a FORTRAN statement to tell the computer to repeat a group of instructions a certain number of times. The DO statement allows us to do this very thing. Not only can we specify in one statement how many times a group of instructions is to be repeated, but we have an index variable which can be used, if desired, inside the loop. The specific initial value of the variable, the test value, and the increment value are all given in the DO statement. The computer carries out the operations required in a counter loop with the parameters found in the DO statement. So let us turn our attention to the DO statement—a powerful command to provide counter loop operations within a single statement.

9.2 The DO Statement

The DO statement is a command to execute a series of statements one or more times. The statements that physically follow the DO statement, down to and including the indicated end (range statement) of the loop are repeated as specified in the DO. The four operations required in loop control—initialization of an index variable, the increment, the test value of the variable, and the exit point in the loop—all are specified in the DO statement. Thus, the DO statement becomes a shorthand version for a counter loop. The general form of the DO statement is presented in figure 9.1 on the next page.

The index variable is established as the control variable or index to be initialized, incremented, and tested. The parameters m_1, m_2, and m_3 are called *indexing parameters* and specify the value to be used for the initialization, testing, and increment of the index variable, respectively.

Program Looping and the DO Statement 143

DO Statement—General Form

     Do n ivar = m_1, m_2, m_3

or:

     DO n ivar = m_1, m_2

where:

  n — is the statement number of the last statement to be included in the range of the DO loop.

  ivar — is called the index variable and is an integer variable used by the computer in the initializing, incrementing, and testing functions.

  = — is a necessary separator.

  m_1 — is an unsigned integer constant or variable used as the initial value of the index variable.

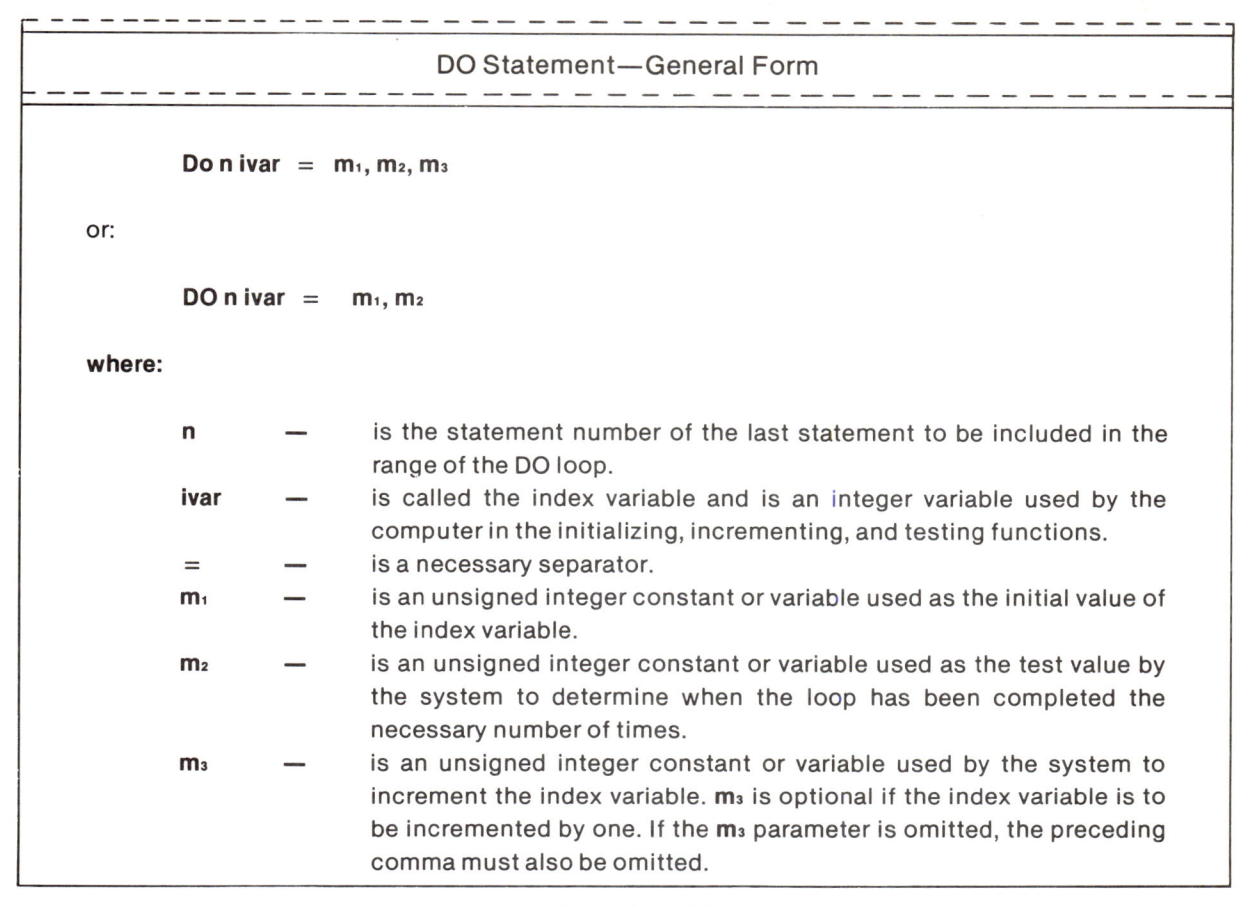

  m_2 — is an unsigned integer constant or variable used as the test value by the system to determine when the loop has been completed the necessary number of times.

  m_3 — is an unsigned integer constant or variable used by the system to increment the index variable. m_3 is optional if the index variable is to be incremented by one. If the m_3 parameter is omitted, the preceding comma must also be omitted.

Figure 9.1. General Notation of the DO Statement

Examples of the DO statement are:

        DO 10 I = 1, 10, 1
        DO 76 N = 1, 100, 2
 168 DO 33 INDEX = 1, 53
        DO 24 K = 1, N
        DO 43 JJ = L, NMAX, M

   The first example uses I as the index variable and repeats a loop through statement number 10, ten times. The second example uses N as the index variable, starts at 1, and increments by 2 until N exceeds 100; thus, the series of statements through 76 are executed 50 times. The third example (with assigned statement number 168) uses INDEX as the index variable, starts at 1, and increments INDEX by 1 until INDEX is greater than 53; thus, the series of statements through statement number 33 are executed 53 times. Remember, the increment is understood to be by 1 if this parameter is not included. The fourth example uses K as the index variable, starts at 1, and increments K by 1 until it is greater than the value of N; the series of statements through 24 are executed the number of times as the value in N. The last example uses JJ as the index variable, starts JJ at the value in L, increments it by the value in M, and tests JJ against the value in NMAX. Statement number 43 is assigned to the last statement in the loop.

A ***DO loop*** consists of all the statements included within the loop (range of the DO). The DO statement sets up the loop and is not included in the loop. All the statements in the range of the DO are repeated in execution as long as the value of the index variable is less than or equal to the test parameter (m_2).

The first time the sequence of statements contained in the DO loop are executed the index variable has the value of m_1. The second time through the loop the index variable has the value of $m_1 + m_3$, the third time $m_1 + 2m_3$, and so on, until the value in the index variable exceeds m_2. The total number of times the DO loop is executed can be quickly calculated by the formula:

$$((m_2 - m_1) / m_3) + 1$$

If the increment value, m_3, is one, the loop will be repeated $m_2 - m_1 + 1$ times. If the value of m_2 is less than or equal to m_1, the loop is executed once.

Following is an example to illustrate the range of a DO in a program which compute the squares and cubes of the numbers from 1 to 100.

```
      DO 11 K = 1, 100
      KSQ = K * K
      KCUBE = KSQ * K         } DO Loop
   11 WRITE (6,95) K, KSQ, KCUBE
   95 FORMAT (1H0,I3,3X,I6,3X,I8)
      STOP
      END
```

The WRITE command with statement number 11 assigned to it is the **range statement** or end of the loop. All statements following the DO down to and including the range statement are in the loop. Remember, the range statement must be an executable statement. The above example also shows how the index variable may be used inside the DO loop.

The DO loop is repeated until the loop is completed the desired number of times. The loop is satisfied when the index variable is greater than the test value. When the loop is completed, transfer of control (the exit) is made to the next executable statement immediately following the range statement.

The DO loop does not have to be fully completed the specified number of times. An IF statement may be included in the loop to test for a certain condition. If the condition is met, a transfer may be made out of the loop, altogether. Consider the following example:

```
      SUM = 0.
      DO 76 K = 1,100,1
      READ (5,99) X
   99 FORMAT (F5.2)
      IF (X .EQ. 0.0) GOTO 19
   76 SUM = SUM + X
      . . .
      . . .
   19 XK = K - 1
      AVE = SUM / XK
```

Other statements such as the Computed GOTO and Arithmetic IF may also be used to transfer control out of the DO loop.

The DO statement provides a structured approach to program loops. The DO statement, therefore, should be used in place of the Assignment and IF statements, whenever possible.

Another helpful technique in following program logic is the indention of all FORTRAN statements in a DO loop. This indention quickly identifies the statements inside the loop and the end of the loop. For example:

```
         SUM = 0.0
         READ (5, 97) N
         DO 10 I = 1, N
           READ (5, 95) X
           WRITE (6, 94) X
   10      SUM = SUM + X
         WRITE (6, 93) SUM
         STOP
   97    FORMAT (I3)
   95    FORMAT (F5.2)
   94    FORMAT (1X, F5.2)
   93    FORMAT (/ 1X, F7.2)
         END
```

One can readily see the three statements that make up the loop because they are indented.

You may have observed that the previous program example incorporated the use of header-card logic. The use of variables for the indexing parameters provides considerable flexibility in the use of the DO statement. The header-card value read from the punched card is used for the value of the test variable. This value then indicates how many times to repeat the loop and also how many data cards to read. Variables should be used for many of the indexing parameters where feasible to provide increased flexibility in program looping.

Flowcharting the DO Statement

There are many techniques used to flowchart the DO statement and its accompanying DO loop. The symbol normally used is the *iteration symbol.* The iteration symbol is a new flowcharting symbol to us. It is a rectangle divided into three areas as follows:

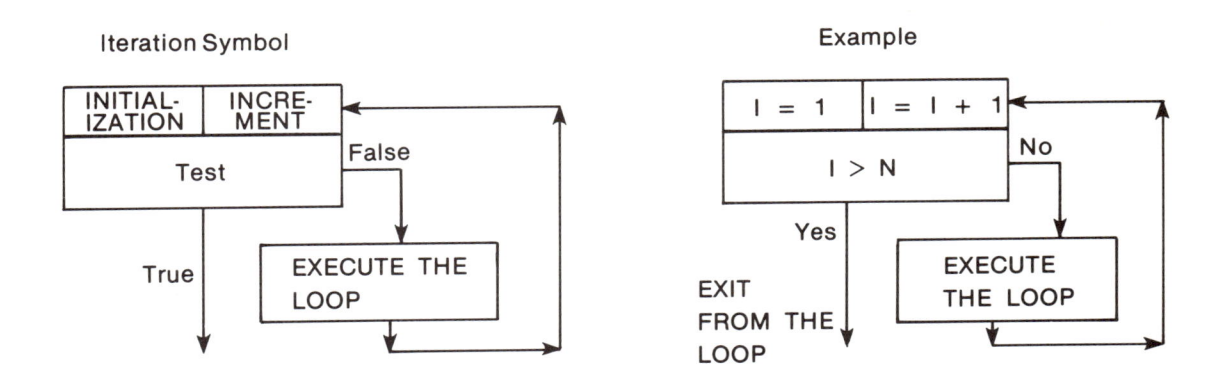

Iteration Symbol

INITIAL-IZATION	INCRE-MENT
Test	

True / False → EXECUTE THE LOOP

Example

I = 1	I = I + 1
I > N	

Yes → EXIT FROM THE LOOP
No → EXECUTE THE LOOP

Each of the three areas represent the three functions—initialization, increment, and testing—of the loop. If the test is true, an exit is made to the next executable statement following the loop. If the test is false, repetition of the loop continues until a true condition is realized or a branch is made out of the loop.

Figure 9.2 presents a flowchart of the last program example. The logic of the program is to read a header card to determine how many data cards to read and sum the input value of X on each card. Note the technique of flowcharting the DO loop.

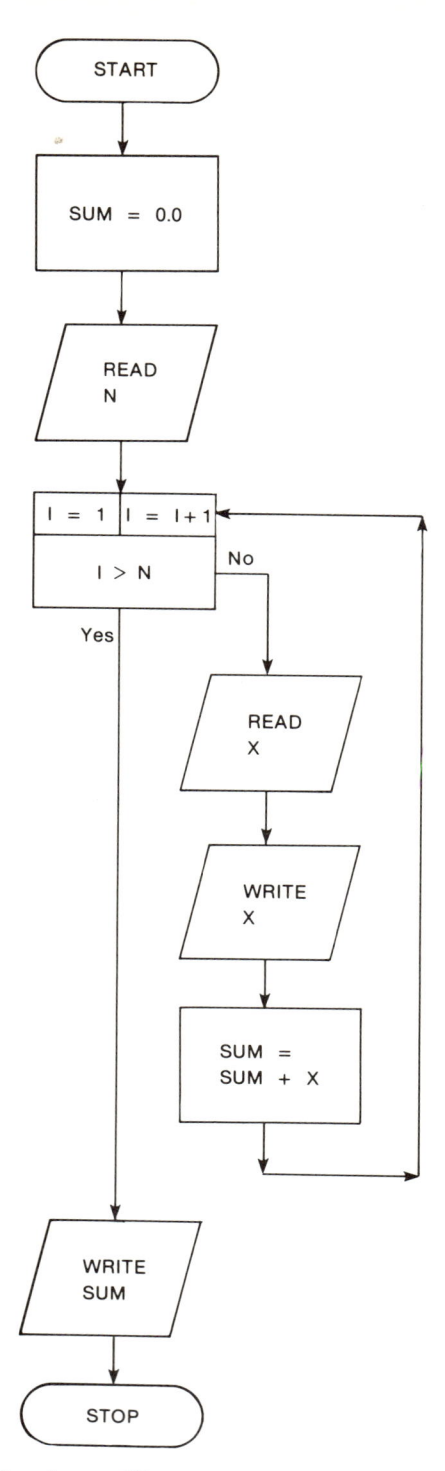

Figure 9.2. Flowchart to Illustrate a DO Loop and Header-card Logic

When the index variable exceeds the test variable, N, an exit is made from the loop.

Some programmers prefer to use a decision symbol at the end of the loop. Inside the decision symbol, a question is asked if the loop is completed. The choice of symbols and techniques used in flowcharting the DO statement and its accompanying DO loop is up to you or your instructor.

The DO statement and DO loops have some unique requirements which must not be violated by the programmer. These rules are presented in the next section.

9.3 Rules in Using the DO Statement

The DO loop created by the programmer must be handled differently from the conditional counter loop with a Logical IF statement. To illustrate this, we cite two mistakes commonly made by many students: (1) attempting to branch from outside the DO loop to a statement in the middle of the DO loop, and (2) not allowing the system to control the incrementing and testing of the index variable. The rules which must be followed are summed up below.

Rule 1

A DO loop may not be entered in the middle. The middle is defined as any statement between the DO and range statement. The loop must always be entered from the top (DO statement). For example (the partial box represents the DO loop):

LEGAL BRANCHES

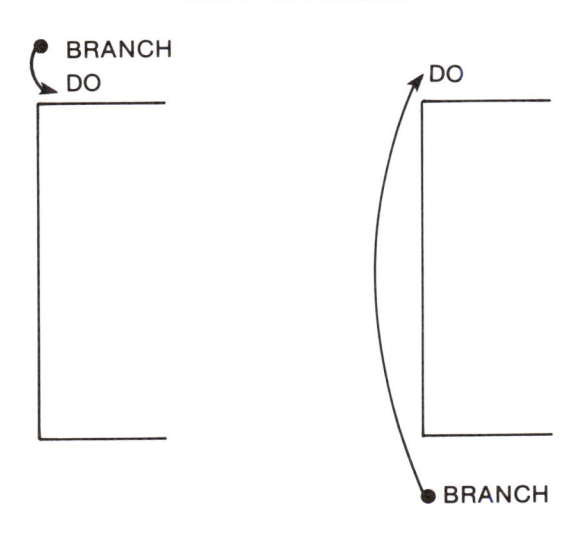

ILLEGAL BRANCHES

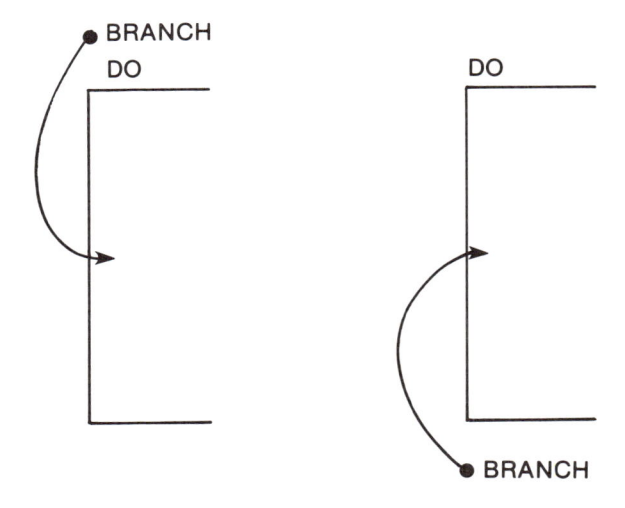

Rule 2

Whenever the DO statement is executed, the index variable is always initialized to the initial parameter and the execution of the loop begins.

Rule 3

A transfer within the loop to another statement inside the loop is always permissible. For example:

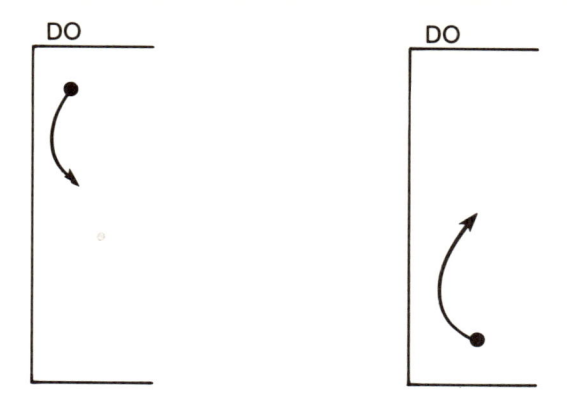

Rule 4

A transfer out of the range of any DO loop is permissible at any time. You do not have to complete all the passes of the loop. For example:

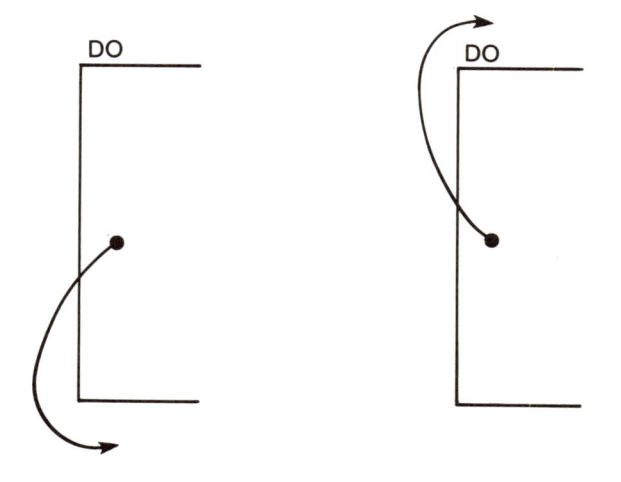

Rule 5

If the DO loop was not completed the specified number of times (i.e., a transfer was made out of the loop), then the index variable has the same value as it had when it was last incremented.

Rule 6

If the DO loop is completed the specified number of times, the value of the index variable is unpredictable and should not be used until assigned another value. The reason for this is that once a DO loop has been satisfied, no attempt is made by the compiler to retain the last value in the index variable.

Rule 7

The first statement immediately following the DO statement must be an executable statement. For example, a FORMAT is not permitted as the first statement following the DO statement.

Rule 8

The last statement in a loop must be an executable statement. For example, a FORMAT statement must not be specified as the range statement in the DO loop.

Rule 9

The last statement in the range of a DO loop cannot be a control statement which includes

    a)   A GOTO statement of any form
    b)   A STOP, PAUSE, or RETURN
    c)   A Logical IF
    d)   Another DO statement

Rule 10

Each indexing parameter must be a positive integer constant or variable.

Rule 11

The index variable and any of the indexing parameters of the DO statement may be used in the range of the DO, but they must not be changed by a statement in the range of the DO. This means that they cannot appear as the assignment variable in the Assignment statement or in the input list of a READ statement.

Rule 12

The DO loop will always be executed at least once. The test is not made until the system completes a pass of the loop. The order in which the parameters are handled in the loop are:

    a)   The index variable is initialized.
    b)   A pass of the loop is made.
    c)   The increment value is added to the index variable.
    d)   The test is made between the index variable and the test value.
    e)   If the index variable is greater than the test value, the loop is complete and control resumes at the next statement after the range statement.
    f)   If the index variable is *not* greater than the test value, the loop is repeated.
    g)   Steps c) through f) are repeated until either step e) is satisfied or a transfer is made out of the loop.

Remember, at the end of each pass through the loop the increment is added, and then the test is made.

Rule 13

The range statement must be executed for the computer to know when to attempt the next pass of the loop. A mistake frequently made by beginning FORTRAN programmers is in testing the index variable and to branch back to the DO statement. If this is done, the loop is reinitialized and begins anew. Thus, in most cases an endless loop will be formed. If some statements are to be skipped in the loop and the next pass begun, the programmer should branch to the range statement. For example,

```
      DO 12 I = 1, 50
      . . .
      . . .
      . . .
      IF (C1.EQ.0.0) GOTO 12
      . . .
      . . .
      . . .
   12 . . .
```

9.4 The CONTINUE Statement

The CONTINUE statement is a "do nothing" statement that may be placed anywhere in the source program where an executable statement may appear. The CONTINUE statement acts as a "dummy statement" and does not affect the sequence of execution in any way. Its sole function is to provide a label which can be referenced by an instruction. In other words, the CONTINUE statement causes no action to take place; when it is encountered the computer continues its execution. Its main use is to end a DO loop that would otherwise end with a forbidden statement (see rule 9). The general form of the CONTINUE statement is given in figure 9.3.

CONTINUE Statement—General Form

---

n CONTINUE

where:

n — is the statement number assigned to the **CONTINUE** statement.

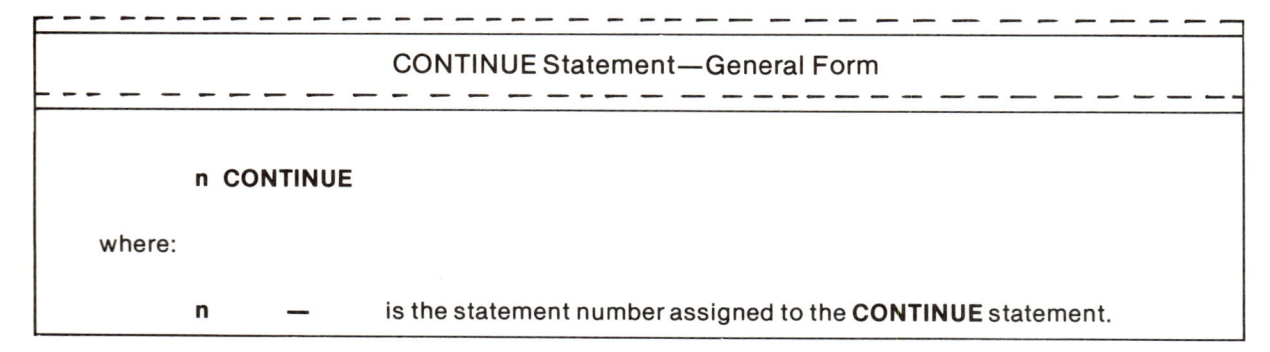

Figure 9.3. General Notation of the CONTINUE Statement

If the CONTINUE statement is used to end a DO loop, its assigned statement number is, of course, the range statement specified in the DO. For example:

```
           ISUM = 0
           DO 34 I = 1, 1000, 1
              ISUM = ISUM + I
              IF (ISUM .GT. 9999) GOTO 7
        34 CONTINUE
         7 WRITE (6,90) ISUM, I
        90 FORMAT (1X,I8,5X,I5)
```

The above example finds the sum of the integers (starting at 1) which first exceeds 9999.

The use of the CONTINUE statement is highly recommended for ending all DO loops. Using the CONTINUE statement allows greater flexibility in the insertion of additional statements at the end of the DO loop or removing the statement normally thought of as the end of the loop. It also serves as an aid in identifying the end of a DO loop.

The CONTINUE is not restricted to use with DO loops. Since it is considered an executable statement, it can be placed at almost any location in a program and used as the target of a branch command.

Often, it is desirable to increment by fractions or by negative quantities. You may wonder how this can be achieved within the DO loop since the DO statement can handle only positive integers. Section 9.5 presents examples of DO loops which illustrate some of their many uses and techniques.

9.5 Examples in Forming DO Loops

The following examples are presented to illustrate the many forms and uses of DO loops.

Example 1:

```
C *** READ 100 CARDS WITH VALUES IN CARD
C *** COLUMNS 1-5 AND SUM THESE VALUES.
      SUM = 0.0
      DO 10 ICOUNT = 1, 100
        READ (5, 99) VALUE
99      FORMAT (F5.2)
        SUM = SUM + VALUE
10    CONTINUE
```

Example 2:

```
C *** ADD ALL THE ODD INTEGERS FROM 1-999
      ISUM = 0
      DO 15 INTEGR = 1, 999, 2
        ISUM = ISUM + INTEGR
15    CONTINUE
```

Example 3:

```
C *** ADD ALL THE INTEGERS FROM -25 TO ZERO
      ISUM = 0
      NEGINT = -25
      DO 8 K = 1, 25
        ISUM = ISUM + NEGINT
        NEGINT = NEGINT + 1
8     CONTINUE
```

Example 4:

```
C *** ADD THE NUMBERS FROM 0 TO 5
C *** IN ONE-QUARTER INCREMENTS
      SUM = 0.
      XNCREM = .25
      DO 12 L = 1, 25, 1
        SUM = SUM + XNCREM
12    CONTINUE
```

Example 5:

```
C *** DECREMENT J BY 1
C *** FROM 9 DOWN TO 1
      DO 76 I = 1, 9
        J = 10 - I
        . . .
        . . .
76    CONTINUE
```

9.6 Nested Loops with Nested DO's

Loops can occur within other loops. We refer to this operation as *nested loops*. The outside loops are called *outer loops*. The loops inside the outer loops are called the *inner loops*.

We may have a DO loop within a loop controlled by a Logical IF statement. Or we can include DO loops within other DO loops. Thus, we have nested DO's or *nested DO loops.*

The DO loops may be nested in any number of ways. Following are pictorial examples of some ways DO loops may be nested.

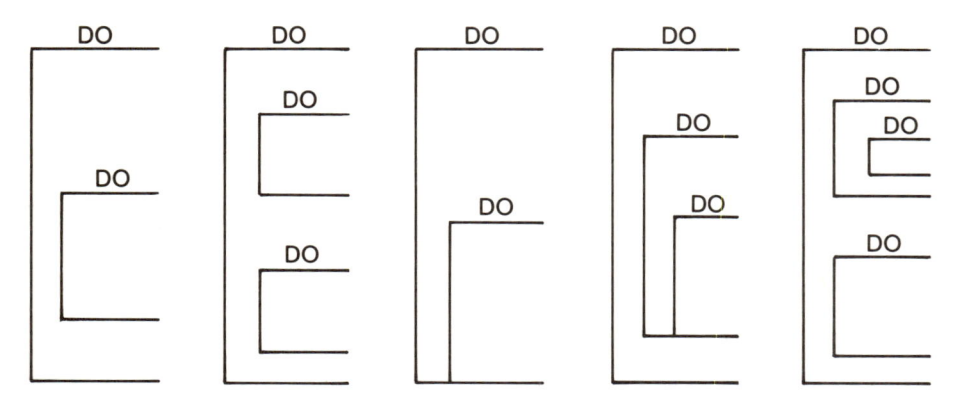

Do loops may be nested to any practical level (number). However, many IBM compilers limit the number of nested DO loops to 25.

Each inner loop must be completed or exited before the next pass of the outer loop is attempted. When the next pass of the outer loop is executed and the inner loop is encountered again, the full iteration of the inner loop will be performed. That is, the inner loop is executed in its entirety upon each pass of the outer loop.

This concept of nested looping is presented in the form of an analogy to walking around a park and riding the merry-go-round. For each trip around the park we make so many rounds on a merry-go-round. The trip around the park is analogous to the outer loop. The rounds on a merry-go-round inside the park is analogous to the inner loop. As you can see, we will complete a required number of rounds on the merry-go-round on each trip around the park. Suppose we make five rounds on the merry-go-round for each trip around the park. If we make five trips around the park, we will have made twenty-five rounds on the merry-go-round.

Following is an example to illustrate nested DO loops.

$$\text{DO 7 I} = 1, 10$$
$$\ldots$$
$$\ldots$$
$$\text{DO 5 J} = 1, 5$$
$$\ldots$$
$$\ldots$$

```
      5    CONTINUE
      7    CONTINUE
```

The outer loop is executed ten times. Since the inner loop is repeated five times for each pass of the outer loop, the inner loop is repeated a total of fifty times.

Nested DO's can get pretty involved. Section 9.7 presents the rules we should follow when using nested DO loops.

9.7 Rules in Using Nested DO's

As with single loops, certain precautions should be taken when forming nested DO loops. All of the rules which apply to a single DO loop also apply to nested DO loops. Additional rules to be followed when using nested DO loops follow.

Rule 1

All statements in the range of the inner DO must be in the range of the outer DO. For example, consider the following illegal construction:

```
      DO 10 I = 1,5
      . . .
      . . .
      DO 15 J = 1,10
      . . .
      . . .
      . . .
   10 CONTINUE
      . . .
      . . .
   15 CONTINUE
```

DO 10 ──┐
DO 15 ──┤──┐
        │ │
        │ │── range of outer DO
        │ │
        ──┘ │── range of inner DO

The end of the inner loop is outside the end of the outer loop so how can we have a loop within a loop? It is permissible in most cases (except in rule 6) for the range statement of the inner loop to be the same range statement as that of the outer loop. For example:

```
      DO 10 I = 1,5
      . . .
      . . .
      DO 10 J = 1,10
      . . .
      . . .
      . . .
   10 CONTINUE
```

This construction is perfectly legal. In fact, many levels of nested DO loops can end on the same range statement. I recommend, however, that different CONTINUE statements be used for the range statements to provide a clearer definition of the DO loops.

Rule 2

You may not branch into the middle of an inner or outer DO loop. For example:

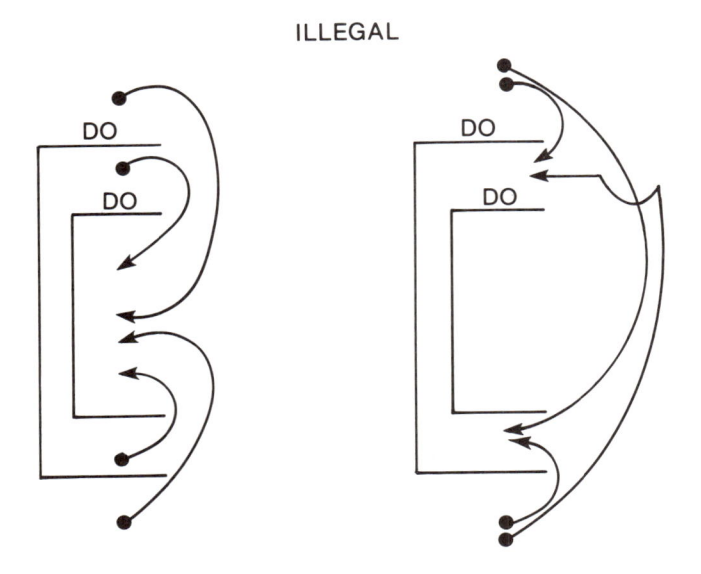

ILLEGAL

Rule 3

You may branch around an inner DO loop or to the inner DO statement from within the outer loop.

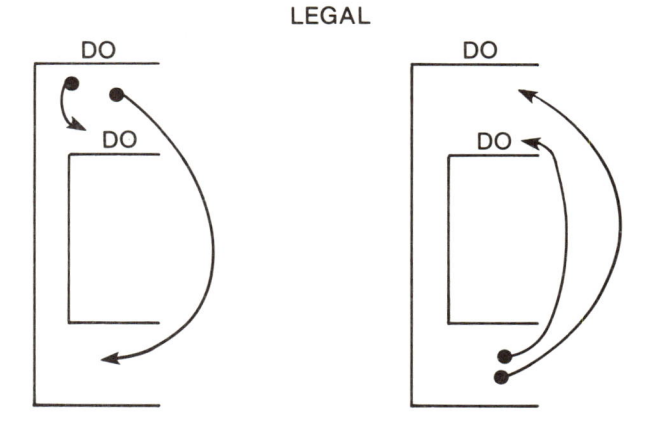

LEGAL

Rule 4

The index variables must be unique in all nested DO statements. For example:

$$DO\ 20\ K = 1, 10$$
$$\ldots$$
$$\ldots$$
$$DO\ 10\ J = 1, 10$$
$$\ldots$$
$$\ldots$$
$$10\quad CONTINUE$$
$$\ldots$$
$$\ldots$$
$$20\quad CONTINUE$$

If K had been used as the index variable for both the outer and inner DO's, the loops would not be repeated the correct number of times.

Rule 5

You may branch out of any inner level nested DO loop to any outer level DO loop at any time. Consider the following legal examples:

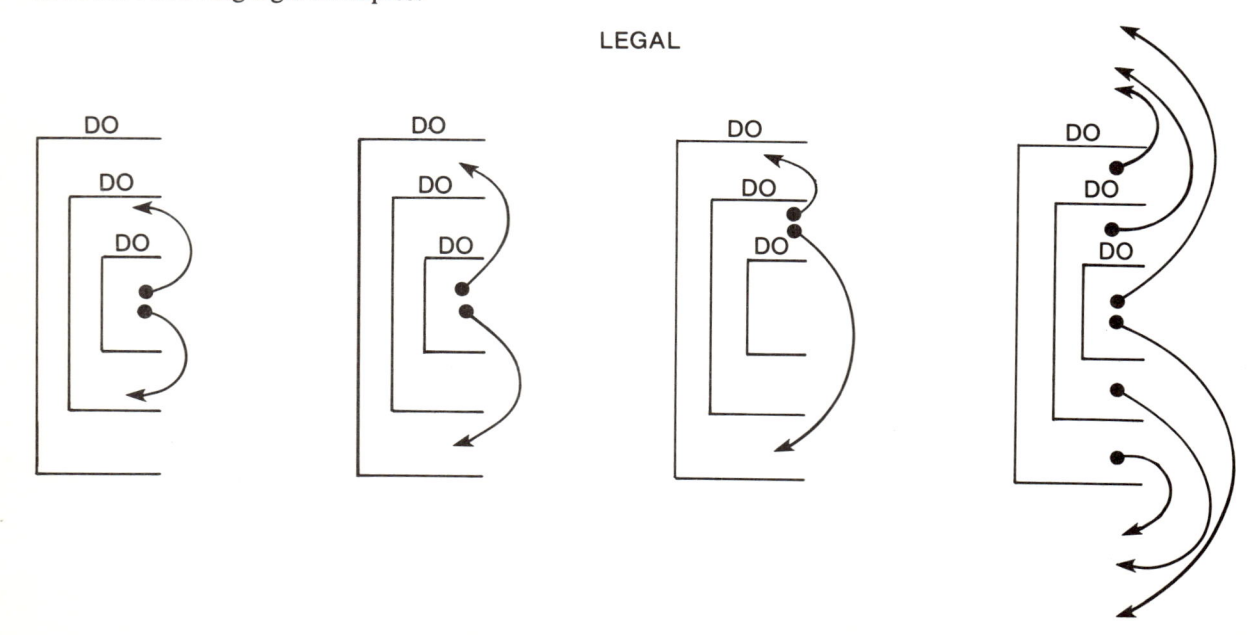

LEGAL

Rule 6

If the same range statement is used to end nested DO loops, the range statement number may be used in a GOTO or IF statement only in the innermost loop. That is, a branch to the range statement number which is used to end multiple nested DO loops is only valid from the most deeply nested DO loop. A branch to the range statement which ends more than one DO loop from any of the outer DO loops will produce invalid results (i.e., the incrementing to the next pass of that respective loop). For example, the following FORTRAN statements violate this rule.

```
           DO 13 I = 1, N
           . . .
           . . .
           DO 13 J = 1, M
           . . .
           . . .
           IF (J .GT. I) GOTO 13
           DO 13 K = 1, 10
           . . .
           . . .
13         CONTINUE
```

Since the Logical IF statement is not inside the inner loop, the branch to the range statement will not provide the correct increments of these DO loops. To overcome this restriction, you must use different CONTINUE statements to end each nested DO loop. For example,

```
           DO 30 I = 1, N
           . . .
           . . .
           DO 20 J = 1, M
           . . .
           . . .
           IF (J .GT. I) GOTO 20
           DO 10 K = 1, 10
           . . .
           . . .
10           CONTINUE
20           CONTINUE
30           CONTINUE
```

Summary of Nested DO Loop Rules

The rules in branching into or out of nested DO loops are summed up in the figure 9.4 diagrams on page 156.

Extended Range of a DO

A question often asked by the student is "Can we jump out of a DO loop, perform some actions, and then branch back into the DO loop where we left off?" Yes, we can, but only from the innermost DO. This concept is referred to as the extended range of the DO.

The *extended range of the DO* is defined as those statements that are executed between the transfer out of the innermost DO of a set of nested DO's and the transfer back into the range of this innermost DO. The index variable and indexing parameters must not be changed in the extended range of the DO. The extended range of the DO must not contain another DO statement that also has an extended range in it.

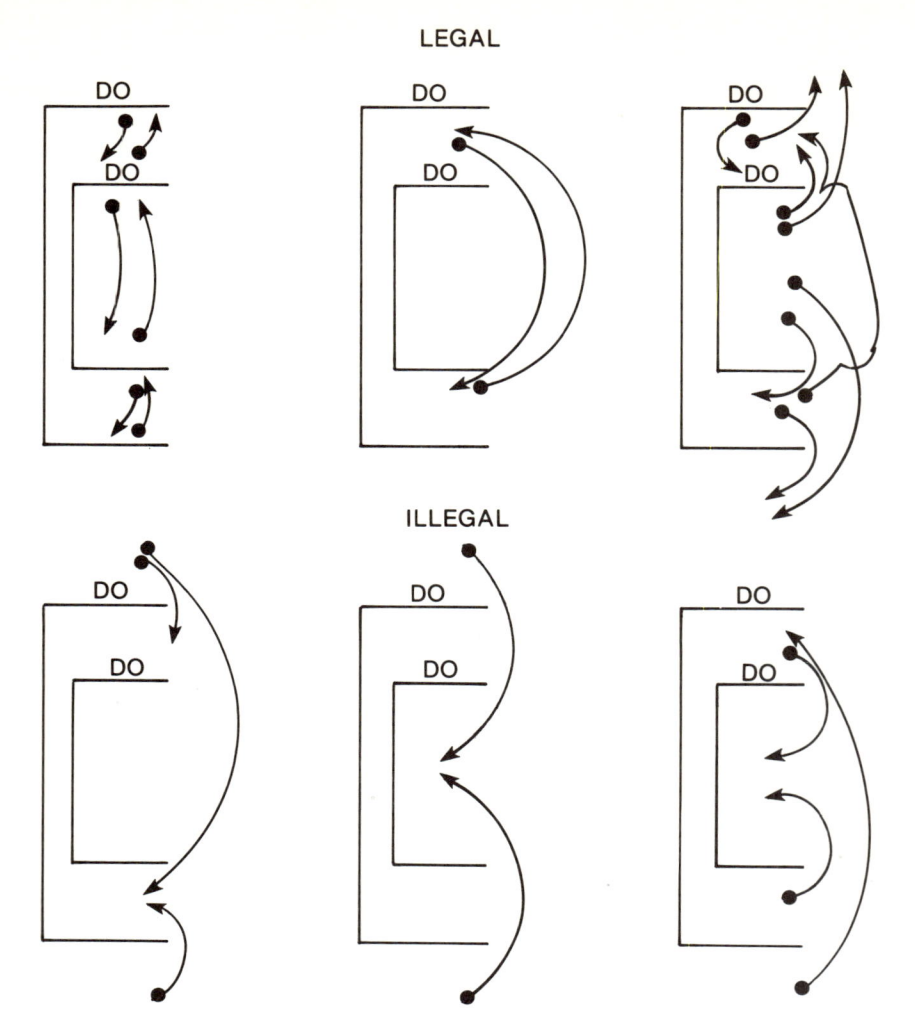

Figure 9.4. Diagrammed Rules in Branching—Nested DO Loops

9.8 A Sample FORTRAN Program to Illustrate DO Loops

Let us develop a FORTRAN program that demonstrates the use and flexibility of DO loops. Suppose we take the problem of producing a table which shows us our savings balance and the amount of compounded interest earned on our savings each year. Say we place a certain amount of money in a saving account, at a specified interest rate, which is compounded so many times per year. How much interest will we earn and what will be our account balance at the end of five years?

Banks and credit unions advertise that they pay so much interest compounded at given intervals during the year. We can compute tables of earned interest and the account balance with the various parameters. These tables will permit us to compare the various interest rates and number of compounded periods, and thus show us which method pays the most interest. In other words, would you prefer to invest your savings in a bank that pays 6 percent interest compounded annually, or in another bank that pays 5.95 percent interest compounded quarterly?

The formula which allows us to compute our new balance and compounded interest is: balance = $p(1 + r / n)^n$ where p represents our invested principal, r represents our percentage interest rate, and n represents the number of times per year that the interest is compounded or computed. For example, if we invest $100 at .06 interest, compounded semiannually (2 periods), what would be our earned interst at the end of one year? Plugging these parameters into our formula we would have: balance = 100

$(1 + .06 / 2)^2$. After carrying out our computations, we find that we will have a new balance of $106.09 at the end of the year and will have earned $6.09 interest.

Program Specifications

We wish to include nested loops and header-card logic in our program to demonstrate these techniques. Based on our requirement to produce multiple tables of compounded interest for various interest rates and different compounded periods, our program specifications are:

1. Produce a desired number of tables which shows invested principal, accumulated balance, and earned interest according to these formulas: new balance = $p (1 + r / n)^n$, and interest = new balance − principal.
2. A header card will be read to specify the number of tables to build, and thus the number of data cards to read.
3. Each data card read will specify the parameters of each table to be produced. These input parameters are principal of original investment, interest rate, number of compounded periods, and length of years for investment.
4. Computations will be made to produce the new balance and earned interest for each year of investment.
5. Each output table will begin on a new page with a page heading that specifies the various parameters used.
6. Column headings will be used to identify the columns of the output results.
7. Each detail printer line will contain the year, beginning balance, new balance, and earned interest for that year.
   With these specs (specifications) let us develop our input and output record formats.

Input and Output Specifications

The input card records and their field descriptions are as follows:

Data Card 1 (Header)

Card Columns	Field Description	Variable Name
1-3	Number of tables to produce	NTAB

Detail Data Cards

Card Columns	Field Description	Variable Name
1-8	Beginning principal (Form = NNNNN.NN)	PRNCPL
10-14	Rate (Form = .NNNN)	RATE
16-17	Number of times interest is compounded per year (Form = NN)	NTIMES
20-21	Number of years principal is left in savings (Form = NN)	NYEARS

158   Program Looping and the DO Statement

We will have a separate data card for each table to be built. If we are to build three tables, we will have three data cards following the header card.

Our output specifications for a single table is presented in printer layout form in Figure 9.5.

NAME: P. COLE   **PRINT LAYOUT**

FORM TITLE: PROGRAM WITH DO LOOPS

```
                TABLE OF COMPOUNDED INTEREST
        FOR INVESTED PRINCIPAL OF NNNNN.NN FOR NN YEARS
        AT N.NNNN INTEREST RATE COMPOUNDED NN TIMES PER YEAR


                BEGINNING        NEW              EARNED
        YEARS   PRINCIPAL        BALANCE          INTEREST

        NN      NNNNN.NN         NNNNN.NN         NNNNN.NN
```

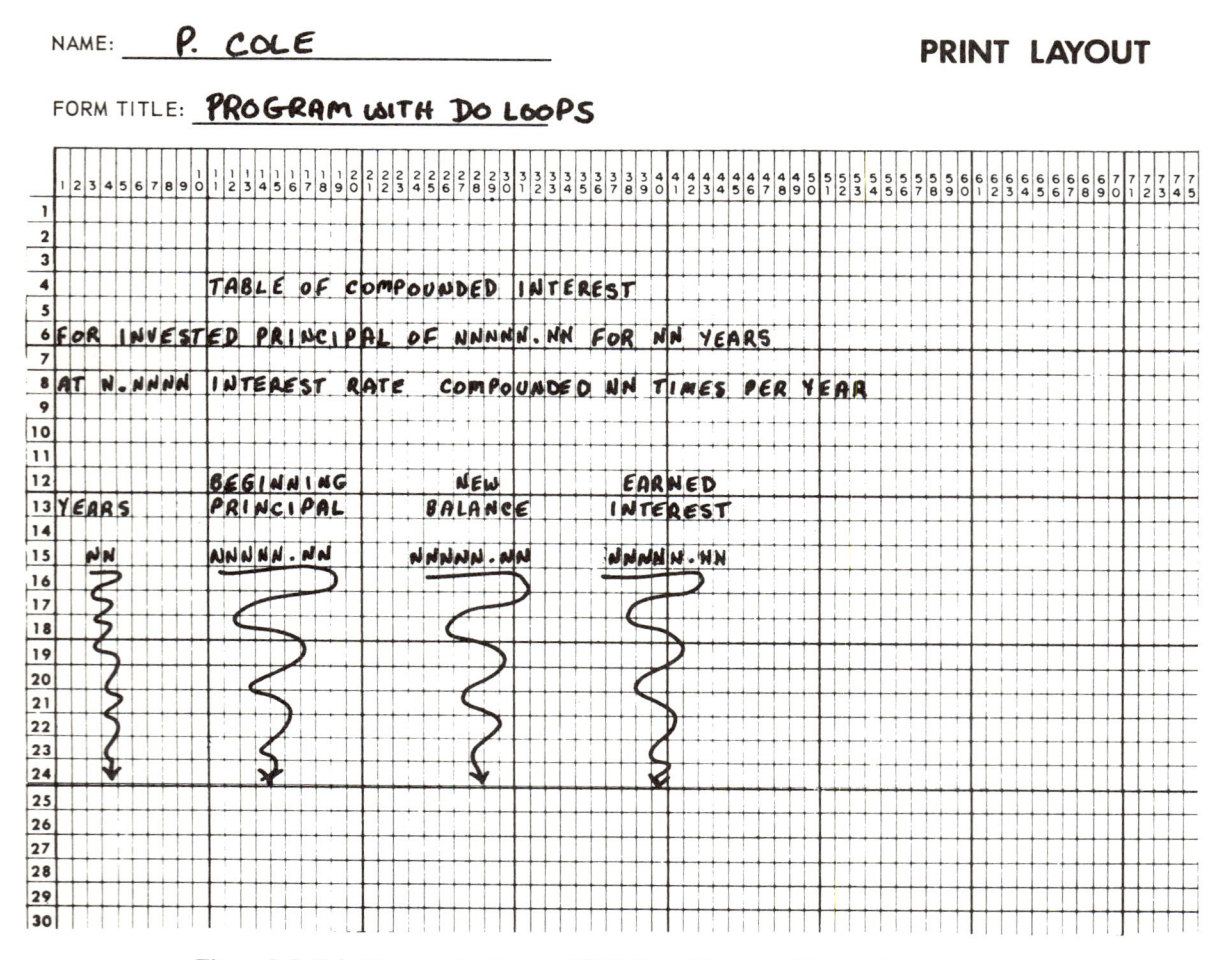

Figure 9.5. Print Layout for Sample FORTRAN Program Illustrating DO Loops

Flowchart and Coded Program

The problem involves header-card logic which reads the header card and determines how many tables to produce. This value in NTAB will be used as the test variable in the first DO statement (outer loop). Each data card read thereafter sets up the test value, NYEARS, for the second DO statement (inner loop) and provides the needed parameter values to produce each table.

Since the factor for computing the balance remains the same, we will compute this factor one time outside the inner loop. This saves the computer from having to recompute it each time through the loop, and thus produces a faster-running program.

The flowchart of the problem logic is presented in figure 9.6 on page 159.

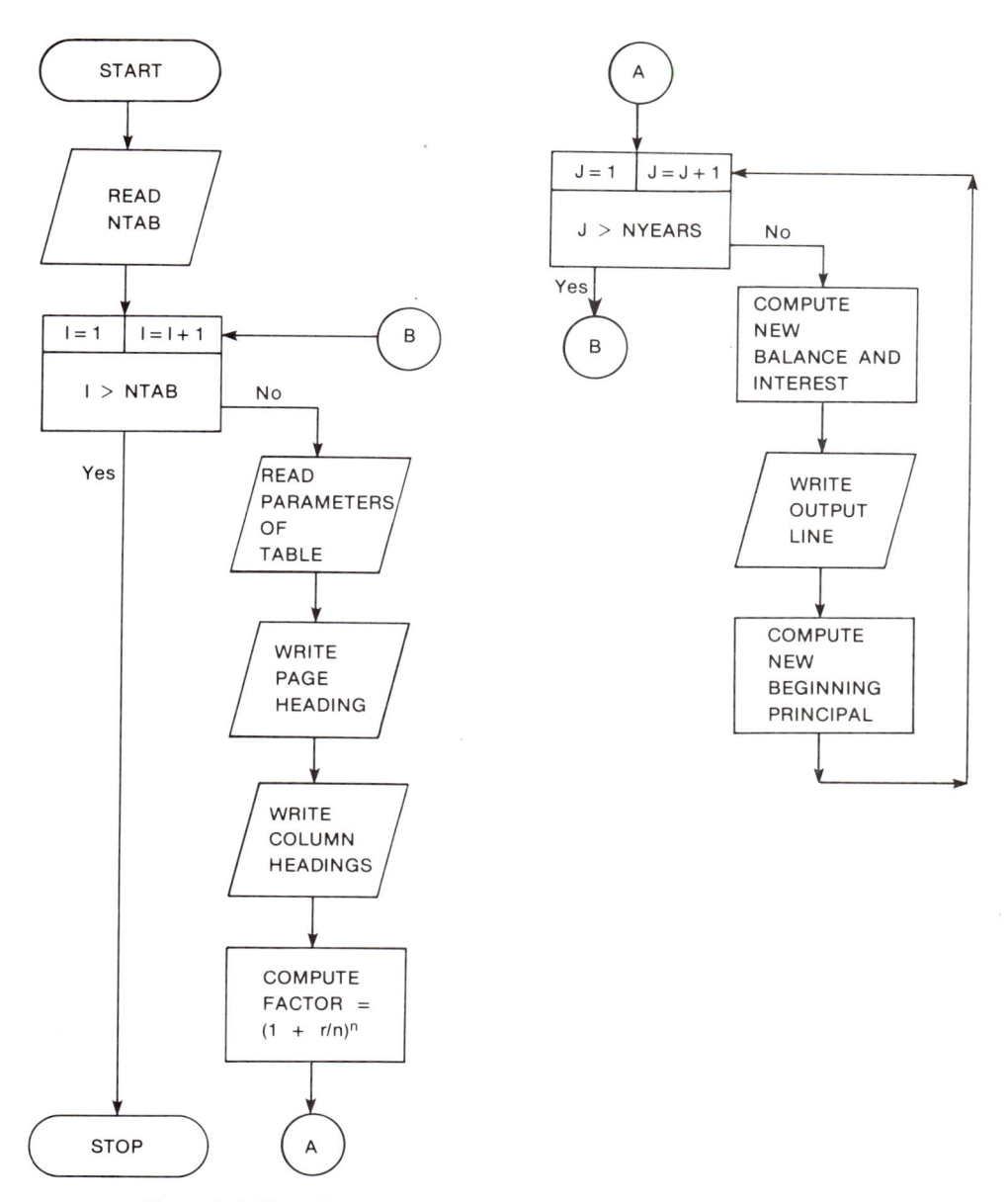

Figure 9.6. Flowchart of Program Logic to Illustrate Nested DO Loops

160   Program Looping and the DO Statement

The coded program is presented in figure 9.7.

```
C *** PROGRAM TO COMPUTE TABLES OF COMPOUNDED INTEREST ON DIFFERENT
C *** PRINCIPALS, INTEREST RATES, AND COMPOUNDING PERIODS.
C
  99      FORMAT (I3)
  98      FORMAT (F8.2,1X,F5.3,1X,I2,2X,I2)
  97      FORMAT (1H1, 10X,28HTABLE OF COMPOUNDED INTEREST/
         1   1H0,26HFOR INVESTED PRINCIPAL OF ,F8.2,5H FOR ,I2,6H YEARS
         2   /1H0,3HAT , F6.4,27H INTEREST RATE  COMPOUNDED ,I2,
         3   15H TIMES PER YEAR//)
  96      FORMAT( 1H0,10X,9HBEGINNING,7X,3HNEW,7X,7H EARNED/1H ,
         1          5HYEARS,5X,9HPRINCIPAL,5X,7HBALANCE,5X,8HINTEREST)
  95      FORMAT (1H0,2X,I2,6X,F8.2,5X,F8.2,5X,F8.2)
C
         READ (5,99) NTAB
C *** OUTER LOOP TO BUILD THE NUMBER OF DESIRED TABLES
         DO 80 I = 1, NTAB
            READ(5,98) PRNCPL,RATE,NTIMES,NYEARS
            WRITE (6,97) PRNCPL,NYEARS,RATE,NTIMES
            WRITE (6,96)
            RTIMES = NTIMES
            FACTOR = (1.0 + RATE/RTIMES ) ** RTIMES
C *** INNER LOOP TO BUILD EACH TABLE
            DO 70 J = 1, NYEARS
               BALNCE = PRNCPL * FACTOR
               XINTRS = BALNCE - PRNCPL
               WRITE (6,95) J, PRNCPL, BALNCE, XINTRS
               PRNCPL = BALNCE
  70        CONTINUE
  80     CONTINUE
         STOP
         END
```

Figure 9.7. Sample FORTRAN Program to Illustrate DO Loops

A page of one table of output results is presented in figure 9.8.

```
               TABLE OF COMPOUNDED INTEREST
       FCR INVESTED PRINCIPAL OF   100.00 FOR 10 YEARS
       AT 0.0595 INTEREST RATE  CCMPOUNDED  4 TIMES PER YEAR
                 BEGINNING       NEW           EARNED
         YEARS   PRINCIPAL       BALANCE       INTEREST
           1      100.00         106.C8          6.08
           2      106.C8         112.54          6.45
           3      112.54         119.39          6.85
           4      119.39         126.65          7.26
           5      126.65         134.35          7.71
           6      134.35         142.53          8.17
           7      142.53         151.20          8.67
           8      151.20         160.40          9.20
           9      160.40         170.16          9.76
          10      170.16         180.51         10.35
```

Figure 9.8. One Table of Output Results from Sample Program to Illustrate DO Loops

The given program example illustrates some of the power and flexibility one can obtain from using the DO statement for program loops. Probably the most powerful use of the DO statement is in table manipulation. Table manipulation involves storing values in a table form and accessing the items by the use of indexes or subscripts. The use of tables (arrays) and the DO with table operations is discussed in the next chapter. How widely the DO statement is used depends upon your ingenuity as a programmer.

9.9 REVIEW QUESTIONS

1. What is the function of the DO statement?
2. List the indexing parameters included in the DO statement and explain their use.
3. If the increment parameter is not included on the DO statement, its value is understood to be _____ .
4. Name the two ways that you may exit from a DO loop.
5. What is the advantage of using variables for the indexing parameters of a DO?
6. Describe the construction of the iteration flowcharting symbol.
7. One may transfer out of a DO loop even though it has not been repeated the specified number of times. (True/False)
8. When is the value in the index variable available for use in calculations outside the loop?
9. What statement is often used as the range statement in a DO loop, especially if a Logical IF statement is needed at the end of the loop?
10. What does the term "nested loops" mean?

9.10 EXERCISES

1. Form a DO statement that will start the index variable, II, at 5, increment by 5's, and stop when II exceeds 500. The range statement is number 56.
2. Form a DO statement that will start the index variable, M, at 3, increment by 3, and stop when M exceeds 99. The range statement is number 38.
3. How many times will the following DO loops be executed and what are the values that the index variable will take on?

    a)   DO 7 I = 1,8
                . . .
                . . .
                . . .
           7 CONTINUE

    b)   DO 13 J = 2,7,3
                . . .
                . . .
                . . .
          13 CONTINUE

    c)   DO 14 K = 4, 9, 2
                . . .
                . . .
                . . .
          14 CONTINUE

4. How many times will the following DO loops be executed and what are the values that the index variable will take on?

    a)   DO 10 J = 1,20
```
            . . .
            . . .
            . . .
         10 CONTINUE
```

    b)   DO 27 L = 2,11,4
```
            . . .
            . . .
            . . .
         27 CONTINUE
```

    c)   DO 18 M = 3,14,2
```
            . . .
            . . .
            . . .
         18 CONTINUE
```

5. How many times will the following inner DO loops be repeated?

    a)   DO 6 K = 1,3
```
            DO 5 J = 1,4
            . . .
            . . .
            . . .
          5 CONTINUE
          6 CONTINUE
```

    b)   DO 9 I = 2,6,2
```
            DO 8 J = 1,5,3
            . . .
            . . .
            . . .
          8 CONTINUE
          9 CONTINUE
```

6. How many times will the following inner DO loops be repeated?

    a)   DO 17 J = 1,5
```
            DO 15 I = 1,4
            . . .
            . . .
            . . .
         15 CONTINUE
         17 CONTINUE
```

    b)   DO 10 MM = 1,9,3
```
            DO 11 JJ = 2,7,2
            . . .
            . . .
            . . .
         11 CONTINUE
         10 CONTINUE
```

7. Locate the errors in the following DO statements.
   a) DO 20 K = 1, K, 1
   b) DO 30, I = 1, 10, 1
   c) DO 40 ICOUNTER = 1, L
   d) DO 50 J = 1, N, J + 1
   e) DO 60 M = 1, 20, −1
8. Locate the errors in the following DO statements.
   a) DO 10 A = 1, 5
   b) DO 20 K = 0, 5, .5
   c) DO LOOP = 2, 20, 2
   d) DO 30 MAXIMUM = 100, 500, 10
   e) DO 40 N = 1, N, 2+J

9.11 PROGRAMMING PROBLEMS

1. Write a FORTRAN program to sum the integers from 1 through 100. Write a literal title that says 'SUM OF INTEGERS 1-100 =' followed by the accumulated sum. Use a DO statement.

2. Write a FORTRAN program to sum the odd integers from 1 through 99. Write a literal title that says 'SUM OF ODD INTEGERS 1-99 =' followed by the accumulated sum. Use a DO statement.

3. Write a FORTRAN program to sum the even integers from 2 through 50. Write a literal title that says 'SUM OF EVEN INTEGERS 2-50 =' followed by the accumulated sum. Use a DO statement.

4. Write a program to prepare a table of simple interest according to the formula i = prt. Let principal = 45000 and rate = 8.5 percent. Use a DO statement to vary time from 1 to 20 years. Write the output results on double-spaced lines with field specifications as follows:

Field Description	Print Positions
Time	1-2 (Format code of I2)
Principal	6-13 (Format code of F8.2)
Rate	17-21 (Format code of F5.3)
Interest	25-32 (Format code of F8.2)

5. Write a program to prepare a table of squares and cubes of the integers 100-110. Use a DO statement to vary the integer number. Write double-spaced output lines with field specifications as follows:

Field Description	Print Positions
Number	5-7 (Format code of I3)
Square	10-14 (Format code of I5)
Cube	20-26 (Format code of I7)

Include appropriate column headings.

6. A student borrows $200.00 at an interest rate of 1 1/2 percent per month on the unpaid balance. If the student pays $10.00 at the end of each month, what is the remaining balance at the end of one year?

Write a program which includes a DO loop to solve this problem. The mathematics involved goes like this. The interest for the first month is the balance ($200) times

.015 = $3.00. Since $10 is paid each month, the balance after the first payment is $200 plus $3.00 less $10.00 = $193.00.

Write output results with appropriate literal titles for the original loan, interest rate, payment, and balance at the end of one year.

7. Write a program to perform the same function as problem six, but also to build a table of balance, amount paid, and interest paid. The interest rate will be 1 1/2 percent per month on the unpaid balance. Read input values for the original loan amount and payment amount in the following card columns:

Card Columns	Field Description
1-7	Loan amount (Form = NNNN.NN)
8-12	Payment amount (Form = NN.NN)

In addition to computing the new balance, accumulate the total interest paid for the year. Use a DO loop.

Write each output line double-spaced with the following values:

Field Description	Print Positions
Month (1-12)	3-4
Balance at beginning of the month	10-16
Interest	22-27
Payment	33-37
Balance at the end of the month	43-49

Print the total interest paid and the total amount of payments at the end of the table. Include appropriate column headings.

8. Write a program to compute a class of students' grades. The first data card serves as a header with the number of students in the class punched in card columns 1-3. The remaining student data cards are punched as follows:

Card Columns	Field Description
1-5	Student ID (Form = NNNNN)
6-8	Score 1 (Form = NNN)
9-11	Score 2 (Form = NNN)
12-14	Score 3 (Form = NNN)
15-17	Score 4 (Form = NNN)
18-20	Score 5 (Form = NNN)

Compute the average grade for each student. Also compute the average score for each of the five test scores. Perform these calculations in the DO loop to read multiple data cards.

Print student output records double-spaced as follows:

Field Description	Print Positions
Student ID	2-6
Score 1	10-12
Score 2	16-18
Score 3	22-24
Score 4	28-30
Score 5	34-36
Sum	40-45 (Form = NNN.NN)
Average	50-55 (Form = NNN.NN)

After all student records have been output, print the average of each of the test scores beneath the respective scores of each student. Include appropriate column headings.

9. Write a program to compute the sum of the reciprocal of the integers from 1 through 10. The reciprocal of a number is simply the number 1 divided by that number. For example, the reciprocal of 2 is 1/2. The sum of the reciprocals from 1 to 3 is $\frac{1}{1} + \frac{1}{2} + \frac{1}{3}$.

Use a DO loop and also real arithmetic. Write the output result preceded by the literal title, 'THE SUM OF THE RECIPROCALS FROM 1-10 IS'.

Subscript Operations and One-dimensional Arrays 10

10.1 Introduction to the Concepts of Arrays

In reading, processing, and outputting data we are often concerned with groups of related items. For example, earlier in the text we developed programs which computed student grades. We assigned unique variable names to each test score and used the selected variable name when we wished to refer to that particular score. These variables that contain a single value are known technically as *scalar variables*. It often becomes convenient to refer to related items by relative position numbers in a group of variables and not by unique variable names. For example, we could refer to the first test score as NTEST(1), the second test score as NTEST(2), the third test score as NTEST(3), and so on. These types of variables are known as *subscripted variables*.

Some programs must be written to store "large" quantities of related data which must be retained in memory for manipulation. This technique of using subscripted variables becomes much more convenient when our list of related items get longer. Say we had 20 test scores that we wanted to read, compute the sum and average, and output the input values and results. We would spend a great deal of time coding the variable names for the READ, Assignment, and WRITE statements if we assigned unique names to each test score. It would certainly be convenient if we could assign one common name to all the test scores and have some shortcut way in which we could refer to each of the scores.

FORTRAN allows us to do this with subscripted array elements. We can specify a common name to any group of related items. Then we can enclose a numeric value within a set of parentheses following the common name. The numeric value within parentheses indicates which item we are referring to, the 1st, 2nd, 3rd, etc. Say we have four test scores of 100, 86, 79, and 90, to which we assign the common name, NTEST. This concept of referring to the values as NTEST(1), NTEST(2), NTEST(3), and NTEST(4) is illustrated below.

Storage Locations

100	86	79	90

NTEST   (1)   (2)   (3)   (4)

We have four memory locations containing the four test scores. The group of four storage locations is called an *array*. An array containing a collection of values in this manner can be simply thought of as a linear *list* or a *row* of items. In mathematical terms, an array such as this is called a *vector*.

The name NTEST refers to the entire array or collection of the four test scores previously described. Each storage location or item within the array is referred to as an *array element*. Each array element can be referenced by the array name, followed by its element number enclosed within parentheses. For example, to refer to the third element or item in the array we would specify NTEST(3). (Read as NTEST sub 3).

Array element numbers simply relate to the position of the elements in the array. We refer to the element number as a *subscript*. So, a subscript value identifies the position of an element within an array, or which item we are referencing.

The term **subscript** in FORTRAN has the same meaning as the subscripts 1, 2, and 3 in the set of mathematical terms X_1, X_2, and X_3. In FORTRAN, however, they are placed inside of parentheses following the variable name to which they apply, i.e., X(1), X(2), X(3). Remember, we must code all characters following each other on the line; we cannot represent subscripts in a FORTRAN statement the way we can in a mathematical notation. So the parentheses are our way of saying a subscript value follows in FORTRAN. Each subscripted reference is known as a *subscripted variable.*

In math we can refer to a subscripted item of array X as X_i, where i can take on any numeric value according to the number of items we have. If we have ten items in a group, the i can take on a value from 1 to 10. We can use this same concept with array items in FORTRAN.

A subscript in FORTRAN can also be a variable as well as a constant. For example, we could refer to NTEST(I). If the variable I had a value of 2, then we would be referring to NTEST(2). If the value of I is altered to contain a 4, then the reference to NTEST(I) would refer to NTEST(4). The use of variables for subscripts is how we derive our brevity, flexibility, and power in using arrays.

To sum twenty test scores in a 20-element array called NTEST could be done in a DO loop as follows:

```
        NSUM = 0
        DO 7 I = 1,20
          NSUM = NSUM + NTEST(I)
     7  CONTINUE
```

Each pass through the DO loop, the value of I is incremented by 1, so the element of the array that I is pointing to is added to the variable, NSUM. This method is much more concise than adding twenty unique variable names in an Assignment statement, and certainly takes far less time to code and keypunch.

You should now begin to see the power and flexibility we can obtain by using arrays. Say we had two 1,000-element arrays called ITEMNR and ITMQTY. ITEMNR contains the part number of 1,000 items in our inventory. The array, ITMQTY, contains the quantity on hand for each of the corresponding part numbers. That is, ITMQTY(1) contains the quantity on hand for ITEMNR(1), etc.

Now if we wished to see if any part number had a depleted quantity on hand (balance of zero), we could do this very easily with the use of arrays. The following statements will check for the quantity-on-hand balance for all part numbers. If a part number has a zero quantity-on-hand balance, then we will write out that part number so we can issue a reorder requisition.

```
        DO 8 I = 1, 1000
          IF (ITMQTY(I) .EQ. 0) WRITE (6,99) ITEMNR(I)
    99  FORMAT (1H0,I7)
     8  CONTINUE
```

Can you guess what we would have to do if we had stored the 1,000 part numbers and their quantities on hand under unique variable names? Yes, we would need 1,000 Logical IF statements. Not counting the input and output operations, we would have lots of coding to do. But with arrays, this can be a very simple operation and requires very few statements.

To permit the operation of subscripting (or indexing, as it is sometimes called), all the elements in an array must be in adjacent memory locations. The DIMENSION statement in FORTRAN tells the compiler to establish an array for the given name and to set up the elements in contiguous storage locations.

An *array,* therefore, is a collection of related items stored in adjacent memory locations under a common name. To refer to a single item in the array, we must give the array name followed by a subscript which denotes the particular element number of the array item.

Arrays representing a list of related items are called a *one-dimensional array.* A one-dimensional array simply means we use only one subscript value to locate any array element.

The use of arrays in computer programming becomes a very powerful tool. With arrays, writing an otherwise long, unwieldly program is made feasible and easy. Thus, arrays are widely used in programming and are indeed an extremely important tool.

Arrays are usually confusing at first for most beginning programmers. Don't get frustrated; many others have had to cross this barrier before. Simply reread the chapter and meditate on the concept being presented. It will help if you draw boxes for the array elements and keep track of the values in each element as we step through some of the examples. It is also important to jot down the beginning value of the subscript used and to alter its value as the subscript value in the examples change. In that way, you can keep track of what is really happening when we work with array items.

Let us first cover the DIMENSION statement to see how we tell the computer we are using an array.

10.2 The DIMENSION Statement

The DIMENSION statement is a specification statement to the FORTRAN compiler to establish arrays. An array name must be assigned for the collection of items. The compiler must know the size (number of elements) of the array in order to set up that many adjacent memory locations. These specifications are given in the DIMENSION statement as illustrated below.

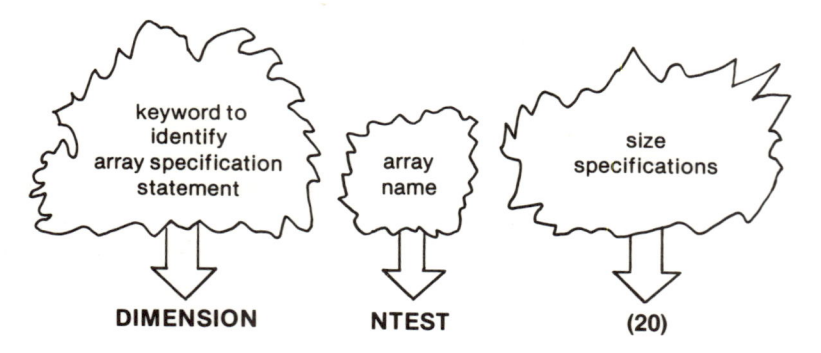

```
   keyword to            
    identify            size
 array specification   array      specifications
    statement          name
       ↓                ↓              ↓
   DIMENSION          NTEST           (20)
```

The general form of the DIMENSION statement is given in figure 10.1.

DIMENSION Statement—General Form

DIMENSION arrayname,(size,), . . . ,arrayname$_n$ (size$_n$)

where:

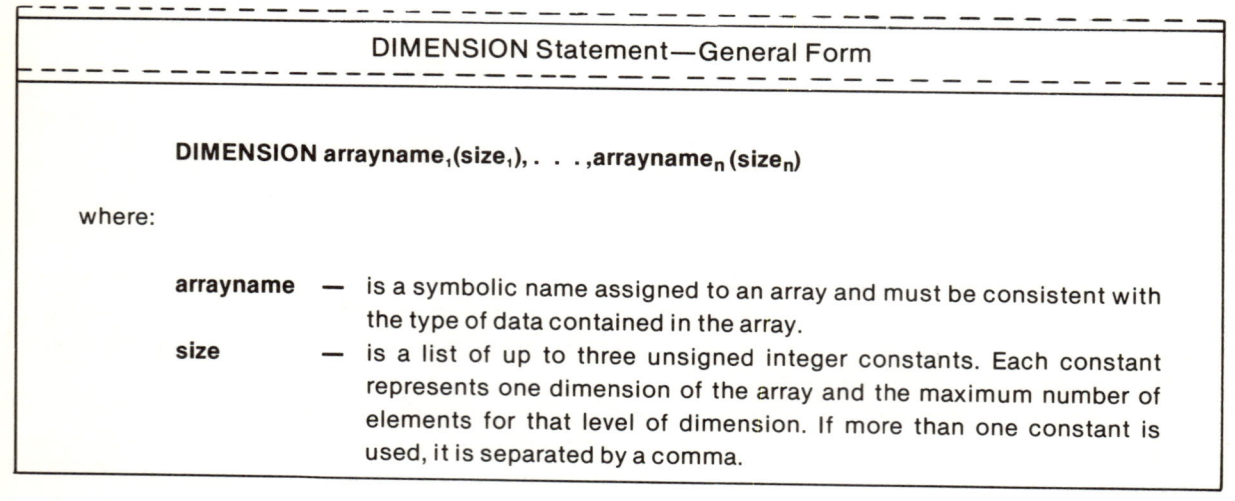

arrayname — is a symbolic name assigned to an array and must be consistent with the type of data contained in the array.

size — is a list of up to three unsigned integer constants. Each constant represents one dimension of the array and the maximum number of elements for that level of dimension. If more than one constant is used, it is separated by a comma.

Figure 10.1. General Notation of the Dimension Statement

Examples of the DIMENSION statement are:

DIMENSION X(10), Y(10)
DIMENSION LISTA(25)

The first example establishes the array X as a 10-element, one-dimensional array and the array Y as a 10-element, one-dimensional array. The second example establishes the name LISTA as a 25-element, one-dimensional array.

If more than one array is used in a program, all may be listed in the same DIMENSION statement, or several DIMENSION statements may be given. Commas must be used to separate each of the array specifications whenever multiple arrays are given in the same DIMENSION statement. The allowance of multiple DIMENSION statements in a program provides flexibility when new arrays may need to be added later. Suppose we had three arrays: A, B, and C. We could specify:

DIMENSION A(5), B(7), C(3)

or

DIMENSION A(5)
DIMENSION B(7), C(3)

or any arrangement you wish. The three arrays could be coded as three separate DIMENSION statements, but this form is discouraged, because of the increased number of cards.

Array names are formed like any other variable name in FORTRAN. The type of array name assigned must always agree with the type of the data to be stored in the array. Thus, an integer array name must be used when the array contains integer values. Likewise, a real array name must be used to store real numeric values. All elements in an array must contain the same type of data, e.g., integer or real.

Information pertaining to dimension, such as the number of dimensions and number of elements in each dimension, is provided inside parentheses following the array name. Dimension information in a "mainline" program must always be given as a numeric integer in FORTRAN. We cannot give a dimension value as a variable and expect to assign a value to the variable at execution to denote the size of the array. For example, the following specification is illegal:

DIMENSION X(N)
READ (5,99) N
99 FORMAT (I2)

We must provide an integer for each dimensional value so the compiler will know what size array to establish at compile time.

Since the DIMENSION statement is a nonexecutable specification statement, it must precede any executable FORTRAN statement. No statement number should be assigned to the DIMENSION statement.

Before we look at the operations with one-dimensional arrays, let us discuss subscripts in more detail.

10.3 Subscripts

A *subscript* is an integer quantity enclosed within parentheses and used to identify an array element. A subscript may also consist of a set of integer quantities separated by commas to identify an array element in a multidimensioned array. The subscript must be written immediately after the array name which contains the data item being referenced.

Do not confuse the term **subscript** with dimension specifications. Dimension specifications are provided within the set of parentheses with the DIMENSION statement. Subscripts are references to subscripted variables (array items) in the executable statements within your program.

The number of values contained in the subscript reference must be the same as the number of dimensions of the referenced array. That is, if A is a one-dimensional array, only one value can be specified in the subscript. For example, the reference of A(I,J) would be invalid.

The value of any given subscript must always lie between one and the number of elements declared in the dimension of the array. If a one-dimensional array X is dimensioned as X(10), then the value of the given subscript must range from 1 through 10.

What happens if a programmer accidently gives an invalid value to a subscript? This can easily happen when a subscript is a variable or an arithmetic expression. If the value of the subscript is zero or negative, most computer systems will detect the error. An error message will be printed and the program terminated. If the subscript is larger than the largest permitted element number, then many FORTRAN compilers will use an address past the end of the array and keep on executing.

Finding the invalid subscript is a very difficult problem. We may use erroneous values both from accessing storage locations past our array and from storing values past our array elements. A printout of all the subscript values and their corresponding array element is usually the best way to debug the invalid subscript problem.

Subscript values may also be other than an integer constant or variable. They may contain arithmetic expressions such as:

$$\text{ITEM } (K + 5)$$
$$\text{LIST } (2*I - 1)$$

The following table illustrates the seven different forms that a legal subscript may take in ANSI FORTRAN:

Seven General Forms of a Subscript	*Example*
Constant	7
Variable	J
Variable + constant	I + 2
Variable − constant	K − 1
Constant * variable	3 * INDEX
Constant * variable + constant	4*I+3
Constant * variable − constant	2*N−1

The restrictions regarding valid subscript forms can easily be circumvented by performing the computations in an Assignment statement just before the statement containing the subscript reference. For example, the subscript reference A(I/M + N) is invalid, but we could compute J = I/M + N and then use A(J) as a valid reference.

The use of arithmetic expressions to specify a subscript value often provides much flexibility in programming techniques. If we wish to compare an array element with the next element, we could write (using array A):

$$\text{IF (A(I).GE.A(I + 1))...}$$

We are now ready to tackle some problems that use arrays. We shall begin with the I/O operations having one-dimensional arrays.

10.4 Input/Output Operations with One-dimensional Arrays

There are three ways we can load (fill) an array with data items or output the items from an array. First, we can specify a subscripted variable as an item in the I/O list of the READ or WRITE statements. Secondly, we can read and write an entire array of items by specifying just the array name by itself in the I/O list. And, thirdly, we can use a special option called the implied DO loop in the I/O list to specify as many subscripted variables as we desire.

Array Input Operations

Let us examine the first technique: reading individually subscripted variables in the I/O list. Assume that we had seven test scores each punched in columns 1-3 on separate data cards. We could write a DO loop to load the array, NTEST, as follows:

```
        DIMENSION NTEST(7)
        DO 10 I = 1,7
           READ (5,99) NTEST (I)
99      FORMAT (I3)
10      CONTINUE
```

The subscript, I, will take on a value of from 1 through 7 from the index variable in the DO statement. Thus, each new test score will be placed in the next element number of the array.

We can also manipulate each subscripted variable as we read it into memory. Suppose we wish to sum all the test scores under the variable, NSUM. The following statements show how we could accumulate this sum inside the same DO loop that is used to load each array item.

```
        DIMENSION NTEST(7)
        NSUM = 0
        DO 10 I = 1,7
           READ (5,99) NTEST(I)
99      FORMAT (I3)
        NSUM = NSUM + NTEST(I)
10      CONTINUE
        AVE = NSUM / 7
```

After the DO loop is completed, we have the sum of the seven array items in the variable, NSUM. Thus, we can compute the average of the seven scores under the variable, AVE.

We are not limited to reading only one subscripted variable in the I/O list. We can include as many as we wish. They can also be interspersed with nonsubscripted variables whenever the need occurs. Consider the following two examples:

```
        DIMENSION X(10), Y(10)
        DO 20 I = 1,10
           READ (5,98) X(I), Y(I)
98      FORMAT (F5.2,3X,F5.2)
20      CONTINUE


        DIMENSION A(15), B(15)
        J = 1
1       READ (5,97) NR1, A(J), NR2, B(J)
97      FORMAT (I3,F4.2,I3,F4.2)
        IF (NR1 .EQ. 999) GOTO 30
        J = J + 1
        . . .
        . . .
        . . .
        GOTO 1
30      . . .
```

172 Subscript Operations and One-dimensional Arrays

The first of these last two examples shows how a subscripted variable of array X and Y are read from the same data card. The second example shows how nonsubscripted and subscripted variables can be intermixed in the I/O list. The variables NR1 and NR2 store the first and third data fields as nonsubscripted variables; the second and fourth data fields, however, are loaded into the Jth element of arrays A and B, respectively.

The second example also illustrates the technique of loading a variable number of items into an array. The variable, J, is used to count the number of items loaded into arrays so we can keep track of how many data items we have in our array. That is, we do not have to load the complete array. We can load any portion of the array elements that we wish.

The second technique for loading arrays is the easiest of all. We can read and store all items in an array by simply giving the array name *without* any subscript references in the I/O list. Let us assume the same problem as in example 1; we wish to load seven test scores punched on separate data cards into the array NTEST. We could simply code:

DIMENSION NTEST(7)
READ (5,99) NTEST
99 FORMAT (I2)

Remember, when we reach the end of a FORMAT statement the computer reads a new data card if the I/O list has not been completely exhausted.

Using only the array name to load an array is simpler. It is also far more efficient in the produced object code than the first technique. The computer begins loading the first data item at the first array element and continues loading subsequent data items at the next higher array element until the array is completely loaded.

The array name by itself means that it is referring to the entire array. It is very important to remember that the array name by itself can only be included in the I/O list of a READ or WRITE statement. We must not use just the array name in trying to manipulate all array items in a single Assignment statement such as LISTA = 0 where LISTA is an array name. In all other FORTRAN statements array items must be referenced as individually subscripted items.

The only drawback to using just the array name to load array items is the fact that all items in the array must be loaded. If we wish to load only part of the items in an array we must not use just the array name. Use of the array name in the I/O list causes the computer to try to load as many items into the array as specified in the DIMENSION statement specifications.

Now let us begin our discussion of the third technique. Assume we have seven test scores punched in the same data card as follows:

```
 89 92 78 86 97 83 90
```

The test scores are to be read into a ten-element array NTEST. We could not use a DO loop as we did in example 1 to read the scores. Each time the READ statement is executed, at least one input card must be read.

To read the seven test scores in as subscripted variables like we did in technique one, we would have to code our input operations as follows:

DIMENSION NTEST(10)
READ(5,96) NTEST(1), NTEST(2), NTEST(3), NTEST(4),
   *   NTEST(5), NTEST(6), NTEST(7)
96 FORMAT(7I3)

If this is to be the case, then we might as well go back to nonsubscripted variables.

The array name technique cannot be used since we are not loading our entire array. Is the long way as shown in the previous example our only way out? No, it is not. We have yet another technique that can be used especially for situations like this.

The third technique used for reading and loading array elements incorporates one or more DO loops in the I/O list. A DO loop in the I/O list is referred to as an *implied DO*. An implied DO can only be used in the I/O list of READ and WRITE statements.

The implied DO option is the most flexible method of reading and loading array items. Array items are treated as subscripted variables, yet all or any portion of the array can be loaded. Input values may be on separate data cards or combined onto one or more data cards.

The function of the implied DO is to provide multiple subscripted variable names in an I/O list. A variable is used as the subscript value of an array reference and the subscript value is varied in our implied DO to reference as many or as few subscripted variables as we wish. Not only is this a very powerful means of referencing multiple array elements in an I/O list, but it also provides a very concise way of doing so.

The implied DO is formed very much like the DO statement except the keyword DO and a range statement number is omitted. An index variable, an initialization value, a test value, and an optional increment value are specified along with a subscripted array name. All these parameters are enclosed within a set of parentheses. A comma must separate the subscripted array name from the index variable.

Now back to our problem: to read seven test scores into the first seven elements of the 10-element array NTEST. This can be accomplished easily by using the implied DO in our I/O list as follows:

DIMENSION NTEST(10)
READ (5,96) (NTEST(I),I = 1,7) ← Implied DO
96 FORMAT (7I3)

In the READ statement:

READ (5,96) (NTEST(I),I = 1,7)
           IMPLIED DO

notice how the indexing parameters in the implied DO correspond to those in the DO statement. The same rules which govern the indexing parameters in a DO statement pertain to the indexing parameters in the implied DO. Remember, the index variable must always be an integer variable.

There are many variations of the implied DO that one can use in an I/O list. More examples of the implied DO with one-dimensional arrays are as follows:

READ (5,98) (A(I),I = 1,N), B, C
READ (5,97) W, (X(J),J = N1,N2,N3), V
READ (5,96) (Y(I),I = 1,5), (Z(I),I = 1,3)

The first example illustrates how we can read N number of values into the array A, followed by the nonsubscripted variables B and C. The second example first reads the nonsubscripted variable W. Then a certain number of data values are read into array X beginning at the position indicated by the variable N1 until the value of J is greater than the variable N2. J is incremented by the variable N3. Next, a value is read into the variable V. The third example reads five data values into the array Y and then reads three data values into the array Z. Of course, the associated FORMAT statements (which are not given) dictate how the items will be read off the input records.

All items in the implied DO must be read before continuing on in the I/O list. In the third implied DO example above, all of the values of array Y are read before the values of array Z. The index variable I varies from 1-5 inside the first implied DO, so the five subscripted variables of Y are loaded. In the second implied DO of the same example, the index variable I is reset to 1 and is incremented by one to read the first three values of the array Z.

174   Subscript Operations and One-dimensional Arrays

We are not limited to only one subscripted variable within an implied DO. Assume that we had twenty values for an array X and twenty values for an array Y. A value of X is punched in columns 1-5 followed by a value of Y punched in columns 10-14. Each value of X (along with a value of Y) is punched in a separate card. We wish to load the twenty values of array X and Y. Our implied DO would be as follows:

```
        DIMENSION X(20), Y (20)
        READ (5,95) (X(I),Y(I),I = 1,20)
    95  FORMAT (F5.2,4X,F5.2)
```

The index variable, I, is used for each subscripted variable in the implied DO before it is incremented. Thus, X(1), Y(1), X(2), Y(2), . . . , X(20), Y(20) are read in that order.

Let us proceed to the output operations with arrays. These same three techniques which allow us to load array elements can be used to write array elements.

Array Output Operations

If we wished to write our seven test scores on separate print lines, we could use a subscripted variable. Using the first technique of individual subscripted variables, our routine is:

```
        DIMENSION NTEST(7)
        DO 35 I = 1,7
            WRITE (6,99) NTEST(I)
    99      FORMAT (1H0,I3)
    35  CONTINUE
```

The subscript, I, will take on a value from 1 through 7 from the index variable in the DO statement. Thus each test score will be taken from the referenced array element and written on a separate double-spaced print line.

The array name technique could be used to accomplish the same output. Using the array name to output an array is easier and also more efficient. Our output routine is:

```
        DIMENSION NTEST(7)
        WRITE (6,99) NTEST
    99  FORMAT (1H0,I3)
```

The implied DO technique to output array items could also be used. Our output routine using the implied DO is:

```
        DIMENSION NTEST(7)
        WRITE (6,99) (NTEST(I),I = 1,7)     ← Implied DO in WRITE statement
    99  FORMAT (1H0,I3)
```

But what if we needed our seven test scores to be printed on the same line? Could we use the subscripted variable technique? Yes, we could. But we would also need to use integer constant subscripts as follows:

```
        DIMENSION NTEST(7)
        WRITE(6,98) NTEST(1), NTEST(2), NTEST(3), NTEST(4),
       *            NTEST(5), NTEST(6), NTEST(7)
    98  FORMAT(1X,7I6)
```

Again, this technique is very awkward and time-consuming.

Our best method is to use the array name technique, and write the entire array. Our routine is:

> **DIMENSION NTEST(7)**
> **WRITE (6,98) NTEST**
> **98 FORMAT (1X,7I6)**

Or we could use an implied DO to produce the same output. Our implied DO routine is:

> **DIMENSION NTEST(7)**
> **WRITE (6,98) (NTEST(I),I = 1,7)**
> **98 FORMAT (1X, 7I6)**

Notice that our format specifications specify the form of our output results in both the array name and implied DO techniques.

But what if we had a ten-element array to hold our test scores and we only wished to output the first seven? The array name could not be used inasmuch as we do not wish to output all the array items. In this case, we must use the implied DO. Our routine is:

> **DIMENSION NTEST(10)**
> **WRITE (6,98) (NTEST(I),I = 1,7)**
> **98 FORMAT (1X,7I6)**

The array name is the easiest to use and the most efficient of the three output techniques. But remember, we can use it only when we wish to output all the elements in our array. As many items will be written as given in the DIMENSION statement specifications.

Nonsubscripted variables may be included in the I/O list along with subscripted variables, array names, and/or the implied DO loop. The format specifications are of the utmost importance since they control the formatting of our output results. Consider the following two examples:

> **DIMENSION NTEST(7)**
> **DO 45 J = 1,7**
>   **WRITE (6,97) J, NTEST(J)**
> **97 FORMAT (1H0,I1,3X,I3)**
> **45 CONTINUE**

This routine writes the value of J along with its associated array element (test score) on each line. The second example is:

> **DIMENSION X(10), Y(10)**
> **WRITE (6,96) X,Y**
> **96 FORMAT (11H0X ARRAY = , 10F8.2/**
>   *****   **11H0Y ARRAY = , 10F8.2)**

This routine writes the ten values of the array X on a line followed by the ten values of Y on a new double-spaced line. Each set of array values is preceded by a literal to identify the array.

The real power and flexibility in outputing array values with or without nonsubscripted variables is found in the implied DO. We must use the implied DO whenever we wish to dump (output) portions of an array. Consider the following examples using the implied DO.

Suppose we want the values in array X to be printed first, followed by the values in array Y. The routine is:

      DIMENSION X(7), Y(5)
      WRITE (6,94) (X(I),I = 1,7), (Y(I),I = 1,5)
  94 FORMAT (1X,7F8.2,5X,5F8.2)

Since there are two implied DO loops, the first one will be completed before the second is started.

We can include nonsubscripted variables inside the implied DO loop, even the index variable itself to be printed. For example:

      DIMENSION X(10), Y(10)
      WRITE (6,95) (I,X(I), Y(I),I = 1,10)
  95 FORMAT (1X,I2,3X,F5.2,3X,F5.2)

This routine writes the value of I followed by the referenced element of array X and Y on the same line. That is, we write 1, X(1), and Y(1) on the first line; 2, X(2), and Y(2) on the second line; and so on.

Nonsubscripted variables may be included in the I/O list with any implied DO loops. For example:

      DIMENSION X(10)
      WRITE (6,94) A, (X(I),I = 1,10)
  94 FORMAT (1X,F5.2,3X,10F8.2)

This routine writes the value of A only once since it is outside the implied DO. Then the ten values in the array X follow on the same line.

There are many variations of the examples we have covered to illustrate the techniques of array I/O operations. There is no one correct technique to use all the time. The technique chosen depends upon the format of your input data and the desired output results.

Some of the techniques we will use to manipulate the elements in an array are discussed next.

10.5 Manipulating One-dimensional Array Items

We can manipulate the values in array elements in many different ways. We can use the array items in various computations. Arrays can be searched for specific values. We can change the values in any or all array elements. Arrays can be initialized or loaded with values that are computed in Assignment statements. There are so many things we can do with arrays that only some of the more common techniques can be covered.

Summing Array Elements

A common operation with arrays is to sum the values in all the elements. Suppose we sum the ten elements in the array NTEST, containing ten test scores. The array items could be easily summed in a DO loop as follows:

      DIMENSION NTEST(10)
      NSUM = 0
      DO 25 I = 1, 10
       NSUM = NSUM + NTEST(I)
   25 CONTINUE

Our index variable, I, varies from 1 through 10 and allows us to add each array item to the variable NSUM.

Initializing Array Elements

Array elements are often loaded by reading and storing input values into them. Sometimes it is required that all or selected items in an array be initialized (given a value) to the same current value. This is especially true with work arrays to which we will be adding values to various elements. The values of all elements are normally set to zero. The routine to initialize the 100 elements in array X to zero is:

```
       DIMENSION X(100)
       DO 25 I = 1,100
          X(I) = 0.0
 25    CONTINUE
```

Creating Duplicate Arrays

There are times when we wish to store the contents of one array into another array. A duplicate array is necessary to retain the values from the original array while we perform manipulations that change the values in the first array. Our routine for storing the contents of one array into another is as follows:

```
       DIMENSION A(50), AHOLD(50)
       DO 85 I = 1,50
          AHOLD(I) = A(I)
 85    CONTINUE
```

It would be nice to simply code

$$AHOLD = A$$

but we can't. Any time an array is used in an Assignment statement, we must include a subscript to refer to a specific element in the array.

Suppose we wish to store the contents of array A into another array, B, in reverse sequence. This can easily be done as follows:

```
       DIMENSION A(100), B(100)
       DO 95 I = 1,100
          B(101-I) = A(I)
 95    CONTINUE
```

The subscript value of 101-I allows us to reference the upper position of array B and work backwards. That is, A(1) is stored in B(100), A(2) in B(99), and so on down to where A(100) is stored into B(1).

Simple Operations with Selected Array Elements

We do not have to manipulate all of the items for an array in a routine. Individual array elements may be referenced as needed. Consider the following examples.

An array element may be used in a calculation to produce a new value for a variable.

$$RESULT = A * X(3) - 5.6 / R$$

A single array item may be changed from a computation in a routine.

$$X(K) = R + S / 7.5$$

An array element may be used in an IF statement to make a decision.

IF (X(J) .LT. 0.0) GOTO 82

In other words, selected array elements may be used in various FORTRAN statements and in any way our program design dictates.

Searching Operations with Arrays

There are times when we wish to find the smallest or largest item in an array. Maybe we want to look for a particular value in an array or even count how many times a certain value occurs in an array. All these routines require a search of the items in an array.

Searching is the process of examining each of the items in an array according to some specific criterion. Our criterion is to find the smallest value, largest value, or a specific value. All of the elements in the array are normally tested during the search operation. Only when we wish to find the first item that matches a specific value do we terminate the searching process.

The simplest search technique is the straight *sequential search.* This search method is also referred to as a serial or a linear search because of the process used. One simply starts at position one and proceeds in a sequential manner through the array. The search is over when a specific item is found or when the end of the array is reached, depending upon our problem requirements.

When the DO loop is used, the sequential search is a very easy routine to write. The initialization parameter is set to one, the test parameter is set to the number of items in our array, and the value one is used for the increment parameter. The array item pointed to by the index variable is, therefore, evaluated on each pass of the loop.

Searching an Array for a Specific Value

Looking for a specific value in an array is a very common requirement in many programs. An array is often searched to locate a specific pay grade, part number, account number, and so on. We may just want to verify its validity or to use the value's element number in accessing other array elements in later processing.

Assume we wish to search a 100-element array, X, for a match on the value contained in the variable, SPECVL. If the value is found, we will branch to the routine at statement number 40. Otherwise, an error message is printed.

The routine to search array X for a match of the value in SPECVL is:

```
       DIMENSION X(100)
       DO 5 I = 1,100
         IF (X(I) .EQ. SPECVL) GOTO 40
    5  CONTINUE
       WRITE (6,99) SPECVL
   99  FORMAT (1H0, 10HTHE VALUE , F10.2, 10H NOT FOUND)
       . . .
       . . .
   40  . . .
```

If the value of SPECVL is found, the location of X(I) is still retained in I, since a transfer is made out of the loop.

Counting the Number of Times a Specific Value Appears in an Array

Often we need to know how many times a certain value occurs in an array. Thus, we search the entire array and compare each item to our specific value. If a hit (match) is found, we add one to a counter. Assume the value we are looking for is in the variable SPECVL. The variable, KNT, will be used to keep count of the occurrences in which SPECVL is found. The routine to count the occurrences of SPECVL in a 100-element array X is:

```
          DIMENSION X(100)
          KNT = 0
          DO 75 I = 1,100
             IF (X(I) .EQ. SPECVL) KNT = KNT + 1
    75    CONTINUE
```

Selecting the Smallest Value in an Array

Suppose your kind instructor announced that he is going to eliminate your lowest test score. You would probably be overjoyed.

There are many ways we can go about finding the smallest value in an array. We might rearrange all the test scores in a descending order, and thus, the last one would be the lowest score. This technique is getting into the subject of sorting which we will discuss later. The main approach we wish to take now is to learn how to perform a search operation on an array to locate the smallest value.

We will begin our search by assuming the first test score to be the lowest and storing it into a variable called LOWEST. Then we will compare each of our array elements containing our test scores against the value in LOWEST. If a test score is greater than or equal to the value in LOWEST, we will go on to the next test score. If our test score is less than the one in LOWEST, we will put that test score in LOWEST. The element number for that test score will be saved in a variable called LOC.

We will continue this process until all test scores have been compared to LOWEST. When we have finished the search of the array, we will have the lowest test score in LOWEST and its element position in LOC. Since we store the first test score in LOWEST, we can begin our search with the second array item. Assuming an array named NTEST, containing ten test scores, our search routine is as follows:

```
          DIMENSION NTEST(10)
          LOWEST = NTEST(1)
          LOC = 1
          DO 55 I = 2, 10
             IF (NTEST(I) .GE. LOWEST) GOTO 55
             LOWEST = NTEST(I)
             LOC = I
    55    CONTINUE
```

How can we program the computer to drop the lowest test score from our average? The routine to eliminate the lowest test score and compute the average of the remaining nine is:

```
          DIMENSION NTEST(10)
          NSUM = 0
          DO 60 I = 1,10
             IF (I .EQ. LOC) GOTO 60
             NSUM = NSUM + NTEST(I)
    60    CONTINUE
          AVE = NSUM / 9
```

Whenever we find a match of the element number with the control variable, we simply jump around the addition operation. So the lowest score is not added into the overall NSUM.

We are now ready to discuss the subject of sorting.

10.6 Sorting One-dimensional Array Items

Sorting is the process of arranging items into an ordered sequence. The items can be sorted into either an **ascending** order with the smallest item appearing first or a **descending** sequence with the largest item appearing first.

Alphabetic data as well as numeric items may be sorted. We will discuss the sorting of only numeric data. All items to be sorted must be items in an array, since we will only cover sorting of data items in memory. This process is referred to as *internal sorting.* Long strings of items and files containing many data records are normally sorted using magnetic tape and disk media. We refer to this type of sorting as *external sorting* which is beyond the scope of this text.

To illustrate the sorting process, let us assume an array, LIST, containing four items. The values of the items are as follows:

$$\begin{array}{ll} \text{LIST(1)} & 10 \\ \text{LIST(2)} & 5 \\ \text{LIST(3)} & 7 \\ \text{LIST(4)} & 3 \end{array}$$

The items are in an unsorted sequence. We wish to rank (sort) the items in an ascending order so the results are 3, 5, 7, and 10 in array elements one through four, respectively.

Three basic operations are needed in the sorting operation. These operations are:

1. A comparison of the items with one another.
2. An interchange of the items between their array elements if they are not in the desired order.
3. Multiple passes/searches through the array to order all the items correctly.

The comparison operation is necessary to determine whether the two items used in the comparison are in order. If the two items are already in order, nothing needs to be done. If they are not, however, then they are placed in order (relative to each other) by the interchange operation.

There are several ways we could do the comparison operation. We could compare the first item with all the others. We could compare the items to each other in pairs, and so on. Since there are many ways to do the comparison, there are many types of sorting techniques. We will discuss only two—the **paired interchange** and the **selection with interchange** sort techniques.

Suppose we use the paired interchange technique to explain the various operations needed in an ascending sort. In the **paired interchange** we simply compare an array item to the next one. We have an IF statement in a DO loop much like a search routine. For example:

```
      DO 25 I = 1, 3
          IF (LIST(I) .LE. LIST(I+1)) GOTO 25
25    CONTINUE
```

When I is 1, we are comparing LIST(1) to List(2). If the quantity in LIST(1) is less than or equal to the quantity in LIST(2), then the items are already in ascending order. The first two items are not in ascending order so we need to interchange the two items to put them into the correct sequence.

Subscript Operations and One-dimensional Arrays   181

The interchange operation simply switches or exchanges the two items between their array elements. But we cannot code:

$$\text{LIST(1)} = \text{LIST(2)}$$
$$\text{LIST(2)} = \text{LIST(1)}$$

This would result in the quantity 5 being placed into both LIST(1) and LIST(2). We need an extra variable to temporarily hold one item and to allow us to properly interchange the two items.

The interchange operation uses a temporary variable to retain one of the items. Then the other items can be properly switched by finally storing the value in the temporary variable into the other item. See the diagram below (LTEMP is used for the temporary variable).

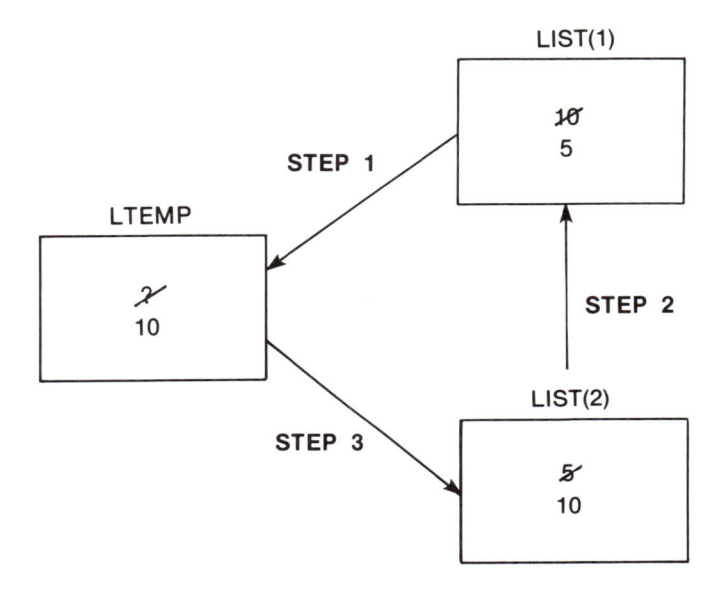

The top number in the boxes (crossed out) represent their original value before the interchange operation. The bottom number is the value after the interchange operation. The step numbers indicate the order of the operations. The value of LIST(1) is first retained in LTEMP. Then LIST(2) is stored into LIST(1) and finally the retained value in LTEMP is stored into LIST(2).

The interchange operation must apply to any two items needed to be exchanged in the loop, so we use the index variable for the subscript. Our coded interchange operation is:

$$\text{LTEMP} = \text{LIST(I)}$$
$$\text{LIST(I)} = \text{LIST(I+1)}$$
$$\text{LIST(I+1)} = \text{LTEMP}$$

If the two items in the comparison are not in order, the interchange operation is performed. Thus, the interchange statements are coded in our original DO loop as follows:

```
      DO 25 I = 1, 3
         IF (LIST(I) .LE. LIST(I+1)) GOTO 25
         LTEMP = LIST(I)
         LIST(I) = LIST(I+1)
         LIST(I+1) = LTEMP
   25 CONTINUE
```

If LIST(I) is greater than LIST(I + 1) we fall through into the interchange statements and exchange the two items.

The comparison loop is processed only three times, since all items would have been compared in pairs. The results of these three loops are:

	LOOP 1 Before	LOOP 1 After	LOOP 2 Before	LOOP 2 After	LOOP 3 Before	LOOP 3 After
LIST(1)	10	5	5	5	5	5
LIST(2)	5	10	10	7	7	7
LIST(3)	7	7	7	10	10	3
LIST(4)	3	3	3	3	3	10

The arrows indicate the two items involved in the comparison on each loop.

The items still have not been completely sorted in ascending order. That brings us to the third necessary operation. Another loop must be included outside the comparison loop to perform additional passes through the list of array items and to get them in final sorted order. This loop is normally performed n − 1 times (where n is the number of items being sorted).

So our complete sort routine is as follows:

```
      DIMENSION LIST(4)
C ***** PAIRED INTERCHANGE SORT
C ***** FIRST THE LOOP TO CONTROL NUMBER OF PASSES
      DO 35 J = 1,3
C ***** NOW THE LOOP TO CONTROL NUMBER OF COMPARISONS
      DO 25 I = 1,3
      IF (LIST(I) .LE. LIST(I + 1))GOTO 25
C ***** THE INTERCHANGE OPERATION
      LTEMP = LIST(I)
      LIST(I) = LIST(I + 1)
      LIST(I + 1) = LTEMP
25    CONTINUE
35    CONTINUE
```

The following diagram shows the order of items after each pass of the outer loop through the list. The before values on passes two and three are the same as the after values on the previous pass.

	PASS 1 Before	PASS 1 After	PASS 2 After	PASS 3 After
LIST(1)	10	5	5	3
LIST(2)	5	7	3	5
LIST(3)	7	3	7	7
LIST(4)	3	10	10	10

We finally have the items in ascending order. I encourage you to follow the interchange operation through each loop within each pass of the list to more clearly see what is happening.

The paired interchange sort is a very inefficient technique to use, especially if the items are already in a partially sorted order. You must always make the maximum number of passes and comparisons loops (unless a technique is used to terminate the sort when no interchange is made in a pass). We discussed this technique first, mainly because it more clearly illustrates the operations that take place in a sort. There are many more efficient sorting techniques we can use. Let us now look at a better sort technique called selection with interchange.

Selection with Interchange Sort Technique

This method searches the array for the smallest value (or largest) and places it in the first element. Then the next smallest value is found and placed in the second element, and so on. The routine is:

```
C *** NUMBER OF PASSES IS N-1 ITEMS
      M = N-1
      DO 35 J = 1,M
      JPLUS1 = J+1
C *** NUMBER OF COMPARISONS TO INCLUDE LAST ITEM
      DO 25 I = JPLUS1,N
      IF (LIST(J) .LE. LIST(I)) GOTO 25
C *** INTERCHANGE OPERATION
      LTEMP = LIST(J)
      LIST(J) = LIST(I)
      LIST(I) = LTEMP
25    CONTINUE
35    CONTINUE
```

Notice the use of the different subscripts in the comparison. One subscript (J) is held constant during the inner loop and indicates which array element is being used to store the smallest item. The other subscript (I) is used to search the remaining items in the list.

The order of the four items in the array LIST after each pass is as follows:

	PASS 1			*PASS 2*		*PASS 3*
	Loop 1	*Loop 2*	*Loop 3*	*Loop 1*	*Loop 2*	*Loop 1*
LIST(1)	5	5	3	3	3	3
LIST(2)	10	10	10	7	5	5
LIST(3)	7	7	7	10	10	7
LIST(4)	3	3	5	5	7	10
Subscript	J=1	J=1	J=1	J=2	J=2	J=3
Values are:	I=2	I=3	I=4	I=3	I=4	I=4

The selection with interchange reduced the number of comparisons by one-third. After the smallest value is found on the first pass, it is placed in element one and element one is never referenced again. On each succeeding pass the next smallest element is found and moved up to its position. After n − 1 (3 in our problem) passes, all items will be in order. If the first n − 1 items are in order, the last item is bound to be in order.

After the sorting operation is completed the items in our array are rearranged according to the type of sort (ascending or descending) we performed. Our original array order is lost. If we are to maintain our original array, the items must be assigned to a duplicate array in which the sorting is usually done.

The real power of a computer can really be appreciated once the use of arrays has been mastered. Arrays have many applications. The ability to use arrays in dealing with a large collection of related data items not only enhances the computer's computational power, but allows us greater ease in developing our programs. The use of arrays is one of the most powerful tools in programming. Learn to use them well and they can save you much work and time.

10.7 Nonstandard Language Extensions

One-dimensional arrays are standard features in every FORTRAN compiler. About the only nonstandard feature is the construct of subscripts. Most compilers allow a more general subscript expression.

Extensions to Subscript Expressions

Many FORTRAN compilers allow any permissible arithmetic expression that results in either an integer or real value. For example, a subscript reference to array A could be:

$$R = A(K/N * L - (5 + 3*M))$$

If a real quantity results from an arithmetic expression, it is converted to an integer form by truncating the fractional part.

The value provided as a subscript to an array may also contain a subscripted variable. That is, we can use values of an array to identify elements in another array. For example, we could write

$$LIST (KTR(J))$$

where the subscripted variable, KTR(J), provides a resulting subscript value for the array, LIST. This is a very complex technique to understand, but can prove very beneficial in accessing arrays. Even though ANSI and IBM Basic FORTRAN do not allow a subscripted variable reference for a subscript, many compilers allow this form of array element reference.

You should stick to subscript expressions of the ANSI standard forms. This will insure compatibility between FORTRAN compilers and generally result in more efficient object coding. If a complicated subscript such as: (K − M) * N / 3 + 1 is needed, it can be achieved easily by computing the subscript value in a prior assignment statement as follows:

$$INDEX = (K - M) * N / 3 + 1$$
$$HOLD = LIST(INDEX)$$

10.8 REVIEW QUESTIONS

1. What is an array?
2. Why is the use of arrays so important to programmers?
3. The statement used to tell the FORTRAN compiler to establish an array and provide the specification information is the _____ .
4. How is an array name formed?
5. The type of array name chosen must agree with the type of numeric quantities it contains. (True/False)
6. The dimension specifications given in the DIMENSION statement may be in the form of variables. (True/False)
7. Only one array name is allowed in each DIMENSION statement. (True/False)
8. During program execution, a _____ is used following the array name to designate a specific array element.
9. Subscript values may be integer constants, integer variables, or a special form of an arithmetic expression. (True/False)
10. Negative subscript values are permitted in FORTRAN. (True/False)
11. Name the three techniques that may be used to read and store values into arrays.
12. The nonsubscripted array name allows us to read/write any portion of an array that we wish. (True/False)
13. The implied DO used in an I/O list allows us to read/write as many or as few array items as we wish. (True/False)
14. The implied DO is always contained in parentheses when used in an I/O list. (True/False)
15. Only one subscripted variable (array name) may be contained in an implied DO. (True/False)

16. The process of examining the items in an array according to some specified criterion is called _____ .

17. The process of arranging the items in an array into a specified order is known as _____ .

18. How does the FORTRAN compiler know that a given symbolic name is an array name?

19. Is it possible to have an array with only one element?

20. What is the importance of DO loops to arrays?

21. Where must the DIMENSION statement be placed in a program?

10.9 EXERCISES

1. Provide the DIMENSION statements for the following specifications.
    a) A group of twelve items called GROUP.
    b) A group of forty items called LISTX.
    c) A group of sixty items called LISTA, and a group of fifty items called LISTB.

2. Provide the DIMENSION statements for the following specifications.
    a) A group of twenty items called K.
    b) A group of two hundred items called XHOLD.
    c) A group of ten items called XARRAY, a group of thirty items called YARRAY, and a group of ten items called ZARRAY.

3. Locate the errors in each of the following statements.
    a) **DIMENSION A(10) B(10)**
    b) **READ (5,99) (LIST(X),X = 1,10)**
    c) **READ (5,98) (A(I),I = 1,7) (B(I),I = 1,5)**
    d) **WRITE (6,97) (GROUP(J,J = 1,8)**
    e) **WRITE (6 96) (HOLD(2*N),N = 1,K,2)**

4. Locate the errors in each of the following statements.
    a) **DIMENSION X(TEN)**
    b) **READ (5,99) (LIST(N)N = 1,20)**
    c) **READ (5,98) ITEMS(J),J = 1,N**
    d) **WRITE (6,97) (ARYZ(Z),Z = 1,9**
    e) **WRITE (6,96) XARRAY, YARRAY ZARRAY**

5. Complete the following exercises on the I/O of array items. Use ARRAY as the array name and use the implied DO.
    a) Write a routine to read eleven data items with the format specifications of 11F5.2.
    b) Write a routine to write the first ten items on a double-spaced line under the specifications 10F7.2.

6. Complete the following exercises on the I/O of array items. Use ARRAY as the array name and use the implied DO.
    a) Write a routine to read sixteen data items with the specifications of 16F4.2.
    b) Write a routine to output the array items read in a) four per each double-spaced line under the specifications 4F7.2.

7. How many data cards will be read by each of the following group of statements?
    a)   **DIMENSION X(10)**
           READ (5,99) X
      99 FORMAT (2F5.2)
    b)   **DIMENSION X(10)**
           READ (5,98) X
      98 FORMAT (10F5.2)

    c)   **DIMENSION X(10)**
              READ (5,97) (X(I),I = 1,10)
          97 FORMAT (2F5.2/F5.2)

8. How many data cards will be read by each of the following group of statements?
    a)   **DIMENSION X(10)**
              READ (5,99) X
          99 FORMAT (4F5.2)

    b)   **DIMENSION X(10)**
              READ (5,98) (X(I),I = 1,10)
          98 FORMAT (F5.2)

    c)   **DIMENSION X(10)**
              READ (5,97) X
          97 FORMAT (F5.2/F5.2)

9. Complete the following exercises on the manipulation of array items.
    a) Write the routine to zero out the first fifty elements of a one hundred-element array named A.
    b) Write the routine to sum all the items stored in odd element numbers of the sixteen-element array called L16.
    c) Write a routine to move all the items in a twenty-element array called X20 down one position. That is, X20(1) is replaced by X20(2), X20(2) is replaced by X20(3), . . . and X20 (20) is replaced by X20(1).

10. Complete the following exercises on the manipulation of array items.
    a) Write the routine to initialize a fifty-element array called L50 with the values of its respective array element numbers. That is, element one contains a 1, element two a 2, . . . , and element fifty the value 50.
    b) Write the routine to sum all the items stored in even element numbers of the sixteen-element array called L16.
    c) A group of high school seniors (509) recently took a college entrance exam. The scores on the exam ranged from one to one hundred. Write a routine that will use a one hundred-element array called NEXAM and create a frequency distribution on each exam score. That is, determine the number of students who made each score. For example, if a student made a score of 76, then add one to the array element 76. Read the student scores from a data card punched in columns 1-3 as I3. Name the input variable NSCORS. Use a trailer card containing the value of 000 for NSCORS to indicate an end of file. When a end of file is reached, branch to statement number 30.

10.10 PROGRAMMING PROBLEMS

1. Write a program to load a twenty-element array called K20 from a single data card. The format specifications for the input items are 20I3.
    Reverse the items in the array. That is, exchange item 1 with item 20, item 2 with item 19, and so on. A duplicate array may be used, but this can be done with the same array by including an interchange operation.
    Write out the original array items on a single print line with format specifications, 20I6. Then write out the reversed array items with the same format specifications.

2. Write a program to read the items for two ten-element arrays named I1 and I2. The items for array, I1, will be read with a 10I2. The items for array, I2, will be read off a separate data card under the same format specifications.

Multiply the items in array, I1, by their corresponding item positions in array, I2. That is I1(1) is multiplied by I2(1), I1(2) is multiplied by I2(2), and so on. The resulting products are to be stored in the corresponding item positions in a ten-element array called I3.

Write out the items for each array on the same print line under the specifications 10I6. Each array will be on a new line.

3. Write a program to find the largest item in a one-dimensional array called NRS. The array, NRS, is a 20-element array and is loaded from a single data card under the format specifications 20I3.

Write the twenty array items on a single line under the format specifications 20I5. Then perform the search.

When the search is completed, write a message that says:

'THE LARGEST ITEM IN ARRAY NRS IS ' mmm
'AND IS LOCATED IN POSITION ' nnn

where mmm is the largest item in the array and nnn is its element number.

4. Write a program to perform a searching operation on a one-dimensional array called NUMBRS. The array, NUMBRS, is a 20-element array that is loaded from a single data card under the format specifications 20I3. A three-digit integer is read from columns 1-3 on the second data card and is used in the search operation. The search operation consists of matching the three-digit input value with one of the array items; the search will begin with the first element. Before the search is begun, write the twenty array items on a single line under the format specifications 20I5.

As soon as a match is found between the input value and an array item, write a message that says:

'THE VALUE ' mmm ' WAS FOUND IN POSITION ' nnn

where mmm is the input value and nnn is the array element number. Then terminate the program.

If the input value is not found in the array, write a message that says:

'THE VALUE ' mmm ' WAS NOT FOUND IN THE ARRAY'

5. Expand the previous problem in number 4 to continue searching the array for additional occurrences of the input value. Each time a match of the input value is found, write the message containing the value and the element number at which the value was found.

6. Write a program to perform a descending sort on ten test scores contained in a ten-element array called NTEST. The values will be read off a single data card in the form, 10I3.

Write the unsorted test scores on a single print line in the form, 10I6. Then write the sorted test scores with the same format specifications.

7. Write a program to read twenty two-digit integer numbers and compute their mean and standard deviation. The input values will be read from a single data card in the form 20F2.0 and stored in the twenty-element array called X.

Their mean will be computed according to the formula:

$$\overline{X} = \sum_{i=1}^{N} X_i$$

The symbol, Σ, means the summation of; the mean is represented by the symbol $\overline{X}$. Thus, the formula says that the mean is equal to the summation of the array elements, X_i, where i goes from 1 to N (N is 20 in our case).

The standard deviation is calculated from the formula:

$$\sigma = \sqrt{\frac{\sum_{i=1}^{N} (X_i - \overline{X})^2}{N-1}}$$

This formula says to first square each of the differences between each element in the array (i.e., X_i) and the computed mean. Each of these differences squared is added to a variable to accumulate their sum. Then divide this summation by the number of items (N) minus one. Finally take the square root of the results to find the value of σ.

For output results write the following:

a) The twenty values in the array X on one print line.
b) The computed mean ($\overline{X}$) on a new line preceded by the literal "MEAN = ".
c) The computed standard deviation (σ) on a new line preceded with the literal "STANDARD DEVIATION = ".

Two-dimensional and Three-dimensional Arrays 11

11.1 Introduction to the Concept of Two-dimensional Arrays

The concept of one-dimensional arrays is indeed a very powerful tool for the manipulation of a group of related items. There are times, however, when multiple groups of related items can be more conveniently handled by using two-dimensional arrays. Two-dimensional arrays may be thought of as a *matrix* or *table* of values.

A two-dimensional array consists of a number of horizontal lists of elements called *rows* (or row vectors), and a number of vertical lists of elements called *columns* (or column vectors). We can refer to any item by using a row vector and a column vector coordinate as we normally do with various tables of data.

An illustration of a two-dimensional array with 2 rows and 3 columns is as follows:

	Column 1	Column 2	Column 3
Row 1	79	83	76
Row 2	91	78	86

A reference to row 1, column 3 would get us the value of 76. If we wished to obtain the item with a value of 91, we would use a reference of row 2, column 1.

This is the same concept as that which we use in mathematics. We could have a matrix X_{ij} with i rows and j columns. If we wished to refer to the item in the second row and third column, we would specify X_{23}.

Assume that we wish to compute the grade averages of five students for three tests. Their averages could be calculated using nonsubscripted variables or a one-dimensional array provided that we did not need to keep all the test scores in memory for later reference. Suppose we do need to retain the test scores for all the students in memory for some later operation, such as searching or sorting. We could have five separate one-dimensional arrays—one for each student. But handing each of these arrays separately would be a very lengthy and time-consuming process.

If we could handle the test scores for all students in a matrix-like arrangement, then the process of manipulating this two-dimensional array becomes quite easy. When the number of students increases, say to thirty, our program remains essentially the same. But if we were to use thirty one-dimensional arrays, this could become quite a burden. Our program would really expand with individual routines to handle each student.

Let us illustrate the concept of two-dimensional arrays by establishing a 5-row by 4-column matrix to contain the data for five students. Each row in the matrix would contain all the information about one student—an identification number and three test scores. Thus, column one would relate to the identification number; column two to the first test score; column three to the second test score; and column four to the third test score.

Our assumed matrix would be as follows:

	Column 1	Column 2	Column 3	Column 4
Row 1	121	79	83	76
Row 2	148	91	78	86
Row 3	137	98	87	94
Row 4	126	81	66	74
Row 5	135	72	75	80

Therefore, student number 121 has the three test scores 79, 83, and 76; the second student with an identification of 148 has test scores of 91, 78, and 86; and so on.

Matrices such as this should not be unfamiliar to you. Not only do we use matrices to hold data for mathematical operations, but matrices are used constantly in our everyday business operations, as tax tables, time schedules, pay tables, etc. To locate a desired item, we normally locate the proper row vector and then read across the columns in that row to find our value. We specify a particular value simply by giving its row and column coordinates, such as row 2, column 4, for example.

A two-dimensional array must be identified in the DIMENSION statement. Two dimensions are indicated for an array by providing two constants in the dimension specifications. For example:

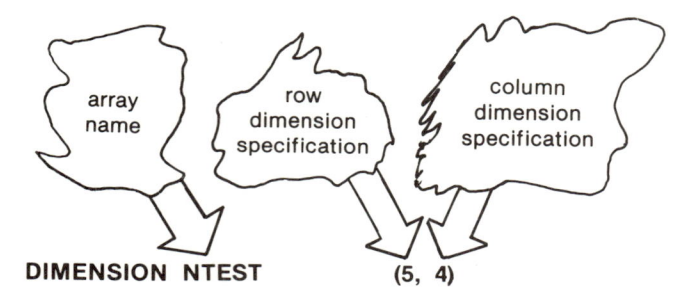

DIMENSION NTEST (5, 4)

(array name / row dimension specification / column dimension specification)

The dimension specifications follow the array name and are enclosed in parentheses, just as they are in one-dimensional arrays. Each dimension specification is separated by a comma. The first specification indicates the number of rows, and the second specification indicates the number of columns. Thus, our DIMENSION statement example tells us we have a two-dimensional array, NTEST, with 5 rows and 4 columns. By multiplying the size of each dimension specification by the others, the compiler determines the total size of the array (number of elements), or how many storage locations to reserve. (5 × 4 = 20, in our example.)

When referring to any array item during the execution of our program, we must provide two subscript values after the array name. For example:

NTEST (1,3)   (i.e., row 1, column 3)

or

NTEST (I,J)   (i.e., Ith row, Jth column)

The first subscript denotes the row vector, and the second subscript denotes the column vector, for a specific element. Each subscript may be any of the seven permissible forms allowed for one-dimensional arrays.

One subscript may take one form and the other subscript may take another form. For example:

NTEST (3,J + 1)

or

NTEST (I,1)

Giving one subscript a constant value and the other a variable can be very handy in some operations.

The same general rules that apply to one-dimensional arrays and their subscripting operations apply also to two-dimensional arrays. But, we have expanded the number of dimension references to two to provide the form of a matrix. To be consistent let us use the variable I for our row subscript and the variable J to denote our column subscript when variables are used as subscripts in our examples. Of course, we can use any integer variable that we want to, but the variables I and J are so frequently used in algebra with references to a matrix, such as X_{ij}. Now let us proceed to discuss the operations of reading and writing array items in two-dimensional arrays.

11.2 Input/Output Operations with Two-dimensional Arrays

We can use the subscripted variable method to read/write individual items or we can use the implied DO. Usage of these techniques can become more complex inasmuch as we may use the index variable in a DO loop to vary one subscript and the implied DO to control the other subscript.

Input Operations with Two-dimensional Arrays

Let us examine each of these two techniques in reading our five students' identification number and three test scores. In loading two-dimensional arrays, all the data items for one row are often punched into the same card. So let us assume each student's data is punched in a separate card as follows:

Card Columns	Input Field
1-3	Student ID (3-digit integer)
4	Blank
5-7	Test Score 1 (3-digit integer)
8-10	Test Score 2 (3-digit integer)
11-13	Test Score 3 (3-digit integer)

We certainly would not want to read the array items as subscripted variables using constants for both subscript references. This would require five separate READ statements in the form:

READ (5,99) NTEST(1,1), NTEST(1,2), NTEST(1,3), NTEST(1,4)
READ (5,99) NTEST(2,1), NTEST(2,2), NTEST(2,3), NTEST(2,4)
etc.

To provide flexibility in expanding the number of student grade cards that can be read, we should vary the row subscript in a DO loop. The routine is:

```
      DIMENSION NTEST(5,4)
      DO 10 I = 1,5
         READ (5,99) NTEST(I,1), NTEST(I,2), NTEST(I,3), NTEST(I,4)
99    FORMAT (I3,1X,3I3)
10    CONTINUE
```

192   Two-dimensional and Three-dimensional Arrays

This is the long way to load the array. What if we had ten test scores to store in each row? So rule this technique out when reading many items stored in the same row vector.

A better technique we could use is the implied DO. All the parameters for subscript control could be placed inside the implied DO as follows:

```
      DIMENSION NTEST(5,4)
      READ (5,99) ((NTEST(I,J),J = 1,4),I = 1,5)
99 FORMAT (I3,1X,3I3)
```

With the implied DO technique, we have an implied DO within an implied DO, or nested implied DO's. Examine the following illustration of the READ.

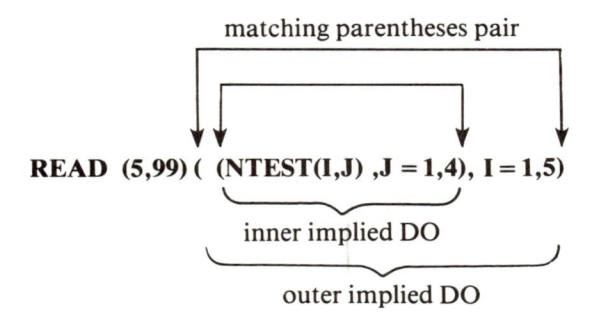

```
                  matching parentheses pair

READ (5,99) ( (NTEST(I,J) ,J = 1,4), I = 1,5)
               inner implied DO
              outer implied DO
```

This produces the same effect as the example at the bottom of the previous page. The inner implied DO (with index variable J) will vary from 1 to 4 while the outer implied DO (with index variable I) holds constant until the index variable J is completed. Then I is incremented by one and the looping is performed again with J varying from 1 to 4. When I is 5 and the inner loop with J is finished, both implied DOs are completed.

The rules for nested implied DO loops are as follows:

1. Each set of implied DO parameters must be enclosed within a pair of parentheses. Hence, in our example we need two pairs of parentheses. One pair will set off the subscripted array name and the end of the parameters for the index variable J. The second pair will include the inner implied DO, and set off the subscripted array name and the end of the parameters for the index variable I.
2. A comma must separate each implied DO. (Note the comma after the inner implied DO.)
3. The implied DO closest to the subscripted array name will be performed first. That is, the subscript corresponding to the innermost implied DO will vary the fastest. Then the next closest implied DO will be performed, etc., until the outermost implied DO is performed. When the outermost implied DO is completed, the implied DO is finished.

The best technique for reading two-dimensional array values punched the way we have is to include a DO loop to control the row subscript reference, and an implied DO to control the column subscript reference. Our revised routine to read the input fields with one implied DO is as follows:

```
      DIMENSION NTEST(5,4)
      DO 10 I = 1,5
         READ (5,99) (NTEST(I,J),J = 1,4)
99    FORMAT (I3,1X,3I3)
10    CONTINUE
```

This routine produces the same results as the nested implied DO, but we can see more clearly what is happening in the subscript references. The subscript, I, will change only when a pass in the DO loop is completed. The subscript, J, is varied and completed in the implied DO on each pass in the DO loop.

Suppose we dimension the array, NTEST, as a 30-row by 7-column array. But, we only want to load the first four columns of each row. The number of students (rows vectors) is read as a header card. Our routine becomes:

```
         DIMENSION NTEST(30,7)
         READ (5,199) N
199      FORMAT (I3)
         DO 10 I = 1,N
         READ (5,99) (NTEST(I,J),J = 1,4)
99       FORMAT (I3,1X,3I3)
10       CONTINUE
```

We can also read two-dimensional arrays with just the array name, as we did with one-dimensional arrays. The array elements are read, however, with the first subscript varying the fastest (i.e., in column-major order instead of row-major order as we have been doing). This means we must either reverse the row-column dimension specifications to column-row order or prepare our input data in column-major order. This technique of using just the array name for I/O operations with two-dimensional arrays is less flexible and tends to produce more errors than the techniques discussed above. This technique does, however, provide more efficiency and will be discussed in section 11.5 (for those interested).

Now let's look at the routines used for writing two-dimensional array items.

Output Operations with Two-dimensional Arrays

The same two techniques for loading our two-dimensional array, NTEST, can be used to write the array elements. Two-dimensional array items are normally written in row-order fashion. That is, all the items for a row vector are written on the same print line.

When writing subscripted variables in the I/O list, we should use a DO loop to control the row subscript. Our routine is:

```
         DIMENSION NTEST(5,4)
         DO 20 I = 1,5
         WRITE (6,98) NTEST(I,1), NTEST(I,2), NTEST(I,3), NTEST(I,4)
98       FORMAT (1H0,4I7)
20       CONTINUE
```

In the I/O list of the WRITE, we can use the implied DO in the same manner that we used it with the READ statement. We can have nested implied DO's or we can have a DO loop to control the row subscript value and one implied DO on the WRITE to control the column subscript value.

Our routine that uses nested implied DO loops in the WRITE is as follows:

```
         DIMENSION NTEST(5,4)
         WRITE (6,98) ((NTEST(I,J),J = 1,4),I = 1,5)
98       FORMAT (1H0,4I7)
```

The nested implied DO's on the WRITE statement work the same as on the READ. The innermost implied DO is completed first and then the outermost implied DO is incremented.

The recommended way to write the array items with an implied DO is to have a single implied DO on the WRITE inside of a DO loop. The routine is:

```
         DIMENSION NTEST(5,4)
         DO 20 I = 1,5
         WRITE (6,98) (NTEST(I,J),J = 1,4)
98       FORMAT (1H0,4I7)
20       CONTINUE
```

This technique is fairly easy to understand and allows greater flexibility.

Let us return to the problem wherein we established a 30-row by 7-column dimension specifications for the array NTEST. Only the first four columns were loaded and the number of students (row vectors) was determined by the value N read from a header card. We want to write only those array items that were loaded. Thus, the implied DO must be used. One routine is:

```
          DIMENSION NTEST(30,7)
          DO 20 I = 1,N
              WRITE (6,98) (NTEST(I,J),J = 1,4)
 98       FORMAT (1H0,4I7)
 20       CONTINUE
```

The implied DO references only the first four columns, even though we have seven column vectors for use.

We can also write all the elements in an array by using only the array name. But the first subscript will vary the fastest to give us the same problems (column-major order) as reading two-dimensional arrays with just the array name. Section 11.5 will discuss how we establish our two-dimensional arrays to provide array I/O operations with just the array name.

We may sometimes want to print only the items in a single row, or a single column, or even a particular element. This is easily done by specifying the desired row or column subscript as a constant and the other subscript as a variable which is varied in a DO loop. For example, the routine to print the second row of a two-dimensional array named X would be:

```
C *** PRINT ROW 2 OF ARRAY X
          DIMENSION X (10,5)
 99       FORMAT (1H0, F5.2)
          DO 30 J = 1,5
              WRITE (6,99) X (2,J)
 30       CONTINUE
```

An implied DO could be used if we wish to print all the items on the same print line.

The routine to print the items in column 3 of array X would be:

```
C *** PRINT COLUMN 3 OF ARRAY X
          DIMENSION X (10,5)
 99       FORMAT (1H0,F5.2)
          DO 20 I = 1,10
              WRITE (6,99) X(I,3)
 20       CONTINUE
```

If we had wanted the items for row 2 or column 3 to be printed on the same line, we would have had to change our FORMAT statement and use an implied DO in our WRITE.

To print a particular element such as row 4, column 1, our statement would be:

```
C *** PRINT A SINGLE ELEMENT OF ARRAY X
          WRITE (6,99) X(4,1)
```

So we see that there are many variations in the way we can read and write two-dimensional array items. It all depends upon what we need. We now turn our attention to the techniques we use to manipulate two-dimensional array elements.

11.3 Manipulating Two-dimensional Array Items

We perform calculations, search operations, and sorting with two-dimensional array items just as we do with one-dimensional arrays. Two-dimensional array operations are a little more involved because we have two subscripts to take care of.

Since there must be two subscript values in order to reference a two-dimensional array element, the normal procedure is to control the two subscript values with nested DO loops. The inner DO loop normally controls the column vector subscript and the outer DO loop controls the row vector subscript.

Summing Array Elements

Suppose we sum the three test scores and compute their average for each of five students. We will use the same array, NTEST, with 5 rows and 4 columns to illustrate this technique. We must have two nested DO loops to control our subscripts and reference all the test scores. Our routine is:

```
         DIMENSION NTEST (5,4)
         DO 20 I = 1,5
           NSUM = 0
           DO 10 J = 2,4
             NSUM = NSUM + NTEST(I,J)
   10      CONTINUE
           AVE = NSUM / 3
           WRITE (6,98) (NTEST(I,J),J=1,4), NSUM, AVE
   98    FORMAT (1H0,5I7,F6.2)
   20    CONTINUE
```

Our inner DO loop begins at 2 to pick up the first test score and goes through 4 to add in the first, second, and third test scores. The variable, NSUM, must be initialized to zero outside the inner loop to reset this counter between each grade calculation. After the average of each student is calculated we print the four array items with their sum and average. The outer loop is processed five times to compute the average for each student. The number of test scores and/or students processed may easily be expanded by changing the indexing variables of the proper DO.

If the sum and average values need to be retained for later array operations (searching, sorting), these values could easily be stored in unused columns of the array. For example:

```
         DIMENSION NTEST(5,6)
         DO 20 I = 1,5
           NTEST(I,5) = 0
           DO 10 J = 2,4
             NTEST(I,5) = NTEST(I,5) + NTEST(I,J)
   10      CONTINUE
           NTEST(I,6) = NTEST(I,5) / 3
   20    CONTINUE
```

The sum is stored into column 5 of each row, while the average is stored into column 6.

Initializing Array Elements

Two-dimensional arrays are often used for work arrays. Thus, their elements must be initialized to some value, normally zero. The routine to initialize a two-dimensional array X to zero is:

```
         DIMENSION X(10,5)
         DO 25 I = 1,10
           DO 15 J = 1,5
             X(I,J) = 0.0
   15      CONTINUE
   25    CONTINUE
```

196 Two-dimensional and Three-dimensional Arrays

The preceding example sets the elements of array X to zero in row order, i.e., the column subscript is varied the fastest. If one wishes to access the array in column order (with the row subscript varying the fastest), the order of the DO loops is simply reversed. For example:

```
         DIMENSION  X(10,5)
       DO 30  J  =  1,5
         DO 20  I  =  1,10
            X(I,J)  =  0.0
20       CONTINUE
30    CONTINUE
```

Accessing the elements in a two-dimensional array in column order as shown above is usually more efficient than in row order. But, there are times we need to access the elements in row order, as in our grade computation examples. For small arrays, the difference in efficiency is usually very minor. So the choice is left to you. If you feel more comfortable in accessing the array elements in row order, then do it that way. Remember, **row order** means we are accessing elements in different columns in the same row. **Column order** means we are accessing the items in different rows in the same column vector.

Creating Duplicate Arrays

There is often a need to retain the original values of an array for later output if the values are changed in processing. A routine for storing the values in array X into a duplicate array called Y is as follows:

```
         DIMENSION  X(10,5), Y(10,5)
       DO 25  I = 1,10
         DO 15  J = 1,5
            Y(I,J)  =  X(I,J)
15       CONTINUE
25    CONTINUE
```

Simple Operations with Selected Array Elements

Single array elements can be referenced in Assignment and IF statements. Consider the following examples.

An array item is used in calculating a value for a nonsubscripted variable:

$$RESULT = A*X(I,6) + 3.2/T$$

A single array item may be calculated in an Assignment statement:

$$X(I,7) = A - B*8.7$$

An array item may be used in an IF statement to make a decision:

$$IF\ (X(I,J)\ .GT.\ 100.0)\ GOTO\ 90$$

Searching Operations with Two-dimensional Arrays

Often we need to find a particular value, the smallest value, or the largest value in a two-dimensional array. The same sequential search technique as that which we used in one-dimensional arrays is used. The main difference is that now we must search each column in every row.

Suppose we wish to count the occurrences of the items in the array X that match the quantity in the variable, SPECVL. The routine is:

```
          DIMENSION X(10,5)
          KNT = 0
          DO 25 I = 1,10
            DO 15 J = 1,5
              IF (X(I,J) .EQ. SPECVL) KNT = KNT + 1
15        CONTINUE
25      CONTINUE
```

The outer loop controls the selection of a row vector while the inner loop controls the column vector being searched.

Suppose we wish to find the largest item and its location in the array X. The routine is:

```
          DIMENSION X(10,5)
          XLARGE = X(1,1)
          ILOC = 1
          JLOC = 1
          DO 25 I = 1,10
            DO 15 J = 1,5
              IF (X(I,J) .LE. XLARGE) GOTO 15
              XLARGE = X(I,J)
              ILOC = I
              JLOC = J
15        CONTINUE
25      CONTINUE
```

The item in X(1,1) is assumed to be the largest value and is placed into the variable XLARGE. Then we search the array and compare each item against the quantity stored in XLARGE. Whenever a larger value is found, it is placed into XLARGE along with its element numbers into ILOC and JLOC.

Sorting Two-dimensional Array Items

A sorting operation with a two-dimensional array is usually done on the values in one specific column. The main difference between sorting a two-dimensional array and sorting a one-dimensional array is in the interchange operation. All items in the two rows must be exchanged and not just the two items in the comparison.

Suppose we sort the averages of the five students used in our previous discussion. Assume the sum and average have been stored in the fifth and sixth columns, respectively. The routine to produce a descending sort in average order using the selection with interchange method is:

```
          DIMENSION NTEST (5, 6)
          N = 5
C ****** LOOP TO CONTROL THE NUMBER OF PASSES
          M = N - 1
          DO 45 I1 = 1,M
            IPLUS1 = I1 + 1
C ****** LOOP TO CONTROL THE NUMBER OF COMPARISONS
            DO 35 I2 = IPLUS1, N
              IF(NTEST(I1,6) .GE. NTEST(I2,6))GOTO 35
C ****** LOOP TO CONTROL THE INTERCHANGES BETWEEN ROWS
              DO 25 J = 1,6
                NHOLD = NTEST(I1,J)
                NTEST(I1,J) = NTEST(I2,J)
                NTEST(I2,J) = NHOLD
25            CONTINUE
35          CONTINUE
45      CONTINUE
```

Three DO loops must be used. The innermost loop is used to exchange all the items in the two rows (referenced by I1 and I2). Thus the control variable, J, is varied from 1 through 6, to exchange the six column values of the two rows. The best way to fully understand what is happening is to play computer. Step through each of the loops and change the values in the array as called for in the routine.

11.4 Introduction to the Concept of Three-dimensional Arrays

Often there is a need to refer to a group of matrices in an array arrangement. That is, we can shorten our program by using a third subscript reference to indicate which matrix we are referencing. Such an array is called a three-dimensional array, which can be considered as being two or more related two-dimensional arrays. We have a *row* subscript to denote the row vector, a *column* subscript to denote the column vector. And we have a *level* subscript to indicate which level number or matrix we are referencing.

For example, we may want to write a program to play Bingo on the computer. Suppose we decided to play with three Bingo cards. This would be an ideal situation in which to use a three-dimensional array. The first two subscripts would be used to indicate the row and column coordinates on a single card. The last subscript would be used to tell us which card we are playing.

The dimension specifications in the DIMENSION statement tells the compiler when a three-dimensional array is to be used in a program. The DIMENSION statement to provide the array specifications for three 5-row by 5-column matrices to be used for our Bingo program would be:

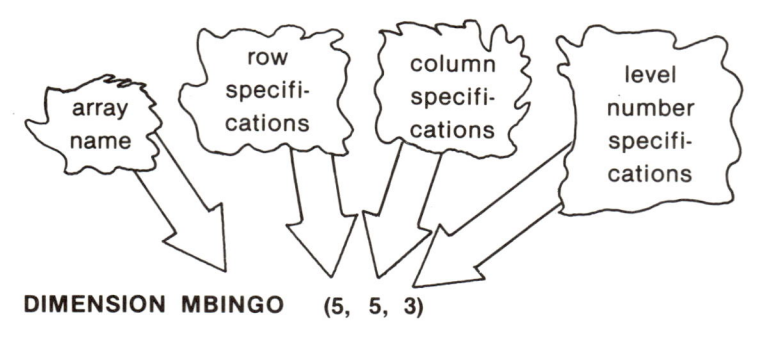

DIMENSION MBINGO (5, 5, 3)

(array name, row specifications, column specifications, level number specifications)

Whenever an element in the array MBINGO is needed, three subscripts must follow the array name. The first subscript refers to the row vector, the second to the column vector, and the third to the respective card vector. Consider in figure 11.1, the three Bingo cards used with our array MBINGO.

BINGO CARD 1					BINGO CARD 2					BINGO CARD 3				
2	16	33	47	63	1	16	32	46	62	1	17	31	47	63
5	17	34	50	67	3	18	34	48	65	2	19	33	49	65
7	20	Free	52	71	4	21	Free	51	68	6	24	Free	51	70
10	22	39	57	72	8	23	38	55	70	10	27	41	53	72
11	27	41	59	74	13	27	42	58	73	14	28	43	57	75

Figure 11.1. Three Sample Bingo Cards

A subscript reference of MBINGO(4,2,1) would get us the value 22. If we wanted to get the value in the second row and first column of the third card, we would use a reference of MBINGO(2,1,3).

We could write a program to play Bingo as three two-dimensional arrays. But, there must be a separate routine to search each card for a hit with a called number (generated from a random number routine) and to check for a BINGO. By using a three-dimensional array, we need only one routine with three nested loops. The inner loop would control the column subscript, the middle loop would control the row subscript, and the outer loop would control the level or card number.

Three-dimensional arrays are generally even more involved than two-dimensional arrays. But the use of a three-dimensional array can shorten our program and make our job of programming less tedious and time-consuming. The one-dimensional and two-dimensional arrays are by far the most frequently used in programming. It is these two forms of arrays on which you should concentrate your learning efforts; you may never have the need to use three-dimensional arrays.

11.5 Achieving Efficiency in Multidimensional Array Operations

This section is intended primarily for those students who, in the hope of achieving greater proficiency in the use of multidimensional arrays, need a deeper explanation of how they are handled. Those students who are planning to pursue the profession of programming, or who might perform research on problems requiring the use of large multidimensional arrays, will benefit greatly from this discussion.

Multidimensional arrays, as used in solving many statistical and application problems, make a powerful and necessary tool. Unless dimensioned and manipulated in the most efficient manner, however, large multidimensional arrays may cause an excessive amount of computer time to be used. First, we must understand how storage locations are allocated for array elements before discussing the most efficient means of manipulating array elements.

Specifications in the DIMENSION statement determine how many memory locations are reserved for an array and their address assignments. All memory locations allocated for an array are assigned ascending, contiguous addresses, no matter how many dimensions an array may have. That is, a two-dimensional array is *not* stored in memory in a matrix form, as we visualize on paper. All the items are stored in one linear arrangement, because memory locations are organized that way.

So memory locations for each array are assigned in an ascending order, with the value of the first element number (subscript) increasing the fastest and the value of the last element number (subscript) increasing the slowest. This is a simple arrangement for a one-dimensional array which only uses one element number for a subscript reference. For example, the one-dimensional array named LIST, with four items would be assigned storage locations as follows:

memory locations				
subscript reference	LIST(1)	LIST(2)	LIST(3)	LIST(4)

There is no problem to reading/writing the entire array when the array name I/O technique is used.

A two-dimensional array uses two element numbers—one for the row and one for the column. When memory locations are assigned, the first element number increases the most rapidly and the second element number increases the slowest. Assuming a 3-row by 2-column dimension specification for an array named X, the six elements would be assigned storage locations as follows:

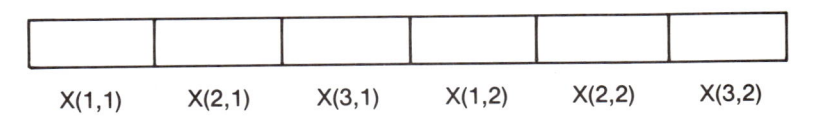

X(1,1)	X(2,1)	X(3,1)	X(1,2)	X(2,2)	X(3,2)

A three-dimensional array uses three element numbers for subscript references—one for the row, one for the column, and one for the level number. The first element number would vary the fastest, the second element number the next fastest, and the third element number the slowest. Assuming a 2-row by 2-column by 2-level dimension specification for an array named Y, the eight storage locations would be assigned as follows:

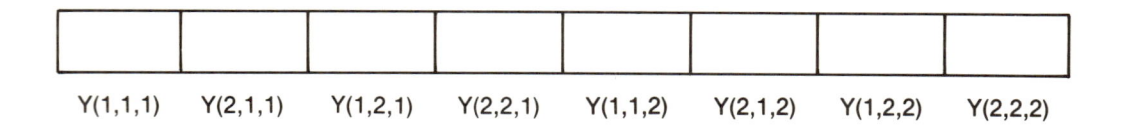

Y(1,1,1)　Y(2,1,1)　Y(1,2,1)　Y(2,2,1)　Y(1,1,2)　Y(2,1,2)　Y(1,2,2)　Y(2,2,2)

Since two-dimensional arrays are more widely used, our discussion on multidimensional array efficiency will be limited to two-dimensional arrays. Two-dimensional arrays are allocated memory locations in FORTRAN in what is known as column-order fashion. That is, the row subscript varies the fastest so that all items for a column are stored next to each other. Whenever large arrays are used by some computers, the array will be stored on disk in order to cut down on the total memory requirements of the program.

Many computers keep only one column of items for a large array in memory at a time. If the array item that is needed by the computer is not in memory, the computer will read the column of items that contains the needed item. Before a new column of items is read into memory, the old column is written to disk. Thus, many computers read/write array items a column at a time from/to disk whenever an item in a different column is needed. This operation is referred to as *segmentation* or *paging* on the various computers.

If we reference the items in an array where the first subscript varies the fastest, all the items in the same column can be used between the I/O of array columns. When the second subscript is altered and a new column is referenced, the column currently in memory is written to disk and the new column read. This is certainly the most efficient way to change the subscript values. Suppose we use a 3-row by 2-column array named X and a routine to initialize all array elements to zero.

```
          DIMENSION  X(3,2)
          DO 20 J = 1,2
             DO 10 I = 1,3
                X(I,J) = 0.0
10           CONTINUE
20        CONTINUE
```

By varying the first subscript (row) the fastest, we will access the array elements in column order—the same way the elements are allocated in memory, i.e., X(1,1), X(2,1), X(3,1), X(1,2), X(2,2), and X(3,2).

On the other hand, if the second (column) subscript is varied the fastest, a disk write and read operation must be performed on the reference of each array item. The disk I/O operation must be done since the next item being referenced is stored at a different column. Can you imagine the increase in time needed to manipulate large arrays referenced in this manner? For example:

```
          DIMENSION  X(3,2)
          DO 20 I = 1,3
             DO 10 J = 1,2
                X(I,J) = 0.0
10           CONTINUE
20        CONTINUE
```

	←——— Column 1 ———→	←——— Column 2 ———→			

Storage Locations: X(1,1)   X(2,1)   X(3,1)   X(1,2)   X(2,2)   X(3,2)

Access Order: 1   3   5   2   4   6

Can you see the inefficiency caused by varying the second subscript the fastest? A different column is accessed on each array element reference.

The largest vector (i.e., the one containing the biggest dimension specifications) should be established as the first subscript (row) whenever logically feasible. For example, if we had a 20 by 100 matrix containing various data items, the 100-element vector should be set up as the row vector and the 20-element vector as the column. If our array was named A, this would be specified as DIMENSION A (100,20). This arrangement would greatly increase program efficiency since more items would be contained in a column and fewer I/O operations would be required.

The array name I/O technique with two-dimensional arrays reads/writes the data in column order. That is, the I/O sequence is down the columns and across the rows. Suppose we read the items of an array named X with just the array name as follows:

DIMENSION X(3,2)
99 FORMAT (6F5.2)
READ (5,99) X

The data items are read and stored in the following order: X(1,1), X(2,1), X(3,1), X(1,2), X(2,2), and X(3,2). All the items for column 1 are stored and then those in column 2.

If we prepare and read our input data in column-order fashion with the first subscript varying the fastest, we can now read the data into the array by using only the array name in the I/O list. We can use simply the array name to output the elements of a two-dimensional array if we write the elements in column order instead of row order. Remember, when we use just the array name to read/write elements in a multidimensional array, we must read or write all of the elements. The array name I/O technique will gain us a little efficiency over subscripting each element in I/O operations.

A problem arises when we wish to read the input data punched in row order and write it the same way using only the array name. For example, suppose we have ten students with an ID number and 12 test scores for each. The logical way to prepare the input data is to punch all the information relative to each student in a separate card. But, if our dimension specification is

DIMENSION STUDNT (10,13)

and we use the array name technique to read the input items, we do not get the data read in such a way that we may reference the elements pertaining to each student in a straightforward, logical manner. That is, the ID number and test scores for the second student would be loaded beginning at array element STUDNT (4,2), the third student's data beginning at array element STUDNT (7,3), etc. (You may wish to draw a 10 by 13 matrix for yourself to confirm this.)

So you might think that the array name technique could not be used, and that you must use the implied DO. If you do wish to use the array name I/O technique for efficiency, to read (or write) the data in row order, there is a simple way. You may reverse the order of the subscripts. That is, the first dimension would refer to the column and the second dimension would refer to the row. Our dimension specifications would then be:

DIMENSION STUDNT (13,10)

If the first subscript (J as the column) referred to the elements of a particular student and the second subscript (I as the row) indicated which student, then we could handle the array elements with no problem. The array could be read and the 12 test scores could be summed as follows:

```
         DIMENSION  STUDNT (13,10)
      99 FORMAT  (I3,2X,12F4.0)
      98 FORMAT  (1H0,I3,2X,12F6.0,F10.0)
         READ  (5,99) STUDNT
         SUM  =  0.0
         DO 20  I  =  1,10
           DO 10  J  =  2, 13
             SUM  =  SUM  +  STUDNT (J,I)
      10   CONTINUE
           WRITE  (6,98) (STUDNT(J,I),J = 1,13),SUM
      20 CONTINUE
```

To sum up the rules for efficiency on two-dimensional arrays: you should

1. Always vary the first subscript the fastest. This accesses the array elements in the same order as their storage assignments.
2. Strive to use the largest vector requirements as the first subscript. This will cut down on disk reads/writes for column vector elements on many computers.
3. When the data must be read/written in a row-order fashion, reverse the order of the subscripts. Thus, the first subscript acts as the column vector and the second subscript acts as the row. When manipulating individual array elements, remember to give the column subscript first and the row subscript last.

The same concepts apply to three-dimensional arrays. You should always vary the first subscript the fastest, the second subscript next, and the last subscript the slowest to obtain the utmost in efficiency. The array elements will be referenced in the same order as that in which they are arranged in memory.

11.6 Nonstandard Language Extensions

Two-dimensional arrays are standard on all FORTRAN compilers. Full ANSI FORTRAN provides for up to three levels of dimension, while Basic ANSI FORTRAN allows only two levels of dimension. The IBM "G" level compiler and WATFIV allow seven levels of dimensions. Some FORTRAN compilers accept more than seven levels of dimensions, while a few compilers allow an unlimited number of dimension levels for an array.

You may wonder why anyone would ever want to use more than three levels of dimension. Multidimensional levels are convenient when we think in logical grouping. We might want to have a four-dimensional array to represent telephone extension numbers for individuals. Each of the four digits in the number could be used as a element number and the name of the individual placed into the array element. For example, the array reference NRTELE(2,4,4,1) might contain the name, COLE.

Suppose someone wished to index the keywords in several volumes of manuals. (Imagine the work!) They might use one subscript for the line number on a page, one subscript for the page number, one subscript for the section number, one subscript for the manual number, one subscript for the volume number, and so on. So you see that having access to a number of dimension levels can be very useful in some applications.

11.7 REVIEW QUESTIONS

1. What advantage does the use of a two-dimensional array have over the use of multiple one-dimensional arrays?
2. How do we reference an item in a two-dimensional array?
3. How does the FORTRAN compiler know that we are using a two-dimensional array in our program?
4. How does the FORTRAN compiler determine how many storage locations to reserve for a two-dimensional array?
5. When using nested implied DO's, the innermost implied DO is always completed first. (True/False)
6. Each implied DO in nested implied DO's must be enclosed within a set of parentheses and separated by a comma. (True/False)
7. When should three-dimensional arrays be used instead of two-dimensional arrays?

11.8 EXERCISES

1. Provide the DIMENSION statements for the following specifications.
    - *a)* A matrix of 7 rows by 5 columns named MATRIX.
    - *b)* A matrix of 20 rows by 11 columns named TABLE.
    - *c)* Four groups of matrices with 6 rows and 4 columns named CLASS.
2. Provide the DIMENSION statements for the following specifications.
    - *a)* A matrix of 11 rows by 6 columns named X.
    - *b)* A matrix of 30 rows by 4 columns named MATX.
    - *c)* Five groups of matrices with 10 rows and 7 columns named SALES.
3. How many data cards will be read by each of the following group of statements?
    - *a)*
      ```
      DIMENSION  Z(3,12)
      DO 10  I = 1,3
         READ  (5,99)  (Z(I,J),J = 1,12)
      99 FORMAT  (6F5.2)
      10 CONTINUE
      ```
    - *b)*
      ```
      DIMENSION  L2(5,30)
      READ  (5,99)  (L2(I,J),J = 1,30),I = 1,5)
      99 FORMAT  (15I2)
      ```
4. How many data cards will be read by each of the following group of statements?
    - *a)*
      ```
      DIMENSION  B(6,20)
      DO 10  I = 1,5
         READ  (5,99)  (B(I,J),J = 1,10)
      99 FORMAT  (10F4.1)
      10 CONTINUE
      ```
    - *b)*
      ```
      DIMENSION  MM(11,40)
      READ  (5,99)  ((MM(I,J),J = 1,40),I = 1,11)
      99 FORMAT  (20I1)
      ```
5. Complete the following exercises on the I/O of two-dimensional arrays. Use X as the array name and use at least one implied DO in the READ/WRITE statements. Assume X is dimensioned as a 6-row by 10-column array.
    - *a)* Load the array X from six data cards punched with ten items per card under the specifications 10F6.2.
    - *b)* Write the first eight items in each row under the specifications 8F10.2.

6. Complete the following exercises on the I/O of two-dimensional arrays. Use Y as the array name and assume dimension specifications of 8 rows and 12 columns. Use at least one implied DO in the READ/WRITE statements.
   - *a)* Load the first nine columns of each row in the array Y from eight data cards. Each data card contains nine values under the specifications 9F4.0.
   - *b)* Write the first five values in each row double-spaced under the specifications 5F7.0.
7. Complete the following exercises on the manipulation of two-dimensional array items.
   - *a)* Write the routine to initialize the elements in a 9-row by 30-column array named K to one.
   - *b)* Write the routine to sum all the items stored at odd-numbered rows in a 6-row by 6-column array named NN.
   - *c)* Write a routine to search a 50-row by 7-column array named D for the value 76.5. As soon as the value is found, branch to statement number 777.
8. Complete the following exercises on the manipulation of two-dimensional array items.
   - *a)* Write the routine to add one to each element in a 4-row by 20-column array named HOLD.
   - *b)* Write the routine to create a second array named B2 from an array named B1. The arrays contain 12 rows and 3 columns.
   - *c)* Write the routine to count all the occurrences of a quantity stored in K with the items in a 40-row by 20-column array named KARY.

11.9 PROGRAMMING PROBLEMS

1. Write a program to find the smallest item in a two-dimensional array named XDATA. The array, XDATA is a 5-row by 25-column array and is loaded from five data cards under the format specifications 25F2.0.

    Write the array in row-order fashion under the output specifications of 25F5.0. Then search the array.

    When the search is completed, write a message that says:

    'THE SMALLEST ITEM IN ARRAY XDATA IS' mmm

    where mmm is the quantity of the smallest item in the array.

2. Expand problem 1 to include the row and column location at which the smallest item was found. When the search is over, write a message that says: 'THE SMALLEST ITEM IN ARRAY XDATA IS' mmm 'AND WAS FOUND AT ROW' ii 'AND COLUMN' jj where mmm is the smallest item in the array, ii is the row location, and jj is the column location.

3. Write a program to search a two-dimensional array named M and count all the occurrences of the items that match the value in a variable, MVAL. The array, M, is a 3-row by 10-column array loaded from three data cards punched with the format specifications 10I2. The variable, MVAL, is read from a fourth data card with its value punched in columns 1-2.

    Write the array M in row-order fashion under the format specifications 10I5. Then search the array. After the search operation is completed, write a message that says: 'THE VALUE' mmm 'OCCURRED IN THE ARRAY M' nnn 'TIMES' where mmm is the quantity in MVAL and nnn is the count of occurrences.

4. Write a program to compute a biweekly gross pay report using a search operation on a pay grade table. A biweekly payroll card is prepared for each employee, containing his

employee number punched in columns 1-4 (integer form), his hours worked (form = NNN.NN) punched in columns 7-12, and his pay grade (form = NN.) punched in columns 15-17. The pay grade field is used to search a pay grade table to locate the hourly wage rate. By using the hourly wage rate from the table and the number of hours worked from the payroll card, the employee's gross pay can be calculated.

The pay grade table is a 9-row by 2-column array named PAYGRD. Nine data cards are used to load the table in row order with each card containing the pay grade punched in columns 1-3 (form = NN.) and the hourly wage rate in columns 5-9 (form = NN.NN).

Next, nine employee payroll cards are read and processed against the pay grade table to compute gross pay. As each payroll card is read, a search is performed in the array, PAYGRD, to match the pay grade (column one) in the payroll card. Then the column two quantity (hourly rate) is multiplied by the hours worked (from the payroll card) to produce gross pay.

As the gross pay is calculated for each employee, write a print record as follows:

Field Description	*Print Positions*
Employee number	7-10
Pay grade	14-16
Hourly wage rate	20-24
Hours worked	30-35
Gross pay	40-46

Also accumulate a running total of the gross pay calculated for all employees.

The last payroll card (record 10) read will contain a value of 9999 for the employee number field. This data card is used as a trailer card to signal the end of the payroll transaction cards. Print the accumulated gross pay after the last printed employee record, with an appropriate literal. Also include appropriate column headings to identify each column of output values.

5. Write a program to process transactions for an inventory control application. The file of ten inventory items is read into a two-dimensional array named INVCTL. The array, INVCTL, is a 10-row by 3-column array loaded in row order from ten data cards in the following format:

Card Columns	*Field Description*
1-5	Item number (Form=NNNNN)
10-12	Quantity on hand (Form=NNN)
13-15	Quantity on back order (Form=NNN)

After the file containing the ten records is loaded into the array, INVCTL, write the array in row order under the output specifications 3I10. Include column headings to identify each column of data. The next eight data cards represent transaction cards that are processed against the items in INVCTL. The transaction cards are punched according to the following format.

Card Columns	*Field Description*
1-5	Item number (Form = NNNNN)
7	Transaction code (1=Receipt of items, 2=Shipment of items)
10-12	Quantity involved (Form = NNN)

As each transaction card is read, search the array, INVCTL, for a match on the item number (column one). When a match is found, update the quantity-on-hand vector (column two) and/or the quantity-on-back-order vector (column three) according to the following requirements.

a) If the transaction code is a receipt of items (code of 1), you should add the quantity in the transaction card to the quantity-on-hand column for that item number.

b) If the transaction code is a shipment of items (code of 2), then you want to subtract the quantity in the transaction card from the quantity-on-hand column for that item number. However, you must first check to see if any items are on hand. If there is a sufficient quantity, do the subtraction. If there is not a sufficient quantity on hand to satisfy the shipment request, then ship as many items as are on hand and add the different (between shipment request and quantity on hand) to the quantity-on-back-order column. If the difference is added to the back-order column, be sure to set the quantity-on-hand column to zero.

Assume that the inventory item numbers in the transaction cards are valid and that the transaction codes are also valid. You should always program for possible error conditions in the transactions for "real-world" applications. To shorten the problem, however, we will assume that they are correct.

A ninth transaction card will be included as a trailer card to indicate the end of the transaction cards. This data card will have 99999 punched as its item number. When this card is read, print the updated array under the same output specifications as before. Also include column headings.

Alphanumeric Data and Compile Time Specifications   12

12.1 Introduction to Alphanumeric Data and Format Specifications

Most data read and processed by FORTRAN programs are numeric in form. But frequently the need to read alphanumeric data arises. The term—*alphanumeric*—is used to describe data in which the characters may be alphabetic, numeric, special characters, or a combination of the three types. Alphanumeric data is treated as character or string data and must not be used in computations. That is, digits read as alphanumeric characters are treated differently in the computer than numeric digits read with an I or F format code.

Data items such as name, part number, item description, course number, and address are typical alphanumeric values used in computer programs. Alphanumeric data must be read with the alphanumeric (A) format code. Any attempt to read alphabetic letters or special characters with any of the numeric-type format codes will normally cause an error condition during execution time.

The Alphanumeric (A) Format Code

The A format code allows us to read alphanumeric data in character form. Any character acceptable to the computer can be read with the A format code. This format code is used to read data items such as name, titles, and other values containing alphabetic letters and special characters. Data items such as student number and part number which may contain a mixture of alphabetic letters and numeric digits must be read with the A format code. A pure numeric field may be read with the A format code; however, any item read with the A format code must *not* be used later in any computations.

The general form of the A format code is:

$$rAw$$

where:

r represents an optional repetition factor that specifies the number of times the format code is to be used. The value of **r**, when used, must be an unsigned positive integer constant whose value is less than or equal to 255. **w** specifies the number of card columns or positions in the input or output field. **w** must be an unsigned positive constant with a value less than or equal to 255. Examples are:

```
 36 FORMAT (5A4, 2X, I3, F6.2)
287 FORMAT (4A4, A3, 10X, 5F8.2)
```

The first example reads the 20 characters in card columns 1-20 as alphanumeric characters. The character field is broken down into five groups of four characters. The second example reads the first nineteen characters of an input record as four groups of 4 characters, plus a group of 3 characters. The need for dividing an alphanumeric field into groups of 4 or less characters is explained below.

We learned that a numeric value must be less than eleven digits (i.e., 2147483647) for it to be read and

207

stored with the I format code on IBM 360 and 370 computers. Likewise, only so many characters can be read and stored as a field with the A format code. The maximum number of characters that can be read and stored as a field is related to the size of a computer word. A computer *word* refers to the size of a storage location that determines the amount of data which can be stored. The size of a computer word varies by type of computer. Since ANSI FORTRAN does not provide specifications for a word size we will assume a computer word that can hold a maximum of 4 characters.

Normally one item or field of information is stored at a storage location. Thus, each numeric value which is read with a format code is placed at a unique memory cell/word. Remember, the format code tells us the form and size of our input data or our output field. It does not dictate the size of the computer word, which is always a fixed size.

If a storage location can hold only 4 alphanumeric characters, then alphanumeric input fields must be broken down into groups of 4 or less characters. Suppose a name field has 16 characters. It could be broken down into four groups of 4 characters each. But what if our name field contains 15 characters? The character string must be broken down into three groups of 4 characters, plus a group of 3 characters. Or the 15-character field could be broken down into five groups of 3 characters each.

Remember that a variable name must be assigned to each group of characters being read to designate their storage location. There must be a different variable for each format code that transmits data to memory. If we have four groups of 4 characters each, we must have four format codes to designate this format of our fields and four different variables. If we have five groups of 3 alphanumeric characters, we must have five format codes and, likewise, five different variables.

The type of variable and array names assigned to store alphanumeric data may be either integer or real. Since no computations should ever be performed with variables containing data in alphanumeric form, the computer will properly process alphanumeric data no matter which type of variable name is used. Integer variable names are preferred in most cases.

Input Considerations

The above discussion concerning the storage of alphanumeric characters may have lead you to believe that a field of more than 4 characters could not be read with an A format code. This is not true; an alphanumeric field of up to 255 characters can be read with a single A format code. The number of characters, however, stored from the input field cannot exceed four. If a 15-character field was read with an A15 format code, only the rightmost 4 characters would be stored.

If the field width specification (**w**) is greater than the maximum number of characters that can be stored at the memory location, then the excess leftmost characters in the field are skipped and the remaining characters read and stored in the variable. Thus, if we read the field DK123C under an A6 format code, only the rightmost 4 characters, 123C, would be stored in memory.

If the field width specification (**w**) is less than the maximum number of characters that can be stored at a memory location, then **w** characters are read and stored in memory. The characters will be stored left-justified in the memory location with the rightmost characters filled with blanks. Suppose the 3-character field, ABC, was read with an A3 format code. The three characters would be stored in the leftmost three positions in the computer word and the fourth position filled with a blank.

If the field width specification (**w**) is equal to the maximum number of characters that can be stored at a memory location, then **w** characters are read and stored to fill the memory word. The 4-character field, COLE, read with an A4 format code would fill a variable with the four characters, COLE.

We can sum up these input considerations by the following rules. (**w** represents the width specification in the A format code)

1. If **w** $>$ variable (word) size, the rightmost characters that can be stored are placed in the memory word. The remaining leftmost characters are lost.
2. If **w** $<$ variable (word) size, the characters are stored left-justified and the remaining positions filled with blanks.
3. If **w** = variable (word) size, then **w** characters are stored in the memory word.

Alphanumeric Data and Compile Time Specifications 209

The field width specification of 4 is recommended for optimum storage utilization, since four characters is the maximum number that can be stored at our memory word. You will use a different value depending on the word size of your computer. You must not fall into the trap of thinking that all alphanumeric fields should be read as a multiple of four characters. Only as many characters as make up the field should be read. If we had a 15-character field, we must not specify a 4A4 field specification. That would read sixteen columns and get us into the next field of data. We should specify a 3A4, A3 or some combination that would read only the 15 characters of data that comprise our total field size.

Output Considerations

Since our computer word holds a maximum of four characters, we can output from 1 to 4 characters from the variable. We do not have to output all the characters contained in the word. But the output characters are selected from the word in a left to right order. For example, if we output a variable containing the 4 characters, COLE, with an A2 format code, we would get the high-order 2 characters, CO, printed.

The following rules apply to the output of alphanumeric data under the A format code.

1. If the width specification (w) is less than the maximum number of characters that can be stored at a variable, the leftmost w characters will be printed. For example, the variable, NAME, containing the 4 characters, JOHN, written with an A3 format code would produce the output results of JOH.
2. If the width specification (w) is greater than the maximum number of characters that can be stored at a variable, the output field will contain the contents of the variable right-justified in the output field. High-order blanks will be inserted in the leftmost positions to make up the size of the output field specifications. For example, the variable, NAME, containing the 4 characters, JOHN, written with an A7 format code would produce the output results of ∅∅∅JOHN (where ∅ represents a blank).
3. If the width specifications (w) is equal to the maximum number of characters that can be stored at a variable, then the output field will contain the contents of the variable. For example, the variable, NAME, containing the 4 characters, JOHN, written under an A4 format code would be printed as JOHN.

The main thing to remember in outputting alphanumeric data is to write the data with the same format specifications that the data was read. Suppose a 15-character field was read as follows:

      READ (5,99) N1, N2, N3, N4, N5
99 FORMAT (5A3)

SMITH MARY LOU

the values would be stored in memory as:

N1	N2	N3	N4	N5
SMI∅	TH∅∅	MAR∅	Y∅L∅	OU∅∅

Suppose the five variables were written as follows:

      WRITE (6,98) N1, N2, N3, N4, N5
98 FORMAT (1H0, 5A4)

The printed output results would be:

      SMI TH   MAR Y L OU

The entire contents, including the padded blanks, would be included in the output field. Data printed as such is hardly legible. What we want to do is to write the variables with the same format specifications that put the characters into memory. Our output statement should be:

WRITE (6,98) N1, N2, N3, N4, N5
98 FORMAT (1H0, 5A3)

Now the printed results would be:

SMITH MARY LOU

This output looks pretty good. If we put only 3 characters into a word during our input operation, we must select only the high-order 3 characters on our output specifications.

Let us illustrate the reading and writing of alphanumeric data with a sample program. Assume again the problem of computing student grades; but instead of a student number, let us include the student's name. Our input items are:

Card Columns	Field Description
1-18	Student name
21-23	Test 1 (Form = NNN)
24-26	Test 2 (Form = NNN)
27-29	Test 3 (Form = NNN)

We wish to compute the sum of the three test scores and their average.

We will read a header card which gives us the number of students in the class. This value is punched in columns 1-2. Our header card will also contain the course number punched in columns 10-21.

We will write a page heading containing the course number. Our individual student output results will be printed as:

Field Description	Print Positions
Student name	1-18
Test 1	25-27
Test 2	33-35
Test 3	41-43
Total	49-54 (Form=NNN.NN)
Average	58-63 (Form=NNN.NN)

Column headings will be printed over each column of data.

The logic involved is basically the same as that for the previous student grade problems we have worked. A flowchart, therefore, is not included. Our program is shown in figure 12.1.

```
 99     FORMAT (1H1,25X,19HSTUDENT GRADES FOR ,3A4/)
 98     FORMAT (1H0,7X,4HNAME, 12X,22HTEST 1   TEST 2   TEST 3,
        *         3X,5HTOTAL, 4X,7HAVERAGE)
 97     FORMAT (I2, 7X, 3A4)
 96     FORMAT (4A4,A2, 2X, 3I3)
 95     FORMAT (1H0, 4A4,A2, 6X, 3(I3,5X), F6.2, 3X, F6.2)
        READ (5,97) NR, TITLE1, TITLE2, TITLE3
        WRITE (6,99) TITLE1, TITLE2, TITLE3
        WRITE (6,98)
        DO 10 I = 1, NR
           READ (5,96) NAM1,NAM2,NAM3,NAM4,NAM5,NTEST1,NTEST2,NTEST3
           TOTAL = NTEST1 + NTEST2 + NTEST3
           AVE = TOTAL / 3.0
           WRITE (6,95) NAM1,NAM2,NAM3,NAM4,NAM5,NTEST1,NTEST2,
        *              NTEST3,TOTAL,AVE
 10     CONTINUE
        STOP
        END
```

Figure 12.1. Sample Program to Illustrate the I/O of Alphanumeric Data

Now let us discuss how we are able to perform decision making (testing) with alphanumeric data.

12.2 Comparing Alphanumeric Data

The question arises, "How does one test a variable for a certain alphanumeric value in FORTRAN?" Of course, the Logical IF statement must be used. But we cannot use a literal constant in the Logical IF statement. That is, we cannot say

IF (CODE .EQ. 'A') GOTO 8

since this syntax is not permitted in ANSI and most FORTRAN compilers. It would be nice if we could assign alphanumeric values to a variable in an Assignment statement such as

LETA = 'A'

but this also is not permitted.

Suppose we wish to process our input data with different routines depending upon a code field in our input record. Maybe we wish to charge different tuition fees depending upon whether the student is a resident or nonresident student. Our input code field might contain the value "R", for resident students and the value, "N", for nonresident students. How can we test the input code field to see what alphanumeric value it contains?

The secret in testing variables for certain alphanumeric values lies in comparing two variables. One variable contains the value being sought and the other variable contains the input value. Then we can use the Logical IF statement such as

IF (KODE .EQ. LETA) GOTO 8

where the variable, LETA, contains the value against which we wish to compare our input value (in variable, KODE).

We can put an alphanumeric value into a variable by reading the value with the A format code. Then the variable, containing the alphanumeric value being looked for, can be compared against our input variable. The statements to determine the tuition rates for our resident and nonresident students might be coded as follows.

```
      C ** READ ALPHANUMERIC VALUES FOR LATER COMPARISON
            READ (5,99) LETR,LETN
      99    FORMAT (2A1)
      C ** READ INPUT RECORD
      1     READ (5,98) ID, KODE,SEMHRS
      98    FORMAT (I6,3X,A1,F4.2)
            IF (ID .EQ. 999999) STOP
            IF (KODE .EQ. LETR) GOTO 10
            IF (KODE .EQ. LETN) GOTO 20
            WRITE (6,97) ID, KODE
      97    FORMAT (1H0,21HINVALID TUITION CODE , I6,3X,A1)
            GOTO 1
      10    . . .
            . . .
      20    . . .
```

One important input rule must be remembered when comparing variables containing alphanumeric values on many computers. The variable names for both variables must be of the same type. Otherwise, an equal condition can never occur in the comparison, even if the contents of the variables are the same. That is, if one name is an integer variable, the other name must be an integer variable; the same holds true for real variable names. This rule does not hold true for all FORTRAN compilers, but is a must for many (especially IBM). You should always try to use integer variable names for alphanumeric data that will be used in comparison operations, since they are more efficient. That is, it takes less time for the computer to compare integer variables than real.

The order of alphanumeric characters as to which ones come before other ones is determined by their **"collating sequence"** (internal representation). The collating sequence of alphanumeric characters is blank ¢ . (+ & ! $ *) : – / , % ? # @ ' = " A thru Z and 0 thru 9. This means that any letter is less than any character digit. The special characters in the collection are less than both letters and digits. The order of the specific character, as to which is less than another, relates to their prescribed order above. That is, a blank is less than a " – "; a " + " is less than a ",".

Don't confuse this collating sequence order with an algebraic comparison. When we compare variables and constants with numeric values, the order is by their algebraic value. When we compare character data, the order is by their internal representation or collating sequence. Whenever variables containing multiple alphanumeric characters are compared, the comparison proceeds from left to right as you would expect. For example "AACD" is less than "ABCE" and "A + " is less than "A* ".

12.3 Introduction to Compile Time Specifications

As you will recall, we learned in chapter 4 that the processing of a FORTRAN program is performed in two phases—compilation phase and execution phase. In the compilation phase the FORTRAN source program you have written is converted into the language of the computer for the second phase.

In chapter 2 we mentioned a category of FORTRAN statements known as specification statements. You probably wondered what this type of statements could possibly be used for, since your knowledge of FORTRAN and computer programming was still quite limited. In chapter 3 we presented the first specification statement for use in every FORTRAN program—the END statement. The END statement was used simply as a signal, in your program, to tell the FORTRAN compiler that was all of your program.

Thus, the END statement is a mandatory specification statement required in every FORTRAN program. The DIMENSION statement (covered in chapter 10) is a mandatory specification statement only when arrays are used to tell the compiler to reserve adjacent storage locations for the given array name. The other specification statements available in the FORTRAN language are generally for optional use in programs. They are available to you primarily for overriding basic default specifications in the FORTRAN language, and to make certain requests to the FORTRAN compiler to perform at compile time.

Wouldn't it be convenient if we could tell the FORTRAN compiler to establish certain initial values in specified storage locations at compile time, rather than having to use Assignment or READ statements at run time? We have this capability with the DATA specification statement. You would probably like to override the default variable types assigned by the initial letter, and to assign instead variable names which have more meaning to you. For example, wouldn't it be clearer to assign the variable name, STKNR, or even, STOCK, to represent an integer data item, stock number, rather than choose a variable beginning with one of the letters I through N? We can do this with the Explicit Type statements. Then there may arise a situation when we wish to equate different variables to the same storage location. The EQUIVALENCE statement permits us to make this equation of different variable names.

So you see, many specification statements are available to help us specify the nature of data in special situations. Or they may be used just to make our job of programming easier. Most programming languages have this capability of making certain requests for the compiler to perform during the compilation of our programs, and to override the basic default specifications built into the language.

Specification statements are often referred to as declarations. They "declare" information to the FORTRAN compiler during the compilation of the program. Specification statements are nonexecutable. That is, they are instructions for the compiler to carry out and not instructions for the computer to perform. Specification statements, therefore, do not generate any machine-language instructions. For this reason they should never be assigned a statement number. Specification statements should be placed at the first part of the program ahead of any executable statement. This placement ensures prior declaration of variables before they are used and also locates them for easy reference. Of course, the END statement must always appear as the last statement in each program.

12.4 The DATA Statement

Up to now you have learned two statements that put values into memory locations—the Assignment statement and the READ statement. Both of these statements are executed during run time of the program. The DATA specification statement tells the FORTRAN compiler to place certain values at the indicated variables during the compilation of the program.

The initial value of a variable or array element is placed within slashes following the variable. If a list of variables are given, followed by a list of values within slashes, there is a one-to-one, left-to-right correspondence between the variables and the given values. In ANSI FORTRAN we cannot initialize values for the entire array by giving only the array name in the DATA statement. Each array element, with an integer constant subscript, and its initial value, must be specifically indicated. Some non-ANSI compilers do allow for an entire array or beginning portions to be initialized by giving only the array name, but not ANSI.

The DATA statement may be used to initial variables with literal (character) values. If fewer than 4 characters are given in the H format specifications for the literal value, the characters will be left-justified and the rest of the storage location filled with blanks.

The general form of the DATA statement is given in figure 12.2.

DATA Statement—General Form

DATA name$_1$/con$_1$/, . . ., name$_n$/con$_n$/

or

DATA name$_1$, . . ., name$_n$/con$_1$, . . ., con$_n$/

where:

name — any **name** represents a variable or subscripted array element (subscripts must be integer constants).

con — any **con** represents any type of valid constant used to assign initial values to the variable or array element. Any **con** can be preceded by **i*** where **i** is an unsigned integer constant that indicates the **constant (con)** is to be specified **i** times.

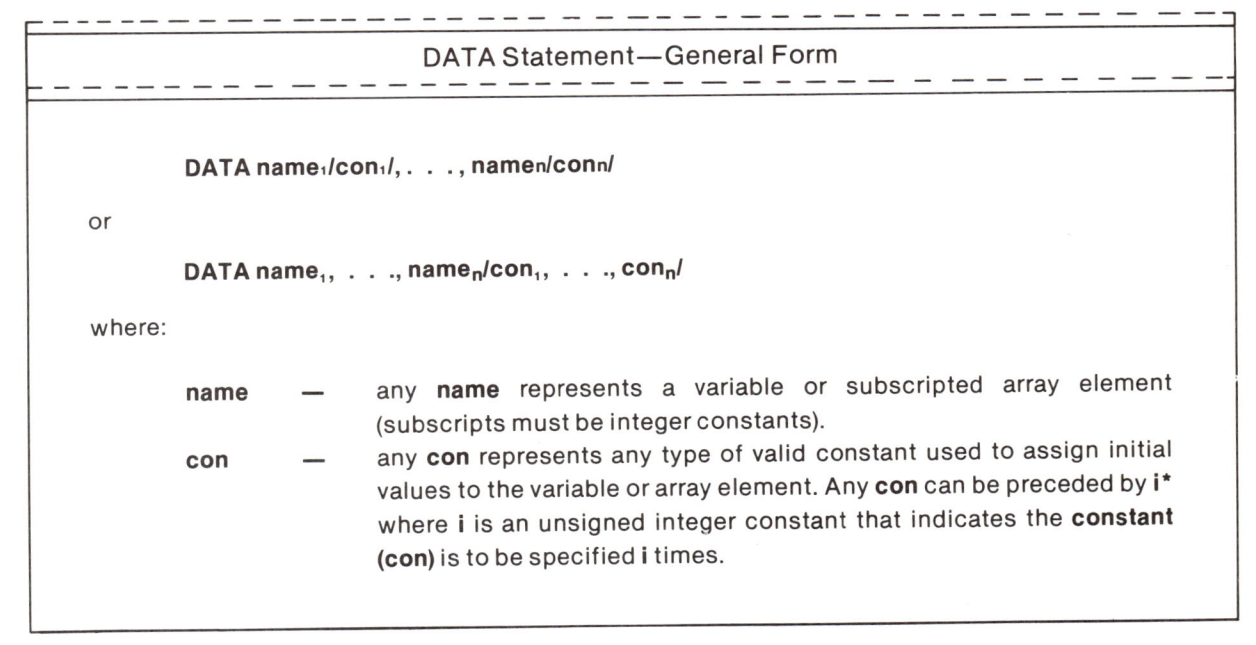

Figure 12.2. General Notation of the DATA Statement

Examples of the DATA statement are:

```
DATA A/1.0/, B/2.0/, C/3.5/, K/3/
DATA D,E/4.0, 5.0/, F/6.0/
DATA A,B,C,D /1.0, 2.0, 3.5, 4.0/, L/1/
DATA ARYX(3)/1.3333/, T(1), T(2), T(3)/3*0.0/
DATA CARD(1)/1HS/, CARD(2)/1HD/, CARD(3)/1HH/, CARD(4)/1HC/
DATA X,Y,Z/3*0.0/
```

The first example places the values 1.0, 2.0, 3.5, and 3 in the variables A, B, C, and K, respectively. The second statement places the values 4.0, 5.0, and 6.0 in the variables D, E, and F, respectively. The third example places the values 1.0, 2.0, 3.5, 4.0, and 1 in the variables A, B, C, D, and L, respectively. The fourth example initializes the third element in array ARYX to 1.3333 and the first three elements in array T to zero. The fifth example stores the literal values of "S", "D", "H", and "C" into the first four elements of the array named CARD. The last example initializes the three variables, X, Y, and Z to zero. (The * says to repeat the constant that follows the * as many times as the constant given before the *.)

The sole function of the DATA statement is to establish initial values in variables and array elements at compile time. Thus, run time statements can be eliminated and a saving in run time can be realized.

The DATA statement is very useful when input items must be compared with alphanumeric values. Suppose various transaction codes in some input records are alphanumeric values. We must decide what routine to use based on the alphanumeric transaction codes. For example, an input code of "A" might mean to add the transaction quantity to our balance on hand and the code of "D" might mean to subtract the transaction quantity from the balance. We must use either the READ or the DATA statement to get an alphanumeric value in a variable. The DATA statement is by far the easiest way to go. For example:

```
      DATA LETA/1HA/,LETD/1HD/
      . . .
      . . .
      . . .
      READ (5,99) KHAR, NQTY
   99 FORMAT (A1,I4)
      IF (KHAR .EQ. LETA) GOTO 10
      IF (KHAR .EQ. LETD) GOTO 20
      . . .
   10 . . .
      . . .
   20 . . .
```

If the input value read under the variable, KHAR, were an "A", a true condition would result in the first Logical IF statement and a transfer would be made to statement number 10.

The DATA statement cannot precede any other specification statement that refers to the same variable or array. If the DATA statement refers to any array name, then the DIMENSION statement for the same array name must appear first.

12.5 The Explicit Type Statements

The Explicit Type statements allow the programmer to declare specific variables to be of a certain type. The Explicit statement is often referred to as the "Type" specification statement, since any variable or array name in FORTRAN can be specified as to a particular type. The Explicit Type statements are very popular and widely used by FORTRAN programmers. Explicitly typing the variables, rather than using the predefined FORTRAN convention of the initial letters, normally gives a program more "linguistic" meaning. Names can be used for variables that tie in more closely with their use in the program, thereby providing greater clarity of meaning. For example, the variable QUANTY may be more meaningful than NR to represent an integer quantity of items in an inventory control program.

The Type statements also provide a simple way of correcting the use of illegal variable names in a program. For example, if an integer type name was used for a real variable a number of times in a program, a single Type statement could be included to tell the compiler to treat the variable as type real instead of type integer. Thus, the programmer is saved the time of changing each of the statements containing the illegal variable name.

Any Type specification statements must appear at the beginning of a program unit and should not be assigned a statement number. A Type statement must precede the DIMENSION statement if the dimension information is not given for the names in the Type statement. Dimension requirements may be provided for array names given in a Type statement. Or the dimension information may be given in a separate DIMENSION statement, but must not be given in both specification statements.

We have been working with integer and real type variables in our programs. There are two Explicit Type statements—known as the INTEGER and REAL Type statements—to declare the variables and array names given in these statements, delineating the particular type. We also have the DOUBLEPRECISION Type statement to declare variables to contain double precision values. There are additional Type statements which must be used to declare the type of variables when we work with some of the other forms of data available in FORTRAN. These Type statements will be covered when we discuss the additional types of constants and data available for use in FORTRAN programs.

INTEGER Type Statement

The INTEGER Type statement tells the compiler to treat all the variables and arrays contained in the list as type integer. The general form of the INTEGER Type statement is given in figure 12.3.

INTEGER Type Statement—General Form

INTEGER name$_1$, name$_2$,..., name$_n$

where:

name — any **name** represents a variable or array name. Any array name may contain its dimension information.

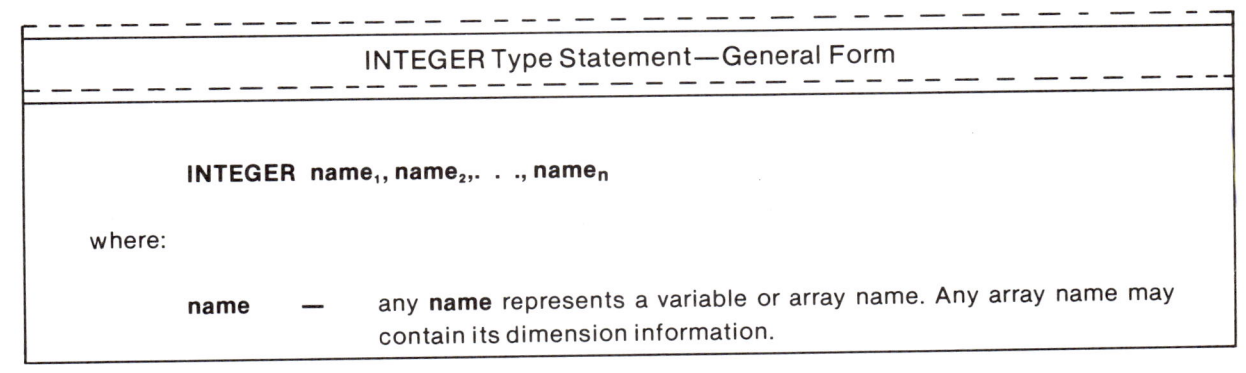

Figure 12.3. General Notation of the INTEGER Type Statement

Examples are:

    INTEGER A, COUNT
    INTEGER X(10), Y(7,5), TOTAL
    INTEGER X1, X2, X3, Z(10,4,2)

The first example declares the variables A and COUNT to be type integer. The second example declares the array named X to be a one-dimensional integer array of ten elements, the array named Y to be a two-dimensional integer array of seven rows and five columns, and the variable TOTAL to be type integer. The third example declares X1, X2, and X3 to be type integer variables and the array named Z to be a three-dimensional integer array of ten rows, four columns, and two levels.

REAL Type Statement

The REAL Type statement tells the compiler to treat all the variables and arrays contained in the list as type real. The general form of the REAL Type statement is given in figure 12.4.

REAL Type Statement—General Form

REAL name$_1$, name$_2$,..., name$_n$

where:

name — any **name** represents a variable or array name. Any array name may contain its dimension information.

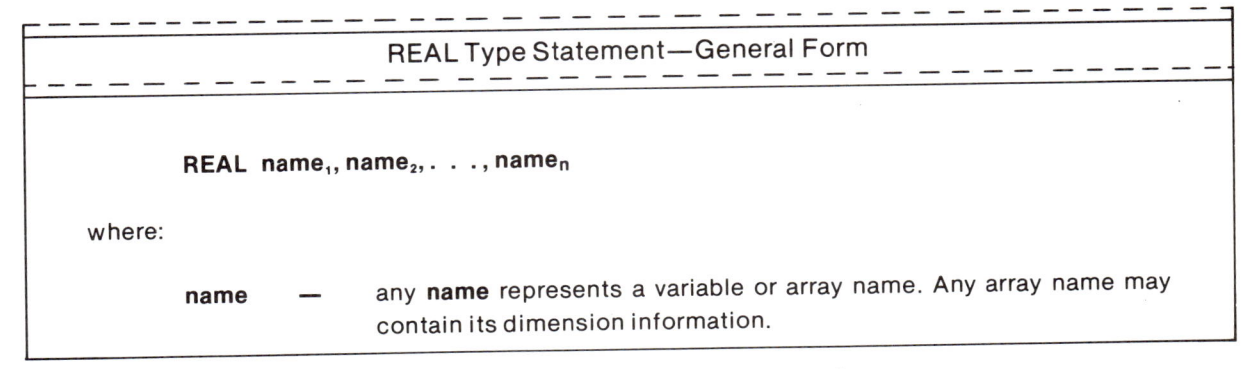

Figure 12.4. General Notation of the REAL Type Statement

Examples are:

> **REAL NTOTAL, J**
> **REAL MATRIX(6,4), LIST(11)**
> **REAL K, L, M(8,3,3), N**

The first example declares the variables NTOTAL and J to be type real. The second example declares the array named MATRIX to be a two-dimensional real array of six rows and four columns and the array named LIST to be a one-dimensional real array of eleven elements. The third example declares K, L, and N to be real type variables and the array named M to be a three-dimensional real array with eight rows, three columns, and three levels.

If the arrays are dimensioned in the INTEGER and REAL type statements as shown in the above examples, this eliminates the need for a DIMENSION statement. This method is the recommended one since it is more documenting and the programmer has to look at fewer statements when checking on the specifications for certain variables or arrays. A separate DIMENSION statement could be used, however, if one so desires. For example:

> **INTEGER X,Y**
> **REAL LIST, MATRIX**
> **DIMENSION X(10), Y(7,5), LIST(11), MATRIX(6,4)**

12.6 The EQUIVALENCE Statement

Suppose a large program was divided among three programmers. One programmer might code the input operations and editing of the input data. A second programmer might code part of the calculation routines. The third programmer might code the rest of the computations and the output operations. When the three programmers attempt to put their routines together, they discover they have used different variables to represent the same data. For example, one programmer may have used L, another LNG, and the third the variable LENGTH to represent the variable for the value of length.

What should they do to make the variables represent the same location in memory? Should they agree on one variable and recode the other two to agree with the first? This operation might involve a considerable amount of time-consuming work. No, they do not have to perform this operation. There is an easy solution—use the EQUIVALENCE statement.

The EQUIVALENCE statement allows the programmer to equate two or more corresponding variables to the same memory location. So no matter which variable is used, you are still referring to the same location in memory. The general form of the EQUIVALENCE statement is given in figure 12.5.

EQUIVALENCE Statement—General Form

EQUIVALENCE (name$_1$,..., name$_n$), (name$_1$,..., name$_n$)

where:

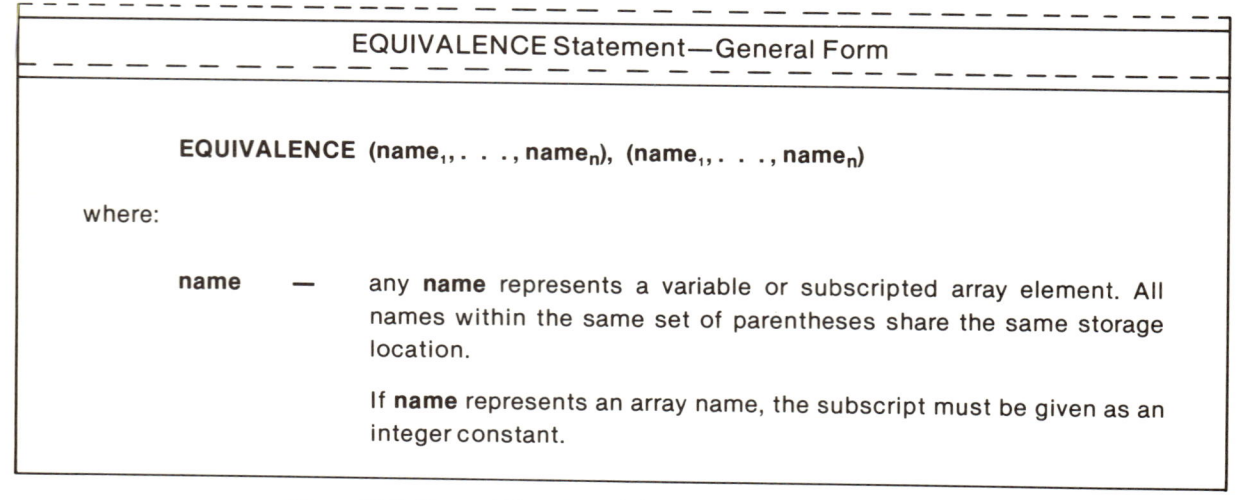

name — any **name** represents a variable or subscripted array element. All names within the same set of parentheses share the same storage location.

If **name** represents an array name, the subscript must be given as an integer constant.

Figure 12.5. General Notation of the EQUIVALENCE Statement

Examples of the EQUIVALENCE statement follow:

> EQUIVALENCE (L,LNG,LENGTH)
> EQUIVALENCE (B,BLK,BLOCK), (A,RA)
> EQUIVALENCE (A(1),A1), (B(5),B5), (C(1),D(1))

The first example equates the variables L, LNG, and LENGTH to the same memory location. The second example equates the variables B, BLK, and BLOCK to the same memory location and the two variables A and RA to the same memory locations. The third example equates the first element in array A with the variable A1, the fifth element in array B with the variable B5, and the start of the two arrays C and D to the same memory location so that these two arrays equate to the same locations.

Remember, equivalence between variables implies storage sharing only, and not mathematical equivalence. The storage location value will change any time a new value is assigned to any one of the equated variables. The EQUIVALENCE statement is used to override the compiler in assigning unique storage locations to each new variable encountered in the program.

Perhaps the same programmer develops a long program. He may unconsciously change the variable name at some point in the program. Whenever this mistake is found, the programmer simply uses the EQUIVALENCE statement to correct his error. One may also use the EQUIVALENCE statement to conserve memory requirements in a large program. The programmer may have used the variables A, B, and C earlier in the program. These variables may cease to be used in later segments of the program. Thus, the variables X1, T2, and S1 which are used in the latter parts of the program may be equated to A, B, and C to save three memory cells.

The saving in memory locations can be significant if large arrays are equated to each other. Suppose the first part of a program used array A of 200 memory cells and was not needed in the latter part of the program; array B of 100 memory cells is used only in the latter part of the program. If these arrays are equated so that they occupy part of the same locations in memory, 100 storage locations are saved.

When using the equivalence statement with elements in multiple arrays, one must be careful that array elements do not extend to the left of the beginning element in the other array. For example:

> DIMENSION A(10), B(7)
> EQUIVALENCE (A(1), B(3))

This is illegal since the first two elements in the array B (B(1) and B(2)) would extend to the left of the first element in array A. This could be disastrous.

You, as a student, may not have the need to use the EQUIVALENCE statement to conserve memory. But on small computers and with large FORTRAN programs, the EQUIVALENCE statement might mean the difference between fitting the program into memory or not.

The EQUIVALENCE specification statement must be placed at the beginning of a program unit ahead of any executable statements and must not be assigned a statement number. This statement must appear after the DIMENSION statement if any array elements are placed in equivalence.

12.7 A Sample FORTRAN Program with Multiple Specification Statements

Since we have introduced the majority of the specification statements, you are probably confused as to their placement in a FORTRAN program. What happens if you use several specification statements in the same program; which one should come first?

Following is a list of the various FORTRAN statements that we have covered so far, and their general sequence in a program unit.

> Explicit Type Statements
> DIMENSION
> EQUIVALENCE
> DATA
> all other executable statements
> END

218 Alphanumeric Data and Compile Time Specifications

FORMAT statements may appear anywhere between the DATA and END statements, but must not appear between continuation statements.

Let us look at a sample FORTRAN program containing some of the specification statements, by way of illustrating their use. Our sample program will read the number and suit of a playing card (poker card, to any gamblers) and interpret its respective rank and suit.

The rank of the playing card is read as a two-character alphanumeric field from card columns 1 and 2. The suit of the card is read as a single alphanumeric character in column 3. For example, a two of hearts would be punched as a 2∅H. The ten of clubs would be punched as a 1∅C. A jack of diamonds would be punched as a J∅D. The ace of spades would be punched as a A∅S. For example:

```
 A S
```

We will interpret the playing card and print its rank and suit on a single print line. If we had read a K, blank, C, then we would print, beginning in print positions one:

RANK AND SUIT OF INPUT CARD IS THE KING OF CLUBS

The coded program is given in figure 12.6.

```
      INTEGER RANKIN, RANK (13), SUIT (4), SUITIN
      INTEGER CR1(13),CR2(13),CSUIT1(4),CSUIT2(4)
      DATA RANK(1)/1H2/,RANK(2)/1H3/,RANK(3)/1H4/,RANK(4)/1H5/,
     *     RANK(5)/1H6/,RANK(6)/1H7/,RANK(7)/1H8/,RANK(8)/1H9/,
     *     RANK(9)/2H10/,RANK(10)/1HJ/,RANK(11)/1HQ/,RANK(12)/1HK/,
     *     RANK(13)/1HA/
      DATA CR1(1)/3HTWO/,CR1(2)/4HTHRE/,CR1(3)/4HFOUR/,CR1(4)/4HFIVE/,
     *     CR1(5)/3HSIX/,CR1(6)/4HSEVE/,CR1(7)/4HEIGH/,CR1(8)/4HNINE/,
     *     CR1(9)/3HTEN/,CR1(10)/4HJACK/,CR1(11)/4HQUEE/,
     *     CR1(12)/4HKING/,CR1(13)/3HACE/
      DATA CR2(1)/1H /,CR2(2)/1HE/,CR2(3)/1H /,CR2(4)/1H /,CR2(5)/1H /,
     *     CR2(6)/1HN/,CR2(7)/1HT/,CR2(8)/1H /,CR2(9)/1H /,
     *     CR2(10)/1H /,CR2(11)/1HN/,CR2(12)/1H /,CR2(13)/1H /
      DATA  SUIT(1)/1HC/,SUIT(2)/1HD/,SUIT(3)/1HH/,SUIT(4)/1HS/
      DATA CSUIT1(1)/4HCLUB/,CSUIT1(2)/4HDIAM/,CSUIT1(3)/4HHEAR/,
     *     CSUIT1(4)/4HSPAD/,CSUIT2(1)/1HS/,CSUIT2(2)/4HONDS/,
     *     CSUIT2(3)/4HTS  /,CSUIT2(4)/4HES  /
   99 FORMAT (A2,A1)
   98 FORMAT (1H0,19HINVALID INPUT RANK ,A2)
   97 FORMAT (1H0,19HINVALID INPUT SUIT ,A1)
   96 FORMAT (1H0,35HRANK AND SUIT OF INPUT CARD IS THE ,A4,A1,
     *        4H OF ,2A4)
  105 READ (5,99) RANKIN, SUITIN
      DO 10 I = 1, 13
         IF (RANKIN .EQ. RANK(I) ) GOTO 20
   10    CONTINUE
      WRITE (6,98) RANKIN
      STOP
   20 DO 25 J = 1, 4
         IF (SUITIN .EQ. SUIT(J) ) GOTO 30
   25    CONTINUE
      WRITE (6,97) SUITIN
      STOP
   30 WRITE (6,96) CR1(I),CR2(I),CSUIT1(J),CSUIT2(J)
   40 STOP
      END
```

Figure 12.6. Sample Program to Illustrate Specification Statements

The beginning programmer may generally find little need for including many of the various specification statements in his program. They can, however, make your job of programming easier, especially with large or complex programs. They also perform a specific function when the need arises in certain problems.

12.8 Nonstandard Language Extensions (IMPLICIT Statement)

Assignment of Literal Constants

The B6700 and CDC 6000 compilers permit the assignment of literal constants in an Assignment statement. On the B6700 compiler one may assign a literal constant to a variable by using the H format code or the quote. For example:

$$\text{LIST1} = \text{3HSUM}$$
$$\text{LIST2} = \text{'TOTAL'}$$

The result is the same as if the literal was read with an A format code. That is, the literal is left-justified in the computer word. If less characters are assigned than the size of the word, the remaining low-order positions are filled with blanks.

On the CDC 6000 compilers one may assign a literal to a variable with the H format code as follows:

$$\text{LIT1} = \text{3HSUM}$$

One may also assign a literal to a variable with the R (Hollerith Right-justified) format code. For example:

$$\text{LIT1} = \text{3RSUM}$$

With the R format code on the CDC compilers the data is stored right-justified with octal (base 8) zeroes in any excessive high-order positions.

The Character (C) Format Code

On the B6700 and H6000 compilers alphanumeric data may be read with the C format code. The C format code stores the input alphanumeric field *right-justified* in a computer word. Any excessive high-order positions are padded with hexadecimal (base 16) zeroes on the B6700 or octal (base 8) zeroes on the H6000. The general form of the C format code is:

$$rCw$$

where **r** is an optional repetition factor from 1 to 255. **w** represents the width of the input or output field. For example, a data field containing the two letters AB read under a C2 format code would store the two letters "AB" in the two low-order positions in a word and fill the excessive high-order positions with hexadecimal (or octal) zeroes.

You may wonder why we need the C format code in addition to the A format code? The reason lies in being able to compare alphanumeric data in FORTRAN on these computers, as is discussed below.

Comparing Alphanumeric Data on B6700 and H6000 Computers

When alphanumeric data is read with an A format code on the B6700 and H6000 computers, the data affects the sign bit in the word. The sign bit is a special bit in the high-order position of the word to determine the sign of the value contained in the word. If we are comparing alphanumeric data for an equal or not equal condition, there is no problem. But if we are comparing for a less than or greater than condition with our alphanumeric data, then we will not get a correct result.

To avoid setting the sign bit incorrectly in the computer word we must read the alphanumeric data with the C format code to store the data right-justified in the word. Also, we must store one less character than can be held by the computer word to avoid filling the entire word with alphanumeric characters. Since the length of a computer word on the B6700 and H6000 is 6 characters, the maximum number of characters read with the C format code should be 5, i.e., C5.

This restriction of reading alphanumeric data with the C format code pertains only to alphanumeric data used in comparison operations. If we do not need to compare the alphanumeric data for a less than or greater than condition, we can use the A format code to read and store the data.

The Right-justified (R) Format Code

The Right-justified (R) format code is used on the CDC 6000 to transmit alphanumeric data. The basic difference between the R and A format codes is that the R format code causes character string data to be right-justified in a variable. The general form of the R format code is:

$$rRw$$

where **r** is an optional repetition factor from 1 to 255. **w** represents the width of the input or output field.

On *input* **w** characters are read from the data field. If the **w** specifications exceed the maximum allowed characters in a variable, the high-order characters are truncated. If **w** is less than the maximum characters allowed, the characters are right-justified in the variable and *high-order zeroes* are added to the left. Remember that the A format code causes the data to be left-justified, and blanks are added to the right for a "short" character string.

On *output,* if the **w** specifications exceed the number of characters stored in a variable, blanks are added in the high-order positions. If the number of characters in a variable exceed **w**, then the rightmost **w** characters are printed. Remember that the A format code extracts the characters from the left in a variable on output.

IMPLICIT Statement

Another specification statement is available on many compilers that allows us to declare the type of variables and arrays by specifying that names beginning with certain letters be of a certain type. This statement is called the IMPLICIT statement. The individual letters are given within a set of parentheses. A range of letters can be specified by giving the first and last letters separated by a minus sign (−). The general form of the IMPLICIT statement is given in figure 12.7 shown on the next page.

Examples of the IMPLICIT statement are:

IMPLICIT INTEGER (A-D,S)
IMPLICIT REAL (J,M)
IMPLICIT REAL (K-N), INTEGER (R-W)
IMPLICIT INTEGER (E,H-J,T), REAL(N)

The first example declares that all variables which begin with the letters A, B, C, D, and S to be integer. The second example declares the variables that start with the letters J and M to be type real. The third example declares that the variables which begin with K, L, M, and N to be type real, and the variables that begin with the letters R through W to be type integer. The fourth example declares that the names beginning with the letters E, H, I, J, and T to be type integer and the variables starting with the letter N to be type real.

IMPLICIT Statement—General Form

        IMPLICIT type (letter, . . ., letter), type (letter, . . ., letter)

where:

    type — specifies any of the permissible type variables. **Type** may be followed by *s where s represents an optional length.

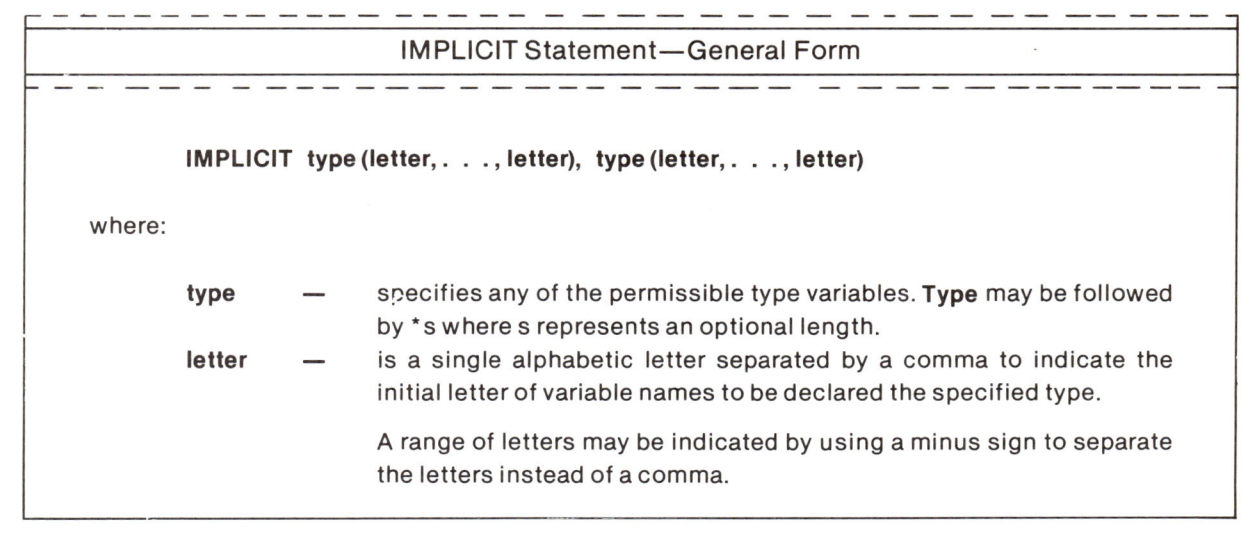

    letter — is a single alphabetic letter separated by a comma to indicate the initial letter of variable names to be declared the specified type.

             A range of letters may be indicated by using a minus sign to separate the letters instead of a comma.

Figure 12.7. General Notation of the IMPLICIT Statement

Only one IMPLICIT statement is allowed in a FORTRAN program or subprogram. When the IMPLICIT statement is used, it must be the *first* statement in the program. The symbol **$** is considered to be the last letter in the alphabet (i.e., after Z) on compilers which allow the **$** to be used as a letter.

The IMPLICIT statement provides a fast and simple way of overriding the standard type and length of FORTRAN variables. But what if a programmer forgets that he has declared a certain letter to be a different type variable than established by the predefined convention of the FORTRAN language? The results could be disastrous, and cause much time-consuming debugging effort. The IMPLICIT statement, therefore, is not recommended for use in changing the type of specific variable names as to initial letters; use the Type statements to specifically identify the variables that are to be a certain type. When both the IMPLICIT statement and an Explicit statement have a conflict as to the same first letter and a specific variable, the Explicit statement overrides the specifications and determines the type of the variable.

Length Options and Initial Values with the Type Statements

Certain compilers permit the programmer to specify different length options and to include initial values of variables/arrays given in the Type statements. The length option refers to different size storage locations such as a halfword or doubleword which can be specified for variables on some compilers. The initial values refer to assigning values to the given variables at compile time as performed in the DATA statement.

The IMPLICIT statement may also specify different length storage locations. But initial values cannot be specified in the IMPLICIT statement since it only denotes the type of variables/arrays by their initial letter.

12.9 REVIEW QUESTIONS

1. The term alphanumeric refers only to letters. (True/False)
2. All characters acceptable by a computer may be read with the A format code. (True/False)
3. The maximum number of alphanumeric characters that may be stored at a variable or array element is determined by the size of a storage location (word) on different computers. (True/False)
4. If more characters are read than can be stored in a word of storage, the excess rightmost characters are lost. (True/False)
5. The comparison of alphanumeric data must be done with two variables. (True/False)

6. What is the purpose of specification statements in FORTRAN?
7. Which specification statement must be in all FORTRAN programs?
8. What is the primary function of the DATA statement?
9. What is the purpose of the Explicit Type statements?
10. What is the function of the EQUIVALENCE statement?
11. Where are specification statements placed in a FORTRAN program?

12.10 EXERCISES

1. Determine the proper A format codes to read the following data fields. Assume all fields begin with the first given letter. The symbol "∅" is used to indicate blanks at the end of the field. Assume a maximum of 4 characters per word.
    a) COLE J W PERRY∅∅∅
    b) GO
    c) L602M3
    d) 60 WATT BULBS∅
    e) SAN ANTONIO, TEXAS

2. Determine the proper A format codes to read the following data fields. Assume all fields begin with the first letter. The symbol "∅" is used to indicate blanks at the end of the field. Assume a maximum of 4 characters per word.
    a) JONES JERRY T, JR.∅∅∅∅
    b) STOP HERE
    c) K0329M7
    d) HEAVY DUTY BATTERIES
    e) COLORADO SPRINGS, CO∅

3. Construct a DATA statement to initialize the variables A, X, and M to 3.75, 8.6, and 12, respectively.

4. Construct a DATA statement to initialize the variable IY to 10 and the three elements in the one-dimensional array, TAB3, to all ones.

5. Construct an Explicit Type statement to declare the variables A and B to be integer, and the array named X to be a one-dimensional integer array of 100 elements.

6. Construct an Explicit Type statement to declare the variables COST, D, and E to be integer, and the array named Y to be a two-dimensional integer array of twenty rows and ten columns.

7. Construct an Explicit Type statement to declare the names I, J, and INTRST to be real variables, and the array named LIST2 to be a one-dimensional real array of seven elements.

8. Construct an Explicit Type statement to declare the names LENGTH, M, and M2 to be real variables, and the array named MATX to be a two-dimensional real array of ten rows and four columns.

9. Form an EQUIVALENCE statement to equate the variables L, L1, L2, and LX to the same storage location.

10. Form an EQUIVALENCE statement to equate the variables X, X1, and the first element of the single-dimensioned array XX to the same storage location.

11. Locate and explain the errors in the following specification statements.
    a) DATA /A/ 5.5,/B/ 6.3
    b) REAL M,N   INTEGER X,Y
    c) INTEGER XXARRAY(20)
    d) INTEGER YARRAY (100)/100*0./
    e) REAL I,J,K /3*9.0/

   f) **EXPLICIT REAL NSUM, NTOTAL**
   g) **EQUIVALENCE A,AX,ZA**
12. Locate and explain the errors in the following specification statements.
  a) **DATA /5.0/A,/7.2/B**
  b) **REAL YMATRIX(17,5)**
  c) **INTERGER T, U, V**
  d) **INTEGER FGLOBAL/10./**
  e) **REAL K/1/,N/2/**
  f) **EQUIVALENCE (B/BA/BB),(K/KA)**
  g) **DATA X,Y/0,1/**

12.11 PROGRAMMING PROBLEMS

1. Write a program to read a single data card with your name punched in columns 1-25 and your date of birth punched in columns 30-46. Your date of birth should be punched as month, day, year (example, December 16, 1937). Print these two values on a double-spaced line with your name beginning in position 10 and date of birth beginning in position 50.

2. Write a program to read and list an inventory control card file. The input values are punched in the following card columns.

Card Columns	Field Description
1-7	Part number (Form is alphanumeric)
10-35	Description (Form is alphanumeric)
40-43	Quantity on hand (Form = NNNN)
44-50	Unit cost (Form = NNNN.NN)

List the input file according to the following print positions.

Field Description	Print Positions
Part number	3-9
Description	15-40
Quantity on hand	46-49
Unit cost	55-61

Terminate the read loop with a trailer card containing a value of 9999999 for part number. Include a report title and appropriate column headings.

3. Write a program to prepare a list of employees who have had 30 or more years of service with XYZ Company. The data for each employee is contained in a punched card in the following layout.

Card Columns	Field Description
1-20	Employee name
21-30	Department (Form is alphanumeric)
31-32	Years of service (Form = NN)

All employees who have had 30 or more years of service with the company will be recognized at the company's annual employee awards banquet. Select all employee records containing a value in the years-of-service field with 30 or greater.

Print the selected records double-spaced according to the following print specifications.

Field Description	Print Positions
Employee name	1-20
Years of service	30-31
Department	40-49

Terminate the loop with a trailer card containing the value 99 for Years of Service. Include a report title and appropriate column headings.

4. Write a program to prepare a list of blood donors with a specific type of blood. The program should read a header card containing the type of blood in columns 1-4 which is compared with each of the donors in the card file. (Examples: APOS—A Positive; ANEG—A Negative; OPOS—O Positive; ONEG—O Negative, etc.)

The donor record is layed out according to the following fields.

Card Columns	Field Description
1-6	Donor number (Form = NNNNNN)
7-26	Donor name
28-31	Donor blood type (Form is alphanumeric)
32-38	Telephone number (Form is alphanumeric)

Read the donor file and compare the blood type of each donor with the blood type specified in the header card. Prepare a list of donors with the specified blood type. The roster will be used to call donors when the supply for that type of blood runs low.

The list of donors with the specified blood type should be printed double-spaced according to the following specifications.

Field Description	Print Positions
Donor number	3-8
Donor name	14-33
Telephone number	40-46

Include a report title indicating the type of blood donors. Include appropriate column headings.

5. Write a FORTRAN program to compute the total, average, and percentile ranking of five test scores for a class of ten students. The input data will be read in the following specifications.

Card Columns	Field Description
1-18	Name
19-20	Blank
21-25	Test 1 (Form = NNN.N)
26-30	Test 2 (Form = NNN.N)
31-35	Test 3 (Form = NNN.N)
36-40	Test 4 (Form = NNN.N)
41-45	Test 5 (Form = NNN.N)
46-80	Blank

Compute the total and average for each student's five test scores. Next, sort the records into descending sequence by average. Compute the percentile ranking of each student by dividing each student's average by the top student's average and multiplying by 100.

Print output results as follows:

Field Description	Print Positions
Name	1-18
Test 1	21-25
Test 2	29-33
Test 3	37-41
Test 4	45-49
Test 5	53-57
Total	61-66
Average	71-76
Percentile	81-86

Print each output record double-spaced. Include appropriate column headings.

Use an implied DO to control the number of column values read on input for each student. You must use three nested loops to perform the sort. The inner loop will control the exchange of items between the two records to be exchanged in the array.

6. Write a program to draw a specific symbol depending upon the specified type read from a punched card. The type of symbols to be drawn will be a line, a square, a rectangle, and a triangle. Each punched card will indicate what symbol to draw as follows:

Field Description	Card Columns	Input Values
Type symbol	1-4	LINE
		SQUA
		RECT
		TRIA

For example, if the value "LINE" is read, you will draw a line such as **********. If the value "TRIA" is read, you will draw a triangle such as

```
       *
      * *
     *   *
    *     *
   *       *
  *         *
 *********
```

Use a DATA statement to specify the four-character literals which are used in the Logical IF statement to be compared with the input variable. For example:

DATA ISQ /4HSQUA /
IF (ISYMBL .EQ. ISQ) WRITE (6,98)

The output results will denote the symbol being drawn followed by the drawn symbol. For example (the value "RECT" is read):

```
RECTANGLE          ***************
                   *             *
                   *             *
                   *             *
                   ***************
```

The denoted type symbol will begin in print position 1. The drawn symbol will begin in print position 20. The dimensions of the printed symbols will be:

Symbol	Dimension
line	15 units long
square	7 units long by 5 units high
rectangle	15 units long by 5 units high
triangle	base will be 9 units long and the height will be 5 units high

Triple-space between each drawn symbol.

7. Write a program to process a file of transaction cards against a 20-element, one-dimensional array named X. The array will be loaded from a single data card (i.e., 20F4.1). The elements in the array are to be summed. Print each value in the array on a double-spaced line in the form: X(i) = n where i is the array element number (1-20) and n is the value of that respective array element. After the array is printed, write the accumulated total of the elements preceded by the literal "TOTALX = " beginning in print position 11.

Next, read a file of transaction cards to update the array values. The record layout of each transaction card is:

Card Columns	Field Description
2-3	Array element to be updated (Form = NN)
10	Type of update character (A=Add S=Subtract M=Multiply D=Divide)
15-20	Numeric value to be used in the update (Form = NNN.NN)

As each transaction card is read, check the update code and apply that arithmetic operation with the input value against the specified input array element. For example, a transaction card:

3   S   50.50

would mean to subtract the value, 50.50, from the value in the third array element. The transaction card:

14   D   100.00

would mean to divide the value in the fourteenth array element by 100.00. The transaction update codes of A and M should be self-explanatory.

After all the transaction cards have been read and applied against the array X, accumulate the new sum of the array elements. Then print each element of the updated array as before on a double-spaced line (i.e., X(i) = n). Then print the literal title "TOTALX = " as before, followed by the new accumulated sum.

Use a DATA statement to initialize the alphanumeric values of A, S, M, and D. Hint: be sure to assign integer variables to each letter (or the same type of variable assigned to type of update) to ensure an equal condition in the Logical IF statements.

Subroutine Subprograms and the COMMON Statement 13

13.1 Introduction to the Use of Subroutine Subprograms

Programs are often written which include various routines that are repeated at various locations throughout a program. It would be convenient if we could write a routine only once and call it whenever we need its use in our program. A subprogram allows us to do just that. We can write our routine as a subprogram and refer to it whenever we wish in our program. This has the same effect as writing the routine at each appropriate location in the program.

A *subprogram* may be defined as a set of instructions contained in a separate program that performs a specific task under the direction of another program. The subprogram is executed when its name is used in a calling program. The subprogram is designed to solve only one small part of a problem, such as sorting, searching, calculations, output operations, or whatever other operation we wish. It is a complete program in itself and plays a subrole in our problem solution. Hence, the name *subprogram* makes a very appropriate title.

There are two basic types of subprograms in FORTRAN:

1. SUBROUTINE subprogram
2. FUNCTION subprogram

This chapter will concern itself solely with subroutine subprograms which can be used for any type of routine needed. Function subprograms are used primarily to compute mathematical functions whereby a single resulting value is produced, and are covered in chapter 15.

The two types of subprograms are alike in many ways, and in many ways they are different. But each type of subprogram must have at least four statements. These statements are:

1. A SUBROUTINE or FUNCTION statement to identify the type of subprogram.
2. At least one executable statement (e.g., READ, WRITE, Assignment).
3. A RETURN statement.
4. An END statement.

An END statement must be included as the last statement in every subprogram to inform the FORTRAN compiler where the subprogram ends. Thus, each subprogram is a separate and complete program within itself. It may contain any number of source statements. A subprogram may also contain any statement that a main problem does with the exception of another subprogram statement (i.e., SUBROUTINE or FUNCTION).

Whenever the term *subprogram* is used in this chapter, we are referring to both types of subprograms in general. When the term *subroutine* is used, we are referring to the specific type of subroutine subprograms.

The program that creates the whole program and normally calls the subprograms is referred to as the **main program** or **mainline program.** Only one main program is allowed in a FORTRAN job. One or more subprograms may be combined with a main program to form the whole program designed to solve a given problem. The subprograms are normally placed in the job deck after the main program.

The term *calling program* refers to the program unit that is invoking (using) a subprogram. The term *program unit* means any type of program (i.e., main program or a subprogram).

The CALL statement in the calling program makes a subroutine subprogram available for use. Whenever the subroutine is referenced, control of execution is directed to the subprogram. After the execution of the subroutine is completed, control is turned back to the calling program. The compiler automatically provides the necessary transfers of control and communication links. You need only to specify the subprogram name, the information you want passed to the subprogram, and the information you want returned.

Assume we have a program that needs a sorting routine at several locations in a program. Suppose we write the sorting routine as a subroutine named SORT. The concept of transferring control to and from the subprogram is illustrated in figure 13.1. The arrows indicate the transfer of execution.

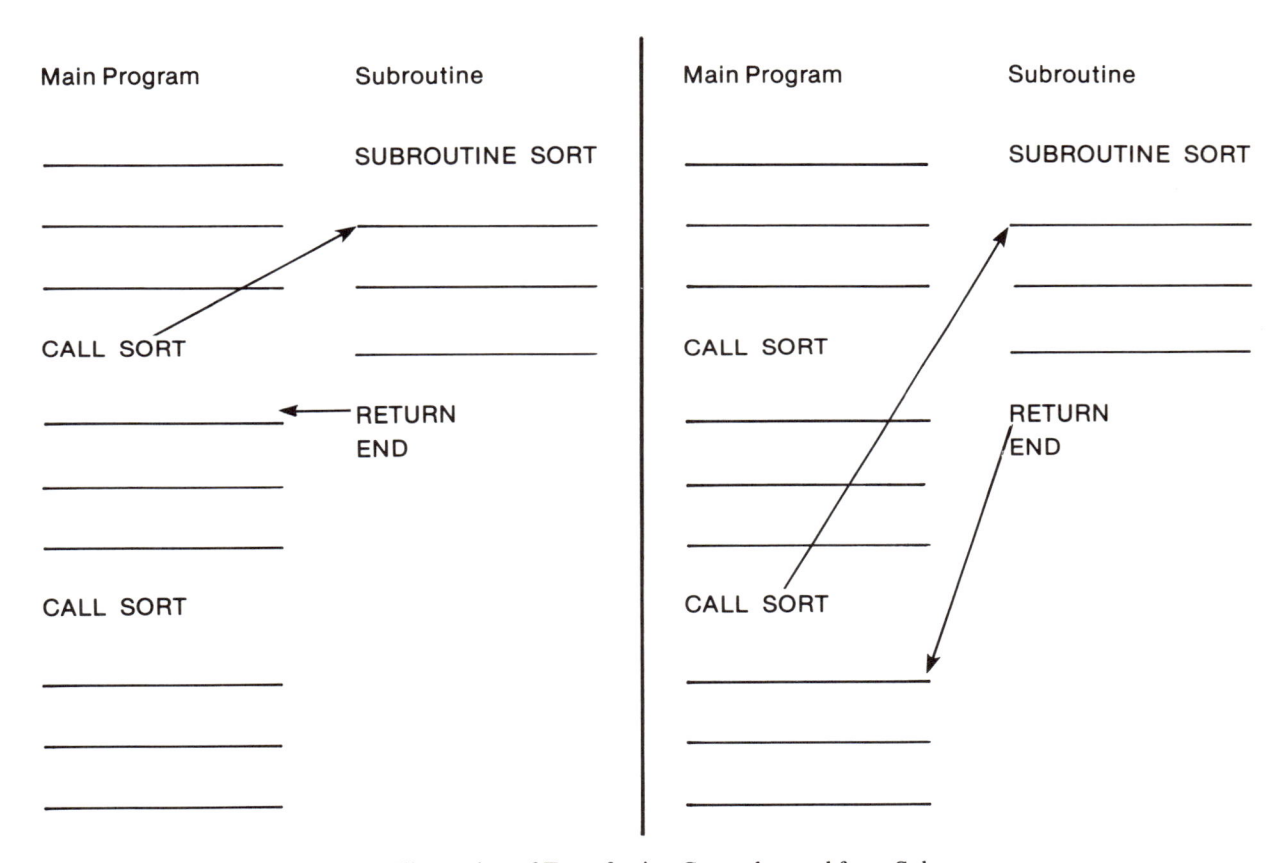

Figure 13.1. Illustration of Transferring Control to and from Subprograms

Control is transferred to the first statement in the subroutine whenever its name is referenced in the CALL statement. Control is turned back to the calling program at the statement after the CALL.

Let us define our problem in a little more detail. Suppose we wish to read the arrays X, Y, and Z and sort the items in each of these arrays. The following "skeleton" programs in figure 13.2 on the next page illustrate the two methods we might use to produce our desired operations.

If we coded our program in the typical in-line (sequential order) fashion, we would have to write our sort routine three times. The subprogram approach keeps us from having to rewrite long or complex program segments. We can simply write our sort routine, SORT, one time and include it in our source deck as a subroutine subprogram. So whenever we need to use it, we simply call it by its assigned name. The sav-

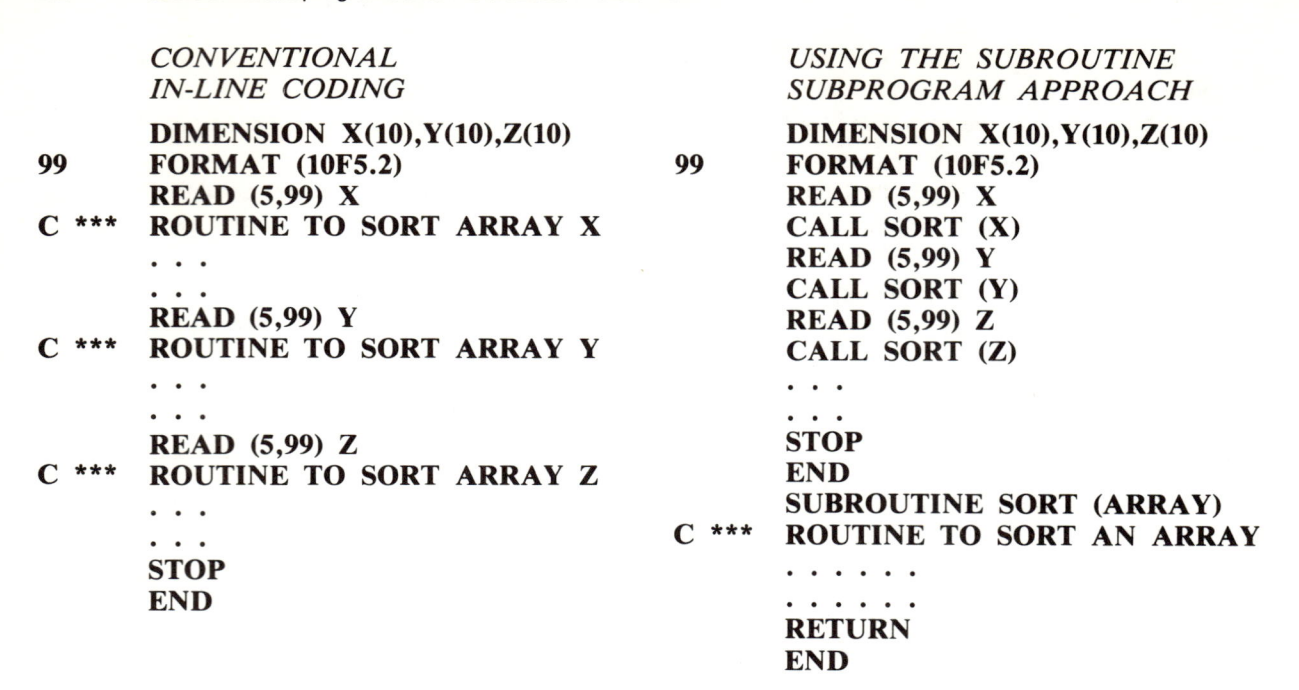

```
              CONVENTIONAL                              USING THE SUBROUTINE
              IN-LINE CODING                            SUBPROGRAM APPROACH
              DIMENSION X(10),Y(10),Z(10)               DIMENSION X(10),Y(10),Z(10)
     99       FORMAT (10F5.2)                   99      FORMAT (10F5.2)
              READ (5,99) X                             READ (5,99) X
     C ***    ROUTINE TO SORT ARRAY X                   CALL SORT (X)
              . . .                                     READ (5,99) Y
              . . .                                     CALL SORT (Y)
              READ (5,99) Y                             READ (5,99) Z
     C ***    ROUTINE TO SORT ARRAY Y                   CALL SORT (Z)
              . . .                                     . . .
              . . .                                     . . .
              READ (5,99) Z                             STOP
     C ***    ROUTINE TO SORT ARRAY Z                   END
              . . .                                     SUBROUTINE SORT (ARRAY)
              . . .                             C ***   ROUTINE TO SORT AN ARRAY
              STOP                                      . . . . . .
              END                                       . . . . . .
                                                        RETURN
                                                        END
```

Figure 13.2. Illustration of Two Approaches in Providing a Sort Routine

ings in coding and keypunch time become increasingly important to us as we increase the number of times we need to refer to the subroutine.

The beginning of our subroutine subprogram is identified as follows:

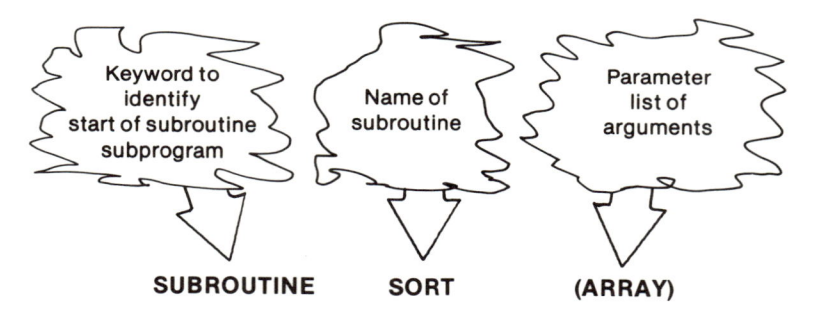

- Keyword to identify start of subroutine subprogram → **SUBROUTINE**
- Name of subroutine → **SORT**
- Parameter list of arguments → **(ARRAY)**

A subroutine name is formed the same as other names in FORTRAN (i.e., 1-6 characters, and beginning with a letter). The type of name chosen does not have to conform to the integer/real convention, since no values are ever associated with the name. A variable or array, however, must not have the same name as any subroutine name used by the program.

You may wonder how we can sort different arrays with the same routine. The answer lies in the ability to pass different values called *arguments* for the subprogram to use. The subprogram is written in a general fashion and sorts whatever array is passed to it. The arguments as shown in our example are placed within a pair of parentheses following the subprogram name. Thus, the values (X), (Y), and (Z) are our different arguments in our calls to the subroutine SORT. The SUBROUTINE statement receives each array and places the items in its array named ARRAY. The sorted items in ARRAY are returned to the respective array that was passed to the subprogram.

The calling program is usually the main program, but may also be another subprogram. The CALL statement invokes (starts execution of) the subroutine identified by its name in the CALL.

Consider the following illustration of the CALL statement.

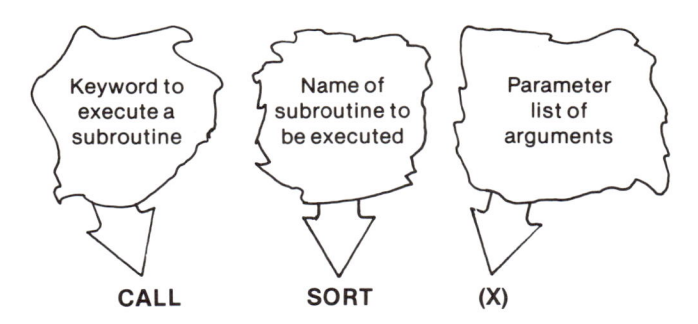

```
      Keyword to      Name of         Parameter
      execute a       subroutine to   list of
      subroutine      be executed     arguments
           ↓               ↓               ↓
         CALL            SORT            (X)
```

This statement calls the subroutine named SORT and passes the array X as an argument to the subprogram. A subprogram may call other subprograms, but cannot call itself. If a subprogram called itself, this would result in an endless loop on many computers. Neither can we have two subprograms which call each other. With these two restrictions, subroutines may include calls of other subroutines to practically any depth we desire.

A RETURN statement in the subprogram tells the computer when to turn control back to the program that called it. A RETURN statement must be used and not a STOP statement. The STOP statement would tell the computer to terminate processing for our job altogether.

The use of subprograms provides many advantages in the designing and developing of a program to solve various problems. These advantages are:

1. Reduction in the amount of required work. Since our routine is included at only one location in our program, there is a substantial reduction in the amount of coding and keypunching performed.
2. Modular programming techniques can be incorporated. Now we can code each task as a different subroutine. Each subroutine can be thought of as a different module in the development of our program. This module technique provides a highly structured program approach to problem solving.
3. Debugging is easier. It is usually far easier to debug an individual subprogram than it would be to debug a routine included in line in the main program. There are fewer statements to look at, thus the logic can be more easily understood.
4. Division of work simplified. Suppose a program must be developed by three people. The major tasks can be identified and coded as subprograms. Thus the tasks can easily be divided among the programmers. In contrast, the development of one main-line program by more than one person becomes quite a hassle.
5. Amount of required storage locations reduced. Every executable statement in a program requires a certain amount of memory. By using a subprogram instead of repeating a routine in a program, fewer statements are required and thus, less memory.
6. Once written and debugged, the subroutine can be used repeatedly. Programming errors in the reuse of the same routine are eliminated.
7. Incorporation of library-supplied subprograms. Those subprograms which are frequently used by many programmers may be included in a library of subprograms. The library of subprograms is normally included with the FORTRAN compiler and is available during compilation of the program. The programmer does not have to code the subprogram, but merely calls it in his program. Thus, the subprogram is copied from the library and included in the program.

The statement numbers, variable names, and array names are completely independent of the statement numbers and symbolic names used in the other program units. Therefore, we may have duplicate statement numbers, variables, and array names between program units without any harm. Statement numbers and variables/arrays are assigned unique locations within each program unit. Thus, the variable X in a subprogram refers to a different storage location than the variable X in another program unit. This is why we must pass arguments in a parameter list when we communicate values between program units.

Subprograms are one of the most powerful tools available to the programmer. Knowledge of their use and capability will greatly increase your competence in programming. Now that the basic concepts of subprograms have been discussed, let us examine the subroutine related statements in detail.

13.2 The SUBROUTINE, RETURN, and CALL Statements

The SUBROUTINE statement identifies the beginning of a subroutine subprogram. Thus, this statement must be the first one in the subroutine. The SUBROUTINE statement begins with the word SUBROUTINE, followed by its assigned name. A parameter list, if used, is enclosed within parentheses and follows the subroutine name. Figure 13.3 presents the general form of the SUBROUTINE statement.

SUBROUTINE Statement—General Form

SUBROUTINE name(darg$_1$, . . ., darg$_n$)

where:

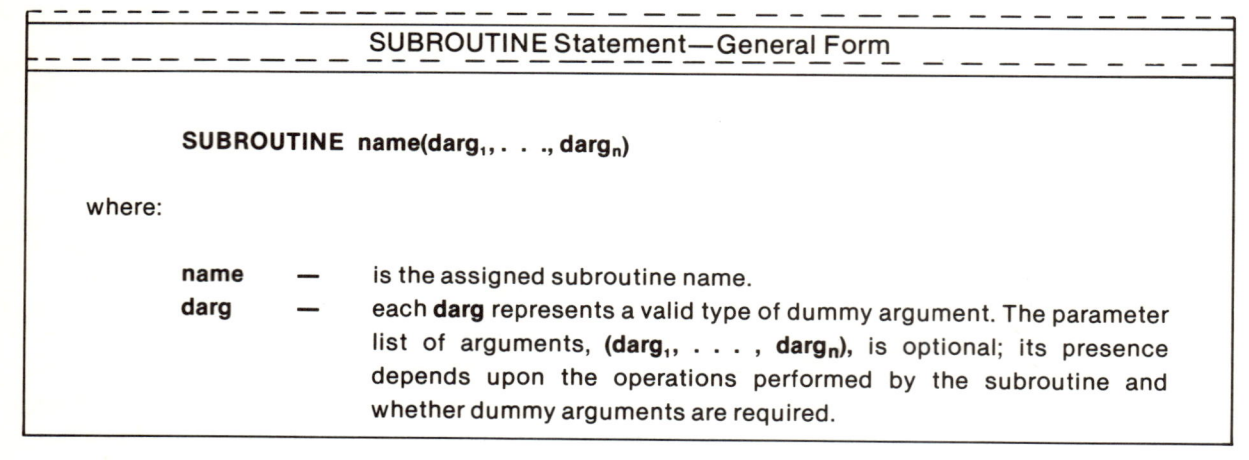

- **name** — is the assigned subroutine name.
- **darg** — each **darg** represents a valid type of dummy argument. The parameter list of arguments, (darg$_1$, . . . , darg$_n$), is optional; its presence depends upon the operations performed by the subroutine and whether dummy arguments are required.

Figure 13.3. General Notation of the SUBROUTINE Statement

The RETURN statement terminates execution of the subprogram and returns control to the calling program. There can be more than one RETURN statement in a subprogram, just as there can be more than one STOP statement in the main program. The RETURN statement consists solely of the keyword, RETURN.

Figure 13.4 gives the general form of the RETURN statement.

RETURN Statement—General Form

RETURN

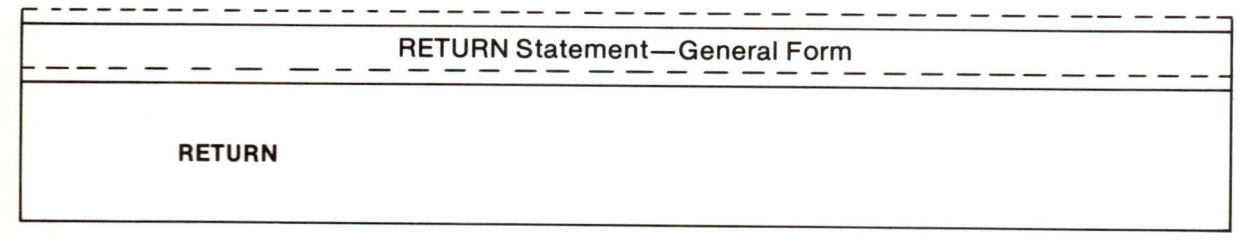

Figure 13.4. General Notation of the RETURN Statement

A subroutine is called into use by the CALL statement. This statement contains the word CALL followed by the subroutine name. A parameter list of arguments, if required, is enclosed in parentheses and follows the subroutine name. The CALL statement may be included in the main program or in any subprogram that requires use of the subroutine.

Figure 13.5 gives the general form of the CALL statement.

```
┌─────────────────────────────────────────────────────────────────────┐
│                    CALL Statement—General Form                      │
╞═════════════════════════════════════════════════════════════════════╡
│                                                                     │
│         CALL  name (arg₁, . . ., argₙ)                              │
│                                                                     │
│   where:                                                            │
│                                                                     │
│         name    —   is the name of the subroutine to be executed.   │
│         arg     —   each arg represents an actual argument to be    │
│                     passed to the subroutine. The parameter list,   │
│                     (arg₁, . . ., argₙ), is optional and is         │
│                     provided only if required by the subroutine.    │
└─────────────────────────────────────────────────────────────────────┘
```

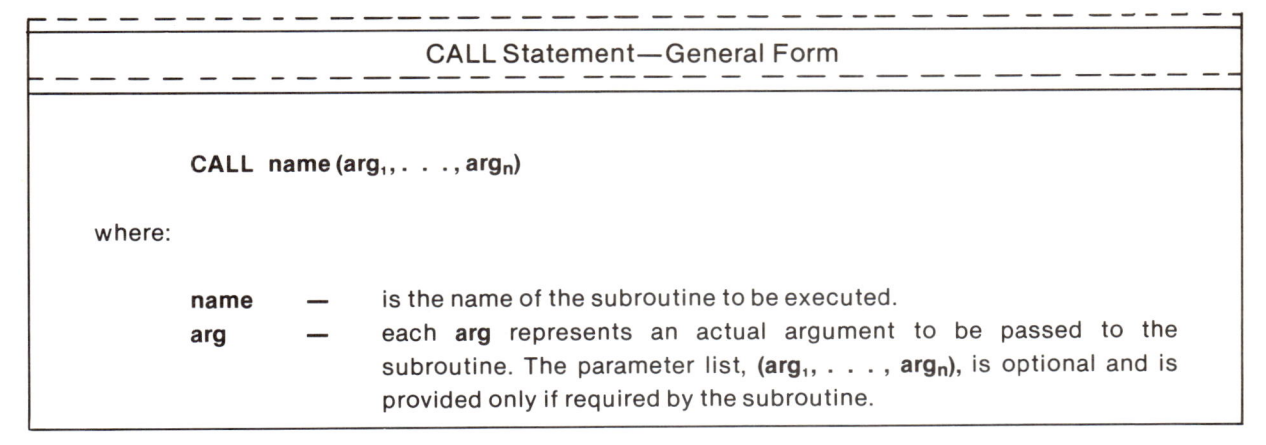

Figure 13.5. General Notation of the CALL Statement

13.3 Passing Arguments in a Parameter List

A *parameter list* is defined as a list of variables, constants, and/or arrays which communicates values between a subprogram and a calling program. Each variable, constant, or array name in the list is called an *argument* which either passes a value to another program unit or takes on a value passed from a program unit. Some programmers refer to the list of arguments as an *argument list* instead of a parameter list.

The parameter list of arguments is our way of communicating values between program units. If a subroutine requires a certain number of arguments to be used in calculations, both the SUBROUTINE and the CALL statements must include corresponding arguments in their parameter list. If a subroutine returns some calculated results, arguments likewise must be included in both parameter lists for the returned values.

The parameter list is an optional part of the SUBROUTINE and CALL statements. That is, if no arguments need to be passed between the subroutine and the calling program, then the parameter list is omitted. We may have, for example, a subroutine named NEWPAG that simply performs a page eject operation and no arguments need be passed.

The arguments included in the parameter list of the CALL statement are referred to as ***actual arguments*** (or formal arguments). The variables and array names contained in the parameter list of the subprogram are called *dummy arguments.* Dummy arguments may be thought of as merely placeholders for the actual arguments. The first dummy argument holds a place for the first actual argument, and so on. So whenever we use the term **dummy argument,** we are referring to an argument in the subprogram parameter list. Likewise, the term **actual argument** will refer to an argument in the CALL.

Dummy arguments may be one of three types:

1. input arguments—that furnish values for the subroutine to use in its operations and are not modified by the subroutine.
2. output arguments—that the subroutine calculates and returns to the calling program.
3. input/output arguments—that furnish values to the subroutine, are modified in the subroutine, and are passed back to the calling program. Input/output arguments are also known as mixed arguments.

When a subroutine is called by a program unit, the values are placed in the input arguments. Then the execution of the subroutine begins. When execution is finished, the calculated or redefined values are placed in the output arguments and control is returned to the calling program.

The name of an actual argument can be the same as that of its corresponding dummy argument, or it can be a different name. For the sake of clarity and documentation, it is wise to use the same names where feasible. If, however, different programmers write different program units, no difficulty is encountered in the use of different names.

Arguments in the parameter list may include values other than numeric constants or variables. They may include a literal constant or even various types of expressions. Following is a list of the different types of values that may be included as actual arguments in a parameter list.

1. A numeric constant of any type
2. A logical constant (a special type of constant to be covered in chapter 16)
3. A literal constant
4. Any type of nonsubscripted variables
5. Any type of subscripted variables
6. Any type of array name
7. Any type of arithmetic expression
8. Any type of logical expression
9. The name of a subprogram

If an array is passed as an argument, then the associated argument in both parameter lists is an array name. The array dimensions *must be included* in both programs. If a single item in an array is to be passed, then it is passed as a subscripted variable by specifying the element number after the array name.

Constants and expressions are used only as actual arguments. It makes no sense to include constants or expressions as dummy arguments. Dummy arguments may only be a variable, array name, or a subprogram name. A subprogram must not change the value of any dummy argument having as its corresponding actual argument a constant or expression. An example of a CALL containing constants, an expression and a subscripted variable is as follows:

CALL SUBR(10,8.7,3HCAT,2*5 + N,RESULT(1))

The first two arguments (10,8.7) are numeric constants. The third argument is an example of a literal constant. The fourth argument is an arithmetic expression that yields an integer value. The last argument is an example of a subscripted variable which passes only the first value in the array RESULT. The use of subprograms as arguments will be discussed in chapter 15 when we cover some of the more exotic subprogram statements.

All arguments in the parameter list of the calling program must agree with the arguments in the subprogram's parameter list in four ways. These four ways are:

1. In number (that is, by count in the parameter list)
2. In order (that is, in a left-to-right progression)
3. In type (that is, integer, real, literal, etc.)
4. In length (that is, single precision, double precision)

The same number of arguments must be included in both parameter lists. If the SUBROUTINE statement contains three arguments, then the CALL statement must contain three arguments in its parameter list.

The arguments must agree in a one-to-one, left-to-right order in both lists. If a subroutine expects to receive a certain value as the first argument in its list, then you must ensure that the corresponding value being passed appears first in the list of the CALL. The positional relationship of all corresponding arguments must be maintained in each list.

Suppose we have a subroutine CALC that sums the items in each of two arrays named A and B. The variables ASUM and BSUM are used to return the sum of array A and B, respectively. Assume the SUBROUTINE statement is coded:

SUBROUTINE CALC (A,ASUM,B,BSUM)

Array A is considered to be the first array (one-dimensional containing ten items), and B is the second array (one dimensional containing twenty items).

We wish to pass two arrays named X and Y to the subroutine to calculate the sum of their items and receive these sums back under the variables XSUM and YSUM, respectively. Array X is a ten-element one-dimensional array and array Y is a twenty-element one-dimensional array. The CALL statement must be coded as follows:

CALL CALC (X,XSUM,Y,YSUM)

Thus, the order of the actual arguments placed in the parameter list agrees with the order of the corresponding dummy arguments in the SUBROUTINE list.

The relationship between the arguments in the CALL and SUBROUTINE statements is illustrated as follows:

CALL CALC (X,XSUM,Y,YSUM)
SUBROUTINE CALC (A,ASUM,B,BSUM)

We see that the array names A and B in the subroutine take on the values in the arrays X and Y, respectively. And the arguments XSUM and YSUM in the CALL take on the values in ASUM and BSUM, respectively. Remember, it is the position of the arguments in the CALL and not their names which determines their use in the subroutine.

We may include the arguments in the parameter list of the SUBROUTINE statement in any order we wish. But the order of the arguments in the list of the CALL must be arranged to agree with the order of arguments in the subroutine list. Suppose we write the subroutine list as follows:

SUBROUTINE CALC (A,B,ASUM,BSUM)

The CALL statement must be coded as:

CALL CALC (X,Y,XSUM,YSUM)

The positional relationships of the corresponding arguments are still maintained.

The common practice is to include the input arguments first in the subroutine parameter list, and the output arguments last. This is only a recommended procedure. The main thing is that the arguments between the two lists agree with one another in the order in which they are put into their lists.

Agreement by type and length means that the corresponding arguments in the two lists must agree in their mode of data. If a dummy argument is type integer, the actual argument must be an integer type also. If the dummy argument is a real type array, then the corresponding actual argument must be a real type array. Agreement by length means that the corresponding arguments between the two lists must agree in precision. If a dummy argument is double precision, then its corresponding actual argument must also be double precision.

If an actual argument is intended to be a different type than the type implied by the first letter of its name, then the argument must be declared in an Explicit Type statement. For example:

Mainline	*Subroutine*
REAL INCH	**SUBROUTINE CONVRT (RINCH,FEET)**
. . .	. . .
. . .	. . .
CALL CONVRT (INCH,FEET)	. . .

Likewise, we must use an Explicit Type statement in a subprogram if a dummy argument is intended to be a different type from that implied by the first letter of its name.

There is generally no limit on the number of arguments that can be included in a parameter list. Thus, the maximum number of arguments permitted is the maximum number that can be contained on allowed continuation cards.

Now to some sample programs.

13.4 Sample Subroutine Subprograms

To illustrate the concepts and statements we have been talking about, let us examine three subroutine subprograms. We will first look at a subroutine that uses no arguments. Then we will examine a subroutine that has both input and output arguments in its parameter list. Finally, we will examine a subroutine that uses an array.

We may have the need to print a page heading and column heading at various points in a long program. This operation is very easily performed in a subprogram, which makes our work easier and less prone to coding errors in complex format specifications.

Let us assume a grading program that needs a page header that reads "GRADES FOR DATA PROCESSING 101." We want to print column heading for a student number, test 1, test 2, test 3, test 4, test 5, and average. Our coded subroutine named HDGS is:

Subroutine　　　　　　　　　　　　　　　　　　　　*Main Program*

```
      SUBROUTINE HDGS                          _____
      WRITE(6,99)                              _____
   99 FORMAT(1H1,30X,                          _____
     *30HGRADES FOR DATA PROCESSING 101)       CALL HDGS
      WRITE(6,98)                              _____
   98 FORMAT(/1H0,5X,4HNAME,14X,               _____
     *6HTEST 1,3X,6HTEST 2, 6HTEST 3,3X,       _____
     *6HTEST 4,3X,6HTEST 5,3X,7HAVERAGE)       CALL HDGS
      RETURN                                   _____
      END                                      END
```

To illustrate the use of a parameter list, let us examine a subroutine that computes the sum, difference, product, and quotient of any two values passed to it. Our subroutine is named CMPUTE and is:

Subroutine Subprograms and the COMMON Statement   237

Subroutine
```
SUBROUTINE CMPUTE(X,Y,S,D,P,Q)
S = X + Y
D = X - Y
P = X * Y
Q = X / Y
RETURN
END
```

Main Program
```
      READ(5,99) A,B
99    FORMAT (2F5.2)
      CALL CMPUTE(A,B,S,D,P,Q)
      WRITE(6,98)A,B,S,D,P,Q
98    FORMAT(1H0,6F10.2)
      READ(5,99)G,H
      CALL CMPUTE(G,H,A1,A2,A3,A4)
      WRITE(6,98)G,H,A1,A2,A3,A4
      CALL CMPUTE(8.3,5.71,S,D,P,Q)
      WRITE(6,98)S,D,P,Q
      STOP
      END
```

The subroutine CMPUTE receives the two input values under the arguments X and Y. The sum, difference, product, and quotient are computed and passed back under the arguments S, D, P, and Q, respectively.

The main program first reads the two variables A and B and passes them to the subroutine. The results are received back under the variables S, D, P, and Q. The same variable names are used for the actual arguments in the CALL parameter list as the dummy arguments in the subroutine parameter list. This is fine, and is recommended where feasible. But the variable names do not have to be the same. In the second CALL the variables A1, A2, A3, and A4 are used to receive the computed values. The third CALL passes two numeric constants as the input values to the subroutine.

To illustrate a subroutine that uses an array, let us assume a subroutine named SUMARY which sums the items in variously sized one-dimensional arrays. The array name and number of elements are passed to the subroutine as the first two arguments, respectively. The subroutine is:

Subroutine
```
      SUBROUTINE SUMARY(X,N,TOT)
      DIMENSION X(20)
      TOT = 0.0
      DO 10 I = 1,N
         TOT = TOT + X(I)
10    CONTINUE
      RETURN
      END
```

Main Program
```
      DIMENSION A(10),B(20)
      READ(5,99)A
99    FORMAT (10F4.2)
      CALL SUMARY(A,10,SUM)
      WRITE (6,98) A,SUM
98    FORMAT(1,H0, 10F7.2/
     *1H0,11HSUM OF A = ,F7.2)
      READ(5,97)B
97    FORMAT (20F3.1)
      CALL SUMARY(B,20,SUM)
      WRITE(6,96)B,SUM
96    FORMAT(1H0,20F6.1/
     *1H0,11HSUM OF B =  ,F7.2)
      STOP
      END
```

The main program passes the array to be summed as the first argument. The second argument is a numeric constant used to specify the number of items in the array. The sum of the array items is received back under the variable SUM. The subroutine receives the values of the array under the array named X. The array X is dimensioned in the subroutine as twenty elements. So we can sum the items in any one-dimensional array up through twenty elements. The size of the array is received under the variable N. The total of all items in the array is accumulated under the argument TOT.

These sample programs should help you understand the many ways in which we can use subroutines. Subprograms are very powerful tools to make your job of programming easier. They should be used wherever feasible to provide a modular approach to programming with an application which makes your program more understandable and easier to debug.

Our next section discusses how we flowchart subprograms when used in our programs, and the card deck setup for programs which include subprograms.

13.5 Flowcharting Subprograms and Job Setup of Subprograms

The *predefined process symbol* is used in a flowchart to indicate a subprogram. This one symbol refers to the entire subprogram, so a separate flowchart must be developed to specify all the logical steps that are included in the subprogram. The predefined process symbol is:

This symbol is simply the rectangular operations symbol with a vertical line drawn down each side. Figure 13.6 illustrates the use of this symbol in our flowcharts. We will flowchart our third sample program in section 13.4, which calculated the sum of the elements in a one-dimensional array.

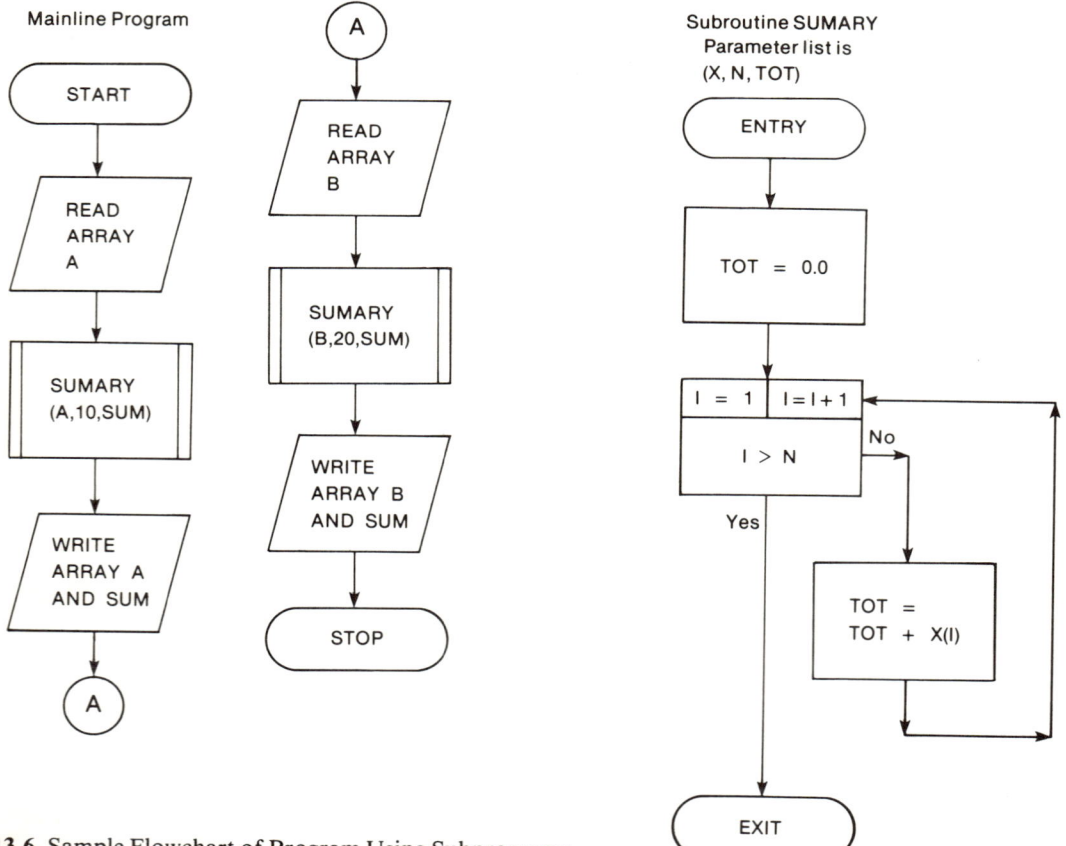

Figure 13.6. Sample Flowchart of Program Using Subprograms

Subroutine Subprograms and the COMMON Statement 239

The predefined process symbol in our main program indicates that the subroutine is called at that point in the program. The symbol includes the name of the subroutine called, since we can have more than one subroutine. The parameter list is optional, but is helpful for documentation purposes. The subroutine, itself, is flowcharted in detail on a separate page.

Job Setup for Subprograms

Subprograms may be included in a job deck either before or after the main program. In fact, you can have some subprograms before and some after the main program. The recommended way is to include all subprograms after the main program. A good technique to aid in locating a subprogram is to put them all in alphabetical order by their name. Figure 13.7 illustrates the setup of a job which includes subprograms.

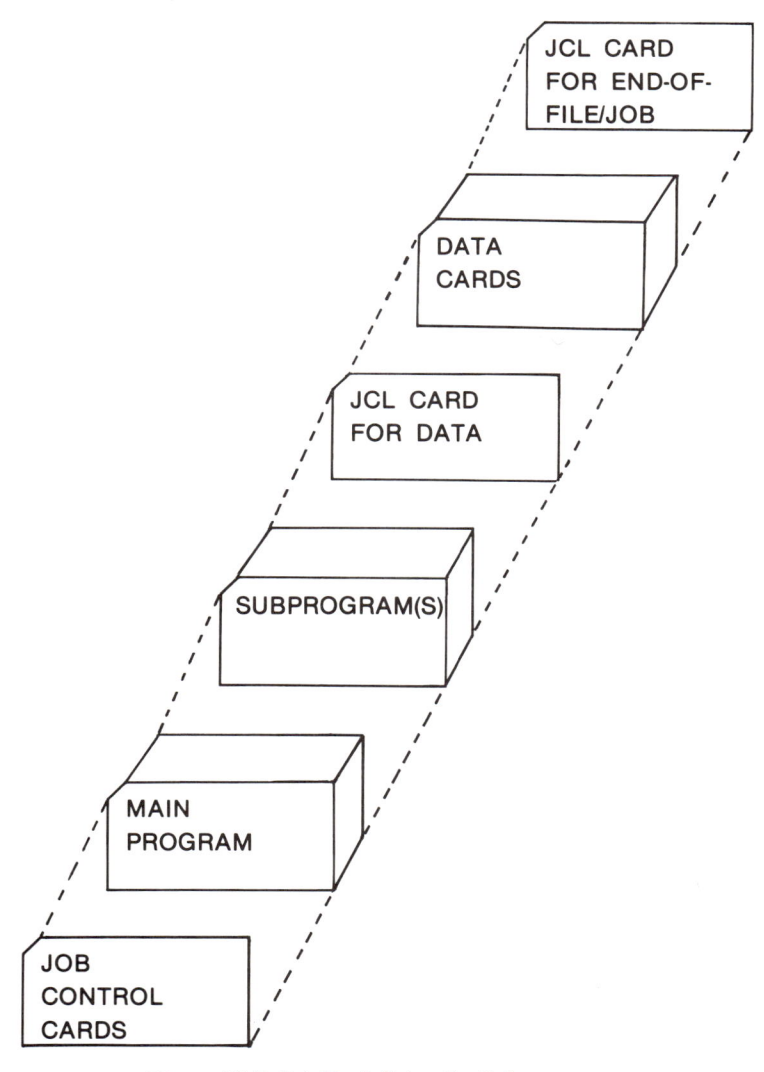

Figure 13.7. Job Deck Setup for Subprograms

No matter how the subprograms are arranged in the job deck, execution is always begun at the mainline program.

Subroutine Subprograms and the COMMON Statement

The COMMON Statement

Up to this point we have passed arguments between the calling program and subroutines in the parameter list. FORTRAN allows us, however, to set aside a number of storage locations that can be shared between program units. The shared storage locations are referred to as "COMMON" storage, meaning common to a number of program units. Thus, arguments can be assigned to the Common storage locations and need not be passed in a parameter list. The group of common storage locations is often referred to as a ***Common Block***.

The COMMON statement is used to establish common storage locations and to identify the variables and/or arrays assigned to these locations. A COMMON statement must be included in *every* program unit that is to share the assigned variables and arrays.

Suppose we wish to assign the three variables X, Y, and Z to the common storage area, and have them share these locations between program units. The COMMON statement

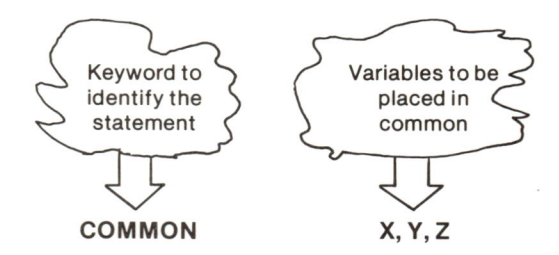

```
        Keyword to                Variables to be
        identify the              placed in
        statement                 common
             ↓                         ↓
         COMMON                     X, Y, Z
```

would be included in each program unit that needs to reference these variables. Three storage locations are set aside in the Common storage area.

These locations contain the quantities of X, Y, and Z, respectively; and these locations are accessible to every program unit that contains this particular COMMON specification statement. The last value assigned to any of the three variables by a program unit becomes its current value. That is, if our main program last assigned the value of 3.5 to X, that would be the current value of X when it is used by a subprogram. If the subprogram altered the value of X to 7.8, then that would be the current value of X when control is returned to the calling program.

This type of Common memory area is called ***Blank Common*** or ***Unlabeled Common*** inasmuch as no name is assigned to the Common Block. Only one Blank Common Block can be established in FORTRAN.

The communication between program units using a COMMON statement is illustrated in figure 13.8. Thus we see that any program unit using the COMMON memory may share variables between any number of other program units.

Use of the COMMON statement not only eliminates the necessity of establishing a parameter list, but is also a more efficient technique. The number of storage locations is reduced by at least 50 percent. Instead of having redundant locations in each program unit, they can share a common location in the Common area. Program units process faster during execution because values do not have to be passed between the program units. The address of the variables in the Common area are established during compile time for each program unit.

A subroutine may receive or pass an argument via the parameter list *or* by the COMMON statement. But the argument must not be passed by both methods. That is, a dummy argument used in a parameter list must not appear in a COMMON statement.

Variables and/or array items are assigned to Blank Common cumulatively in the order of their appearance within the COMMON statement. That is, the first variable found in the COMMON statement is assigned to the first storage location, the second variable is assigned to the second location, and so on. If multiple COMMON statements are used, the variables/arrays in the second and successive COMMON

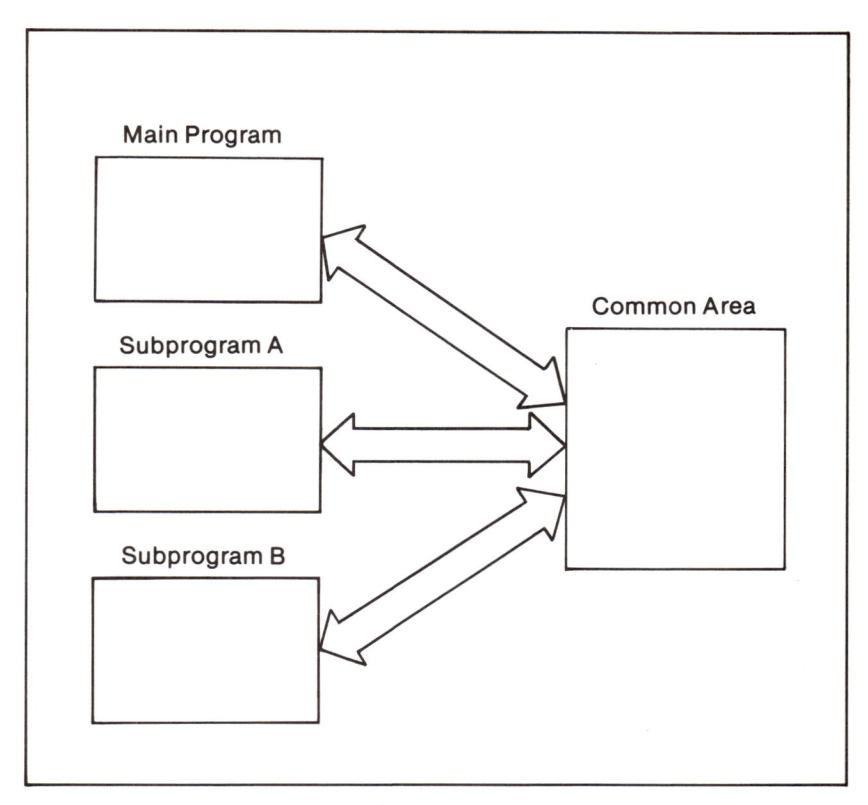

Figure 13.8. Communications Between Program Units Using COMMON

statements are appended to the end of the Common Block where the other variables leave off. For example:

        COMMON A,B
        COMMON D,E,F
        COMMON G

The variables A and B are assigned to the first and second locations, respectively. Variables D, E, and F and next assigned to the third, fourth, and fifth locations, respectively. And the variable G is assigned to the sixth location in Common memory.

The same variable may not be placed in a Blank Common Block more than once. For example, the following statements are illegal:

        COMMON A,B,C
        COMMON I,J,A

The variable A appears in two COMMON lists which is illegal.

The variables in the COMMON statements between program units need not be the same variable/array names. The concept works the same as with the actual arguments in a calling program and the dummy arguments in the subprogram. The variables/arrays in COMMON statements within *each* program unit are assigned beginning with the first location in the Common Block. For example:

    Main Program: **COMMON A,B,C**
    Subprogram: **COMMON X,Y,Z**

These variables are assigned to Common memory as follows:

Main Program:	A	B	C
Common Block:			
Subprogram:	X	Y	Z

The variables X, Y, and Z in the subprogram related to the variables A, B, and C, respectively in the main program.

It is best to use the same symbolic names and identical COMMON statements in all program units which make use of the Common Block. You must take extreme care when a change is made to a COMMON statement to ensure that the corresponding changes are made in all program units using COMMON.

A Blank Common Block does not have to be the same length (i.e., contain the same number of variables) in all program units. A programmer must ensure the proper arrangement of arguments in Common storage by maintaining agreement as to (1) order, (2) type, and (3) length of variables. A COMMON statement may include "dummy" (extra) variables/arrays in order to provide the desired arrangement as to the locations of arguments.

Just as the position of arguments is important in a parameter list, so the positional correspondence of variables/arrays is important in using COMMON statements. If positional correspondence is violated, confusing results as well as hard-to-locate errors may occur.

Whenever the types of arguments are mixed as to length, it is good practice to list the arguments in descending order as to length. So double-precision variables and arrays should precede single-precision variables and arrays in the COMMON statement. This ensures proper "boundary" alignment of storage locations.

Whenever array names are contained in the COMMON statement, they may be supplied with dimension specifications in the COMMON statement itself. For example:

COMMON A, B(10), C(5,3)

where B is a ten-element one-dimensional array and C is a five-row by three-column, two-dimensional array.

Or the dimension specifications may still be given in the DIMENSION statement and only the array names given in the COMMON statement. For example,

DIMENSION B(10),C(5,3)
COMMON A,B,C

These statements provide the same specifications as the previous example. Either way to provide dimension specifications for arrays in the Common memory area is correct, but the preferred way is to supply the dimension specifications within the COMMON statement, itself. This technique provides clearer documentation and requires fewer statements. If the dimension information is given in a DIMENSION statement, it must precede the COMMON statement containing the array names.

In situations where Common storage is needed, it is often useful to divide the Common area into subdivisions of smaller blocks. Labels (names) are assigned to these subdivisions to identify the "labeled" Common Block. Whereas there can be only one Blank Common Block, multiple *Labeled Common Blocks* can be established.

The second type of Common Blocks to which names are assigned is called *Labeled* or *Named COMMON*. The term **Labeled Common Blocks** will be used to refer to this second type of Common memory.

The name of a Labeled Common Block is enclosed within slashes following the keyword COMMON. The names of the variables and/or arrays to be included in the block are given after the block name. For example:

> name assigned to a Labeled Common Block

> list of variables/arrays assigned

COMMON/LIST1/ A,B,C(10)

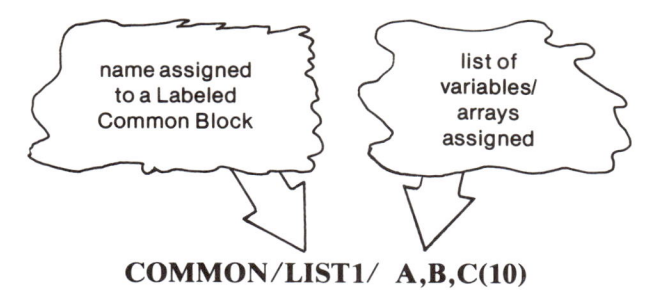

A Labeled Common Block name is formed the same as other symbolic names in FORTRAN. No meaning is given to the first letter of the name (i.e., integer or real) since different types of data may be assigned to a Labeled Common Block. The name used as a label in the COMMON statement must not be used as a variable/array name elsewhere in the program.

Breaking up the Common area into labeled blocks makes it possible for program units to communicate with each other directly without concerning the other program units. For example:

Main Program:	**COMMON/LIST1/A,B,C/LIST2/X,Y,Z**
Subprogram 1:	**COMMON/LIST1/A,B,C**
Subprogram 2:	**COMMON/LIST2/X,Y,Z**

Thus, the Labeled Common Block named LIST1 is used between the main program and subprogram 1. The Block named LIST2 is only used between the main program and subprogram 2.

Now that both types of Common Blocks have been introduced we can give the general form of the COMMON statement. See figure 13.9.

COMMON Statement—General Form

$\quad$ **COMMON /comnam/var$_1$, . . ., var$_n$**

where:

$\quad$ **comnam** $\quad$ — is an optional Common area name. If no name is used, the variables are assigned to a Blank Common area. If this option is used, the name identifies the labeled Common area.

$\quad$ **var$_1$, . . ., var$_n$** — are variable or array names which identify the locations assigned to a Common area.
$\quad$ Multiple labeled Common areas may be included on the same COMMON statement.

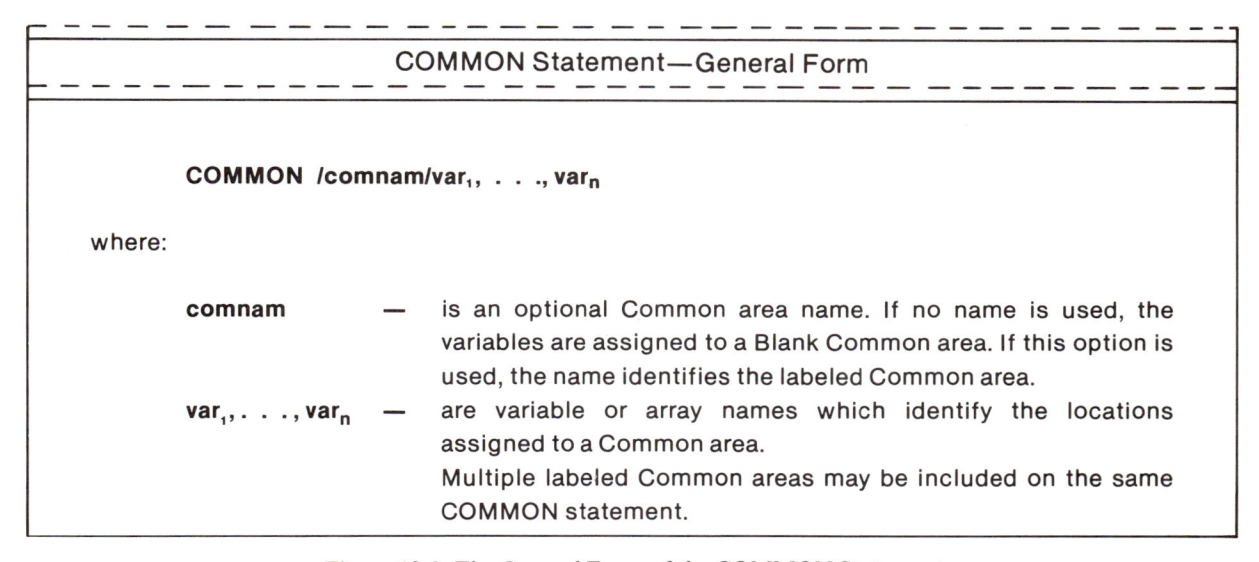

Figure 13.9. The General Form of the COMMON Statement

The COMMON statement is a specification statement. It must appear before any executable statements and must not be assigned a statement number. Any other specification statement referring to variables or arrays listed in the COMMON statement must precede the COMMON statement.

Both Blank Common items and Labeled Common Blocks may be listed in the same COMMON statement. For example:

COMMON A,B/COM1/X,Y,Z

If the Labeled Common Block is given first, two consecutive slashes must be used to indicate the end of a Labeled Common Block and the beginning of Blank Common items. For example:

COMMON /COM1/X,Y,Z //A,B

A Labeled Common Block may appear more than once in the same or additional COMMON statements. For example:

COMMON /COM1/ X,Y /COM1/ Z

or:

COMMON /COM1/ X,Y
COMMON /COM1/ Z

Additional items in the same Labeled Common Block are appended to the end of the list. This has the same effect as if the Labeled Common Block had been coded as:

COMMON /COM1/ X,Y,Z

So the FORTRAN compiler will continue to assign items at the end of a Blank Common Block or Labeled Common Blocks as it translates each COMMON statement.

The same block size must be maintained in all program units using the same Labeled Common Block. That is, the same number of storage locations assigned to a Labeled Common Block must agree in all program units.

The names assigned to a Labeled Common Block may be different in each program unit. But the block name and the total number of locations in the block must agree. For example:

Main Program: **COMMON /GRP1/ A,B,C**
Subprogram: **COMMON /GRP1/ A1,B1,C1**

The variables A1, B1, and C1 equate to the same locations in the Labeled Common GRP1 as A, B, and C, respectively.

The EQUIVALENCE statement must *not* be used to declare two variables or arrays which appear in COMMON statements within the same program unit equivalent to each other. It is permissible to declare a quantity in COMMON equivalent to another variable which does not appear in a COMMON statement. This has the effect of placing the second variable into Common storage also, but at the same location as the first quantity.

No values can be predefined in a DATA or Explicit Type statement for any value in a Blank Common Block. The DATA and Explicit Type statement can be used to preassign values in a Labeled Common Block. But, this preassignment of values to Labeled Common must be done in a special program unit called a BLOCKDATA subprogram. Section 13.7 will cover the construction of BLOCKDATA subprograms.

The COMMON statement can provide quite a shortcut when many calls are made to subprograms. It also eliminates the tedious construction of a long parameter list which can often cause difficult-to-find errors arising from misarranged arguments. But, the COMMON statement must be used with caution to ensure that the proper storage location contains the value you think you have.

To illustrate the use of a COMMON statement, let us recode the second sample program in section 13.4 using Blank Common. This sample program computed the sum, difference, product, and quotient of two values. Our subroutine and main program with a Blank COMMON statement are:

Subroutine	*Main Program*
SUBROUTINE CMPUTE	COMMON A,B
COMMON X,Y,S,D,P,Q	COMMON A1,A2,A3,A4
S = X + Y	READ (5,99) A,B
D = X - Y	99 FORMAT (2F5.2)
P = X * Y	CALL CMPUTE
Q = X / Y	WRITE (6,98)A,B,A1,A2,A3,A4
RETURN	98 FORMAT (1H0,6F10.2)
END	READ (5,99) A,B
	CALL CMPUTE
	WRITE(6,98)A,B,A1,A2,A3,A4
	STOP
	END

The subroutine CMPUTE used six variables in Blank Common to compute and store its results. The main program also used the first six variables in Blank Common to store the input values and write the output results.

There are several important features of this program we must examine more closely. First, the variables in the COMMON of the main program did not use the same names as in the COMMON of the subroutine. The same basic rules hold true for variables in the COMMON as for arguments in the parameter list. They must agree in order, type, and length. That is, the variables in COMMON statements within each program unit are assigned at the beginning of the Common area. The variables in COMMON between our program units share memory illustrated as follows:

BLANK COMMON

Subroutine CMPUTE Variables	X	Y	S	D	P	Q
Storage Locations						
Main Program Variables	A	B	A1	A2	A3	A4

We could use the same variable names in the COMMON statements for both program units, but it is not necessary. The variables will be matched in a one-to-one, left-to-right order as they are assigned to Blank Common. I recommend, however, that you use the same variable names assigned to Common memory for documentation purposes and easier-to-follow programs.

The second thing to notice in our program is that we did not use the variables G and H to read the values from the second data card. These variables could be used if we had equated them by an EQUIVALENCE statement to the locations used for A and B. This statement would be:

EQUIVALENCE (A,G),(B,H)

Thus the locations for G and H would be the same for A and B, respectively. A third COMMON statement which included the variables G and H would not work since these variables would have been placed at the end of Blank Common after the variable A4.

Sometimes it becomes advantageous for us to include part of our arguments in COMMON and pass part of them via a parameter list. Normally, those variables which do not change should be put into Blank Common or even Labeled Common. Then those arguments which have different names and should not be equated to other variables are passed in a parameter list. This technique must be used in our sample program if the values in the variables A and B needed to be saved and should not be equated with G and H. That is, the input values should be placed in a parameter list and the calculated variables put into a COMMON statement to provide the flexibility we need.

The EQUIVALENCE statement cannot be used to extend a Blank Common Block to the left. That is, we must not equivalent the first variable in a Common Block with other than the first element of an array. For example:

```
DIMENSION X(10)
COMMON A
EQUIVALENCE (A,X(5))
```

Blank Common Blocks may, however, be extended to the right with the EQUIVALENCE statement. The following example is a valid use of the EQUIVALENCE statement.

```
DIMENSION A(10),B(20)
COMMON A
EQUIVALENCE (A(5),B(1))
```

Now let us look at the special type of subprogram called BLOCKDATA.

13.7 BLOCKDATA Subprograms

A BLOCKDATA subprogram has as its sole purpose the assigning of initial (compile time) values to storage locations in Labeled Common Blocks. It is the only way in which one is permitted to initialize variables and array elements in Labeled Common Blocks. The BLOCKDATA subprogram, however, may not be used in ANSI FORTRAN to provide initial values for variables/arrays assigned to a Blank Common Block.

The BLOCKDATA subprogram begins with the BLOCKDATA statement and ends with the END statement. No executable statements are included in the subprogram. The only statements allowed in the subprogram are specification statements which assign compile time values to storage locations and provide declaratory information. The statements are limited to the COMMON, DIMENSION, EQUIVALENCE, DATA, and Explicit Type statements. The subprogram may also include comment cards.

Only the DATA statement may be used in ANSI FORTRAN to assign initial values. Other FORTRAN compilers also allow the Explicit Type statements to be used to assign initial values. The COMMON statements must precede any DATA statements in the subprogram.

When the elements of an array are to be initialed in a BLOCKDATA subprogram, each element must be listed individually with its subscripted name. Some compilers, other than ANSI, permit the use of the array name followed by a constant with the "*s" specification to initialize the entire array.

An example of a BLOCKDATA subprogram is as follows:

```
BLOCKDATA
INTEGER R,T
REAL K,L,L2
DIMENSION A(5), B(3)
COMMON /LIST1/SUM,A,R,K
COMMON /LIST2/T,L,B
EQUIVALENCE (L,L2)
DATA SUM/0.0/,R,T/1,2/
DATA A(1),A(2),A(3),A(4),A(5)/5*0.0/,L/3.5/
DATA B(1),B(2),B(3)/3*1.0/
END
```

This BLOCKDATA subprogram initializes two arrays and four variables in two Labeled Common Blocks. Note the fact that all variables in a Labeled Common Block must be included in the COMMON statement whether or not they are initialized. For example, the variable K in our sample subprogram is not given an initial value.

Initial values may be assigned to more than one Labeled COMMON Block in a single BLOCKDATA subprogram. For example, our sample program includes two Labeled Common Blocks, LIST1 and LIST2. All values in a Common Block that are to be initialed must be located in the same BLOCKDATA subprogram. That is, part of the values in a COMMON Block cannot be initialed in one BLOCKDATA subprogram and other values in the same COMMON Block cannot be initialed in another BLOCKDATA subprogram.

13.8 Nonstandard Language Extensions (Multiple RETURN n Option)

Control of execution, ordinarily, returns from the RETURN statement in a subroutine to the next executable statement following the CALL in the calling program. An extra option is available in the IBM "G" level compiler, WATFIV, and other FORTRAN compilers to allow the return of program control to other statements in the calling program as well.

Returns to Alternate Points in a Program (RETURN n)

The alternate return points in the calling program are specified in the CALL statement parameter list by an actual argument of &n, where the & identifies a statement label as an actual argument and the n represents a valid statement number. The statement number must, of course, be assigned to an executable statement. Since our arguments in both parameter lists must be the same number, an asterisk (*) is included as a dummy argument in the Subroutine parameter list to be matched with each alternate return point.

A number is included in the RETURN statement to indicate which alternate return point control should be returned to in the calling program. That is, a RETURN 1 means to return control to the first alternate return point given in the parameter list, a RETURN 2 means the second alternate return point, and so on. The regular RETURN statement, by itself, still means to return control to the statement after the CALL.

The following routines illustrate the use of the alternate return point option.

```
        Calling Program                          Subroutine
        CALL ASUB(X,Y,&40,&50,&60)               SUBROUTINE ASUB(X,Y,*,*,*)
        ADD = X + Y                              IF (X.GT.Y) GOTO 4
        . . .                                    IF (X.EQ.Y) GOTO 5
        . . .                                    IF (X.LT.0.0) GOTO 6
    40  SUB = X - Y                              RETURN
        . . .                                  4 RETURN 1
        . . .                                  5 RETURN 2
    50  XMUL = X * Y                           6 RETURN 3
        . . .                                    END
        . . .
    60  DIV = X / Y
        . . .
```

The calling program provides for three alternate return points by the actual arguments &40, &50, and &60 (i.e., statements numbered 40, 50, and 60 respectively). Three asterisks must be placed in the Subroutine parameter list to match these three alternate return point arguments.

The subroutine evaluates each of the Logical IF statements and makes the proper return of control. If the quantity X is greater than the quantity in Y, control is returned to statement number 40 in the calling program. The RETURN 2 returns control to statement number 50 and the RETURN 3 returns control to statement number 60. If the value of X is less than Y, but not less than zero, the normal RETURN is taken. The integer constant included on the RETURN statement must not exceed the number of alternate return points given in the parameter list.

13.9 REVIEW QUESTIONS

1. List three advantages in using Subroutine subprograms in programming.
2. How is a Subroutine subprogram identified in a program?
3. What four statements must be included in every Subroutine subprogram?
4. The statement used in a Subroutine subprogram to return control of execution to the calling program is the _____ statement.
5. How is a Subroutine subprogram referenced in the calling program?
6. How are values passed to and returned from a Subroutine subprogram back to the calling program unit?
7. Where is control of execution returned to in the calling program?
8. What is the minimum number of arguments that can be passed to or from a Subroutine subprogram?
9. What are the four ways that the arguments in a Subroutine reference must agree with the arguments in the SUBROUTINE statement?
10. What is the purpose of the COMMON statement?
11. Name the two types of Common areas.
12. What is the purpose of a BLOCKDATA subprogram?
13. Can any executable statements be included in a BLOCKDATA subprogram?

13.10 EXERCISES

1. Determine the error in each of the following pairs of statements.

    Main Program   *Subroutine*

    a) CALL MPY(A,B,C)   SUBROUTINE MPY(X,Y)
    b) CALL CAL(3,X)   SUBROUTINE CAL(A,B)
    c) CALL PAGERT   SUBROUTINE PAGERN

2. Determine the error in each of the following pairs of statements.

    Main Program   *Subroutine*

    a) CALL CAT(X(1),Z)   SUBROUTINE CAT(X(1),Z)
    b) CALL DOG(G)   SUBROUTINE DOG(5.7)
    c) CALL MOUSE   SUBROUTINE MOUSE(M)

3. Write the SUBROUTINE statement for each of the following requirements.
    a) A subroutine named PAGE that prints page and column headings. No arguments are passed.
    b) A subroutine named PAGELN that performs a line count operation and prints a page heading. The dummy argument is named LINCNT.
4. Write the SUBROUTINE statement for each of the following requirements.
    a) A subroutine named CALC that produces the sum of three input arguments named A, B, and C. The computed sum is passed back under the dummy argument T.
    b) A subroutine named COUNT that counts the number of occurrences of an item in an array argument named X. The computed count is returned under the dummy argument KNT.

5. Write the CALL statement for each of the following requirements.
    a) Call a subroutine named SUBRTN that prints column headings. No arguments are needed.
    b) Call a subroutine named SUB1 that performs a line count operation and prints a page heading. The argument to be passed is named LINCNT.
6. Write the CALL statement for each of the following requirements.
    a) Call a subroutine named COUNT which counts the number of occurrences of a value in the array argument named A. The computed count is received back under the argument K.
    b) Call a subroutine named SUM to compute the sum of the four arguments named, A, B, C, and D. The computed sum is received back under the argument named TOT.
7. How many storage locations are reserved in Blank Common by each of the following sets of statements?
    a) **COMMON A,B**
       COMMON C,X,Y
    b) **COMMON G,H(11),I,J**
    c) **DIMENSION Z(15),N(5,5)**
       COMMON R,S,Z
       COMMON N
    d) **DIMENSION B1(100), B2(50,2)**
       COMMON T,V
       EQUIVALENCE (B2(1,1),B1(1),V)
8. How many storage locations are reserved in Blank Common by each of the following statements or sets of statements.
    a) **COMMON A,B,C,D**
       COMMON I,J,R
    b) **COMMON X,Y(5),Z(7),W**
    c) **DIMENSION G(10)**
       COMMON G,H,I
    d) **DIMENSION X(75),Y(30)**
       COMMON X,Z
       EQUIVALENCE (X(1),Y(1))
9. Write a BLOCKDATA subprogram to initialize the items of A, X, and J in a Labeled Common Block named GROUP1 to 5.3, 8.7, and 3, respectively.
10. Write a BLOCKDATA subprogram to initialize the items of X, Y, and Z in a Labeled Common Block named SET1 to 1.7, 86.5, and 3.87, respectively. Also initialize the four elements of a one-dimensional array named KNT to zeroes.

13.11 PROGRAMMING PROBLEMS

1. Write a FORTRAN subroutine named SQR which will square each item in a one-dimensional array named M. The subroutine must also accumulate the total of all newly squared items. This total is to be returned to a main program under the argument MSUM.

A mainline program will read the values of a twenty-element, one-dimensional array named M20 under the format specifications 20I2. The main program will first write the array M20 under the format specifications 20I5. Then the array will be passed to the subroutine SQR for the squaring of each item. Upon return of control to the main program, the updated items will be written under the specifications 20I5. The total of the updated items will be printed with an appropriate literal.

2. Write a subroutine named SORT to sort the items in a one-dimensional array into ascending sequence. Include a mainline program to read ten values under the format specifications 10F4.1 and load them into a ten-element array named A. Write the array A under the format specifications 10F7.1. Then pass the array A and the number of items to be sorted to the subroutine SORT. After the items are sorted and control returned to the main program, write the sorted array under the specifications 10F7.1.

3. Write a subroutine named SMALL to find the smallest value in a two-dimensional array. Include a mainline program to read twenty values under the format specifications 20I2 and load them into a four-row by five-column array named MATRIX. Pass the array, the number of rows, and number of columns to the subroutine SMALL. After the subroutine has returned the smallest value, write the array SMALL in row-order and the smallest value found.

4. Write a subroutine named MATADD to perform a matrix addition operation on two symmetrical two-dimensional arrays. The two input arrays named A and B each contain five rows and five columns. The output matrix is named C and is also a 5 by 5 two-dimensional array. Matrix addition is accomplished by adding corresponding elements in the two input arrays and storing the sum at the same location in the output array. That is, A(1,1) is added to B(1,1) and stored at C(1,1), A(2,1) is added to B(2,1) and stored at C(2,1), and so on.

Include a mainline program to read the values of A under a 25F3.1 specifications and the values of B from a second data card under the same specifications. After the subroutine has performed the matrix addition write each of the three matrixes in row-order fashion under a 5F6.1 specification. Include a title over each output matrix to identify the matrix.

5. Write a subroutine named MATSUB to perform a matrix subtraction operation on the two input arrays described in problem 4. Matrix subtraction is performed by subtracting corresponding elements in the second array (B) from the first array (A) and storing the sum in the same location in the output array (C). That is, B(1,1) is subtracted from A(1,1) and stored in C(1,1), and so on. Include a mainline program to read and print the arrays as described in problem 4.

6. Write a subroutine named QUAD which solves a quadratic equation of the form $ax^2 + bx + c = 0$. Two real roots are possible. The formulas for computing the roots are:

$$R_1 = \frac{-b + \sqrt{b^2 - 4ac}}{2a}$$

$$R_2 = \frac{-b - \sqrt{b^2 - 4ac}}{2a}$$

Include a mainline program to read the values of A, B, and C under a 3F5.2 format specifications. After the two roots are returned, print the values of A, B, C, ROOT1, and ROOT2 under the specifications 5F10.2. Include column heading to identify each printed value.

7. Select any FORTRAN program that you have previously written and convert it into a subroutine subprogram. Include a main program to read the input values and pass them to the subroutine by a parameter list. Return any computed values by arguments in the parameter list and output these values in the main program.

8. Revise problem 1 to incorporate the use of Blank COMMON instead of using a parameter list to pass the arguments.

9. Revise problem 3 to incorporate the use of a Labeled COMMON in place of a parameter list to pass the arguments.

Structured Programming and Debugging Techniques

14

14.1 Introduction to the Concepts and Use of Structured Programming

The programs that have been assigned and that you have written so far have probably been relatively straightforward as to the logic required. You have generally been able to comprehend the problems and to keep most of the needed logic in your head. Therefore, designing, writing, and debugging relatively small, but important programs should not have been a "major" or impossible task once you learned what programming is all about. This is not meant to belittle the effort you have put forth. To the beginning student programmer a problem of any size is a complex task and requires much work. We who are experienced programmers and instructors sometimes forget how difficult the simplest problem can be to one just learning to program. Until the student crosses the hurdle of learning how to express the logic of a problem in a computer program, programming is a difficult, frustrating job.

But what about truly complex programs, or those large problems that require hundreds or perhaps even thousands of program instructions? Programs that require more than one page of instructions are often difficult to accomplish by the professional programmer, much less by the student. How can we approach such problems so that they become relatively easier to solve, and we know the output results to be correct? Better yet, how can we avoid making logical errors in writing our program? The concepts and techniques employed in top-down structured programming can make the task of developing a program of any size easier to accomplish.

Structured programming is a method of designing and writing programs so that they are easier to code, easier to understand, and contain less chances for logic errors the first time around. Thus, structured programming is a style of programming that, by following a set of rules, aids one in the development of logic, and provides a lessened probability for error. Although a relatively new method of developing programs, it has time and again proven its merit.

The rising cost of software development and maintenance led to the increasing demand for greater precision and productivity in the field of data processing. This demand has, in turn, led to the development of structured programming.

During the first ten to twenty years in the data processing field, computer hardware required the greatest percentage of the data processing budget. But with the development of solid-state technology and mass production of computers, hardware cost began a downward trend in the mid-sixties. Meanwhile, the cost of software development continued to grow at a significant rate. (*Software* is the collective term for computer programs; both for applications and for system programs such as compilers, operating systems, etc., which are used to enable the computer to process applications programs.) By the late-sixties, software development cost had reached the point at which it equaled the cost of hardware expenditures. By the early seventies the cost of software development had increased to such an alarming percentage that a "software crisis" became apparent. It is estimated that presently the ratio of current software cost to hardware cost is roughly 65:35.

The traditional approach to programming was to strive for efficiency. Programmers would use hours and even days to develop "tricks" that made their programs run faster. The techniques used in these tricks were often known only to the programmer who wrote the original program. The original programmer was often replaced and another person was assigned the task of maintaining the programs. When the program had to be modified by the new programmer, difficulty frequently ensued because the new programmer could not understand the logic of the program which he had not written. Thus, many futile attempts might have to be made to modify the program. Many times a *new* program was written if the logic could not be quickly grasped. So we can see how the problem of swelling software costs has come about.

Today the emphasis is on clear, logical structuring of code that lends itself to an understanding of what is being done. The goals of understandability, reliability, and maintainability take precedence over efficiency. Therefore, programmers should write their programs in a straightforward manner, emphasizing clarity and reliability. Programs should be "modularized" so that an individual module that is used quite frequently might be replaced by an even more efficient module when the programmer has time. After the program is working correctly, then optimize the code used most often (usually within inner loops).

Traditionally, only a small portion of the programmer's time was spent on actual coding. About 25 percent was spent in program planning and design, 25 percent was spent in actual coding, and the remaining 50 percent was spent in program checkout (testing and debugging) and other miscellaneous chores. Since structured programming requires the programs to be written in a clearer, more straightforward manner, the resulting code contains fewer logic errors. Thus, the programs are more accurate and require less time in program checkout. Of course, more time is required in the planning and design phase, but the end result is a more reliable and maintainable program. Some people have reported an increase of over 500 percent in the amount of debugged source statements written in a person-day. The two-fold objective of program reliability and programmer productivity, therefore, go hand in hand. If the majority of the bugs can be avoided to start with, the programming process will naturally be cheaper.

Structured programming is an outgrowth of formal correctness studies. Dr. Edsger W. Dijkstra proposed that we take a "constructive" approach to the problem of program correctness by controlling the process of writing programs.[1] We should establish a correct program as we are designing and coding it. Our main concern is with the flow of control within the program. Böehm and Jacopini in a 1966 paper first showed that (1) simple statement sequencing, (2) IF-THEN-ELSE conditional branching (i.e., selecting the next statement[s] to be executed from the result of a test), and (3) DO-WHILE conditional iteration would suffice as a set of control structures for expressing any flowchartable program logic.[2] The restriction to these three types of control structures is not arbitrary. They are based upon mathematical theorems which prove program correctness. Dijkstra introduced the concept of structured programming in various publications, beginning in 1968, based on the theorems established by Böehm and Jacopini. Structured programming, therefore, was introduced to provide a proof-of-correctness approach to programs which would be better understood and more accurate.

The actual use of structured programming methods and techniques by various data processing organizations have resulted in the following benefits:

1. Programs easier to read and comprehend
2. Significant reduction in program logic errors
3. Increased programmer productivity
4. Significant reduction in program maintenance and enhancement costs

Because of these increased dividends, structured programming has become a practice widely accepted by professional programmers and the academic environment.

A structured program consists of blocks. These blocks are formed according to six basic control constructs (expanded from the three theorems given by Böehm and Jacopini) which govern the transfer of con-

---

1. E.W. Dijkstra, "A Constructive Approach to the Problem of Program Correctness," *BIT8* (1968): 174-86.
2. C. Böehm, and G. Jacopini, "Flow Diagrams, Turing Machines and Languages with Only Two Formation Rules," *Communications of the ACM,* 9, no. 5 (May 1966): 366-71.

trol between the different blocks of a structured program. Structured programming, then, is the combining of these statement blocks according to the six prescribed control constructs.[3]

The idea behind the block structure is that a block of statements may be substituted for a simple statement at any place in a program where an executable statement may appear. A block consists of one or more statements which are combined in some fashion and can be treated as a simple statement. Thus, a block may contain other blocks as part of its structure, or it may be constructed from one or more simple statements. A subroutine is a block. A program is the entire overall block. A structured program becomes a block which contains other blocks built only from the six basic control constructs.[4]

The following paragraphs discuss the six control structures used in structured programming. The symbol **exp** denotes the expression included in the IF statement that provides the condition for the test. Each of the letters **a, b,** and **s** represent a simple statement or a block of simple statements. A simple statement may be any of the available statements in the programming language used to implement structured programming. Simple statements include the arithmetic Assignment statement, READ, WRITE, subroutine CALL, etc. They may not include any form of the GOTO statement, since this would cause a transfer of control structure other than those allowed in the structures. However, it may be necessary to include the GOTO statements in order to implement the control structures in a language such as FORTRAN, which is not block structured. Therefore, the GOTO is not considered a simple statement, but an integral part of the control structure.

Sequence: The statements are executed in the sequential order in which they appear. Control is passed unconditionally from one statement to the next one in sequence. The *Sequence* control structure is so basic to programming that it needs no elaborate explanation. Yet this structure is necessary for the construction of a block from statements that are to be executed sequentially. The logic diagram for this structure is:

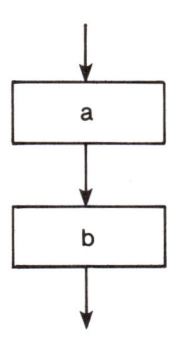

where **a** and **b** represent a simple statement or a block of statements.

IF exp THEN a: The condition represented by **exp** is tested. If **exp** is true, then the statement **a** is executed and controls continues at the next statement. If **exp** is not true, **a** is not executed and control is passed to the next statement. The statement **a** may be a simple statement or a block. The logic diagram for this structure is:

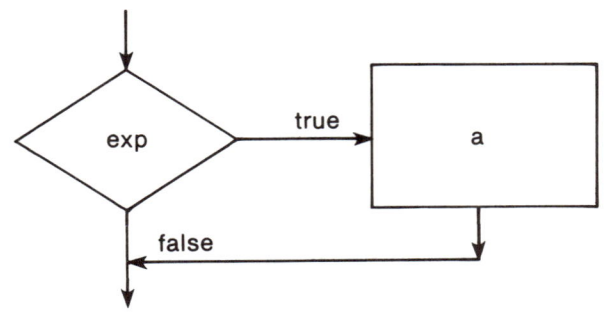

---

3. Ted Tenny, "Structured Programming in FORTRAN," *Datamation* 20, no. 7 (July 1974): 110-15.
4. Ibid., p. 110.

IF exp THEN a ELSE b: The condition represented by **exp** is tested. If **exp** is true, then statement **a** is executed and statement **b** is skipped. Control is resumed at the next statement after **b**. If **exp** is not true, statement **a** is skipped and statement **b** is executed. Control passes to the next statement after **b**. The logic diagram for this structure is:

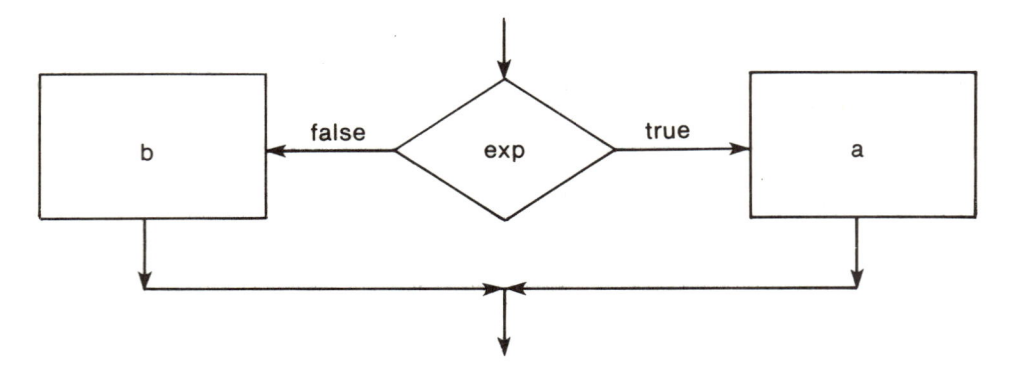

While exp DO a: The condition represented by **exp** is tested. If **exp** is true, the statement **a** is executed and control returns to the beginning of the loop for another test of **exp**. When **exp** is false, then statement **a** is skipped and control is passed to the next statement after **exp**. The logic diagram for this structure is:

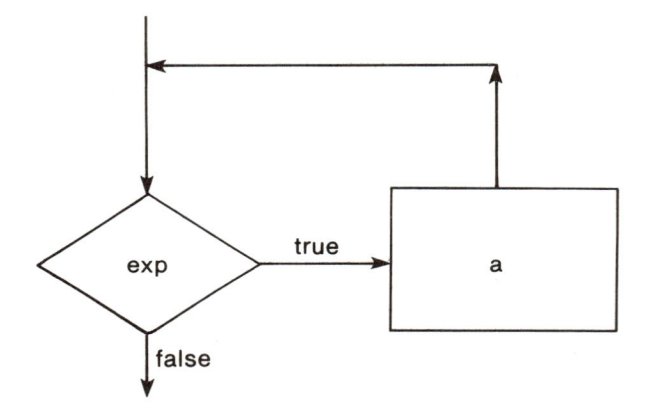

Repeat a UNTIL exp: The statement **a** is executed and then the condition represented by **exp** is tested. If **exp** is *true,* control continues at the next statement after **exp** (since the structure means to continue statement **a** until the condition **exp** is true). When **exp** is false, controls returns in the loop to statement **a** for another iteration. The logic diagram for this structure is:

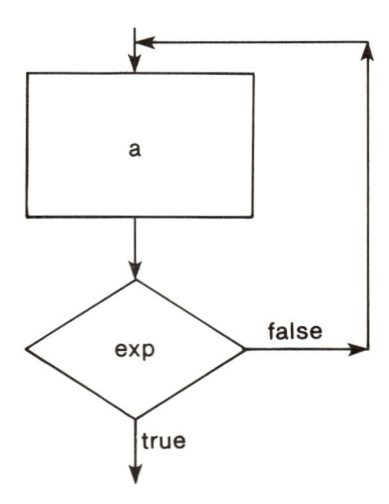

Select Case i OF $(s_1, s_2, \ldots, s_n)$: The ith statement of the set $(s_1, s_2, \ldots, s_n)$ is executed and all other statements of this set are skipped. That is, the value of **i** determines which statement from among the set is to be executed. Control continues at the next statement after the last one in the set (following s_n). The logic diagram for this structure is:

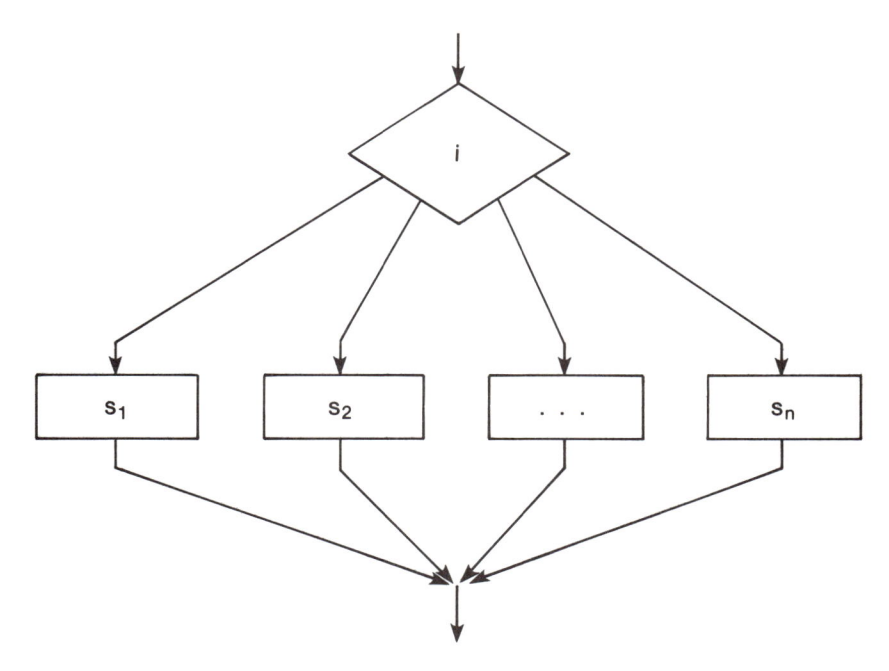

Just because we must use these six basic control constructs does not mean that we must use them individually or separately from one another. Control blocks may be included within other control blocks to achieve the desired logic (see fig. 14.1 on page 256).

Each of the mentioned control structures has only one entry and one exit. When these control structures are transformed into source code, we have a program that can be read from top to bottom and without jumping around in the logic. It is much easier to comprehend what a function does if all the statements that influence its action are physically close by. Thus, structured programming also denotes a group of control structures used for developing programs that are written with respect to a collection of one-entry-point-in and one-exit-point-out structures.

Top-down readability is one consequence of using only the six control structures and avoiding the GOTO statement, except in very special cases, such as the simulation of a control structure in a programming language that lacks it. Structured programming, therefore, is often referred to by some people as "GOTO-less" programming. GOTO's make the logic in a program hard to follow since the logic is heavily interlocked. Interlocked nonlinear segments are increasingly difficult for the human mind to comprehend and follow.

Consider the following code to select the largest value of three variables. Assume the variables A, B, and C have already been read.

```
   IF (A .GT. B) GOTO 10
   IF (B .GT. C) GOTO 20
   BIG = C
   GOTO 40
10 IF (A .GT. C) GOTO 30
   BIG = C
   GOTO 40
20 BIG = B
   GOTO 40
30 BIG = A
40 WRITE (6,98) BIG
```

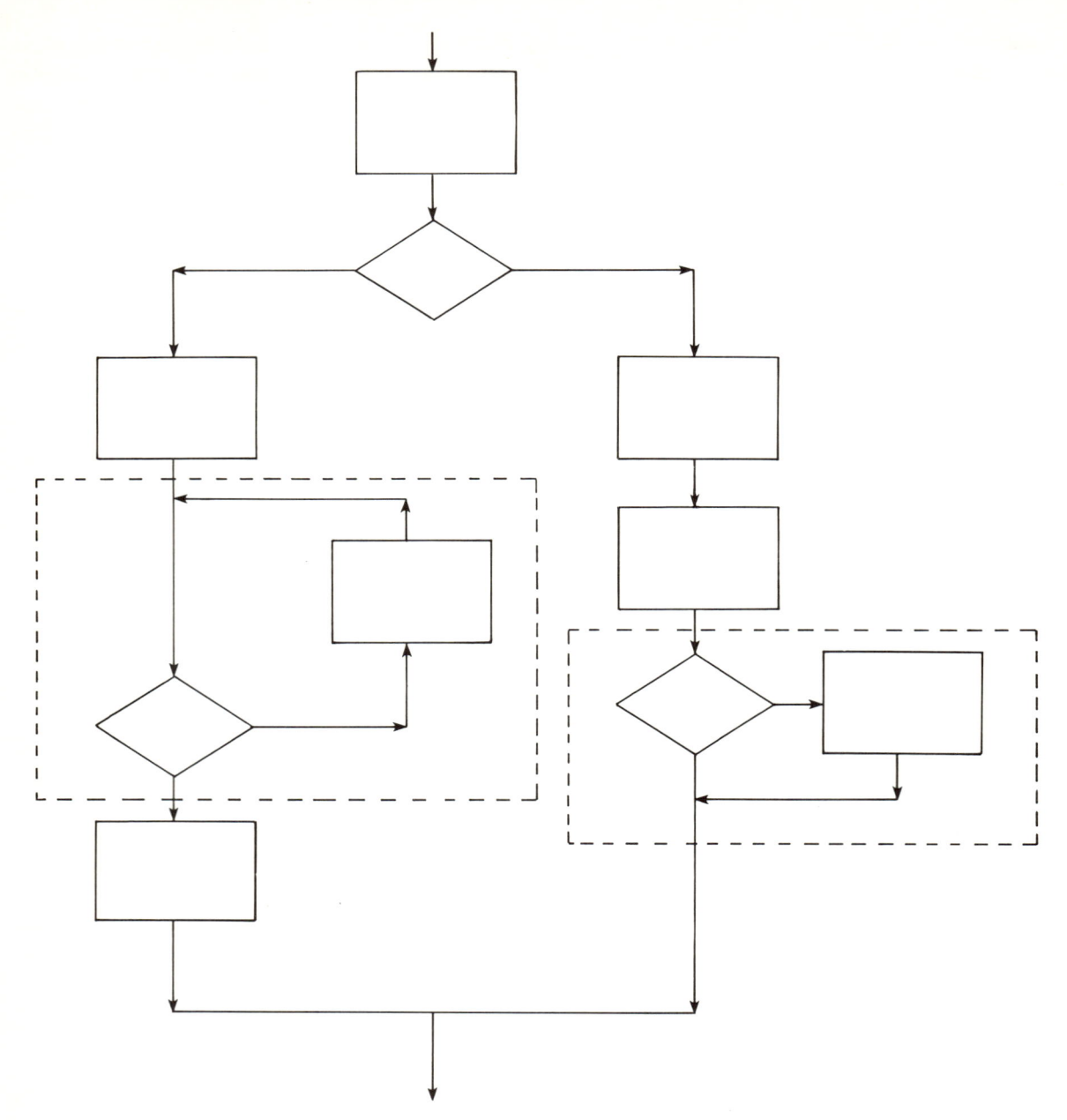

Figure 14.1. A Block of Statements Containing Various Control Structures

The logic is hard to follow because of the numerous GOTO statements.

Now consider this same logic written in a structured form:

```
BIG = A
IF (B .GT. BIG) BIG = B
IF (C .GT. BIG) BIG = C
WRITE (6, 98) BIG
```

The logic in this structured form is very easy to follow. If we modified the problem to find the largest of four variables, this modification could easily be incorporated into the logic.

It is often very difficult to write a complex program without a single GOTO statement. Some people say that GOTO statements should be avoided altogether. Others agree that GOTO statements are okay within each segment of a program that provides a certain routine. But these program segments or modules should have one entry point and one exit point in the logic. That is, it is okay to branch within a module of instructions, but you should not jump into the middle of another module and transfer back again to the module you were originally in. The latter position—using a limited number of GOTOs within an individual program segment—will be our accepted principle.

Some programming languages, such as COBOL, PL/I, and ALGOL, are quite adaptable to structured programming. These languages allow a block (group) of instructions to be executed in an IF statement whenever the given condition is met. These languages also have an ELSE option with the IF statement that can provide an instruction or block of instructions to execute whenever the condition in the IF is not met.

FORTRAN, however, does not have this capability. Only one statement may be executed with the Logical IF statement when the given condition is met. When the condition is not met, control proceeds to the next executable statement. Thus, structured programming is harder to achieve in FORTRAN than in some other languages because of the limited capability of the Logical IF.

The complete elimination of GOTO's, in a language like FORTRAN, is practically impossible. But their use can be reasonably restricted by permitting the FORTRAN programmer to use them only to build the standard control structures which are not in the language. Structured programming, therefore, is regarded as being a programming language independent methodology. It is a set of highly organized, well-disciplined techniques for developing programs in any language.

Our next section will illustrate how the six structured programming constructs are implemented in FORTRAN. Section 14.4 presents some guidelines for writing better programs. Section 14.5 provides techniques for debugging logical errors occurring during the execution of your program.

14.2 FORTRAN Implementation of the Six Control Structures

Structured programming in FORTRAN requires that the six control structures be implemented by the use of the simple statements available in the FORTRAN language. This section illustrates the implementation of these control structures with the available FORTRAN statements. The same symbols will be used to denote expression and statement blocks as given in section 14.1.

Sequence

Implementation of this control structure is accomplished by any of the simple statements in FORTRAN. These statements include the:

> Arithmetic Assignment
> READ
> WRITE
> CALL

This control structure is also implemented by using any of the other control constructs in a sequential fashion.

IF-THEN

The implementation of this structure is accomplished by the Logical IF statement. If **exp** is true, the statement given with the Logical IF is executed and control continues at the next statement. If **exp** is not true, control passes to the next statement. For a simple statement to be executed when the condition is met, the following construct of the Logical IF is used.

```
      C *** THE IF-THEN CONSTRUCT
            IF (exp) a
```

For example:

$$\text{IF } (X \text{ .GE. } 1.) \; Y = 2. * X$$

or

$$\text{IF (SUM .LT. 100.0) CALL SUBRTN}$$

Whenever multiple statements need to be executed for a true condition, we must use a different construction of the Logical IF, since only one statement can be executed if the condition is met. The logical operator .NOT. is placed before the given condition to transfer control around the statements for block **a** which are placed after the Logical IF. This construct is:

```
C *** THE IF-THEN WITH MULTIPLE STATEMENTS
      IF (.NOT. (exp)) GOTO 10
        a
10    CONTINUE
```

For example:

```
      IF (.NOT.(X.GE. 1.)) GOTO 10
         Y = 2. * X
         A = B + C
10    CONTINUE
```

One may wish to reverse the logic of the relational operators and eliminate the logical operator .NOT. to produce the same results. Following is a list of the relational operators with their opposite condition to provide the same results by omitting the .NOT.

Relational Operators	Relational Operators for the Opposite Condition
.EQ.	.NE.
.NE.	.EQ.
.GT.	.LE.
.GE.	.LT.
.LT.	.GE.
.LE.	.GT.

Thus, our Logical IF equivalent for the given example would be:

```
      IF (X.LT. 1.) GOTO 10
         Y = 2. * X
         A = B + C
10    CONTINUE
```

IF-THEN-ELSE

This control structure is also implemented with the Logical IF. If the expression **exp** is true, the block represented by **a** is executed and control proceeds to the statement following block **b**. If the expression **exp** is false, the block represented by **b** is executed and control continues at the next statement. The logical operator .NOT. is used to provide a true condition when **exp** is false in order to transfer around block **a**. A GOTO must be included as the last statement in block **a** to transfer around block **b** when the statements in block **a** are executed. The construct is:

```
C *** THE IF-THEN-ELSE CONSTRUCT
      IF (.NOT. (exp)) GOTO 10
          a
          GOTO 20
10    CONTINUE
          b
20    CONTINUE
```

For example:

```
      IF (.NOT. (X .EQ. 1.0)) GOTO 10
      Y = 2. * X
      GOTO 20
10    CONTINUE
      Y = X ** 2
20    CONTINUE
```

As with the IF-THEN construct, the relational operator may be reversed and the .NOT. omitted.

WHILE-DO

This control structure is implemented by the Logical IF at the head of a loop. The loop is executed as long as **exp** is true. The construct is:

```
C *** THE WHILE-DO CONSTRUCT
10    CONTINUE
      IF (.NOT. (exp)) GOTO 20
          a
          GOTO 10
20    CONTINUE
```

For example:

```
10 CONTINUE
   IF (.NOT. (K.LT. 50)) GOTO 20
   K = K + 1
   SUM = SUM + ARRAY(K)
   GOTO 10
20 CONTINUE
```

REPEAT-UNTIL

This construct is implemented with a loop controlled by the Logical IF. The loop is performed until **exp** becomes true. The construct is:

```
C *** THE REPEAT-UNTIL CONSTRUCT
10    CONTINUE
          a
      IF (.NOT. (exp)) GOTO 10
```

For example:

```
      10 CONTINUE
         K = K + 1
         SUM = SUM + ARRAY(K)
         IF (.NOT.(K.EQ.50)) GOTO 10
```

SELECT-CASE

The select-case construct is implemented with the Computed GOTO statement. An Unconditional GOTO must be included at the end of each statement block to transfer control to the end of the blocks. The construct is:

```
C *** THE SELECT-CASE CONSTRUCT
         GOTO (10,20, . . . ,80), i
  10     CONTINUE
             a
         GOTO 100
  20     CONTINUE
             b
         GOTO 100
             . . .
             . . .
  80     CONTINUE
             h
 100     CONTINUE
```

For example (assume the value of I has been previously assigned):

```
         GOTO (10,20,30), I
  10     CONTINUE
             Y = 2. * X
         GOTO 40
  20     CONTINUE
             Y = X ** 2
         GOTO 40
  30     CONTINUE
             Y = X - 5.0
  40     CONTINUE
```

Additional concepts and techniques are also connected with structured programming. The concept of top-down program design is discussed in the next section.

14.3 Top-down Program Design and Development

The beginning programmer is often quick to surmise that all there is to solving a problem on the computer is to jot down the FORTRAN statements, keypunch them, and run them on the computer. This may be true for small simple programs. But, for more complex programs, this represents only a fraction of the total time required to get a program running and producing the correct results. The majority of one's time and effort goes into program planning and debugging.

The student who hurriedly writes a program without complete planning usually ends up with a "bucket of worms." By this, I mean that there is little organization to the program. It is difficult to read and understand, much less to debug. Programs that are carefully designed and developed with a logically sound approach are far easier to debug and to modify if the need ever arises.

Structured Programming and Debugging Techniques   261

The top-down approach to program development provides an orderly approach to the solution of a problem and program design. The major tasks are first identified so they can be developed in a logical order and "we can see the way we are headed." That is, complicated problems can be divided into smaller units that can be comprehended more readily. These smaller units can in turn be subdivided into smaller parts to a level at which the logic can be understood. This principle may be thought of as "divide and conquer" in the development of large and/or complex problems.

Consider the problem of writing a text book. The author does not sit down and start writing about each of the statements and features of the FORTRAN language. He, or she, first makes a table of contents to delineate what information is to be included and in what sequence the material is to be covered. That is, the organization of the text is first decided, and then the details follow. Likewise, in programming, you should first develop an overall view of the logical parts of the problem, then you can visualize the major tasks required of a program.

With the top-down approach to programming, the first thing to be done is to describe a generalized structure or set of logic modules for the problem solution. A *module* is defined as a segment of code that performs some necessary function. These first level modules may then be divided into succeeding levels of modules to form a tree-like pattern for the program development process. Given modules are expanded to as detailed a level of functional definition as is necessary to produce a relatively easily understood and manageable breakdown of the tasks required in the problem solution.

A **hierarchy chart** illustrates the major tasks needed in the program. These tasks are divided into subtasks and even sub-subtasks to further refine the logic that is required to solve the problem. Figure 14.2 provides an example of a hierarchy chart that shows the tasks and subtasks in solving a payroll problem.

The problem definition is included as the program requirement at level 1. The major tasks of computing gross pay, deductions, and net pay are identified at the second level. These tasks are further sub-

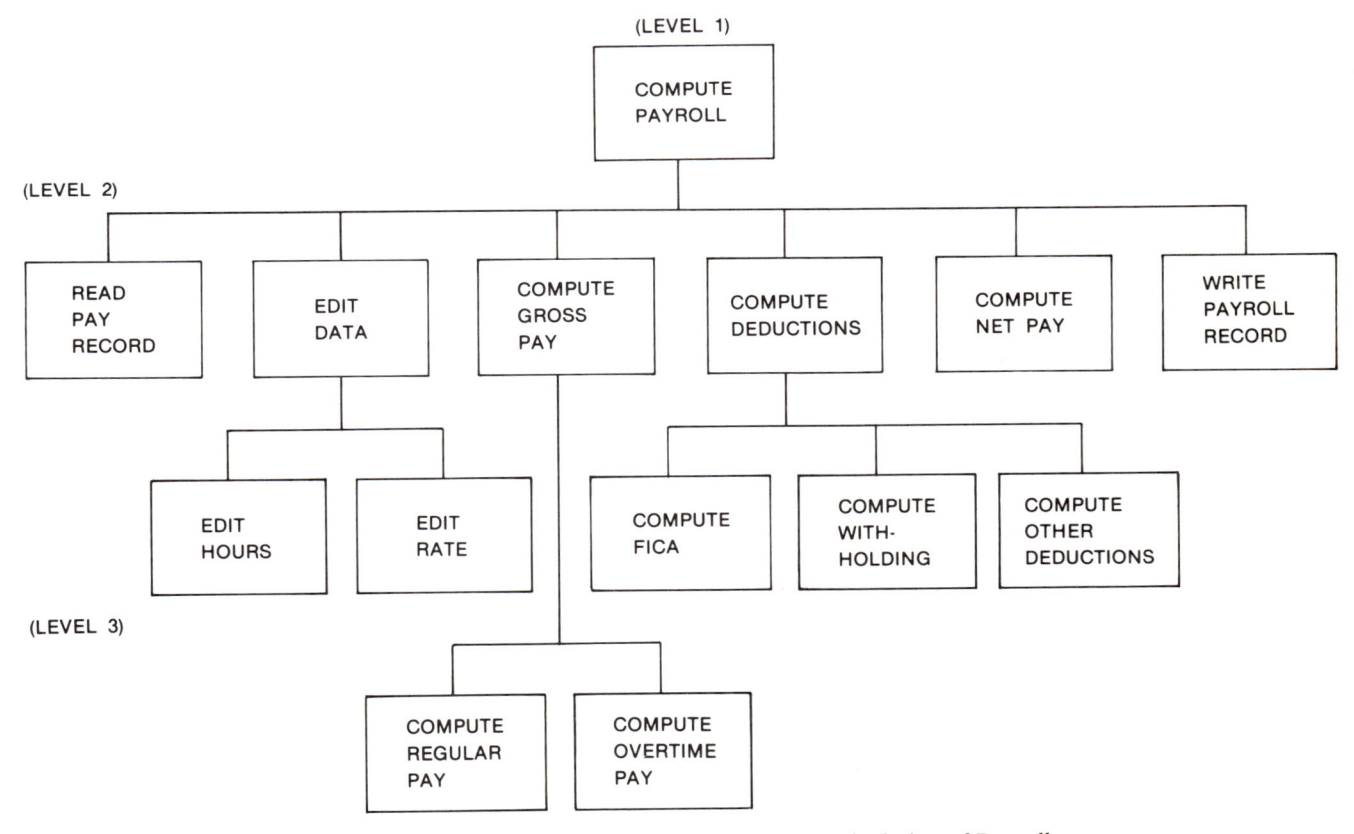

Figure 14.2. Top-down Hierarchy Chart Showing Calculation of Payroll

divided into the functions required to accomplish the task under which it appears. These subtasks appear in a hierarchical or tree-fashion to show they are subrelated to some higher task.

In the top-down process, program development is carried out beginning at the top level. The highest level tasks of the program are written and tested first. The main tasks acting as "driver" routines are first coded with program stubs standing in for the subtasks. The program stubs are simply a line or two of code to indicate that the subtask was indeed invoked, even though the code has not yet been developed. For example, the major tasks may be coded in our example as:

>   READ . . .
>   CALL EDIT
>   CALL GRSPAY
>   CALL DEDUCT
>   CALL NETPAY
>   WRITE . . .

Each of these major tasks may be subroutines or modules of code that will call other subtasks in the hierarchy of logic.

The functional expansion process of each subtask can be carried out in a page of code at a time, in which new sub-subtasks are denoted simply by assigned names, which will eventually contain the code for the next level of expansion. Such a page of code is called a **segment.** A segment is identified by a name which indicates its function at the next higher level segment in the program. The segments of a program, therefore, form a tree structure as shown in figure 14.2.

Each program segment has only one entry—at the top, and one exit—at the bottom. If other segments are named within it, they are, in turn, entered at the top and exited at the bottom, back to the calling segment.

The problem of proving the correctness of any additional sublevel tasks is thereby reduced to proving the correctness of a program of at the most one page (segment), in which there may exist various named subtasks. The named subtasks will subsequently be verified, possibly in terms of even more detailed sublevels of specifications, until segments with the "action" code are reached and verified.

Thus, the top-down programming process provides a vehicle for maintaining the integrity of the program/system step by step. The code is produced in a top-down fashion rather than bottom-up. Integrating and control code is produced before functional code, and no unit checking of detail modules occurs until the modules are needed. This top-down approach provides the benefits of giving the critical top segments the most testing, of giving earlier warning of problems with segment interfaces, and spreading the debugging and testing over a greater part of the development cycle.

Another principle of top-down structured programming is a limit on the number of statements that make up a segment or routine. Easy program readability requires that it not be necessary to turn many pages to understand how a routine works. A practical rule is that a segment should not exceed a page of code—about fifty lines. But segmentation should be more than just breaking up a program into page-sized pieces. Program segmentation should reflect the division of the program into parts that relate to each other in a hierarchical fashion. Well-designed segments should represent some natural function of those operations which are closely related to each other. This makes the segment easier to understand and to modify if the need ever arises.

One of the best techniques for providing program segments and promoting clarity is the intelligent use of subprograms. A subroutine may be used to contain short, well-defined tasks that can be understood and checked out separately. Subprograms, therefore, provide an easy way of segmenting a program and providing modules which serve as building blocks in the design of large programs.

If a structured program has been properly designed, program checkout is greatly facilitated. Most of the bugs can be localized to a particular segment of code, inasmuch as structured programs are highly modular and the flow of control parallels the listing. That is, if you get correct results down to point X in the program, then the program has correctly performed those statements prior to point X, and it hasn't done what is required after this point. This is generally not true of nonstructured programs because of the numerous transfer statements and the skipping around in the program.

Structured programming and top-down design go hand in hand and complement each other. The combination of these two techniques tends to generate a modularized, clearly organized description and development of a computer program.

14.4 Techniques for Better Designed Programs

The design of one's program is often built around various techniques used to make a program more understandable and readable. Seven rules recommended for a more readable FORTRAN program are:[5]

1. Comment cards should be used freely, but wisely. Each program and subprogram should include an initial group of comments that completely describes the program, what it does, what the inputs are, and what outputs are produced. Comments should be used at the beginning of "logical blocks," that is, the logical grouping of statements for a major task/block in the program. Several blank comment cards should be used to set off the logical sections so that they will stand out as such.

    Comment cards should normally include a few special characters such as asterisks, at the beginning, so that the cards will not be confused with FORTRAN statements. Remarks contained in the comment card should tell *what* is logically being done, not how, as given in the FORTRAN statement. For example, writing a comment that says add 1 to K before the statement K = K + 1 is no particular help. Use a remark like bump records processed counter by 1. Again, tell *what* is being done in terms of the problem and not in terms of the statement. Express your comments in a way that they will help the reader know what is going on.

2. Eliminate unnecessary GOTO's. The GOTO statement should be used sparingly. Programs containing excessive GOTO's are difficult to follow, understand, and document, since they are heavily interlocked. Situations that may require the heavy use of GOTO's may be better programmed as subprograms. In some cases, the excessive use of GOTO's can be avoided by the use of multiple Logical IF statements. For example:

    IF (N .EQ. 1) K = 0
    IF (N .EQ. 1) A = B + C

3. Every DO statement should have a CONTINUE as the range statement in the loop. Each DO statement should also refer to a separate CONTINUE statement. The statements contained in the DO loop should be indented several spaces under their respective DO. Consistent indentation enhances readability so that the finished program will exhibit in a pictorial way the relationships among the statements.

4. When an IF statement contains compound conditions, place each simple condition within a set of parentheses to make the order of the logical operators (.AND., .OR., and .NOT.) crystal clear. When there are three or more simple conditions, use continuation cards to put each simple condition on a separate card and align the conditions vertically.

5. Assign statement numbers in some systematic order that will facilitate understanding in following the program. For example, assign the statement numbers in ascending sequence in some increment, say by tens. Thus, new statement numbers can be inserted between the existing numbers. The main point here is that a program is easier to follow when a reference is made to a statement number that occurs elsewhere in the program. FORMAT statement numbers should be assigned a statement number beginning with a 9, to clearly indicate this type of statement.

---

5. Daniel D. McCracken, and Gerald M. Weinberg, "How to Write a Readable FORTRAN Program," *Datamation* 18, no. 10 (October 1972), pp. 73-77.

6. All FORMAT statements should be placed at the beginning of the program. This facilitates the search for a particular FORMAT statement that is used by several I/O statements in the program. The FORMAT statements should be placed in a sequential order by their statement number. Thus, one has merely to flip back to the beginning of the source listing to find any FORMAT statement. The removal of the FORMAT statements from the coded logic also improves the readability of the program.
7. Assign meaningful symbolic names. Don't be afraid to use all six characters to form a symbolic name. This is especially helpful with variables, where the name helps identify what value is being used. Never use the same variable name to represent different values in the same program, just to save space. This can cause a debugging nightmare.

The use of these suggested techniques can help you write better designed programs which are more meaningful to other people. We now turn our attention to some techniques that can be used to debug execution time errors.

14.5 Debugging Techniques for Execution Time Errors

Programmers are not infallible, you can be sure. The existence of program bugs is almost guaranteed, especially with large and/or complex programs. The programmer must devise ways to verify his output results and to locate the errors in his program when they occur. This section discusses various techniques that may be used in locating execution time errors.

Execution errors usually fall into one of two types. The first type results in *logic errors*—the output results are not correct. The computer produces some output, but not what was expected. The second type of error results when the computer is told to do something that it cannot do and the program "blows up." Such an error is called a *system error* and is caused by some illegal operation.

First, we must never assume our output to be correct, therefore we should always check our output results. As mentioned in chapter 4, the best way to verify our answers is to make hand calculations of some of the input values and determine what our results should be. When our program has several paths it can take in the logic, an input value should be used for each alternative path. If the calculations for the input values involve fractions and are highly complex, prepare some simple numbers and test data. Then run the program with this simple set of input values and compare the results with the easy-to-calculate values.

Locating the Logic Error

Once we have discovered that a bug exists in our program, there are numerous techniques we can use to help us locate the logic error. Among the many techniques available are:

1. Desk-check the program.
2. Echo-check the input data.
3. Print intermediate values from calculations.
4. Include check-point messages.
5. Rewrite the segment of code in error.
6. Use various control card (JCL) options.

Desk-check the Program

You are probably already familiar with the desk-checking technique to locate logic errors. You play the role of the computer and manually trace the execution path through the program. As you follow the logic, jot down the computed values and see if you have told the computer the proper order and way to solve the problem. When an error is discovered, the logic can be corrected. You should not stop as soon as you have found the first error. Continue through the program and correct as many errors as you can find. Resist the temptation to rush back to the computer after finding only one error. Sometimes we can't figure out why the computer doesn't give the same answers that we get in the desk-check. Now we need to turn to other techniques to find out what the computer is really doing and where we went wrong.

Echo-check the Input Data

Too often the beginning programmer assumes the error is in the program and spends fruitless hours trying to find the bug, when in reality the error lies in the data. A data field may be punched in the wrong columns, in the wrong format, or have incorrect values. The computer does not question what values were given to it to manipulate. It simply reads the data as told. Perhaps you should be using an F7.2 format code when you are actually using an F6.2 specification. Maybe you should be skipping only three blank positions when you are actually skipping four blank columns.

During the first test run you should echo-check (print) your input items to verify their validity. Echo-checking your input data means to print the input values you have just read. This way you can be certain that the input values were read correctly and that you have the proper values stored in memory. After you are certain the input values are read correctly, the WRITE statements can be removed from your program.

Print Intermediate Values from Calculations

Sometimes we compute a long list of values or use a long, "hairy" formula such as the computation for a linear regression. It is best to break up long formulas into shorter ones. When the correct output is not produced, we can print the resulting values in each of the Assignment statements. Each computed value can be examined to see where we went wrong in the calculations.

Include Checkpoint Messages

Temporary WRITE statements can be inserted in a program to trace the flow of execution. You may wonder how you got to a certain statement and how you could possibly get the output results you did. Perhaps you thought a certain condition would be met, but you didn't take into account a special situation. Maybe you have an arithmetic truncation error that always produces a false condition in a comparison.

Checkpoint messages inserted throughout your program will allow you to follow the execution sequence and see exactly what paths the computer is actually taking. Checkpoint messages also tell you exactly how far the computer got in your program before blowing up, in case you encounter a system error.

Different messages must be printed to follow the flow of execution if GOTO's are used in the program. Consider the following example.

```
            . . .
            . . .
         1  . . .
            WRITE (6, 99)
         99 FORMAT (1H0,7HCHECK 1)
            . . .
            . . .
            IF (A .EQ. 3.0) GOTO 2
            WRITE (6,98)
         98 FORMAT (1H0,7HCHECK 2)
            . . .
            . . .
         2  . . .
            IF (B .EQ. 0.) GOTO 1
            WRITE (6, 97)
         97 FORMAT (1H0,7HCHECK 3)
```

With different checkpoint messages as illustrated in the previous example, we can accurately trace the flow of execution in the program to see what branches are taken in our logic.

We may also combine the use of checkpoint messages and the output of intermediate values for certain variables to provide useful trace routines. When the program is free of errors, at least in certain segments, these WRITE statements are easily removed. The combination of printing checkpoint messages and desired variables is perhaps the best technique for a beginning student to use in finding logic errors in complex programs.

Rewrite the Segment of Code in Error

When all else fails, one should rewrite the routine in which the error occurs. Sometimes rewriting the same statements reveals the hidden error. Perhaps the formula was coded incorrectly in the Assignment statement. Maybe an extra set of parentheses is needed to change the order of computations. Or maybe a needed variable or constant was overlooked in the formula. If rewriting the same routine does not work, try a different approach in the logic of the routine. Often a different algorithm or method of attack will correct the problem.

Use Various Control Card Options

Most compilers allow the use of certain control card options to provide extra information from the compilation. One common option on IBM compilers is the MAP option to provide the type and location in memory of each variable name used by the program. The generated machine language may also be obtained, but has little value for the beginning student (and may cause volumes of output).

System Errors

The second type of execution errors is one that causes the computer to abnormally abort your program. An arithmetic calculation may result in an attempt to compute a number exceeding the maximum value that a specific computer is designed for. This type of error is called an *overflow*. The calculated value may truly be too large for a storage location, or the error may be caused by a logic error in the program. For example, on IBM compilers, if a student reads a data value under an integer format code and stores it at a real variable, with a subsequent multiplication, an overflow error will occur. The reverse of an overflow error is an *underflow* error. This means that one tried to compute a number exceeding the smallest value allowed.

Other types of system errors may be caused by a divide-by-zero, an invalid operator, an invalid subscript, trying to read input data with the wrong type of format specifications, etc. The computer aborts your program and prints an associated system error message as to what caused the termination. With these systems messages and printed results produced up to program termination, you may have a clue as to the statement or area of your program that caused the error. Otherwise, you may have to resort to some of the previously discussed debugging techniques. The best way to find system errors is the use of checkpoint messages.

A memory dump can be taken on many computers when a program aborts. The **memory dump** displays the entire contents of memory used by the program. This dump can be extremely helpful in finding the most obscure errors. However, one normally must have a thorough knowledge of the computer's machine language and addressing procedures in order to profit from a memory dump. System programmers are quite familiar with a memory dump and can provide assistance as to what and where the error is that caused the abnormal termination. Beginning programmers will normally waste their time trying to comprehend the large output generated from such a dump.

The WATFOR and WATFIV compilers are extremely helpful in locating system execution errors. In addition to a clear execution fault error message, a message saying "PROGRAM WAS EXECUTING LINE n IN ROUTINE M/PROG WHEN ERROR OCCURRED" will be printed. The line number n in this message refers to the line sequence number assigned to the statements in your program by the compiler. You may still have to echo-check the variables used in the statement to find which value caused the error. These compilers can also detect many logic errors such as undefined variables, invalid subscripts, changing the index variable and indexing parameters in a DO loop, and many more.

Guidelines to Aid in Debugging

The following suggestions are given as aids to debugging execution errors and also to help you prevent these types of errors.

1. Know the program. Know where each routine that performs a particular function is in the program. This can save considerable time in determining what the program was doing when an error occurred.
2. Check the initial values of variables. Programmers often fail to initialize a variable to the proper beginning value. When a routine does not work, it is difficult to understand why. Make sure all counters and variables are initialized to their proper values. Make sure all variables used in an expression to the right of the equals sign have been defined. Make sure that index variables which are later reused are reinitialized to their necessary values.
3. Plan for errors. The old saying "an ounce of prevention is worth a pound of cure" is certainly true in programming. You must realize that errors will occur and plan to avoid them. Some suggestions are:
    a) Write your program in a structured form so that errors can be localized to certain modules. Use subprograms where feasible.
    b) Check for all possible conditions in your program. If there are three possible valid input values for an item and, thereby, three possible conditions, don't check for two and assume the third. Check for all three and execute an error routine if the condition is not one of the three valid ones.
    c) Use straightforward logic so that everything you are doing is clear and obvious—not tricky and hidden, and thereby likely to be difficult to debug.
    d) Plan for debug traces and printouts in the development of the program. Assume that you will eventually need them and include these precautionary measures from the start. Thus, additional compiles and test runs can be avoided. The debug statements can be removed after various routines have been proven correct.
    e) Save all test decks and test results. Rerunning these same test decks can save considerable programmer checkout effort if any modifications are made to the program.
4. Check the values of constants in the statement, especially those with more than seven digits. Also consider the possibility of truncation or rounding errors.
5. Check the variable names for misspelling or for type. Output values of zeroes are often the result of conflicting format specifications on some compilers.
6. Check the order of computations in the Assignment statement. You may need an extra set of parentheses in the expression to cause one operation to take place before another.
7. Check for integer division. This may provide a truncated result which leads to a bad value.
8. Check the subscript value and array dimensions for a subscripted item. A bad subscript value of less than one or greater than the dimension size can produce addressing errors.
9. Make sure that the arguments passed between program units are correct and agree in number, order, and mode. Upon entering a subprogram, print a checkpoint message that you entered the subprogram and the values of the dummy arguments. This ensures that the correct values are passed for the arguments and prevents the hard-to-find errors resulting from incorrect arguments being used.
10. After you have found the error in a routine, check the entire program to make sure you did not make the same error elsewhere.

Always try to locate the error yourself, first by trying a specific or combination of the debugging techniques discussed in this section. If you are finally unable to locate the bug, then consult your instructor. Bring the results of the debugging attempts with you, as your instructor will often need these in assisting you to find your problem.

14.6 REVIEW QUESTIONS

1. Structured programming techniques may be used to write any size program. (True/False)
2. Structured programming has been brought about by the soaring cost of hardware development. (True/False)
3. Structured programming stresses program readability, reliability, and maintainability over efficiency. (True/False)
4. List three benefits that usually result from structured programming.
5. List the six control structures which may be used in structured programs.
6. FORTRAN is a very popular language to use in structured programming because of its powerful block structure capabilities. (True/False)
7. Structured programming techniques can be used for developing programs in any language. (True/False)
8. The top-down program design technique uses a chart which shows a hierarchical structure relationship between the required major tasks and their subtasks in a program. (True/False)
9. In top-down structured programming the lowest level or functional subtasks are coded and tested first. (True/False)
10. An important feature of program segments in a top-down structured program is that they have only one entry and one exit point. (True/False)
11. Name the two types of errors that can occur during the execution of a program.
12. The process of manually following the logic in a program to determine its operations is called _____ _____ .
13. The printing of input values as they are read to verify their correctness is known as an _____ .
14. The display of various messages to trace the flow of execution through a program or various segments is called _____ messages.
15. Name three types of system error messages that may occur.
16. Name five things that could cause a logic error in a program.

Function Subprograms and Additional Subprogram Statements 15

15.1 Introduction to Function Subprograms

Function subprograms serve the same purpose as subroutines, in FORTRAN programs. That is, they provide a subprogram that can be coded once and executed at different locations in other program units. *Function subprograms* are used where a single calculated value is needed.

In mathematics the term *function* refers to an operation or rule for computing a quantity from a set of values called arguments. For example, we may want to compute the function of Y according to the formula: $Y = X^2 + Z$. The values X and Z are input arguments to the function of Y. The value of Y can be computed by using the input arguments according to the given formula. If $X = 3$ and $Z = 1$ then Y is calculated as 10.

Thus, the concepts of function subprograms closely match those of mathematical functions. One or more arguments are passed to the function subprogram which calculates a single resulting value. The returned value by the function subprogram is known as the *function value.*

A function subprogram is a separate program unit just as a subroutine is. Since a function subprogram is compiled as a separate program unit, there is no conflict between duplicate variable names and statement numbers in other program units. There must be at least four statements in a function subprogram just as in a subroutine, except that the first statement is the FUNCTION statement to identify the start of the subprogram.

A function name is assigned to the subprogram and follows the keyword FUNCTION. A parameter list is *always* included after the function name. For example:

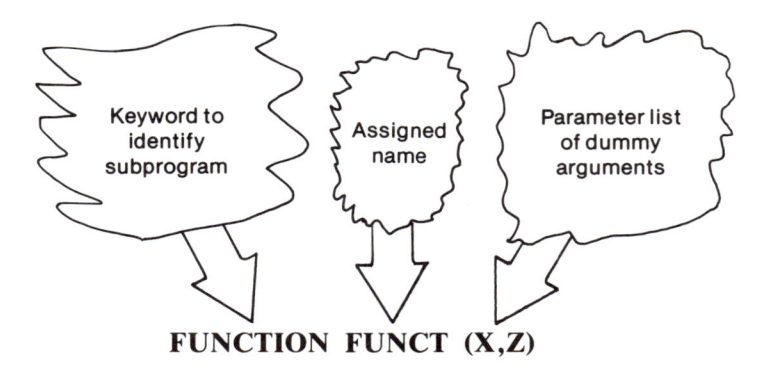

```
          Keyword to        Assigned       Parameter list
           identify           name         of dummy
          subprogram                       arguments
              ↓                ↓                ↓
            FUNCTION    FUNCT   (X,Z)
```

A function subprogram must contain at least one executable statement, which normally includes an Assignment statement that assigns a value to the subprogram name. The RETURN and END statements serve the same purpose as in subroutines.

The function subprogram name is formed the same way as other symbolic names. However, the initial letter of the function name determines the type of value returned. That is, if we assign a real symbolic name as the function subprogram name, a real value is returned. Likewise, an integer value is returned if an integer symbolic name is assigned.

270   Function Subprograms and Additional Subprogram Statements

The function subprogram can contain any number of statements and any kind of statement except another FUNCTION, SUBROUTINE, or BLOCKDATA statement. So in many respects, a function subprogram is constructed and operates the same way as a subroutine. The three major differences are:

1. The way it is invoked (i.e., called).
2. The number of values that may be returned and the method used.
3. The location in the calling program to which control is returned.

A function subprogram is invoked, not by a CALL statement, as in a subroutine usage, but by using its name in an arithmetic expression. We will use the term *invoke* or *reference* to indicate the way a function subprogram is executed—not the term "called"—to prevent confusion with a subroutine execution. Although, many people use the term "call" to refer to the execution of either a subroutine or a function subprogram.

The function subprogram reference is normally made in an Assignment statement such as:

$$\text{RESULT} = \text{FUNCT}(X, Z)$$

where FUNCT is the name of the subprogram; X and Z are arguments. A function subprogram reference, however, may be used at any place in an expression where a variable or constant of the same type could be used. For example:

$$\text{IF (FUNCT}(X, Z) \text{ .GT. } 0.0) \text{ GOTO } 20$$

Here the function subprogram is invoked to obtain a value which is used to determine a logical condition in an IF statement.

The second major difference in function subprograms is the argument list and the way the computed value is returned to the calling program. Normally only one value is returned from the function subprogram and it is returned under the subprogram name. The calculated value is placed in the function name by an Assignment statement such as:

$$\text{FUNCT} = X ** 2 + Z$$

where FUNCT is the function subprogram name.

In ANSI FORTRAN the function subprogram may return more than one value by assigning new values to any of the dummy arguments. If any dummy argument is modified, the values of corresponding actual arguments in the calling program are also changed. In many other compilers it is forbidden in a function subprogram to modify any of the variables representing the dummy arguments. That is, the dummy arguments may be used in calculations, decision making, or whatever, but they must not be given a new value in the function subprogram. The modification of a dummy variable in a function subprogram is not a good programming practice. This technique often results in hard-to-find errors during debugging.

The function subprogram must *always* include a parameter list. The parameter list may contain more than one argument; but at least one argument must always be given in the parameter list. ANSI FORTRAN generally allows an unlimited number of dummy arguments. Some smaller compilers like Basic FORTRAN IV limits the number of dummy arguments to fifteen.

The parameter list of a function subprogram and the parameter list of the invoking statement may contain any type of arguments that can be used with subroutine parameter lists. However, the actual arguments must agree with the dummy arguments in the same four ways—number, order, type, and length. Actual arguments may or may not be the same name as their corresponding dummy arguments. They must always keep positional correspondence in a left-to-right order. For example:

```
mainline
function
reference:    Y = FUNCT (A,B)

FUNCTION
statement:    FUNCTION FUNCT (X,Z)
```

If an array name is passed as an argument, then the array must also be dimensioned in the function subprogram.

The third major difference in using function subprograms is the location in the calling program to which control is returned. Control is returned to the *same* statement that invoked the function subprogram. It is *not* returned to the next statement as with the CALL and subroutines. The reason for this is that the execution of the statement that invoked the subprogram must be completed. For example:

$$VALUE = FUNCT(X,Z) + 3.7 * S$$

The reference to the function subprogram named FUNCT supplies only one value in the expression. The rest of the expression must be completed before the one resulting value is assigned to the variable VALUE. The same principle holds true even if we had an Assignment statement such as:

$$VALUE = FUNCT (X,Z)$$

Control must be returned to the same statement to check for additional operations and to assign the computed value to the assignment variable.

In an arithmetic expression that contains a function subprogram reference, the function reference takes precedence over any arithmetic operator and is, therefore, performed first. Of course, parentheses can be used to override this hierarchy and cause other operations to be performed before the function reference. For example:

$$Y = (3.0 + R) * FUNCT (X,Z) - T$$

First, R is added to 3.0 because of the use of parentheses. Secondly, the function FUNCT is invoked because it takes precedence over arithmetic operations. Next, the result of the function is multiplied by the result of the first step. Lastly, T is subtracted from the result in step three to obtain the final value for Y.

The returned value in the function subprogram name can never be accessed by using the subprogram name other than where an expression is allowed. For example, it is forbidden to write:

WRITE (6,98) FUNCT (X,Z)

where FUNCT is a function subprogram reference. The returned value via the subprogram name must be assigned to a new variable in an Assignment statement if the calculated value is to be printed or used in later calculations.

Function subprograms may invoke other function subprograms or call subroutines. But a function subprogram must not invoke itself or invoke another subprogram that references the function subprogram. So basically the same concepts that apply to subroutine subprograms hold true for function subprograms, except for the three noted differences.

To illustrate the operation of function subprograms let us code our problem of computing the function of $Y = X^2 + Z$. Our program units are as follows:

Function Subprogram

```
FUNCTION FUNCT(X,Z)
FUNCT = X ** 2 + Z
RETURN
END
```

Mainline Program

```
99 FORMAT(2F5.2)
98 FORMAT(1H0,F7.2)
   READ (5,99) X,Z
   Y = FUNCT(X,Z)
   WRITE (6,98) Y
   READ (5,99) A,B
   Y = FUNCT (A,B)
   WRITE (6,98) Y
   STOP
   END
```

In our mainline program we first read two values under the variables X and Z. The function subprogram is invoked using these input values. The subprogram FUNCT receives the input values X and Z under the dummy arguments of X and Z, respectively. The function is computed according to the formula $X^2 + Z$ and the calculated value placed in the function subprogram name. Control is returned to the mainline program where the calculated value is assigned to the variable Y and printed. This same process is also performed with two more input variables A and B.

Table 15.1 summarizes the differences between function and subroutine subprogram operations. Study this table carefully to avoid confusing the two types of subprogram usage.

TABLE 15.1
COMPARISON OF FUNCTION AND SUBROUTINE SUBPROGRAMS

Function	Subroutine
Invoked by using its name in an expression.	Invoked by means of a CALL statement.
Name takes on a value and may be used in an expression. Dummy arguments may be modified if more than one value is returned.	Name does not take on a value. All values are returned through the arguments.
The type of the value returned depends upon the type name assigned.	Subroutine names are not typed. The type of values returned depends upon the type of the arguments.
Must have at least one dummy argument.	Does not require any dummy arguments.
Control is returned back to same statement.	Control is returned to the next executable statement after the CALL.

You may now wonder where we do use a function subprogram in place of a subroutine. You can always use a subroutine subprogram to perform any type of operation. Function subprograms should be used when a single calculated value is needed from a routine.

Function subprograms are flowcharted by the same symbol and technique as subroutines are. The job setup for programs which include function subprograms also remain the same as with subroutines. Whenever function subprograms are used in conjunction with subroutines, it makes no difference which type of subprogram is included first. In fact, you may have a subroutine followed by a function subprogram followed by another subroutine, and so on.

There are two kinds of Function subprograms available in FORTRAN.

1. Function Subprograms—a complete subprogram similar to subroutines.
2. Statement Functions—a one-statement function subprogram represented in an arithmetic Assignment statement.

First, we will continue to discuss the function subprogram.

15.2 The FUNCTION Statement and a Sample Subprogram Using Arrays

The FUNCTION statement identifies the beginning of a function subprogram. It normally begins with the word FUNCTION, and is followed by its assigned name. A parameter list is enclosed within parentheses and follows the function subprogram name.

A type option may be given before the keyword FUNCTION if one wishes to explicitly type the resulting value of the subprogram name. For example:

INTEGER FUNCTION FSUB (I,K,M)

Even though the function subprogram is assigned a real type name (FSUB), the resulting value is integer because of the Explicit INTEGER type option used. The type specification can be INTEGER, REAL, DOUBLE PRECISION, LOGICAL, or COMPLEX to explicitly type the resulting value in the function subprogram name.

Figure 15.1 gives the general form of the FUNCTION statement.

FUNCTION Statement—General Form

---

      type FUNCTION name (darg$_1$, . . ., darg$_n$)

where:

type	—	is an explicit type specification which denotes the type value of the function name. **Type** may be REAL, INTEGER, DOUBLE PRECISION, LOGICAL, or COMPLEX. Its inclusion is optional.
name	—	is the assigned function name.
(darg$_1$, . . ., darg$_n$)	—	represents a parameter list of dummy arguments.

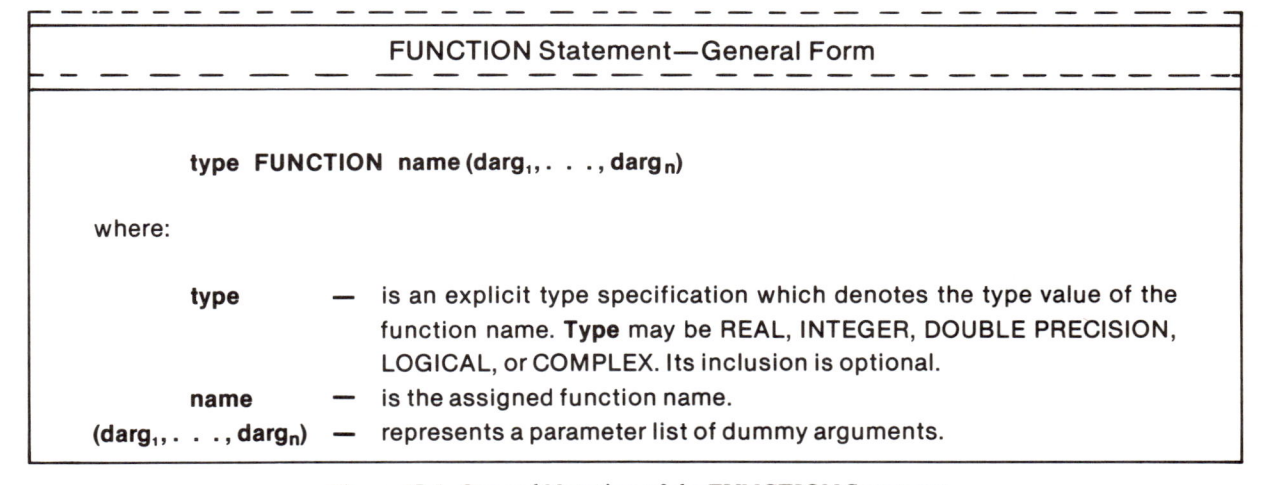

Figure 15.1. General Notation of the FUNCTION Statement

Now let us examine a function subprogram that uses an array as a dummy argument. Suppose we write a function subprogram named NSUM to sum the elements in a one-dimensional integer array. The dummy arguments will include an array name for a one-dimensional array and the number of elements to sum. Our coded program units are:

Function Subprogram

```
       FUNCTION NSUM(LIST,N)
       DIMENSION LIST(20)
       NSUM = 0
       DO 10 I = 1,N
           NSUM = NSUM + LIST(I)
10     CONTINUE
       RETURN
       END
```

Mainline Program

```
       DIMENSION L1(10),L2(20)
99     FORMAT (10I3)
98     FORMAT (20I2)
97     FORMAT (1H0,10I5,I6)
96     FORMAT (1H0,20I4,I5)
       READ (5,99) L1
       L1SUM = NSUM(L1,10)
       WRITE (6,97) L1,L1SUM
       READ (5,98) L2
       L2SUM = NSUM(L2,20)
       WRITE (6,96) L2,L2SUM
       STOP
       END
```

Our mainline program first reads a ten-element array named L1. The array name and the number of elements are passed as actual arguments when the subprogram is invoked by the statement

$$L1SUM = NSUM (L1,10)$$

The subprogram receives the input arguments under the dummy arguments of LIST and N, respectively. The sum of the ten items are computed and assigned to the function subprogram name. The calculated sum is returned under the function name NSUM and assigned to the variable L1SUM in the main program. Then the array L1 and its sum are printed.

Next, the mainline program reads a twenty-element array named L2. The array and its number of elements are passed in the statement

$$L2SUM = NSUM (L2,20)$$

Again the array and number of elements are received under the dummy arguments LIST and N, respectively. The sum of the items is calculated and returned under the function name. The mainline program assigns the returned value to the variable L2SUM. Then the array L2 and its sum L2SUM are printed.

The FORTRAN compiler also includes many preprogrammed or "canned" function subprograms which are available to the programmer. The next section discusses FORTRAN Library-supplied function subprograms available for your use.

15.3 FORTRAN Library-supplied FUNCTION Subprograms

Dozens of function subprograms are available with the FORTRAN language. For that reason many people refer to them as *built-in* functions. The terms Library-supplied functions and built-in functions are used interchangeably in this section. All you have to do to invoke the function is to use its name and appropriate arguments in the proper FORTRAN statement. User-developed function subprograms may also be kept on a user Library which are stored on magnetic disk and be available for programmer use.

The FORTRAN Library-supplied function subprograms provide a wide range of calculations. Many compute trigonometric functions such as sine, cosine, tangent, etc. For example, to compute the trigonometric sine of a 90° angle (which is $\pi/2$ in terms of radians) we simply write:

$$VAL = SIN (3.14 / 2.0)$$

Other functions convert data from one type to another (i.e., real to integer and integer to real). For example, if we wish to convert the real variable X to an integer value to be used in an expression, we could use the function **IFIX** and code:

$$\text{IRESLT} = \text{IFIX}(X) + 7 * K$$

The computer would provide an integer quantity for X to provide integer mode arithmetic for the expression. If we wanted to convert the variable I to real mode for an expression we could use the function **FLOAT** and code:

$$\text{RESULT} = 7.5 * \text{FLOAT}(I)$$

Still others compute the square root of a real quantity, find the absolute value of a quantity (the positive value), and find the smallest or largest value in a group of variables. Some of the more common built-in functions are described in examples below.

SQRT   to compute the square root of a real argument ($\sqrt{X}$). For example:
$$\text{RESLT} = \text{SQRT}(X)$$

ABS   to find the absolute value of a real argument (|X|). For example:
$$\text{RESLT} = \text{ABS}(X)$$

IABS   to find the absolute value of an integer argument (|K|). For example:
$$\text{RESLT} = \text{IABS}(K)$$

AMIN1 to find the smallest real quantity in a group of real arguments. For example:
$$\text{RESLT} = \text{AMIN1}(A,B,C,D)$$

AMAX1 to find the largest real quantity in a group of real arguments. For example:
$$\text{RESLT} = \text{AMAX1}(A,B,C,D,E)$$

ALOG   to find the natural log of a real argument. For example:
$$\text{RESLT} = \text{ALOG}(X)$$

MOD   to obtain the remainder from the division of two integer values by dividing the first argument by the second. For example:

$$\text{IREM} = \text{MOD}(17,3)$$

The result is 2 (17/3 produces quotient = 5 and remainder = 2)

Some functions require only one argument, whereas others require more than one. Each function also requires a specific type of argument. For example, the SQRT function requires that its argument be type real. If we tried to take the square root of an integer argument using SQRT, we would get an error message. The arguments of the function ALOG must not be equal to zero.

An argument of a built-in function may be a constant, variable name, subscripted variable, or an arithmetic expression. An arithmetic expression may also include built-in functions. Under *no* condition may a built-in function be an argument of itself. The following statement would be invalid.

$$\text{RESLT} = \text{SQRT}(3.5 * \text{SQRT}(X))$$

Examples of valid arguments with Library-supplied functions are:

$$C = \text{SQRT}(A**2 + B**2)$$
$$\text{RES} = \text{ALOG}(987.4)$$
$$\text{AMX} = \text{AMAX1}(A(1),A(2),A(3),B)$$
$$\text{RV} = \text{FLOAT}(I + 3)$$
$$X = \text{SQRT}(A) + \text{ABS}(B)$$
$$\text{RSLT} = \text{SQRT}(\text{FLOAT}(L))$$

TABLE 15.2
LIBRARY FUNCTIONS USED IN BUSINESS AND SCIENTIFIC PROBLEMS

User Reference Name	FORTRAN Name	Mathematical Definition	Number of Arguments	Type of Arguments	Type of Returned Value
Square Root	SQRT	$\sqrt{a}$	1	Real	Real
Absolute Value	ABS	$|a|$	1	Real	Real
	IABS	$|k|$	1	Integer	Integer
Exponential	EXP	e^a	1	Real	Real
Conversion from integer to real	FLOAT	change integer mode argument to real mode.	1	Integer	Real
Conversion from real to integer	IFIX	change real mode argument to integer mode.	1	Real	Integer
Modulo Arithmetic (Real remainder)	AMOD	y = Modulo(a,b)	2	Real	Real
Modulo Arithmetic (Integer remainder)	MOD	y = Modulo(i,j)	2	Integer	Integer
Natural Logarithm	ALOG	$\log_e(a)$	1	Real	Real
Common Logarithm	ALOG10	$\log_{10}(a)$	1	Real	Real
Largest Value	AMAX0	$\max(a_1,...,a_n)$	≥ 2	Integer	Real
	MAX0			Integer	Integer
	AMAX1			Real	Real
	MAX1			Real	Integer
Smallest Value	AMIN0	$\min(a_1,...,a_n)$	≥ 2	Integer	Real
	MIN0			Integer	Integer
	AMIN1			Real	Real
	MIN1			Real	Integer
Truncation	AINT	eliminate the fractional portion of the argument.	1	Real	Real
	INT			Real	Integer
Transfer of sign	SIGN	sign of a_2 times $|a_1|$	2	Real	Real
	ISIGN			Integer	Integer
Positive Difference	DIM	$a_1 - \min(a_1,a_2)$	2	Real	Real
	IDIM			Integer	Integer

Table 15.2 provides a list of the Library-supplied functions commonly used in business and scientific problems. Table 15.3 provides a list of the Library-supplied trigonometric functions. Remember that you must always supply the correct number and type of arguments required by the function. The Type-of-Returned-Value column indicates the type of value returned as to a real or integer mode quantity.

TABLE 15.3
LIBRARY OF TRIGONOMETRIC FORTRAN FUNCTIONS

User Reference Name	FORTRAN Name	Mathematical Definition	Number of Arguments	Type of Arguments	Type of Returned Value
Trigonometric Sine	SIN	sin(a) a in radians.	1	Real	Real
Trigonometric Cosine	COS	cos(a) a in radians.	1	Real	Real
Arctangent	ATAN	arctan(a) a in radians.	1	Real	Real
	ATAN2	arctan(a_1/a_2) a in radians.	2	Real	Real
Hyperbolic Tangent	TANH	tanh(a) a in radians.	1	Real	Real
*Trigonometric Tangent	TAN	tan(a) a in radians.	1	Real	Real
*Trigonometric Cotangent	COTAN	cotan(a) a in radians.	1	Real	Real
*Arcsine	ARSIN	arcsin(a) a in radians.	1	Real	Real
*Arccosine	ARCOS	arccos(a) a in radians.	1	Real	Real
*Hyperbolic Sine	SINH	sinh(a) a in radians.	1	Real	Real
*Hyperbolic Cosine	COSH	cosh(a) a in radians.	1	Real	Real

*Indicates nonstandard ANSI function, but included since they are found in most FORTRAN Libraries.

Many computer centers have hundreds of subprograms available for a wide variety of application problems. You should check with the computer center to find out what subprograms are already available before you attempt a time-consuming crash project of writing a complex routine.

15.4 The Statement Function Subprogram

A statement function is a one-statement function subprogram. It is not coded as a separate subprogram in the same way that subroutines and regular function subprograms are. There is no keyword to identify the statement function, nor are any RETURN or END statements associated with it. The statement function can compute only one value.

A "modified" arithmetic Assignment statement is used to define the mathematical function (hence the term **statement function**). The modified Assignment statement that defines the statement function looks like this.

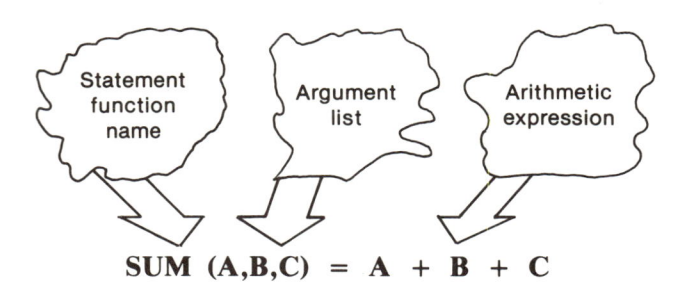

```
       Statement      Argument       Arithmetic
       function        list          expression
        name
          ↓              ↓              ↓
      SUM (A,B,C)  =  A  +  B  +  C
```

To the left of the equal sign is the statement function name followed by its parameter list of dummy arguments. The right-hand part of the statement contains the arithmetic expressions that define the calculations. The statement function must not be assigned a statement number.

The statement function name is formed the same way as a function subprogram name. Its initial letter determines the type of value returned. Of course, the chosen name of the statement function must not be the same as another variable or array name in the program.

We may have any number of dummy arguments in the statement function, but it must contain at least one. (Basic FORTRAN limits the number of dummy arguments to 15.) The same dummy arguments may be used in more than one statement function and may be the same as other variable names used in the program. A dummy argument may not be a subscripted variable; however, subscripted variables may be used as actual arguments that are passed to the statement function. The same argument rules that apply to function subprograms also apply to statement functions. The actual and dummy arguments must always agree in number, order, type, and length.

The arithmetic expression appearing to the right of the equals sign may contain

1. numeric constants
2. numeric variables
3. built-in functions
4. other user-defined function subprograms
5. other statement functions

The expression must *not* contain any subscripted variables. The variables may be the same as the dummy arguments, and also additional variables that do not appear in the argument list. Those variables used in the arithmetic expression, but not appearing in the argument list, contain their current value. When variables of this nature are used, their values may be changed between references to the statement function. The expression must contain all of the dummy arguments.

This one statement subprogram is often called the *statement function definition*. This statement is nonexecutable and only defines the statement function; thus, it does not cause any computation to take place. It is limited to one statement and can return only one value.

The statement function must immediately precede the first executable statement in the program. It must also be placed after all specification statements. Thus, its position in a program is between the last specification statement and the first executable statement.

The general form of the statement function is:

$$\text{name}(darg_1, \ldots, darg_n) = \text{expression}$$

where:
- **name** — is the name assigned to the subprogram.
- **darg₁, . . . , dargₙ** — represent a parameter list of one or more nonsubscripted dummy arguments.
- **expression** — is an arithmetic expression involving the dummy arguments and optionally additional constants, variables, and various types of function subprogram references.

Following are valid examples of statement function definitions.

DIF(X,Y) = X − SQRT(Y) * 2.0
TOT(A,B,C,D) = A * (B−C) / D + E
IRSLT(J,K,L) = IFUNCT(J,K) / L − 3 * L

A statement function is invoked in the same way that we reference a function subprogram. Its name may be used in any form of arithmetic expression. To invoke the statement function named SUM which computes the sum of three arguments, we would code:

RESULT = SUM(A,B,C)

or:

IF (SUM(A,B,C) .LE. 0.0) RSLT = 0.0

The actual arguments are passed in a left-to-right correspondence to the dummy arguments. For example:

SUM (X,Y,Z) = X + Y + Z

R = SUM(A,B,C)

The statement function must not refer to itself. The statement,

SUM(A,B,C) = A + B + C * SUM(A,B,C)

would be invalid. A statement function may, however, include another statement function as part of its arithmetic expression. When this is true, the statement function included in the arithmetic expression must appear before the definition of the statement function in which it is used.

Since this type of function subprogram appears in the program and is not a separate program unit, it is compiled along with the program unit it is in. Therefore, the statement function can only be referenced by the program unit it is in and not by any other program unit. Any program unit, however, mainline, subroutine, or function subprogram can contain its own statement functions.

Now let us construct a sample program to illustrate the use of statement function subprograms. We will use the same example given for our function subprogram—to compute $Y = X^2 + Z$.

Our program is:

```
C *** MAINLINE WHICH INCLUDES A STATEMENT FUNCTION
      FUNCT(X,Z) = X ** 2 + Z
   99 FORMAT (2F5.2)
   98 FORMAT (1H0,F7.2)
      READ (5,99) X,Z
      Y = FUNCT(X,Z)
      WRITE (6,98) Y
      READ (5,99) A,B
      Y = FUNCT (A,B)
      WRITE (6,98) Y
      STOP
      END
```

Our statement function along with our mainline achieves the same results that were produced by our sample function subprogram on page 272. Statement functions are very beneficial when one has complex calculations that are used many times in a program.

The remaining sections discuss the additional statements and concepts that may also be used with function and subroutine subprograms.

15.5 The EXTERNAL Statement and Subprogram Names Passed as Arguments

We learned in section 13.3 that a subprogram name can be used as an actual argument in the parameter list of the program calling or invoking a subprogram. The subprogram name can be passed as an actual argument in one of two ways: (1) as a function subprogram name followed by its required parameter list, or (2) as a function or subroutine subprogram name by itself.

When the actual argument is a function subprogram name followed by its required parameter list, the subprogram is executed and its resulting value is passed to the invoked/called subprogram as the actual argument. For example:

$$\text{CALL SUBR (A,FUNCT(X,Z),RESLT)}$$

A and RESLT are variables and FUNCT is the name of a function subprogram. However, the function FUNCT has its arguments included with the function subprogram name. Thus, the function subprogram FUNCT is executed before the arguments are passed to the subroutine SUBR. The resulting value of FUNCT(X,Z) is then passed as the second actual argument along with A, and RESLT.

Suppose our function subprogram FUNCT and subroutine SUBR are:

Function

```
FUNCTION FUNCT(X,Z)
FUNCT = X ** 2 + Z
RETURN
END
```

Subroutine

```
SUBROUTINE SUBR(A,F,R)
R = A + F
RETURN
END
```

If the value of A is 10.0, the value of X is 3.0, and the value of Z is 5.0, then the statement

$$\text{CALL SUBR (A,FUNCT(X,Z),RESLT)}$$

would produce a value of 24.0 for the argument RESLT. When a function subprogram name with its required parameter list is passed as an actual argument, there is no problem.

When we pass just the subprogram name as an actual argument, it presents the name of a subprogram that is to be executed during the execution of the called/invoked subprogram. That is, the subprogram name by itself is used as an actual argument which specifies which subprogram is to be executed when the corresponding dummy argument is used in the called/invoked subprogram.

We are discussing the flexibility we can obtain by changing the subprogram that we execute within a subprogram simply by using a subprogram name as an actual argument. That is, we can pass a different subprogram name in the actual argument list, but the called subprogram logic still remains the same.

Suppose we have a function subprogram called MINMAX. If we pass it the built-in function AMIN1, then the subprogram will use the function AMIN1 in its calculations. If we pass it the built-in function AMAX1, then the subprogram will use the function AMAX1 in its calculations. Our coded function subprogram is:

```
      FUNCTION MINMAX (SUBPGM)
      READ (5,99) A,B,C
   99 FORMAT (3F5.2)
      MINMAX = SUBPGM(A,B,C)
      RETURN
      END
```

Suppose we invoked the subprogram MINMAX with the following statement.

RESLT = MINMAX (AMIN1)

Since we pass the built-in function AMIN1 as an actual argument, the function subprogram MINMAX would return the smallest value of the three input variables.

Now suppose we invoked the subprogram MINMAX with the following statement.

RESLT = MINMAX (AMAX1)

Since we pass the built-in function AMAX1 as an actual argument, the function subprogram MINMAX would return the largest value of the three input values. Thus, the statement in the subprogram MINMAX that we coded

MINMAX = SUBPGM (A,B,C)

could be used to execute *any* function subprogram that used three arguments.

A problem arises when we pass just the function or subroutine subprogram name by itself. How does the computer know that the actual argument is a subprogram name and not a variable or an array name? It would know that it is not an array name if it was not dimensioned in a proper dimensioning statement. Otherwise, the computer assumes that the actual argument is a variable.

When we use only the subprogram name as an actual argument, the computer has no way of knowing that it is a subprogram name and not a variable unless we tell it, somehow. The statement needed to tell the computer that an actual argument is indeed a subprogram name is the EXTERNAL statement. When we code:

EXTERNAL AMIN1, AMAX1

(Keyword to denote that external subprograms are used in program)

(Subprogram names)

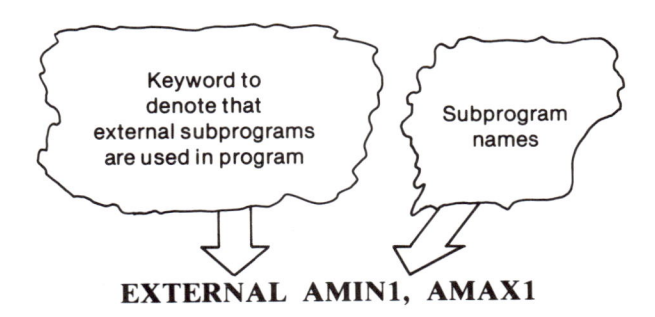

then the computer knows the symbolic names AMIN1 and AMAX1 to be names of subprograms that are external to this program unit.

282 Function Subprograms and Additional Subprogram Statements

The EXTERNAL statement, therefore, specifies that a symbolic name is to be regarded as the name of a subprogram, rather than as a variable, when it is included as an actual argument in a parameter list. The general form of the EXTERNAL statement is given in figure 15.2.

EXTERNAL Statement—General Form

   EXTERNAL name$_1$,name$_2$,...,name$_n$

where:

   name$_1$,...,name$_n$ — represent names of subprograms when included as actual arguments without a parameter list are interpreted to be subprogram names rather than variables.

Figure 15.2. General Notation of the EXTERNAL Statement

The EXTERNAL statement is a specification statement and should not be assigned a statement number. It must precede any statement function subprograms and executable statements in the program unit.

A single EXTERNAL statement may be used to contain all the names of external subprograms. Or we may use multiple EXTERNAL statements in a program unit to declare this information.

So to make our program correct when we use a subprogram name as an actual argument, we must include the EXTERNAL statement with the names of all subprograms used as actual arguments. Our program unit that invokes the Function subprogram MINMAX should be:

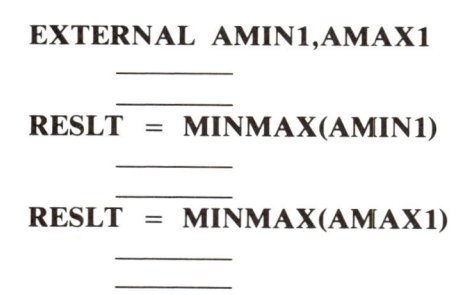

   EXTERNAL AMIN1,AMAX1
   _____

   RESLT = MINMAX(AMIN1)
   _____

   RESLT = MINMAX(AMAX1)
   _____
   _____

The same concepts hold true when we call subroutines. An actual argument may be the name of a function or subroutine subprogram. For example:

Mainline	*Subroutine*
EXTERNAL SQRT,SIN	**SUBROUTINE SUBF (X,FX,R)**
CALL SUBF (20.5,SQRT,R)	**R = FX(X)**
. . .	**RETURN**
. . .	**END**
CALL SUBF (3.14, SIN,R)	
. . .	
. . .	

The first time the subroutine SUBF is called, the built-in function SQRT is passed. So the subroutine calculates the square root of the input value in X. The second time SUBF is called, the built-in function SIN is passed. So the subroutine calculates the trigonometric sine of the input value in X.

Although our examples have shown only function subprogram names, subroutine names may be passed as actual arguments. So in the CALL statements we can have the same flexibility in changing the subroutine being called as one of the actual arguments.

15.6 Object Time Dimensions in Subprograms

Object time dimensions means that we can provide the dimension size of arrays used in subprograms at execution time. As you will recall from chapter 10 on Arrays, we must specify the dimension size of arrays in our main program as a numeric constant, so that the proper amount of adjacent storage locations are assigned during compilation. But within subprograms we can say, for example:

DIMENSION X(N)

where the value of N is provided during the execution of the subprogram.

The dimension size of an array provided by the value of variables must be equal to or less than the actual size specifications of the array passed as an actual argument. That is, if we pass a fifty-element, one-dimensional array named A as an actual argument, then the size of the corresponding dummy array argument must never exceed fifty. For example (assume the arrays A and B have been read),

Main Program	*Subprogram*
DIMENSION A(50),B(30)	SUBROUTINE SUBR(X,N,S)
CALL SUBR(A,50,SUM)	DIMENSION X(N)
. . .	S = 0.0
. . .	DO 10 I = 1,N
CALL SUBR(B,30,SUM)	S = S + X(I)
. . .	10 CONTINUE
. . .	RETURN
	END

A COMMON or a Data statement must not be used by the subprogram to assign a value to the variable of the DIMENSION statement specifications.

Object time dimensions may also be used with two- and three-dimensional arrays. Extreme caution is advised, however, when using object time dimensions with two- and three-dimensional arrays. Serious problems have been encountered with multidimensional arrays because of the way the storage locations are assigned for multidimensional arrays and are passed to subprograms.

When a two-dimensional array is used as a dummy argument, the first (row) subscript must have the same dimension as the corresponding actual array argument in the main program. The size of the second (column) subscript in the subprogram need not agree with the size of the second subscript in the main program. The same principle holds true for three-dimensional arrays. For object time dimensions of three-dimensional arrays in a subprogram, the first two (row and column) subscripts must be the same size as the array being passed by the calling program. The general rule is that the first $n - 1$ (where n is the number of dimensions) subscripts of the subprogram's DIMENSION statements for object time dimensions must correspond with the dimensions in the main program.

15.7 Nonstandard Language Extensions (ENTRY statement)

There are times when we may wish to enter a subprogram at a different point than at the first statement. This capability is available through the use of the ENTRY statement, the purpose of which is to provide multiple entry points at different locations in a subprogram. The IBM 360-370, WATFIV, and other FORTRAN compilers provide this capability.

The ENTRY statement begins with the keyword ENTRY followed by an assigned name. The name is constructed like other subprogram names. A parameter list enclosed within parameters may follow the ENTRY statement name. For subroutines the parameter list with the ENTRY statement is optional. With function subprograms the parameter list is mandatory. Figure 15.3 gives the general form of the ENTRY statement.

ENTRY Statement—General Form

ENTRY name (darg$_1$, . . ., darg$_n$)

where:

    name — is the name of the entry point into the subprogram.
    darg$_1$, . . ., darg$_n$ — represent a parameter list of dummy arguments. This parameter list is optional for subroutines.

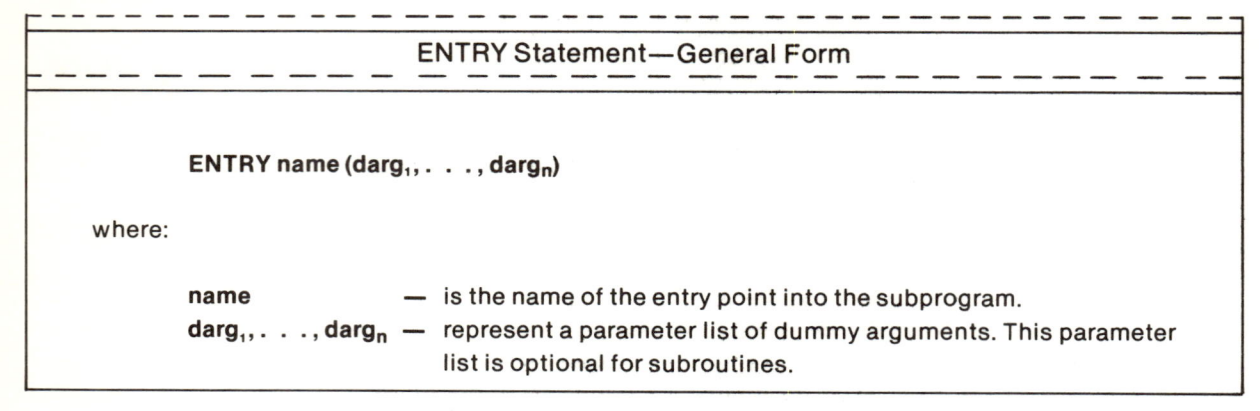

Figure 15.3. General Notation of the ENTRY Statement

If the subprogram is a function, the ENTRY statement name takes on the resulting value, just as a function subprogram name would. Thus the chosen ENTRY statement name must be of the same type as that which would pertain if the function name was used. If the subprogram is a subroutine, the ENTRY statement name is treated the same as a subroutine name. Different type names may be assigned since the values are returned under the dummy arguments in the parameter list included with the ENTRY statement.

Suppose we wish to enter a subprogram at the beginning the first time it is called in order to initial some variables or an array. From then on, we wish to enter the subprogram at a different point whenever we call it again. The following example illustrates this technique on a subroutine named BEGPAG.

Subroutine

```
   SUBROUTINE BEGPAG
   NP = 0
   ENTRY NEWPAG
   NP = NP + 1
   WRITE (6,99) NP
99 FORMAT(1H1,120X,4HPAGE,I6)
   RETURN
   END
```

Mainline

```
CALL BEGPAG

CALL NEWPAG

CALL NEWPAG
```

The first time the subroutine is called (CALL BEGPAG), it is entered at the top (beginning). The variable NP is initialized to zero. Then control continues at the statement: NP = NP + 1. So the variable NP is bumped by one and we get a 1 as the page number from the print operation of the subroutine. The succeeding times that the subroutine is called (CALL NEWPAG), it is entered at the point named NEWPAG. Thus, the page number counter is incremented by 1 and we get our next page number printed at the top of a new page.

Arguments are usually passed to provide intial values. Consider the following function subprogram and its references.

Function Subprogram

```
FUNCTION FSUM(A,B,C)
D = B + C
E = B * C
ENTRY FSUM2 (A)
FSUM = A + D + E
RETURN
END
```

Mainline

```
T = FSUM(X,Y,Z)

T = FSUM2(X)
```

The first time the function FSUM is invoked it is entered at the top with the input arguments of A, B, and C. The input arguments B and C are used to calculate D and E. The variables D and E will retain their values through the use of the subprogram unless they were changed later. This is possible since any variable which is not a dummy argument and is changed in a subprogram retains its last assigned value. The sum of A, D, and E are calculated and stored under the function name FSUM.

The next time the function FSUM is invoked, it is entered at the first executable statement following the ENTRY point named FSUM2. Only the argument A is received and used to calculate the sum of the three variables A, D, and E. The result is stored in the function name FSUM. However, the result is also stored in the ENTRY point name FSUM2. So the resulting value is passed back in both the names FSUM and FSUM2. Since the reference was to FSUM2, the value will be obtained from that name.

From the two illustrated examples we see that the parameter list of an ENTRY statement does not have to agree with the parameter list of the FUNCTION or SUBROUTINE statement. Neither do any two ENTRY statement parameter lists have to be the same. Actual arguments in a parameter list must only agree with the dummy argument's parameter list between the calling/invoking statement and the corresponding ENTRY statement.

The ENTRY statement is considered to be a specification statement and thus is nonexecutable. As such, it does not affect the execution sequence of the statements in the subprogram when control is entered at the beginning. An ENTRY statement, however, must not be placed within the range of a DO loop. The use of an ENTRY statement does not change the rule requiring that all specification and statement functions in the subprogram must appear at the beginning, before any executable statements. Any number of ENTRY statements may be used in a subprogram.

15.8 REVIEW QUESTIONS

1. When do we normally include a function subprogram in the design of our program?
2. How is a function subprogram identified in a program unit?
3. How is a function subprogram referenced by a program unit?
4. What is the minimum number of arguments that must be included in the parameter lists to communicate between the function subprogram and invoking program?
5. What is the normal number of values that are returned from a function subprogram to its invoking program?
6. How is the value normally returned from a function subprogram to its invoking program?
7. What determines the type of value returned from a function subprogram?
8. How is a value generally assigned to the name of a function subprogram?
9. Where is control of execution returned to in the invoking program?
10. Arguments in a function subprogram reference must agree in what four ways with the dummy arguments in the FUNCTION statement?
11. What is the advantage of using a statement function subprogram?
12. Must a statement function have dummy arguments?
13. What is the difference between a statement function and a function subprogram?
14. Where must the statement function subprogram be placed in a program?
15. How is a statement function subprogram referenced?
16. Where does control of execution return to from a statement function subprogram?
17. What is the purpose of the EXTERNAL statement?
18. What is the advantage of object time dimensions in a subprogram?
19. What is the purpose of the ENTRY statement?

15.9 EXERCISES

1. Write the FUNCTION statement for each of the following:
   a) FUN1 with arguments consisting of the variables A, I, and P.
   b) FY with only one dummy argument consisting of the array named Y.
   c) CALC with the following arguments: variables X, Z, and TOTAL.
2. Write the FUNCTION statement for each of the following:
   a) CAT with arguments consisting of the variables MOUSE and RAT.
   b) FUN with the array named GRP1, and the variables S, T, U, and V.
   c) PLOT with the arguments X, Y, and K.
3. Find the error in each of the following statements.
   a) R = FLOAT (A)
   b) FUNCT1 = FUNCT1(X + Z)
   c) AMN = AMIN1(A)
   d) SUBROUTINE SUBA(A,B,C
   e) FUNCTION FOFX
   f) R = SQRT(K)
4. Find the error in each of the following statements or group of statements.
   a) K = IFIX(L)
   b) FUNCT(X,Z) = Y
   c) MAX = AMAX1(X)
   d) FUNCTION CALC /X,Y,Z/
   e) DIMENSION X(10)
      R = SQRT(X)
   f) NMAX = (AMAX0)X,Y,Z
5. Given the following statements:

   Mainline Program   **Subroutine**

   X = 3.0
   CALL SUB1(X,XSQ,XCUBE)   SUBROUTINE SUB1(Y,YS,YC)
   . . .   YS = Y * Y
   . . .   YC = YS * Y
                                      RETURN
                                      END

   a) What is the value of Y in the subroutine?
   b) What is the returned value of XSQ?
   c) What is the returned value of XCUBE?

6. Given the following statements:

   Mainline Program   **Function Subprogram**

   EXTERNAL SQRT   FUNCTION FUNCTX (X,FX)
   A = 25.0   FUNCTX = FX(X)
   R = FUNCTX(A,SQRT) + 7.2   RETURN
   . . .   END

   a) What is the value of X in the Subprogram?
   b) What is the returned value in FUNCTX?
   c) What is the value of R?

7. Determine whether a function or a subroutine is the best type of subprogram to accomplish the following operations:
   a) Initialize an array to zeroes.
   b) Locate the smallest value in an array.
   c) Count the occurrences of an item in an array.
   d) Sort an array into descending sequence.

8. Determine whether a function or a subroutine is the best type of subprogram to accomplish the following operations:
   a) Find the largest value in an array.
   b) Square all the items in an array.
   c) Perform a matrix addition operation on two matrices.
   d) Compute the value of $2.0 * X^3 + Y$.

15.10 PROGRAMMING PROBLEMS

1. Write a FUNCTION subprogram named NSUM to find the sum of the integers from 1 to N. The value of N is read in a mainline program under the specifications I3 and passed to the subprogram via a parameter list. Upon return from the subprogram write the following output:

   'SUM OF THE INTEGERS FROM 1 to' n 'IS' nn

   where n is the input value and nn is the computed sum.

2. Many mathematical and statistical solutions are derived for problems using a factorial series. N factorial (represented by N!) is derived from the product of all the integers from 1 through N. That is, 3! means $3 \cdot 2 \cdot 1 = 6$ (or $1 \cdot 2 \cdot 3 = 6$). However, 1! and 0! are both equal to one.

   Write a function subprogram named NFACTL to compute the factorial of the value contained in the variable N. A mainline program will read the value of N under the specifications I2. The variable N will be passed to the subprogram and the value of N! returned. Then the value of N and N! are to be printed under the specifications 2I10. Include appropriate column headings.

   The following algorithm is given to describe the calculation of N!.
   a) Set NFACTL to 1.
   b) If N $\leq$ 1 RETURN
   c) For values of I from 2 to N multiply NFACTL by I.
   d) RETURN

3. Write a Function subprogram named LARDIV that computes the largest common division of the number N. Include a mainline that reads the value of N under the specifications I3.

   Write an output line that reads 'THE LARGEST COMMON DIVISOR OF' n 'IS' l where n is the input number and l is the computed largest common divisor of N returned from LARDIV.

   Hint: Use the built-in function MOD in the Function subprogram and check for a remainder of zero. Include a DO loop which varies I from 1 to N. Use N as the first argument of MOD and N-I (N minus I) as the second argument of MOD.

4. Write a FORTRAN function subprogram named MEDIAN to compute the median score in a group of n number of test scores. A mainline program is used to read the value of N and the array of test scores named NTEST. The variable N and array NTEST will be

passed to the subprogram named MEDIAN. The returned median score is printed by the mainline upon the return of control to the main program.

The value of N is read from the first data card under the specifications I2. The array NTEST is read from the remaining data cards under the specifications 10I3.

After the array NTEST is read, print the message:

'THE ARRAY NTEST WITH' n 'VALUES IS'

where n is the input value of the variable N. Next print the items in the array NTEST on a new line with the specifications 10I5. When the median score is returned to the mainline, print the calculated median score on a new line preceded by the literal that reads:

'THE MEDIAN SCORE IS' m

where m is the computed median score.

    Hint: The median value of a group of items may be calculated as follows:
- *a)* First, sort the items into an ascending sequence.
- *b)* Next, determine if the value of N is an even or odd value. This may be determined by the formula:

$$N / 2 * 2 - N$$

    If the resulting value is zero, then the value of N is an even number. If the resulting value is -1, then the value of N is an odd number.

- *c)* (1) If N is an odd number, the median value is the middle item and is located at the position $N/2 + 1$.
- (2) If N is an even number, the median value is the average of the two middle items. This median value may be obtained by adding the item at location $N/2$ to the item at location $N/2 + 1$ and taking the average of the sum.

Object time dimension may be used in the problem if you so desire.

5. Write a function subprogram named NBINRY to convert an integer to its corresponding binary integer. The decimal integer is positive and is less than 100. Read the decimal integer under the variable INT with an I3 format specifications. After the decimal integer is converted to its equivalent binary number, print a message that reads:

'DECIMAL NUMBER' dn 'IS = TO BINARY NUMBER' bn

where dn is the input decimal number and bn is the converted binary number.

    Hint: A decimal number is converted to a binary number by repetitive division by 2. The remainder after each division is used to form the binary number. The first remainder becomes the first binary digit, the second remainder becomes the second binary digit, and so on. The division process is complete when a quotient of zero is obtained.

6. Write a function subprogram to accomplish any problem you have previously programmed. Use a mainline program to read any input values and to print the output results.

7. Write a statement function named AVE to compute the average of three variables passed to it. The three input values will be read under the specifications 3F5.2 and named A, B, and C, respectively. The dummy arguments A1, B1, and C1 will be used in the statement function subprogram.

After the computed average is returned, print the three input values and the computed average on the same line. Include appropriate column headings.

Additional FORTRAN Constants and Format Specifications

16

16.1 Introduction to the Additional Types of FORTRAN Constants and Data

The types of data that we have been manipulating in programs so far have been restricted to numeric values of type integer and single-precision floating point, and to alphanumeric values. Most FORTRAN compilers, however, include the capability of working with other forms of data. You may wonder what other types of data there can be. From business application and social science points of view, numeric and alphanumeric data are the basic forms of data that we work with. But from a math and engineering standpoint there are several other types of data.

A primary form used for representing numeric constants is in exponential notation. The *exponential notation* uses an exponent appended to a numeric value to represent the position of the decimal point and thus, the true value of the number. This exponential form of constants is patterned after the *scientific notation* for expressing numeric values. **Appendix D** explains the scientific notation and the basic exponential form for expressing numbers to those who are unfamiliar with this notation.

We can express numeric quantities to the compiler in an exponential form by using the E or D symbol preceding the exponential value. Numeric data fields expressed with an exponential value can also be read or written in FORTRAN with the E and D format codes. Sections 16.2 and 16.3 explain the use of these format codes for the input/output of numeric values.

Mathematicians also represent data in a boolean form. A boolean type constant may only have a value of either true or false. Boolean type constants are also widely used by engineers in expressing inputs and outputs to logic circuits. In FORTRAN boolean values are referred to as logical data. Section 16.4 discusses this type of data, which can be effectively used in computer programs.

In mathematics the use of complex numeric values is very important. The use of complex type data which consists of a real part and an imaginary part is discussed in section 16.5. A Generalized (G) format code for reading and writing any type of numeric value is explained in section 16.6. The use of a P scale factor to override the decimal point location in an input or output value is explained in section 16.7.

System programmers, those who work with compilers and operating systems, often need to display data in a number base other than base 10. The Octal (O) and Hexadecimal (Z) format codes are used to read and write data in base 8 and base 16, respectively. These format codes are explained in section 16.8 since they are not standard format codes. Their availability in a FORTRAN compiler is dependent upon the number base used by a computer to represent its data internally in memory.

The business and social science students studying FORTRAN may never need to use any of these other data types. They are, however, extremely useful in specialized applications and programs.

16.2 Single-precision Exponential Type Constants and the E Format Code

The precision of most real quantities can be expressed by six digits or less. Based on this consideration, computers have been designed to allow internal floating-point calculations with numbers having at least seven decimal digits and an exponent. Externally, a floating-point number may be constructed with or without the exponent. The use of one to seven digits to represent a real constant is known as ***single-precision*** mode. The basic decimal form (i.e., a string of decimal digits with a decimal point) is the way most of us represent numbers with fractions in our everyday life. This is the form we have been using thus far in the text to express real constants.

A single-precision real constant may also be represented as a real or integer number followed by a decimal exponent. The *exponent* consists of the letter E followed by a signed or unsigned 1- or 2-digit integer constant. The exponent represents a power of 10 by which the number is multiplied to obtain its true value. The exponent can range from a power of -78 to $+75$ on IBM 360 and 370 computers. Other computers may have a different range. Thus, extremely large and small numbers may be represented in an abbreviated form.

Single-precision examples of *real* constants expressed with a decimal exponent are:

   432.1E+00   (Has a value of 432.1)
   −8769.2E−2 (Has a value of −87.692)
   65.423E3   (Has a value of 65423.)
   −24.638E−01 (Has a value of −2.4638)

Single-precision examples of *integer* constants followed by a decimal exponent are:

   −87E+0   (Has a value of −87.)
   147E−03   (Has a value of .147)
   2E4   (Has a value of 20000.)

The real type variables that you have been using all along may be used to store single-precision real constants expressed with exponents. The same Assignment statement used to compute and assign real values is also used with this exponential form of real constants. For example:

   PI = .31416E01
   R = 7203E−2
 HAREA = (PI * R ** 2) / .25E+1

Data fields may be read or written with their values in exponential form. The F format code may be used to read input values expressed with an exponent. The E format code, however, is available to read data with exponents and must be used to write data in an exponential form.

The Exponential Floating-point (E) Format Code

The E format code is used to transmit numeric fields normally represented in scientific notation which contain a numeric quantity with an appended exponent. The numeric data fields, expressed in scientific notation, are formed in exactly the same way as numeric constants expressed with an exponent.

The general form of the E format code is:

      rEw.d

where:

r represents an optional repetition factor that specifies the number of times the format code is to be used. The value of **r**, when specified, must be an unsigned positive integer constant whose value is less than or equal to 255. **w** specifies the number of card columns or record positions in the input or output field. **w** must be an unsigned positive integer constant with a value less than or equal to 255. The decimal point (.) is a necessary separator between the **w** and **d** parameters. **d** represents the number of digits in the fractional part of the number. **d** must be an unsigned integer constant between 1 and 255.

All input rules for the F format code apply for the E format code. Any permissible form of constructing an integer or real constant may be read by the E format code.

Input examples to illustrate the input rules when using the E format code are as follows (⌱s represent blanks):

Field Size and Value	E Format Code to Be Used
12.345E+2	E9.3 (Value of 1234.5)
1234.5E−03	E10.1 (Value of 1.2345)
−123.45E00	E10.2 (Value of −123.45)
⌱⌱+12.3E1	E9.1 (Value of 123.)
12.34	E5.2 (Value of 12.34)
12345	E5.3 (Value of 12.345)
123E01	E6.0 (Value of 1230.)
+12.3E3	E7.1 (Value of 12300.)

The E format code is normally used to read data fields expressed in scientific notation with an appended exponent, such as 1.0E04. The above examples show that this format code can also read numeric quantities that we normally express in our everyday decimal form.

Consider the following input example:

READ (5,99) V1,V2,V3,V4
99 FORMAT (E7.2,1X,E6.2,1X,E8.2,1X,E9.1)

```
  12.34E5 − 56.27 18735.10 − 438.1E−2
```

The values written with an E format code will always contain an appended exponent. The following paragraphs discuss the format of the output fields printed with the E format specifications.

Output Considerations

The output rules which apply to the F format code also apply to the output fields when using the E format code. Data fields output under the E format are however, formatted in a different manner, than those formatted with the F format code.

An output field specified with an E format code will consist of an optional sign (required if negative), a decimal point, the number of digits specified by **d**, and an exponent part which denotes the power of 10 to be multiplied with the numeric quantity to obtain the true value of the number. The exponent part always consists of the four positions set out as follows:

1. the letter E
2. the sign of the exponent (a blank is output if the exponent is positive or zero)
3. a two-digit integer constant representing the exponent as a power of 10.

The numeric quantity is expressed in a pure fractional form (called ***normalization***). That is, a decimal point is placed before the first significant digit (a nonzero digit). As many digits are included in the fraction as are specified by the **d** parameter of the E format specification.

Examples of the E format code to illustrate the output form of the data fields are as follows (⌱s represent blanks in the high order field positions):

Value in Memory	E Format Code	Output Field
12.345	E10.5	.12345E 02
12.345	E10.4	0.1235E 02
12.345	E9.5	*********
12.345	E10.3	⌱0.123E 02
−12.345	E10.6	**********
−12.345	E12.4	⌱−0.1235E 02
−12.345	E10.4	−.1235E 02
.0001	E9.3	0.100E−03

Real constants expressed with an exponent are normally used by mathematicians and engineers inasmuch as they are accustomed to expressing numeric quantities in a scientific notation. The main advantage comes from being able to represent extremely large or small numeric values in a shorthand form.

16.3 Double-precision Constants and the D Format Code

Seven digits are hardly enough for accumulating large sums and solving problems that use values containing a substantial quantity. Thus, most computers allow for an extended precision of the basic single- precision seven-digit numbers. This extended form of precision is **called** double-precision, since the number of significant digits is more than double those in single-precision real numbers.

A *double-precision* floating-point number can contain eight through sixteen digits on most computers. A sign, if present, is an allowable character beyond the maximum possible length of sixteen digits. A double-precision floating-point constant in decimal form is formed in the same way and must adhere to the same rules as a single-precision real number in basic decimal form. The double-precision constant must, however, contain more than seven decimal digits, or the quantity will be formed in memory as a single-precision number. Examples of valid double-precision floating-point numbers in basic decimal form are:

$$245.467193$$
$$-.546683207$$
$$+468902043.$$

Double-precision *real* constants may also be expressed with an appended decimal exponent. The exponent may be represented with the letter D followed by a signed or unsigned one- or two-digit integer constant. The value of the exponent can range from -78 to $+75$ on IBM 360 and 370 computers. The range of the exponent value varies on different computer systems.

Examples of double-precision floating-point constants represented with an appended exponent are:

$$.1234567892D+00$$
$$-991.257643D9$$
$$.8765428312D-3$$

A double-precision real constant may also be represented as an *integer* number followed by a decimal exponent. The exponent normally consists of the letter D followed by a signed or unsigned 1- or 2-digit integer constant. Valid examples of double-precision real constants, each formed as an integer constants and each followed by an exponent are:

$$-47D-3$$
$$93D+06$$
$$1D-9$$

While double-precision numbers have the advantage of greater precision, they have a negative effect on computer storage and processing speed. Use of double-precision numbers over single-precision has two main disadvantages. First, double-precision numbers occupy twice as much computer storage space as do single-precision numbers. Secondly, double-precision numbers take more time to process. For these reasons, the programmer should only specify double-precision constants and variables where they are mandatory.

Double-precision Variables

Variables that may be assigned double-precision real values are referred to as double-precision variables. A double-precision variable may be any valid symbolic name, but must be declared a type double-precision in an Explicit type statement. The general form of the DOUBLEPRECISION Type statement is given in figure 16.1.

DOUBLEPRECISION Type Statement—General Form

      DOUBLEPRECISION name$_1$,. . ., name$_n$

where:

      name$_1$,. . ., name$_n$ — each **name** represents a variable, array, or function subprogram name. If the **name** is an array name, the dimension information may be specified following the array name.

Figure 16.1. General Notation of the DOUBLEPRECISION Statement

An example of the DOUBLEPRECISION specification statement to declare double-precision variables and array elements is:

DOUBLEPRECISION A,I,X(10),Y(5,3)

where the variables A and I are declared to be double-precision variables. The array named X is declared to be a one-dimensional array with ten double-precision elements. The array named Y is declared to be a two-dimensional array with fifteen double-precision elements.

    Array dimensions may be specified in the DOUBLEPRECISION statement or may be declared in a separate DIMENSION statement that precedes the DOUBLEPRECISION statement. The DOUBLEPRECISION statement cannot specify initial values for variables or array elements.

Double-precision Expressions and Assignment Statements

A double-precision expression is one that provides a resulting double-precision value to be assigned to a double-precision variable. The expression may contain double-precision constants, variables, and array elements. Following are examples of double-precision expressions:

      10.89D17
      A
      3.5D12 + A * (B − .8671D − 7)

where the variables A and B are declared as double-precision variables.

    The double-precision Assignment statement is used to assign double-precision values to variables. The general form of this Assignment statement is:

double-precision variable = double-precision expression

Examples of the double-precision Assignment statement (the variables W, X, and Y are assumed to be double-precision variables and Z is assumed to be a double-precision, one-dimensional array) are as follows:

      W = 1.345876924
      X = .1D − 12
      Y = W + X * 10.0D + 14
      Z(1) = Y / (345.672983 + W)

Input/Output of Double-precision Fields with the D Format Code

The D format code is used to transfer numeric fields expressed as double-precision quantities. Like the E format code, the numeric fields read with the D format code are normally represented in scientific notation. The appended exponent part normally contains a D to denote double-precision floating-point values.

The general form of the D format code is:

rDw.d

where:

r represents an optional repetition factor that specifies the number of times the format code is to be used. The value of **r**, when specified, must be an unsigned positive integer constant whose value is less than or equal to 255. **w** specifies the number of card columns or record positions in the input or output field. **w** must be an unsigned positive integer constant with a value less than or equal to 255. The decimal point (.) is a necessary separator between the **w** and **d** parameters. **d** represents the number of digits in the fractional part of the number. **d** must be an unsigned integer constant between 1 and 255.

Examples in the use of the D format code are:

```
18 FORMAT (D12.8,2X,2D14.9)
21 FORMAT (1H0,5X,3D24.16)
```

The first example reads a 12-position double-precision numeric quantity in columns 1-12 and two 14-position double-precision values beginning in column 15. The second example (an output specification) prints three 24-position, double-precision fields beginning in print position six.

All input and output rules that apply to the E format code apply to the D format codes. Output fields written with the D format specification will be printed in the same form as with the E format specifications, except that the letter D will be used in place of the letter E.

One unexpected thing happens when single-precision variables are printed with the D format code. If the variable is single-precision, then the computer prints the output results as expressed with an E format specification, even if the D format code is used. The same principle holds true for double-precision variables written with the E format code. The computer prints the output results as expressed in the E field descriptor, but replaces the E with a D.

Double-precision Function Subprograms

Many double-precision built-in Functions are available to manipulate double-precision values. For example, the built-in function DSQRT takes the square root of a double-precision value. Double-precision built-in functions included in FORTRAN are (all functions require one real double-precision argument unless otherwise indicated):

Function Name	*Function Description*
DABS	Absolute Value
DEXP	Exponential
DLOG	Natural Log
DLOG10	Common Log
DSQRT	Square Root
DMAX1	Maximum Value (more than one argument)
DMIN1	Minimum Value (more than one argument)
IDINT	Truncation
DMOD	Modulo Arithmetic (2 arguments)
DSIGN	Transfer of Sign
DBLE	Precision Increase
DSIN	Sine
DCOS	Cosine
DATAN	Arctangent
DATAN2	Arctangent (2 arguments)

Each of these functions returns one double-precision value.

A user may also write double-precision-valued Function subprograms by specifying the type option to be DOUBLEPRECISION. For example:

DOUBLEPRECISION FUNCTION SUM (X,N)

Thus, the resulting value via the Function name SUM will be a double-precision value. User supplied or built-in Double-precision Function subprograms must always have their names declared in a DOUBLEPRECISION statement in the invoking program to inform the compiler that the referenced function subprogram is to return a double-precision value.

16.4 Logical Constants and the Logical (L) Format Code

There are only two logical constants. They are expressed in FORTRAN as .TRUE. and .FALSE.. The period before and after each constant must be used to prevent the constants being taken as variable names.

Logical Variables

A variable which may be assigned a logical value is said to be a logical variable. A logical variable can be any valid symbolic name, but must be declared in an Explicit type statement to be type logical. The general form of the LOGICAL statement is given in figure 16.2.

LOGICAL Statement—General Form

LOGICAL name$_1$,..., name$_n$

where:

name$_1$,..., name$_n$ — each **name** represents a variable, array, or function subprogram name. If the **name** is an array name, the dimension information may be specified following the array name.

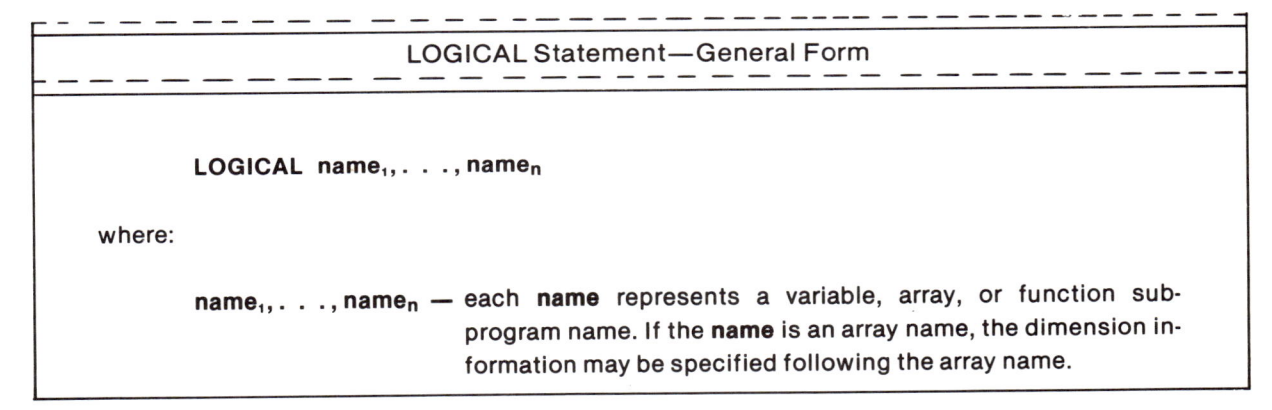

Figure 16.2. General Notation of the LOGICAL Statement

An example of the LOGICAL Type statement to declare logical variables is:

LOGICAL X, Y, Z(7), A(5,2)

The variables X and Y are declared to be logical type variables. The array name Z is declared to be a one-dimensional array with seven logical elements. The array name A is declared to be a two-dimensional array with ten logical elements.

Logical Expressions and Assignment Statements

A logical expression is one that provides a resulting value of .TRUE. or .FALSE.. The expression may contain logical constants, variables, or array elements. The three logical operators may be used within logical expressions to form compound expressions. The arithmetic operators must not be used with logical constants and/or variables since we are not dealing with arithmetic values. Relational operators, however, can be used with numeric constants and/or variables to provide a resulting logical value. Some examples of logical expressions are:

.TRUE.
.FALSE.
X .AND. Y
X .OR. Y
.NOT. X
A .GT. 10.0
B .EQ. C

where the names X and Y are logical variables, and the names A, B, and C are numeric variables. The resulting value of X .AND. Y is true, only if both X and Y have true values. The resulting value of X .OR. Y is true, if either X or Y has a true value. The .NOT. X produces the opposite value of X. The arithmetic expressions with relational operators produce a true or false value depending upon the outcome of the comparison.

The logical Assignment statement is used to assign a logical value to a logical variable. The general form of this type of Assignment statement is:

logical variable = logical expression

The logical expression is evaluated and the resulting logical value is assigned to the logical variable.
Examples of the logical Assignment statement are:

X = .TRUE.
Y = .NOT. X
L1 = .NOT. (X .AND. Y)
L2 = (X .AND. .NOT. Y) .OR. .FALSE.
L3 = (X .OR. Y) .AND. A .LE. 100.0

In the first example, the logical constant .TRUE. is assigned to the logical variable X. In the second example, a false value results from the expression .NOT. X and is assigned to the logical variable Y.

You will remember from chapter 8 that relational operators take precedence over the logical operators, and the logical operators have a precedence of .NOT., and .AND., and .OR. order. Parentheses are used to override these precedences. Thus, the third example says to produce the logical result of X .AND. Y which is false, and then the opposite of this value which is true. So a value of true is assigned to the logical variable L1. The fourth example says to first find the value of (X .AND. .NOT. Y) which is true. This value is "ORed" with the constant .FALSE. which produces a true result and is assigned to L2. The fifth example first evaluates the expression (X .OR. Y) to produce a value of true. Then A .LE. 100.0 is evaluated. Assuming a value of 10.0 for A, the result is true. Thus, the "ANDing" of the first true result and the second true result assigns a value of true to L3.

The Logical (L) Format Code for Reading and Writing Logical Data

The Logical (L) format code is used to read and write logical data. The general form of the L format code is:

rLw

where **r** represents an optional repetition factor from 1 to 255 which may be used to repeat the same width logical data fields in adjacent card columns. The letter **w** is an unsigned integer constant specifying the **number of characters in the field width**. w must be an unsigned integer constant between 1 and 255.

Additional FORTRAN Constants and Format Specifications 297

On input fields the first T or F encountered in the w characters causes a value of .TRUE. or .FALSE., respectively to be assigned to a variable. If the input field contains all blanks or does not contain a T or F, the value of .FALSE. is assigned. Consider the following input example:

```
   LOGICAL  X1,X2,X3,X4,X5,X6,X7
   READ (5,99) X1,X2,X3,X4,X5,X6,X7
99 FORMAT (L4,L5,1X,L3,1X,L4,1X,L3,1X,L1,L2)
```

```
TRUEFALSE CAT FROG DOG T
```

The variable X1 is assigned a value of .TRUE. from the four characters TRUE. The variable X2 is assigned a value of .FALSE. from the input field FALSE. The variable X3 is assigned a value of .TRUE. from the input field CAT read with an L3 format specifications. The variable X4 is assigned a value of .FALSE. from the input field FROG. The variable X5 is assigned a value of .FALSE. from the input field DOG since no letter T or F is found. The variable X6 is assigned a value of .TRUE. from the input field T. The variable X7 is assigned a value of .FALSE. inasmuch as two blanks are read under an L2 format code.

On output a T or F is inserted in the output field depending upon the value of the output variable. The single character is right-justified in the output field and is preceded by w − 1 blanks. For example,

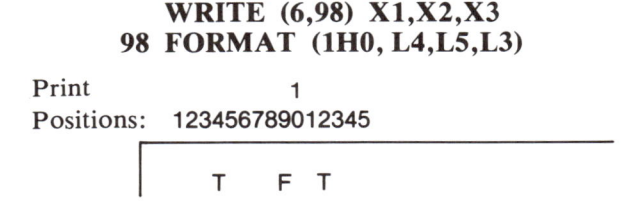

```
   LOGICAL  X1,X2,X3
   WRITE (6,98) X1,X2,X3
98 FORMAT (1H0, L4,L5,L3)
```

Print   1
Positions: 123456789012345

```
   T   F  T
```

We assumed the values of X1, X2, and X3 were .TRUE., .FALSE., and .TRUE., respectively from our input example.

Logical Function Subprograms

A Function subprogram which returns a logical value can be written by the user. The FUNCTION statement must include the LOGICAL type option to specify a logical Function subprogram. For example:

LOGICAL FUNCTION CHECK (N)

If a program unit invokes a Logical type Function, a LOGICAL statement must also be given in the invoking program. This LOGICAL statement must include the name of the Logical type Function subprogram so the compiler will know that the referenced function subprogram is to return a logical value.

The use of logical type constants and data are very helpful to mathematicians, engineers, and logisticians. Logical data can be useful in boolean algebra, hardware circuitry design, scheduling problems, and a wide variety of other applications.

16.5 Complex Floating-point Data

Complex numbers, used mainly in mathematical applications, consist of a pair having two values. The first value in the pair represents the *real part* of the complex number. The second value represents the *imaginary part* of the complex number. This second value is known as the imaginary part because of the resulting value when the term is reduced to its simplest value. For example, the second term may be $\sqrt{-25}$. Since it is impossible to take the square root of a negative number, we must replace this item with $\sqrt{-1} \cdot \sqrt{25}$ which would then give us 5i. The term $\sqrt{-1}$ is known as i.

In math textbooks, complex numbers are often written in the form a + bi where a and b are real numbers and i = $\sqrt{-1}$. Complex numbers are also written in vector form as (a,b) where a represents the real part and b represents the imaginary part. We use the latter form in FORTRAN. Thus, a complex constant is represented in FORTRAN as an ordered pair of signed or unsigned real constants separated by a comma and enclosed in parentheses. For example:

(4.6,−1.37)	(Has the value 4.6 −1.37i)
(−8.91,6.2)	(Has the value −8.91 +6.2i)
(14E+02,−1.6E−1)	(Has the value 1400 −.16i)
(.27E−2,.86E+1)	(Has the value .0027 +8.6i)

Illegal forms of complex constants are:

(A,3.29)	a variable must not be used
(5,7)	integer constants must not be used

Either of the two real constants in a complex number may be a positive, zero, or negative quantity. If either of the two real constants is unsigned, it is considered positive. Each of the two real constants must be a valid real constant formed from any of the permissible types. Both parts, however, must occupy the same number of storage locations (i.e., single-precision or double-precision).

Complex Variables

A variable which may be assigned a complex number is considered to be a complex type variable. A complex variable can be any valid symbolic name, but must be declared a complex type variable. This declaration is accomplished in an Explicit type statement with the keyword COMPLEX. Figure 16.3 shows the general form of the COMPLEX statement.

COMPLEX Statement—General Form

COMPLEX name$_1$,..., name$_n$

where:

name$_1$,..., name$_n$ — each **name** represents a variable, array, or function subprogram name. If the **name** is an array name, the dimension information may be specified following the array name.

Figure 16.3. General Notation of the COMPLEX Statement

An example of the COMPLEX type statement is:

COMPLEX A, X(10), J

The variables A and J are declared to be complex variables and X is declared to be a one-dimensional array of ten complex elements.

Complex Expressions and Assignment Statements

An expression which produces a complex resulting value is said to be a complex expression. Any arithmetic operator may be used with complex expressions and the established rules of operator precedence apply.

A complex variable can be assigned a value using the complex Assignment statement. Its form is

$$\text{complex variable} = \text{complex expression}$$

Following are examples of complex Assignment statements and expressions.

```
COMPLEX A,B,C,D(5)
A = (3.5,-7.24)
B = (-8.21,5.67)
D(1) = A + B
C = D(1) * (3.6,9.2) - A / B
```

Format Specifications for the Input/Output of Complex Numbers

One must remember that a complex number always consists of two parts. So when we read or write a complex value, two format codes must be given. The first format code is used for the real part and the second format code is used for the second part.

There is no special format code for reading or writing complex numbers. The F, D, or E format codes may be used depending upon the precision of the input values or the output form of the printed values. An example of reading a complex number is:

```
      COMPLEX N
      READ (5,99) N
99    FORMAT (2F7.2)
```

```
3214.738610.92
```

An example of writing a complex number is:

```
      COMPLEX N
      WRITE (6,98) N
98    FORMAT (1H0,2F10.2)
```

Complex Function Subprograms

Many built-in function routines are available to manipulate complex numbers. For example, CSQRT(A) will compute the square root of the complex argument A. The result is a complex number. Of course, the argument A must be declared a complex variable.

Complex built-in functions include:

Name	Mathematical Definition
CABS	Absolute value
CEXP	Exponential
CLOG	Natural Log
CSQRT	Square Root
CMPLX	Convert two real values to complex
REAL	Obtain real part of complex number
AIMAG	Obtain imaginary part of complex number
CONJC	Obtain conjugate of a complex number
CSIN	Sine
CCOS	Cosine

A user may write complex Function subprograms that return a complex value. The type option in the FUNCTION statement must be used to specify a complex type function. For example:

COMPLEX FUNCTION CMPLX (A)

The Complex type Function name (CMPLX) must be declared in a COMPLEX statement in the invoking program unit. Any user or built-in complex function subprogram invoked by a program must have their names declared in a COMPLEX statement to inform the compiler that the function subprogram returns a complex value.

16.6 The Generalized (G) Format Code

The G format code is a generalized field specification that can be used to read and write numbers expressed in a decimal (F) or exponential (E) form. The general form of the G format code is:

rGw.d

where **r** represents an optional repetition factor which may range from 1-255. The **w** specifies the width of the field being read or written. The **d** indicates the number of digits to be included in the fractional part of an input field or the number of significant digits to be printed for an output field. The . is a necessary separator. **w** and **d** must be unsigned integer constants between 1 and 255.

On *input fields* the form of the transferred value to memory depends upon how it is represented in the data card.

Consider the following input example.

READ (5,99) A, B
99 FORMAT (G7.2,1X,G5.1)

```
12.31E3 543.2
```

Since the input value for A has an exponent, it will be stored as if read with an E7.2 format code. The variable B is stored as if read with a F5.1 format code.

On *output fields* the computer will select the most suitable format specifications (i.e., E or F) depending upon the size of the number. The **d** specification in the format code determines the number of digits to be printed and, thus, the form of the output value. If the absolute value of an output variable, say X is in the range $.1 \leq |X| < 10^{**}d$, then the F form is used. If the absolute value of X is outside this range, the E form is used.

Thus, the **Gw.d** format code when used to output real values says to use **w** output positions for the field, which includes **d** positions for the number, one position for a decimal point and four positions for an exponent. The **d** specifies the number of significant digits that are to be printed; they may be on *either* side of the decimal point depending upon the output value and its form. The number will appear in F or E form depending upon its magnitude (i.e., within the range of $.1 \leq |\text{value}| < 10^{**}d$).

The location of the value in the output field depends upon the output form used. Values written in the E form will be right-justified in the field and formatted as if written with an E format code. Values written in an F form will be printed with the four rightmost positions left blank, since there is no exponent. There will be **d** significant digits printed with the location of the decimal point determined by the value of the number.

Consider the following output examples:

```
        A = -.103E02
        B = -.005
        C = 485.7263
        D = 391285.7E10
        WRITE (6,99) A,B,C,D
     99 FORMAT (4G11.4)
```

```
Print                 1         2         3         4         5
Positions:   12345678901234567890123456789012345678901234567890
            ┌─────────────────────────────────────────────────
            │ -10.30    -0.5000E-02  485.7      0.3913E 16
```

The G format specifications of G11.4 says to output four significant digits in an eleven-position field with four spaces used for an exponent. Since **d** is 4, the output value will be printed in F form if its absolute value is in the range $.1 \leq |\text{value}| < 10^{**}4$ or $.1 \leq |\text{value}| < 10000$. Since the value of A is $-10.$ and its absolute (positive) value is 10., then A is within the range $.1 \leq |10.| < 10000$. So the value of A is printed in F form. The absolute value of B (.005) is not within the range $.1 \leq |B| < 10000$ so B is printed in E form. Note the rounding result in D caused by the truncation operation.

Additional output examples are as follows (b̸ represents a blank):

Memory Value	G Format Code	Printed Value
99.	G10.2	b̸b̸b̸99.b̸b̸b̸b̸
25.3	G11.4	b̸b̸25.30b̸b̸b̸b̸
−1.396	G11.3	b̸b̸−1.40b̸b̸b̸b̸
.4873	G9.3	0.487b̸b̸b̸b̸
87645.29	G11.4	b̸0.8765Eb̸05

To sum up the output form used by the G format code, the computer attempts to first output the value in F form if it can be represented in **d** positions. Otherwise, the E form is used. If the magnitude of the values to be printed are not known, it is wise to make the **w** specifications at least seven larger than the **d** so that sufficient positions are available should the E form be used.

16.7 The P Scale Factor Specifications

You learned in chapter 6 that the decimal point in the input data field overrode any fractional specifications given by the **d** parameter in the **Fw.d** format code. It is sometimes desirable to override this decimal point location in our data field and specify a new location which gives us a new value. The P scale factor allows us to move the decimal point to the left or right a specified number of positions.

The P scale factor is specified as the first part of a D, E, F, or G format code. The number of positions that the decimal point is to be moved is expressed as a signed or unsigned integer constant before the letter P. For example:

              97 FORMAT (F7.2, 3PF6.1, −2PF5.3)

On input the P scale factor affects the value of the data field as follows.

1. A positive scale factor causes the decimal point to be moved to the left. For example:

              READ (5,99) RATE
              99 FORMAT (2PF4.1)

                  7.5

The value stored at RATE is .075 since the decimal point is moved two positions to the left. If the P scale factor had not been used, the value would have been stored as 7.5.

2. A negative scale factor causes the decimal point to be moved to the right. For example:

              READ (5,98) HOURS
              98 FORMAT (−3PF5.4)

                  .1605

The value stored at HOURS is 160.5 since the decimal point is moved three positions to the right. If the P scale factor had not been used, the value would have been stored as .1605.

The P scale factor is ignored when reading a number with an exponent. Once a P scale factor is established for any field, it applies to all subsequent D, E, F, or G format codes in the same FORMAT statement unless a new scale factor is encountered. A P scale factor of 0 (zero) must be used to discontinue the effect of the previous established scale factor. Consider the following example.

              READ (5,97) A,B,C,D
              97 FORMAT (−2PF6.2,1X,F7.3,1X,E8.3,1X,0PF5.2)

```
                         1         2         3
             123456789012345678901234567890
   card      ─────────────────────────────
   columns    72.583 123.456 4.572E01 82.39
```

The value for A is stored as 7258.3. The scale factor is still in effect for B so the value for B is stored as 12345.6. C remains unchanged, since the number has an exponent. The value of Z is stored as 82.39 since the scale factor is discontinued with the use of 0P.

On output the effect is basically the same except that the decimal point is moved in the opposite direction. For output the effects are:

1. A negative scale factor causes the decimal point to be moved to the left.
2. A positive scale factor causes the decimal point to be moved to the right.

Once the P scale factor is established for an output field, it applies to all subsequent fields in the same FORMAT statement until a new scale factor is encountered. A 0 P scale factor must be used to discontinue the effect of a previous scale factor.

The form of the output value affected by a P scale factor varies as to the format code used. For F format codes, the decimal point is moved the specified number of positions in the output value. For G format codes, the effect of the scale factor is suspended when the number is printed in F form. Thus, the scale factor affects a format code, only if the value is written in E form. For E and D format codes the decimal point is not moved, but the exponent is adjusted to reflect the new value. If the decimal point is to be moved to the left, then the exponent is decreased by the number given in the P scale factor. If the decimal point is to be moved to the right, then the exponent is increased by the number for the scale factor. A repetition code may also precede the D, E, or F format codes when the P scale factor is used. For example, 2P3F6.3 is valid.

The P scale factor can be very useful in converting input and output numbers to represent a new value. It can be used to convert decimal values to percentages, grams to kilograms, etc. This capability of moving the decimal point in data fields during the reading and writing of variables saves us from having to compute the new values, and also allows us to retain the original values.

16.8 Nonstandard Language Extensions (Hexadecimal and Octal Constants and Data)

This section discusses the addition forms of constants and format codes found on the IBM, CDC, Honeywell, and Burroughs compilers. The ability to output the internal representation of values in a number base corresponding to the base used by the computer is put to considerable use by system programmers. The two most common number bases for internal data representation are hexadecimal (base 16) and octal (base 8).

Hexadecimal Constants and Data

The hexadecimal number base consists of sixteen digits which range from 0-F. These hexadecimal digits and their equivalent decimal and binary values are:

Hexadecimal Digit	Decimal Number	Binary Number	Hexadecimal Digit	Decimal Number	Binary Number
0	0	0000	8	8	1000
1	1	0001	9	9	1001
2	2	0010	A	10	1010
3	3	0011	B	11	1011
4	4	0100	C	12	1100
5	5	0101	D	13	1101
6	6	0110	E	14	1110
7	7	0111	F	15	1111

Hexadecimal constants and data fields are only available on base 16-oriented computers such as the IBM 360 and 370, and the Burroughs 6700 and 7700 computers. The number of hexadecimal digits that can be stored in a variable depends upon the computer and the length of the variable used. For a fullword the IBM 360 and 370 computers can hold eight hexadecimal digits. The Burroughs 6700 and 7000 word size can hold twelve hexadecimal digits. Different length variables affect the maximum allowed digits. Each byte (8 bits) contains two digits. A byte is the collection of contiguous bits to represent a character.

A *hexadecimal constant* is a string of hexadecimal digits preceded by the letter Z. Hexadecimal constants may only be given as data initialization values in a specification statement. They cannot be used in expressions in the Assignment statement. An example of providing a hexadecimal constant is:

DATA V1/ZC3D6D3C5/, V2/Z0000FFFF/

The variable V1 is given the value C3D6D3C5. The value 0000FFFF is stored in the variable V2. We are assuming a four-byte word.

The hexadecimal digits are stored right-justified in the word. If the hexadecimal constant is greater than the maximum allowed digits, the high-order digits exceeding the word size are truncated. If the hexadecimal constant is less than the maximum allowed digits, high-order hexadecimal zeroes are added to the left to fill the word. Any type of variable name may be used to hold hexadecimal constants and data.

The Hexadecimal (Z) Format Code

The Z format code is used to transmit hexadecimal data fields. The general form of the Z format code is:

rZw

where **r** is an optional repetition factor that can range from 1 to 255. **w** specifies the width of the field being read or written and must be an unsigned integer constant between 1 and 255. Each hexadecimal digit uses one card column.

On *input* the data field is scanned from right to left. Any leading, trailing, or embedded blanks in the field are treated as hexadecimal zeroes. If the number of digits in the field is greater than the maximum allowed, the leftmost hexadecimal digits are truncated. If the number of digits is less than the maximum allowed, hexadecimal zeroes are added on the left.

Consider the following input example (assume a four-byte word which can contain a maximum of eight hexadecimal digits).

```
       READ (5,99) A,B,C
   99  FORMAT (Z2,1X,Z8,1X,Z10)
```

```
FF D45EC102 B7205E6000
```

The values 000000FF, D45EC102, and 205E6000 are stored at the variables A, B, and C, respectively.

On *output,* if the number of hex digits stored in a variable is less than **w**, then the leftmost print positions of the field are filled with blanks. If the number of digits in a variable is greater than **w**, then only the rightmost **w** digits are printed. If **w** is equal to the maximum number of digits that can be stored in a variable, then **w** hexadecimal digits are printed.

Consider the following output example (assume the same values as used in the input example).

```
       WRITE (6,98) A,B,C
   98  FORMAT (1H0, Z4,1X,Z8,1X,Z11)
```

```
Print                1         2
Positions:  12345678901234567890123456789

            00FF D45EC102    205E6000
```

Octal Constants and Data

The octal number base consists of eight digits which range from 0-7. These octal digits and their equivalent decimal and binary values are:

Octal Digit	Decimal Number	Binary Number
0	0	000
1	1	001
2	2	010
3	3	011
4	4	100
5	5	101
6	6	110
7	7	111

Octal constants and data fields are only available on base 8-oriented computers, such as the CDC 6000, Honeywell 6000, and Burroughs 6700 and 7700. The number of octal digits that can be stored in a variable depends upon the computer and the length of the variable used. On the CDC 6000 a maximum of twenty octal digits may be stored in a word. On the Honeywell 6000 a maximum of twelve octal digits may be stored in a word. On the Burroughs 6700 and 7700, a maximum of sixteen octal digits may be stored in a word. Different length variables affect the maximum allowed digits. Each byte (6 bits) contains two octal digits.

An *octal constant* is a string of octal digits preceded by the letter O. Octal constants may only be given in compile time initialization statements such as the DATA statement. They cannot be used in expressions in the Assignment statement. For example:

DATA V1/O23363325/,V2/O07777/

The octal digits are stored right-justified in the word. If the octal constant is greater than the maximum allowed digits, the high-order digits exceeding the word size are truncated. If the octal constant is less than the maximum allowed digits, high-order octal zeroes are added to the left. Any type of variable name may be used to hold octal constants and data.

The Octal (O) Format Code

The O format code is used to transmit octal data fields. The general form of the O format code is:

rOw

where **r** is an optional repetition factor from 1 to 255. **w** is the width of the field and must be an unsigned integer constant between 1 and 255. Each octal digit uses one card column.

On *input,* any leading, trailing, or embedded blanks are treated as octal zeroes. If the number of digits in the field is greater than the maximum allowed, the leftmost digits are truncated. If the number of digits is less than the maximum allowed, octal zeroes are added to the left. The input operations for octal values are the same as for hexadecimal data except that octal digits are read.

On *output,* if the number of octal digits in a variable is less than **w**, then the leftmost print positions of the field are filled with blanks. If the number of digits in a variable is greater than **w**, then only the rightmost **w** digits are printed. If **w** is equal to the maximum number of digits that can be stored in a variable, then **w** octal digits are printed. The output operations for octal values are the same as for hexadecimal values except that an octal digit is printed.

Complex Constants

On IBM and Burroughs compilers, complex constants and variables may also be double-precision in length. Each constant part of a complex number is expressed in double-precision decimal form or exponential form. Complex variables may be specified to be double-precision by using the DOUBLEPRECISION or REAL * 8 type statements.

16.9 REVIEW QUESTIONS

1. The E format code is primarily used to read and write what type of numeric fields?
2. Describe the general form of a constant in exponential form.
3. Describe the general form of the output field printed with an E format code.
4. How does the computer recognize a double-precision constant?
5. A double-precision variable must be declared in a _____ type statement.
6. The D format code is primarily used to read and write what type of numeric fields?
7. What are the two logical constants available in FORTRAN?
8. A logical variable must be declared in a _____ type statement.
9. The _____ format code is used to read and write logical data.
10. What are the two parts of a complex number?
11. How is a complex constant represented in FORTRAN?
12. A complex variable must be declared in a _____ type statement.
13. What format codes may be used to read and write complex values?
14. The G format code is used to read and write what type of data?
15. The computer attempts to output data written with a G format code in F form unless the **d** specifications are too small. (True/False)
16. What is the purpose of the P scale factor?
17. How is the P scale factor specified with a format code?

16.10 EXERCISES

1. Construct the E format codes necessary to read the following input fields.

	Value	E Format Code
a)	12.3E2	
b)	123E−1	
c)	1.23E+2	
d)	123	

2. Construct the E format codes necessary to read the following input fields.

	Value	E Format Code
a)	.1E−5	
b)	45.6	
c)	456.E+3	
d)	45.6E1	

3. Show the output results produced when printing the following values with their respective E format code. Show all blank positions with the symbol ƀ.

	Memory Value	Format Code	Output Results
a)	12.345	E10.3	
b)	−12.345	E12.5	
c)	123.45	E14.6	
d)	123.456	E14.5	
e)	−12.3456	E14.5	
f)	−.0012	E12.4	

4. Show the output results produced when printing the following values and their respective E format code. Show all blank positions with the symbol ƀ.

	Memory Value	Format Code	Output Results
a)	4.5126	E10.4	
b)	4.5126	E10.3	
c)	−4.5126	E15.7	
d)	−4.5126	E14.4	
e)	.45	E10.3	
f)	−.00016	E11.3	

5. Determine the resulting logical value from each of the following logical expressions. Assume A and B to be .TRUE. and C to be .FALSE.

   a) .NOT. C
   b) .NOT. C .AND. A
   c) .NOT. A .OR. .NOT. C
   d) B .AND. C .OR. A
   e) .NOT. (A .AND. B)

6. Determine the resulting logical value from each of the following logical expressions. Assume A and B to be .TRUE. and C to be .FALSE.

   a) A .AND. .NOT. C
   b) A .AND. B .AND. C
   c) .NOT. (A .AND. C) .AND. B
   d) A .OR. B .OR. C
   e) A .OR. .NOT. (B .OR. C)

7. Show the output results produced when printing the following values with the given Gw.d format code. Show all blank positions with the symbol ƀ.

	Memory Value	Format Code	Output Results
a)	99.0	G9.2	
b)	.1	G9.2	
c)	146E−10	G10.3	
d)	−574.3E−1	G10.3	

8. Show the output results produced when printing the following values with the given Gw.d format code. Show all blank positions with the symbol ƀ.

	Memory Value	Format Code	Output Results
a)	−99.1	G10.3	
b)	1.2E−1	G9.2	
c)	+67.3E−07	G11.4	
d)	3.1416	G12.5	

9. Give the P scale factor to change the following input fields to the new value.

	Input Field	New Value	P Scale Factor and F Format Code
a)	8.75	.0875	
b)	.31416	3.1416	
c)	125.0	.1250	
d)	1.	1000000.	

10. Give the P scale factor to change the following input fields to the new value.

	Input Field	New Value	P Scale Factor and F Format Code
a)	873021.	873.021	
b)	.1	.000001	
c)	34.8762	3487.62	
d)	.1	1000.0	

16.11 PROGRAMMING PROBLEMS

1. Write a program to read the value of RADIUS off a punched card and use the value to compute the volume of a sphere. The input value is punched in exponential form in columns 1-8 and should be read with an E format code.

After reading the value for RADIUS, compute the volume of its associated sphere according to the formula (use 3.14 as π):

$$V = \frac{4}{3} \pi r^3$$

Print the value of RADIUS in positions 1-12 and the computed volume in positions 20-34. Output values are to be written using the E format code. Include appropriate column headings.

2. Write a program to read and sum two complex variables named X and Y. The two complex numbers are to be read with 4F6.2 format specifications. Write the variables X, Y, and SUM under a 6F10.2 format specification. Use column headings to identify each number.

3. Write a logical-valued function subprogram named EVEN to determine whether a numeric integer input is odd or even. If the input value is an even number, the function EVEN returns a logical value of .TRUE.; otherwise, the function returns a value of .FALSE. if the input value is an odd number.

A mainline program is used to read the input value and print the results. The input value must be a positive integer between 1 and 100 and is read with the variable name INTGR under a I3 format specification. After the function EVEN has determined whether the number is even or odd, write an output message as follows:

'THE NUMBER' n 'IS' l

where **n** is the value of the input number, and l is the literal 'ODD' or 'EVEN' depending upon the evaluation of the number.

Hint: Determine if the number is even or odd by the formula:

INRSLT = INTGR/2*2 − INTGR

If the result in INRSLT is 0, then the variable INTGR contains an even integer. If the result is −1, then the input value is an odd integer number.

4. A real estate agency wanted a listing of all property sold during the year. The agency also wanted a total of the sale prices and commissions on all the properties sold. The totals were to be stated by the type of property sold.

Write a program to read a card file containing real estate transactions for the Smiley

Cole Real Estate Agency. Each transaction card is punched according to the following record layout:

Card Columns	Field Description
1-6	Sales number (Form=NNNNNN)
7-15	Sales price (Form=NNNNNN.NN)
16-23	Commission (Form=NNNNN.NN)
24	Type property (P=Private residence, C=Commercial)

Accumulate the amount of sales and earned commissions by each of the two respective type of properties. Also compute what percentage of the sales price is commission (i.e., total of commissions divided by total of sales prices).

For output results:
a) Write a report heading, beginning on a new page.
b) Write appropriate column headings.
c) Write double-spaced detail records with the proper information in its respective columns as described below:

Print Positions	Field Description
3-8	Sales number
15-23	Sales price of private residence
31-38	Commission from private residence
46-54	Sales price of commercial property
62-69	Commissions from commercial property
77	Type property
85-91	Commission percentage of sales price (Form=NNN.NNN)

d) Write the accumulated totals beneath their respective columns (Form = NNNNNNNNNN.NN).

Also use a DATA statement to initialize the letters P and C. Use double-precision variables for sales price and the four counters for
  a) the accumulated sales price for private residences,
  b) the accumulated sales price for commercial property,
  c) the accumulated commissions for private residences, and
  d) the accumulated commissions for commercial property.

Magnetic Tape Statements and Operations 17

17.1 Introduction to the Use and Concepts of Magnetic Tape

Increasing internal speeds of computers calls for high-speed input and output devices to ensure that program execution is not held back by having to wait for input, or by an inability to get processed data out to an output medium. Magnetic tape units, with their dual capability of input and output, provide a very high data transfer rate into and from computer memory. Modern day tape drives can transfer data at the rate of over 200,000 characters per second. This means that a full reel of data can usually be read in several minutes, whereas the equivalence of 400,000 punched cards would take more than eight hours to read through a card reader.

Another prime virtue of magnetic tape is its ability to store an enormous amount of data on an easy-to-handle reel at a very low cost. The equivalent of about 400,000 fully punched cards can normally be stored on one 2400-foot reel of magnetic tape. This is over 32,000,000 characters of data stored on a single reel. When large data files are processed, magnetic tape is often used as the storage medium.

The techniques of magnetic tape recording are very much like those used in the making of tape recordings in the home. A long thin ribbon usually one-half inch wide, made of a plastic base and coated with a metallic oxide, is wound on a reel. This reel is referred to as the *file reel* or *data reel.* A take-up reel, known as the *machine reel,* is used to hold the tape that is unwound from the data reel. The machine reel and file reel are mounted on the tape unit. The tape from the file reel is threaded through the transport mechanism (see fig. 17.1 on the next page).

The recording of data onto the tape takes place by moving the tape under a stationary *read/write head.* Data from the computer memory is transferred to the read/write head, which magnetizes spots in parallel tracks along the length of the tape to record the proper information. A coded pattern of 0 and 1 bits across the width of the tape represents data received from the computer. This data is coded according to a coding scheme which varies by computer and vendor. A different coding scheme is also used depending upon whether a 7- or 9-track tape unit is used (see fig. 17.2). Data previously recorded on the tape is automatically destroyed by overwriting it with new data. As with the home tape recorder, reels of magnetic tape may be used hundreds of times for recording data.

Information that has been recorded onto the tape may be read by passing the tape over the read/write heads, but with the tape unit instructed to read data instead of to write or record data. Reading is nondestructive; the same information can be read time and again from a reel of tape.

Magnetic tape must have some blank space at both the beginning and the end of the reel to allow threading through the feed mechanism of the tape unit. Markers, called reflective strips, are placed on the tape (about twenty feet from each end) to enable the magnetic tape unit to sense the beginning and end of the usable portion of tape. Photoelectric cells in the tape units sense these markers and stop the tape.

The reflective strip at the beginning of the tape is known as the *load-point marker* and indicates where reading or writing of the first record of data is to begin. The tape must be located at this point at the start of processing before the tape unit can begin the reading or writing of data correctly. A computer operator,

Magnetic Tape Statements and Operations 311

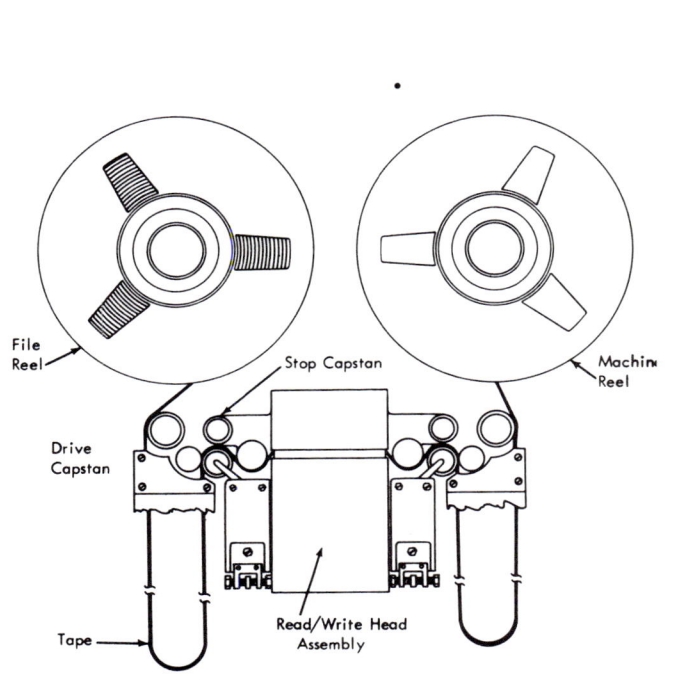

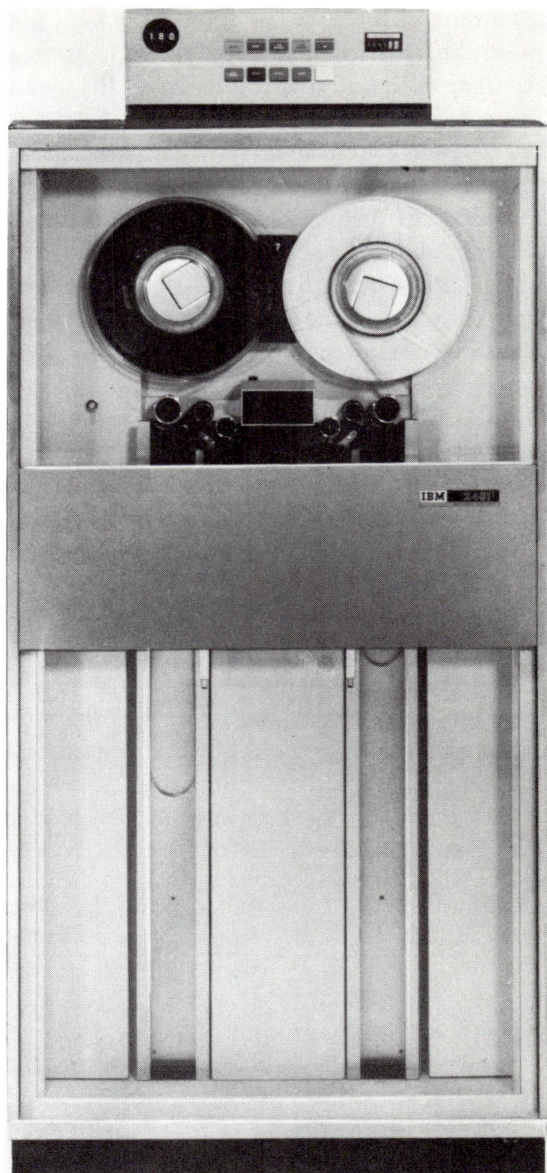

Figure 17.1. A Magnetic Tape Unit and Its Operation.
Courtesy of International Business Machines Corporation.

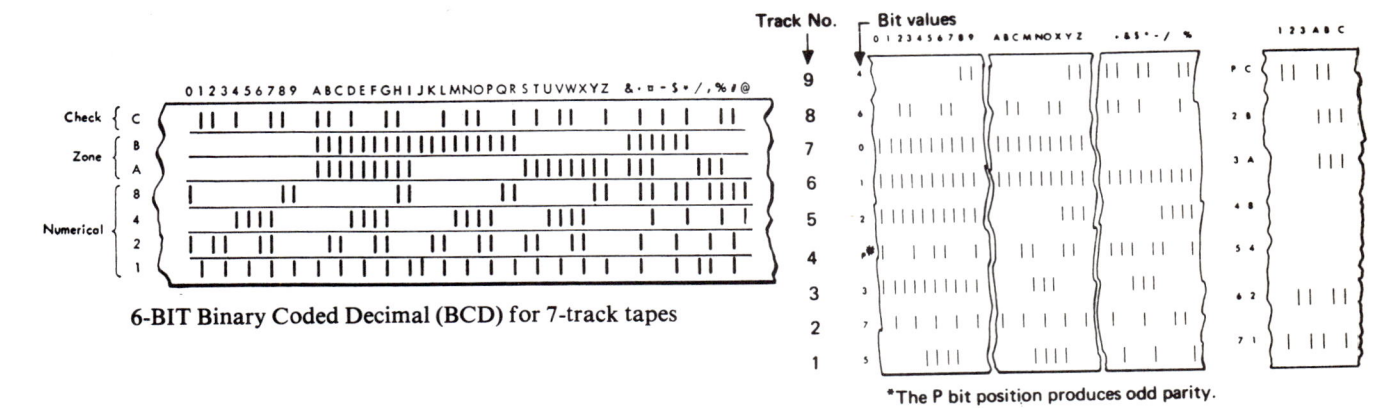

6-BIT Binary Coded Decimal (BCD) for 7-track tapes

Figure 17.2. Magnetic Tape Coding Schemes.
Courtesy of International Business Machines Corporation.

*The P bit position produces odd parity.

8-BIT Extended Binary Coded Decimal Interchange Code (EBCDIC) for 9-track tapes after mounting the reel of data, will press the proper buttons on the tape drive to position the tape at load point. This load point is also referred to as the beginning-of-tape (BOT) marker.

The reflective strip at the end of the tape is known as the *end-of-reel marker* or *end-of-tape (EOT)*. If this reflective marker is sensed during a write operation, the last record will continue to be written and then the tape will be rewound.

The tape unit does not recognize the end-of-reel marker when reading the tape. To indicate the end of the valid data belonging to the last recording, an end-of-file record is written on the tape. The *end-of-file record* is simply a special character that is written in a separate record after the last block of recorded data on the tape. This end-of-file mark (also known as a tapemark) is a must when reading the tape to make sure that one stops at the end of his set of data. Without this marker, one may continue reading data past his file. A technique of writing a record with all 9's or some other chosen character might be used as an alternative to the system's end-of-file record. One would have to check for this special record after reading each record to determine an end-of-file condition.

The recording of data onto magnetic tape is performed in physical records. We refer to a *physical record* as a tape block or simply a *block*. Magnetic tape records are not restricted to any fixed record size as are punched cards. Records may be any practical size within the limits of internal storage capacity. A physical record can consist of a single logical record or a multiple of logical records grouped together. The number of logical records in each physical record is called the *blocking factor.*

Each physical record is separated on the tape by an interblock gap (referred to as an **IBG**). The *interblock gap* is a length of blank tape which varies from 0.60 to 0.75 of an inch depending upon the model of tape drive used. This interblock gap is automatically produced by the computer system at the end of each block of records, during a write operation. During reading, the block begins with the first character sensed after the interblock gap and continues without interruption until the next gap is reached. The entire block of data is read into internal memory for use in processing. Thus the interblock gap allows for starting and stopping the tape between blocks of physical records.

Figure 17.3 illustrates the concept of unblocked and blocked records.

UNBLOCKED RECORDS

BLOCKED RECORDS

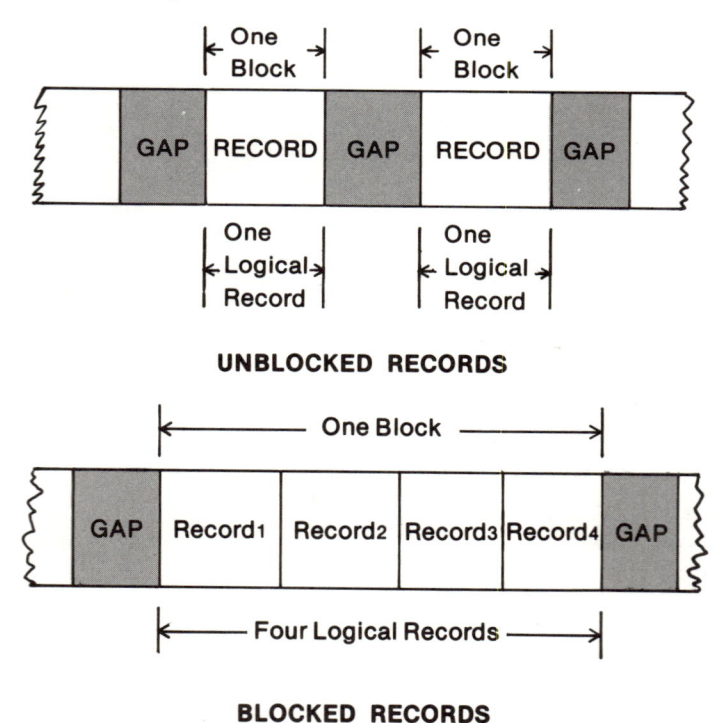

Figure 17.3. Unblocked vs. Blocked Magnetic Tape Records

The tightness at which data can be recorded onto magnetic tape is referred to as its *density*. Modern tape units record data at 800 or 1600 characters per inch. Since characters are recorded in a pattern of parallel vertical bits, the terms bits per inch (BPI), bytes per inch, and characters per inch are synonymous. If we recorded one 80-character logical record on a tape block at a density of 800 BPI, only 0.10 of an inch would be required to hold the 80 characters of data. We can see that for unblocked logical records on tape, most of the tape is blank since so much of the tape is used by the IBGs to separate the records. Both for economy in storage saving and for efficiency in I/O operations, records should normally be blocked in production processing of magnetic tape files.

Not only must we be able to read and write magnetic tape files, but we need also to be able to perform operations such as rewinding the tape, putting an end-of-file marker on the tape, and backspacing to previous records. The seven FORTRAN statements that provide I/O operations on magnetic tape are (1) formatted READ, (2) unformatted READ, (3) formatted WRITE, (4) unformatted WRITE, (5) REWIND, (6) ENDFILE, and (7) BACKSPACE. We will now look at the construct of these statements that permit us to perform these operations on magnetic tape.

17.2 Formatted READ/WRITE Statements with Magnetic Tape Files

Magnetic tape is considered to be a sequential storage medium. Records are written in a serial manner, usually in sequence by a certain *key,* such as SSAN, account number, stock number by warehouse, or some other sequencing field. To get to a specific record, all records falling before the desired one must be read. For example, to retrieve the sixth record, the first five records must have been previously read.

The sequential formatted READ and WRITE statements explained in chapters 3 and 6 apply to magnetic tape files as well as to card and printer files. The general form of the formatted READ statement is the same as for the READ statement you have been using throughout the text.

Valid examples of the formatted READ statement for tape files are:

    3 READ (8,97) A,B,C,D
      READ (9,84) ITEMNR, NRONHD, NRONBO

Any file unit other than 5 (and the standard output units of 6 and 7) may be used, provided that it is an acceptable unit number on the system. When these other file units are used for magnetic tape or disk files, they must be included in the JCL (Job Control Language) to identify which physical peripheral units are actually being used.

The general form of the formatted WRITE statement for tape operations is the same as for the WRITE statement you have been using all along. The output file unit must be other than a 5, 6, or 7. The appropriate JCL card must be included in your Job deck to assign the chosen file unit to a tape unit. Valid examples of the formatted WRITE statement for tape files are:

    15 WRITE (10,37) I,X,Z
      WRITE (8,77) NAME, NUMBER, GRADE

The formatted WRITE statement for magnetic tape functions the same way as the formatted WRITE statement for printer output files. However, no carriage control is required in the FORMAT statement for data written to tape.

The amount of data generally transmitted to tape by a write statement is called a *logical record* or simply a record. The size of a record is determined by the number of list items and the format specifications. If the number of transmitting format codes in the format statement agree with the number of list items in the WRITE statement, one logical record is written. If the number of format codes are less than the number of list items in the WRITE statement, multiple logical records are written. Records are written onto tape under the control of the FORMAT statement in the same way that format codes control the

314   Magnetic Tape Statements and Operations

generation of printer lines. The same principle is true for reading records on tape. An entire record is read or written each time the computer scans from the beginning left parenthesis to the last (ending) right parenthesis in a FORMAT statement. The following I/O statements illustrate the number of characters and records that are read/written with their respective FORMAT statement.

```
   READ (8,78) X,Y,X        (One record with 3 fields of data is read)
78 FORMAT (2F6.1,F4.2)
   READ (10,20) A,B,C       (Three records, each with 7 characters, are read)
20 FORMAT (F7.2)
   WRITE (9,30) X,Y,Z       One record with 3 fields totalling 30 characters is written)
30 FORMAT (3F10.3)
   WRITE (10,40) I,J,K      (Three records, each with 50 characters are written. Note
40 FORMAT (I5,45X)          the last 45 characters in each record are blanks)
```

In addition to reading/writing magnetic tape records with the formatted READ and WRITE statements, unformatted READ and WRITE statements are available for reading and writing magnetic tape files.

17.3 Unformatted READ/WRITE Statements with Magnetic Tape Files

Occasionally, a tape file may need to be read several times by a program or by different programs. Or a tape file may be created by a FORTRAN program to be read by another FORTRAN program in a later processing run. Reading and writing of formatted records in FORTRAN is a very slow operation since individual input fields must be converted from an external form to an internal binary word form in memory, and vice versa on output.

To overcome the inefficient speed of the formatted READ and WRITE operations, two FORTRAN statements are available to allow us to perform unformatted I/O operations. Using unformatted READ and WRITE statements, data is transmitted to and from memory at a highly increased speed. No conversion of the data from an external to an internal form, or from an internal to an external form takes place. Data is read and written in machine word format as it is required to look in internal storage. In addition to the efficient speed of unformatted I/O operations, a secondary advantage is the elimination of writing a FORMAT statement with each I/O statement.

The unformatted READ statement, therefore, allows efficient speed in reading input records already in machine word format. The general form of the unformatted READ statement is given in figure 17.4.

Unformatted READ Statement—General Form

READ (unit) list

where:

unit — represents the file unit being read.
list — represents the I/O list of variables, array names, and/or subscripted variables.

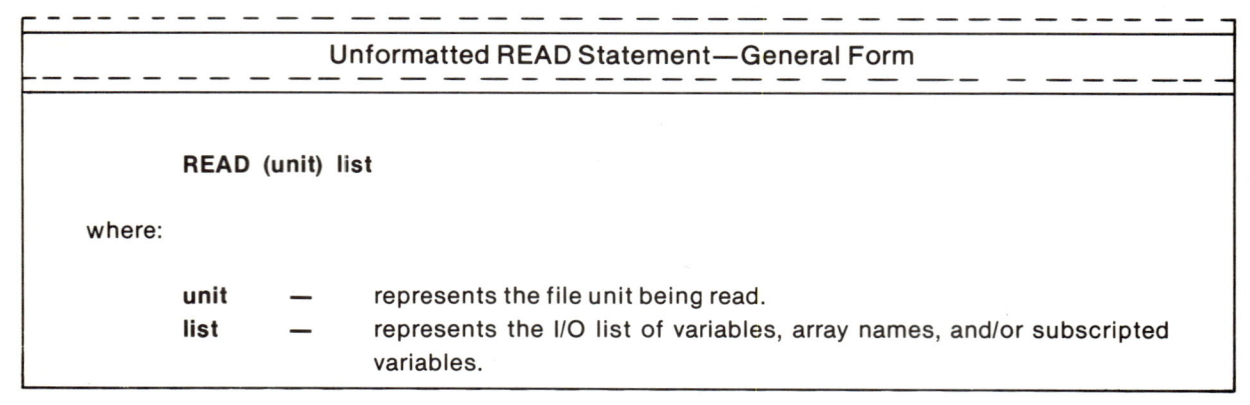

Figure 17.4. General Notation of the Unformatted READ Statement

Valid examples of the unformatted READ statement are:

```
24 READ (8) NRSTUD, GRADE1,GRADE2,GRADE3
   READ (10) T,U,V
```

Magnetic Tape Statements and Operations

The unformatted READ statement allows data to be read and transmitted to memory without any character or field conversion. Data already in binary word format is stored in word locations in the identical form it had when it previously resided in memory. Data read under an unformatted READ statement must have been *previously recorded* by an unformatted WRITE statement.

Since data fields are transmitted in word form under nonformat control, each data field represents one full word. The following READ statement causes one record composed of five words to be read from a magnetic tape and transferred to memory.

READ (8) A,B,C,D,E

When the unformatted READ statement is used, the number of words comprising the tape record being read must be equal to the number of variables and/or array elements given in the list portion of the statement. An unequal match causes a program abort. Special care must also be taken to ensure that the type and length of the variables and/or array elements located in the list of the READ statement are the same type and length as the variables and/or array elements located in the list portion of the WRITE statement that wrote the record on magnetic tape.

Data may be written to tape in a binary word form with the unformatted WRITE. The general form of the unformatted WRITE statement is given in figure 17.5.

```
┌──────────────────────────────────────────────────────────────────┐
│            Unformatted WRITE Statement—General Form              │
├──────────────────────────────────────────────────────────────────┤
│                                                                  │
│       WRITE (unit) list                                          │
│                                                                  │
│  where:                                                          │
│                                                                  │
│       unit   —   represents the file unit being written.         │
│       list   —   represents the I/O list of variables, array     │
│                  names, and/or subscripted variables.            │
│                                                                  │
└──────────────────────────────────────────────────────────────────┘
```

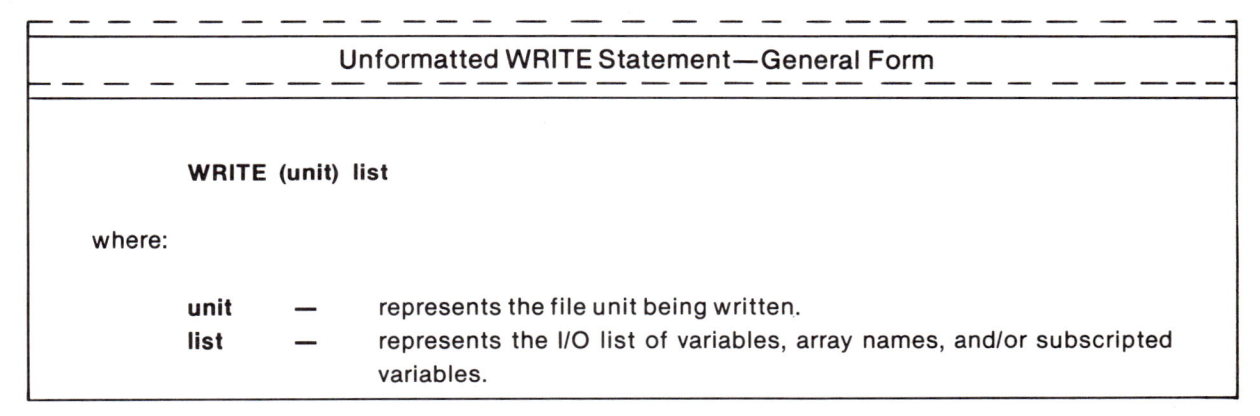

Figure 17.5. General Notation of the Unformatted WRITE Statement

Valid examples of the unformatted WRITE statement are:

7 WRITE (9) DATA1,DATA2,DATA3,DATA4
WRITE (3) A,B,C,D,SUM,AVE

No conversion of data fields occurs. A tape written with an unformatted WRITE statement must be read with an unformatted READ statement when it is used for an input file.

Let us look at the three statements used in manipulating magnetic tape files other than reading and writing records.

17.4 The REWIND, ENDFILE, and BACKSPACE Statements

REWIND Statement

The REWIND statement is used to position a magnetic tape file back to the load point in such a way that the first record of the file is ready to be read or written. This operation is desired after a tape file has been written so that the operator will not have to manually cause the tape to rewind. The general form of the REWIND statement is given in figure 17.6.

316   Magnetic Tape Statements and Operations

```
┌─────────────────────────────────────────────────────────────┐
│              REWIND Statement—General Form                   │
├─────────────────────────────────────────────────────────────┤
│                                                             │
│        REWIND  unit                                         │
│                                                             │
│   where:                                                    │
│                                                             │
│        unit    —    represents the file unit to be rewound. │
│                                                             │
└─────────────────────────────────────────────────────────────┘
```

Figure 17.6. General Notation of the REWIND Statement

Valid examples of the REWIND statement are:

REWIND 8
39 REWIND 2
REWIND JTAPE

Only one file unit may be rewound with a single REWIND statement. The statement REWIND 8, 9 would be invalid in attempting to rewind both units 8 and 9 in one REWIND statement. If one wishes to rewind more than one file unit, separate REWIND statements must be given for each.

ENDFILE Statement

The ENDFILE statement defines the end of a file by causing an end-of-file record to be written to the tape. The general form of the ENDFILE statement is given in figure 17.7.

```
┌─────────────────────────────────────────────────────────────┐
│              ENDFILE Statement—General Form                  │
├─────────────────────────────────────────────────────────────┤
│                                                             │
│        ENDFILE  unit                                        │
│                                                             │
│   where:                                                    │
│                                                             │
│        unit    —    represents the file unit to which the end-of-file marker is written. │
│                                                             │
└─────────────────────────────────────────────────────────────┘
```

Figure 17.7. General Notation of the ENDFILE Statement

Valid examples of the ENDFILE statement are:

ENDFILE 10
200 ENDFILE 8
ENDFILE INTAPE

Only one file unit may be used with each ENDFILE statement. The statement ENDFILE 8, 10 would be invalid in attempting to place an end-of-file mark on both units 8 and 10 with a single ENDFILE statement. If one wishes to place an end-of-file marker on more than one file unit, separate ENDFILE statements must be given.

If the end-of-file record has been written on the tape file and is checked for in the READ statement by the END option (for those compilers that allow this option), then the end-of-file record when read will cause the transfer of control to the statement number indicated by the END option.

BACKSPACE Statement

The BACKSPACE statement is used to position the tape file back one logical record, i.e., to the beginning of the previous record. If the file is already positioned at the load point, the statement has no effect. The form of the BACKSPACE statement is given in figure 17.8.

```
------------------------------------------------------------
                 BACKSPACE Statement—General Form
------------------------------------------------------------

            BACKSPACE unit

   where:

        unit   —    represents the file unit being repositioned.
```

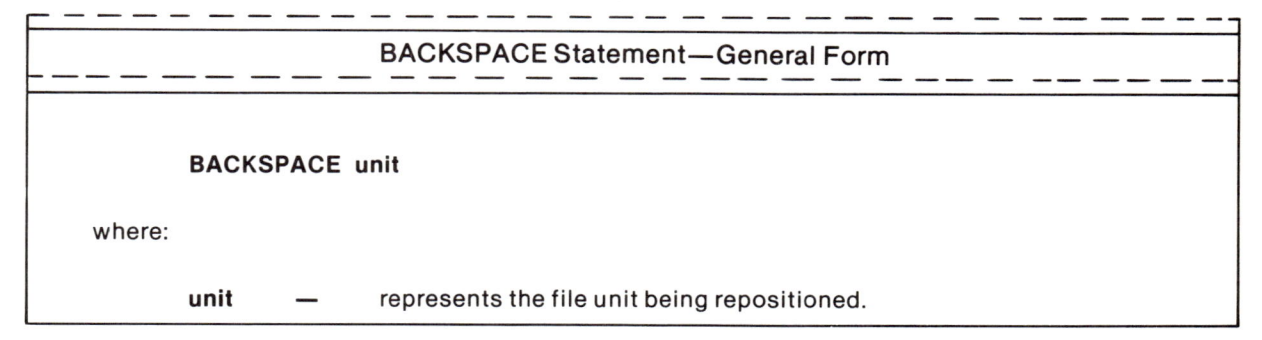

Figure 17.8. General Notation of the BACKSPACE Statement

Valid examples of the BACKSPACE statement are:

        BACKSPACE 9
  27 BACKSPACE 8
        BACKSPACE IFILE

Only one file unit may be backspaced with a single BACKSPACE statement. The statement BACKSPACE 8, 10 would be invalid in attempting to backspace both units 8 and 10 with a single BACKSPACE statement. If one wishes to backspace more than one logical record, separate BACKSPACE statements must be given for each logical record. For example, if it is desired to back up to the third previous record on unit 9 one would code three separate statements as follows:

        BACKSPACE 9
        BACKSPACE 9
        BACKSPACE 9

The BACKSPACE statement is intended to be used for input files only to provide a reread capability of an input record. One must not use the BACKSPACE statement on an output file and rewrite a record. Using the BACKSPACE statement on an output file may produce unpredictable results. On input files a backspaced record can only be reread; it must not be rewritten to the input file. To update a tape file, the output records must be written to a different file.

The following example is to read a record under one set of format specifications and then to reread the record under a different set of specifications.

```
      DIMENSION  TREC(20),SCAN(66), PGMID(2)
      READ (8,91) TREC
   91 FORMAT (20A4)
      BACKSPACE 8
      READ (8,90) ISEQNR, CONT, SCAN, PGMID
   90 FORMAT (I5,A1,66A1,2A4)
```

Now let us look at a complete program to see how we program magnetic tape files.

A Sample Magnetic Tape Program

Figure 17.9 presents a sample program to illustrate the concepts and various statements used in magnetic tape manipulation. The program reads a card file and puts it on tape. The tape file is then rewound, read, and printed.

```
            INTEGER RECD
            DIMENSION RECD(20)
            DATA NINES /4H9999/
   99       FORMAT (20A4)
   98       FORMAT (1H0,20A4)
C ***   CREATE TAPE FILE FROM CARDS
    1       READ (5,99) RECD
            WRITE (9,99) RECD
            IF (RECD(1) .NE. NINES) GOTO 1
   10       ENDFILE 9
            REWIND 9
C ***   READ AND PRINT TAPE FILE
   20       READ (9,99) RECD
            IF (RECD(1) .EQ. NINES) GOTO 30
            WRITE (6,98) RECD
            GOTO 20
   30       REWIND 9
            STOP
            END
```

Figure 17.9. Sample Program to Create a Magnetic Tape File

17.5 Use of Sequential Tape Files in Business Applications

Magnetic tape is a widely used storage medium for sequential files in business data processing. Sequential file processing on magnetic tape is most desirable when the file size is large and the activity (number of transactions against the master file) is considerable. Business files may relate to personnel, payroll, accounts receivable, accounts payable, inventory control, etc., applications.

The primary file of related records belonging to a specific function is known as the *master file.* The master file contains information of a relatively permanent type, relating to the function, as, for example, customer accounts, inventory items, and personnel data. The master file may be read by retrieval systems and various application programs to produce periodic reports. This master file must be updated at certain intervals or periods of time to reflect the latest data. The updating of the master file is known as *file maintenance* or *file update.* A majority of computer time in business applications is normally spent in file maintenance.

File maintenance involves two input files of information. One file is the master file; the other file is known as a transaction file. The *transaction file* contains current and changing information about the records on the master file. The transaction file records are then, applied against the master file to update the master file with the latest information.

Sequential file update requires the input transactions to be batched together and sorted into the same sequence as the master file. The master file records will be sequenced by social security, account number, stock number, customer account number, etc., depending upon the application. By having the records sorted in the master and transaction files according to specific items selected as *sort keys,* the computer does not have to waste time in searching for the desired record. A record is read from the master file and compared with the transaction record. Various decisions are made depending whether the master key is less than, equal to, or higher than the transaction record key. Thus the master file is completely read only once during the file maintenance operation.

File maintenance of the master file involves the updating of information in selected records. In an accounts receivable application the transaction file would contain payments and also new charges against a customer's account. The customer account number would be included in a transaction record with other fields of data. Both the master file and transaction file would be read until the account number in the transaction record matched the account number in a master file record. Then the payment amount field in the transaction record would be subtracted from the account balance in the master file record or the new charge amount field in the transaction record would be added to the account balance in the master file record depending upon the type of transaction. After the transaction had been applied against the proper master file record, the master file would then reflect the latest data on that particular record. The customer's billing might also have been produced at the time his record was updated.

New customers may be added to the accounts receivable master file and old customers removed or deleted from the master file. Thus, in addition to updating a record with current information (normally known as a *change* transaction) new records may be *added* and old records may be *deleted* from the master file.

Errors in a transaction record (such as an invalid account number or amount field) may result in not being able to properly update the master record. *Editing* (checking for invalid fields) should always be performed in order to reject those transaction records with an error. These rejected transaction records with errors are normally written to an error file which is later printed. One reviews the error listing, locates the cause of the reject, and corrects the record which is later resubmitted for processing.

File maintenance with sequential files is illustrated in figure 17.10. Note the output file labeled "Updated Master Tape." In sequential file maintenance, the updated master file is always written to a new tape file. Additions and deletions of records prohibit the writing of records back to the old master file.

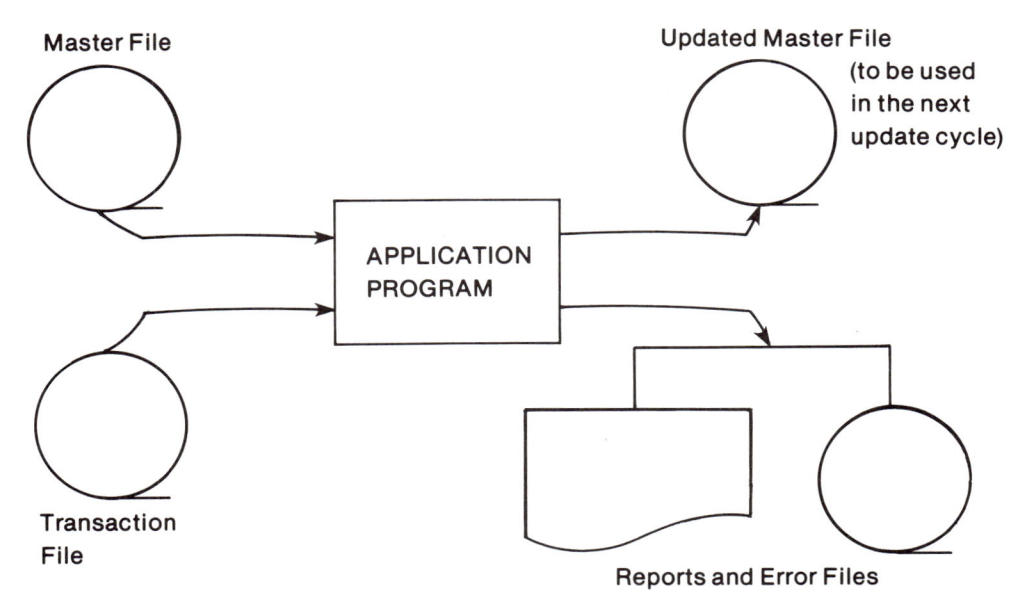

Figure 17.10. System Flowchart for a File Maintenance

The general procedures for updating a master file are as follows:

1. Compare the key in the master file record with that of the transaction file record.
2. If the key in the master file record is less than the key in the transaction file record, then no action is taken against that master record. The current master record is written to the new master file; the next master file record is read, and we return to step 1 to repeat our compare.
3. If the key in the master record matches the key of the transaction record, we apply the required operation indicated in the transaction code. If the transaction code is a change, the appropriate fields in the master record are updated and the updated master record is written to the new master file. A new master record and transaction record are read and we return to step 1. If the transaction code is a delete, a new transaction record is read and we return to step 1. The old master record is not written to the new master file.

4. If the key in the master file record is greater than the transaction record key, we check the transaction code. If the transaction code is an add, the current transaction record is written to the new master file, a new transaction record is read, and we return to step 1. If the transaction code is not an add, the record is in error and an error message is written.

These procedures are continued until all transaction records have been processed. The above type of update does not allow for multiple transaction records with the same key. If multiple transaction records with the same key are allowed, the updated master record is not written nor is another master record read until a less than condition is found against the transaction record key.

17.6 Nonstandard Language Extensions and Additional Remarks

ERR Option in the READ Statement

IBM "G" level, WATFIV, and most other FORTRAN compilers include the ERR = option in the sequential READ statement. This option is available for use in both the formatted and unformatted READ statements.

The ERR = option allows the programmer to check for parity errors. A *parity* error means that an invalid read operation occurred during the I/O of a record. That is, the various checks in the computer system indicate that the data was not read correctly—a bit may have been dropped or an extra bit picked up in transmitting the characters.

The general form of the READ statement that allows the ERR = option is:

READ (unit, n, END = a, ERR = b) list

where **b** represents a statement number to which control is transferred whenever a parity error occurs.

Occasionally dust or smoke particles stick to the recording surface of the tape, or the ferric oxide has flaked away. The area of tape where these problems occur cannot be read by the tape unit. An I/O error (i.e., a parity error) occurs. If the tape unit cannot read the area on tape after several retries, the program will be terminated by the system. The **ERR** option, when available in the READ statement, can be used to prevent an abort and allow the programmer to code his own error routine in his program. The following statements illustrate use of the **ERR** option that can be of value in tape processing.

```
16  READ (8,37,END = 52, ERR = 94) A,B,C
37  FORMAT (3F6.2)
    . . .
    . . .
94  WRITE (6,76) A,B,C
76  FORMAT (21H0PARITY ERROR ON TAPE ,3F10.2)
    GOTO 16
    . . .
52  . . .
```

There is usually no way to get around parity errors. You simply try to figure out the record containing the error and reconstruct it in a later update operation. If an abort of the program can be prevented, the remaining records in the file can be read and are available for use. But the record with the parity error cannot be used and must be skipped.

Additional Remarks

There are other aspects of magnetic tape files to be aware of in a production environment. Tapes may have system header and trailer labels that provide identifying and other useful information about the file. FOR-

TRAN tape files, however, generally do not have these labels. Multiple files can be put on one reel of tape. These files are called multifile reel tape. Tape units may record data in an "even" or "odd" parity mode. Tapes may be protected against accidental erase by removing the file protect ring which is used to allow writing on a tape.

Many details in hardware operation and systems programming involving sequential tape files are left unmentioned here inasmuch as they are beyond the scope of this text. Many of these details usually do not need to be known by the student programmer.

17.7 REVIEW QUESTIONS

1. Give two advantages of using magnetic tape as a storage medium.
2. The most common size reel of magnetic tape contains _____ feet of tape.
3. The reel of tape containing the data to be read is called the _____ reel.
4. Data is recorded onto tape or read from tape by moving the tape under a _____ _____.
5. A magnetic tape must be positioned at its _____ _____ for the first record to be read/written.
6. Recorded data may be read from a magnetic tape many times. (True/False)
7. A magnetic tape may be used to record data over and over again. (True/False)
8. The last record on a reel of tape to indicate an end-of-file condition is called an _____ record.
9. The collection of logical records that form a physical record is also called a _____.
10. Each tape block is separated by an _____ _____.
11. Formatted READ and WRITE statements used with magnetic tape files work the same way as with sequential card and printer files. (True/False)
12. Unformatted READ and WRITE statements read and write data in a pure binary form and, thus, provide a much *slower* I/O operation. (True/False)
13. No FORMAT statement is required with unformatted READ or WRITE statements. (True/False)
14. The unformatted WRITE statement should be used to create an output tape file when the output file is to be read by the same or different programs. (True/False)
15. The statement used to position a magnetic tape back to the load point is the _____.
16. The statement that causes an end-of-file record to be written is the _____.
17. The statement that causes the tape unit to position the tape file back one logical record is the _____.
18. The BACKSPACE statement may be used successfully on both input and output files. (True/False)
19. The data file in a business application that is used to contain all the records relating to a specific function is known as the _____ file.
20. The data file containing various transactions in a business function is known as a _____ file.
21. The process of updating a master file from a given transaction file is known as _____ _____.
22. Sequential file update requires the records in a master file and its associated transaction file to be sorted in a specified sequence according to certain keys. (True/False)
23. List the three types of update actions that a transaction file may contain.

17.8 EXERCISES

1. Construct the following READ statements.
   a) Construct a READ statement to read the variables X, Y, and Z according to the format specifications provided in a FORMAT statement numbered 98. File unit 11 is used to denote a tape file.
   b) Construct an unformatted READ statement to read the same data in a binary form as specified in 1a above.
2. Construct the following READ statements.
   a) Construct a READ statement to read the variables I, J, G, and H according to the format specifications given in a FORMAT statement numbered 97. Use 10 as the file unit number.
   b) Construct an unformatted READ statement to read the same variables in a binary form as specified in 2a above.
3. Construct the following WRITE statements.
   a) Construct a WRITE statement to write the variables M, N, and O according to the specifications given in a FORMAT statement numbered 96. Use 11 as the file unit number.
   b) Construct an unformatted WRITE statement to write the same data in a binary form as specified in 3a above.
4. Construct the following WRITE statements.
   a) Construct a WRITE statement to write the variables A, B, and L according to the specifications given in a FORMAT statement numbered 95. Use the variable NOUT as the file unit designator.
   b) Construct an unformatted WRITE statement to write the same data in a binary form as specified in 4a above.
5. Construct the following statements.
   a) Write the statement to rewind a tape on unit 12.
   b) Construct the statement to write an end-of-file record on unit NTAPE.
   c) Construct the statement to backspace the file on unit 3 to the previous record.
6. Construct the following statements.
   a) Write the statement to rewind a tape on unit INF.
   b) Construct the statement to write an end-of-file record on unit 4.
   c) Write the statements to backspace the file on unit 8 two previous records.

17.9 PROGRAMMING PROBLEMS

1. Create a Master Student Directory on magnetic tape: Prepare a data deck with the following card layout:

CC 1-20	Student Name (Last, First, Middle Initial)
CC 21-36	Student Street Address
CC 37-49	City (Alphanumeric)
CC 50-51	State Code (Alphanumeric)
CC 52-56	ZIP Code (Form = NNNNN)
CC 57-59	Area Code (Form = NNN)
CC 60-66	Telephone Number (Form = NNNNNNN)
CC 67-78	Major area of study (Alphanumeric)
CC 79-80	Reserved

The card deck is arranged in student name sequence. The card deck will contain a trailer card with a value of 99999 for ZIP code to indicate an end-of-file trailer record.

Write a program which reads the data deck, writes it onto a tape with the same format specifications, rewinds the tape, and prints it (double-spaced) for verification. This tape, which we will call the Student Directory Master File, has a record for each student, and is in student name sequence.

The printer output specifications are:

Print Positions	Field Description
1-20	Student name
25-40	Student street address
47-59	City
65-66	State code
70-74	ZIP code
80-82	Area code
84-90	Telephone number
99-110	Major area of study

Include appropriate column headings and a report heading.

2. Write a program that reads the Student Directory Master File created in problem 1 and prepares an output listing of all out-of-state students. An out-of-state student will be one with a state code other than the one chosen to represent an in-state student (i.e., not equal to TX). Use the DATA statement to provide the value for the variable used to identify an in-state student. Use the same printer output specifications as given in problem 1.

3. Write a program to read the Student Directory Master File created in problem 1, and to prepare an output listing of all students majoring in a specific area of study such as COMP SCI (Computer Science). Read the major study area from a punched card in the form 3A4. Use the same printer output specifications as given in problem 1.

4. Suppose a student changes his address or telephone number. Prepare a program to read data change cards in student name sequence which also have all the same fields as the master records. When a match in student name is found on the Master File, replace the current record on the Master File with the values in the new change record. If a change record does not match a record on the Master File (change record name will be less than Master File name due to being out of sequence or not being in the master file) write the change record on a printer file as an error and halt processing.

Use only the first four characters of the name field to determine a match. Also use a trailer card with the value of 99999 for ZIP Code in the card update file to signal an end-of-file. Print the card update records as each is processed to maintain a listing of processed card changes. To verify the correct changes, write the new master file to the printer. Use the same printer output specifications as given in problem 1. Note: A new file must be written for the updated Student Directory Master File.

5. Write a program to read a transaction file of changes to the Student Directory Master File and to update the Master File. Transaction data cards will be punched the same format as the original master data deck cards, but will carry a type transaction code punched in CC 80. If a "1" is punched in CC 80, the transaction record is to be added to the Master File. If a "3" is punched in CC 80, the transaction record is to replace the corresponding record on the Master File. If a "2" is punched in CC 80, the corresponding student record is to be deleted from the Master File. Use only the first four characters in the name field to determine a match.

Write the transaction update records to the printer as each is processed. Write the new Master file to the printer. Use the same printer output specifications as given in problem 1. Remember, a new tape file must be written for the updated Master file.

Magnetic Disk Statements and Operations   18

18.1 Introduction to the Use and Concepts of Magnetic Disk

Today, nearly all third-generation computers have magnetic disk units to store required system program modules, compilers, and data files. The primary value of magnetic disk is that programs and data files may be kept on-line to the computer system and accessed in a fraction of a second. The random-access approach allows direct access to records without extensive searching through a data file.

Modern disk packs may contain over 100 million characters of data. Data transfer rate is double that of the fastest tape unit. Over 400,000 characters of data can be transferred to and from memory in a second.

High capacity, randomly accessible disk storage devices make real-time (or in-line) processing feasible. Real-time processing provides the capability for processing an event as soon as it occurs. Intermixed and nonsequential input data can, therefore, be processed as it becomes available and does not have to be batched and sorted as in sequential processing. The latest up-to-date information is available at a moment's notice. Remote devices can be hooked into the computer through data communication networks and, with the use of large on-line direct-access storage units, make teleprocessing possible. A good example of an efficient on-line teleprocessing system is the airlines reservation system.

Disk storage is extremely effective in a high-activity application with a relatively small master file (for example, 5000 records) involving a comparatively small number of records updated frequently. An on-line disk file is commonly used with applications which require an immediate response. For time-sharing systems, magnetic disk is essential; main storage could not hold all the various users' programs at the same time. Magnetic disk is, therefore, used to temporarily store programs until others have finished execution.

Magnetic disk units are known for their ability to provide random access (or direct access) to on-line files. Random access processing is characterized as being able to skip around within a file and reading or writing specific data with no particular regard to the sequence.

Disk units may contain only one disk drive or they may contain multiple disk drives on a unit. See figure 18.1 for an example of a disk storage unit with multiple drives. A magnetic **disk pack** is mounted on each *drive* (or *spindle,* as it is sometimes called). See figure 18.2 for a picture of an individual disk pack (with a protective plastic case).

Magnetic disk consists of thin circular, metal plates coated on both sides with a ferrous oxide recording material (much like that used on magnetic tapes). This circular disk looks very much like an LP record used on the stereo. Normally, eleven circular plates, called *platters,* are mounted on a spindle to form a disk pack. Data is stored along *circular tracks* on both sides of each platter by magnetic fields recorded by a read/write head. Information is represented by a coding scheme of magnetic spots (the same as used on magnetic tape) caused by the read/write head. Read/write heads are mounted on movable access arms that move them in and out among the platters to the desired track (see fig. 18.3 on page 326).

All the tracks that are available for reading or recording data at one position of the access arm are said to make up a cylinder. A *cylinder* on magnetic disk, therefore, consists of all the tracks in a vertical line throughout the disk pack. See figure 18.4 on page 326 for the concept of a cylinder.

Magnetic Disk Statements and Operations 325

Figure 18.1. A Magnetic Disk Unit with Multiple Disk Packs. *Courtesy of International Business Machines Corporation.*

Figure 18.2. A Magnetic Disk Pack with a Protective Plastic Cover. *Courtesy of International Business Machines Corporation.*

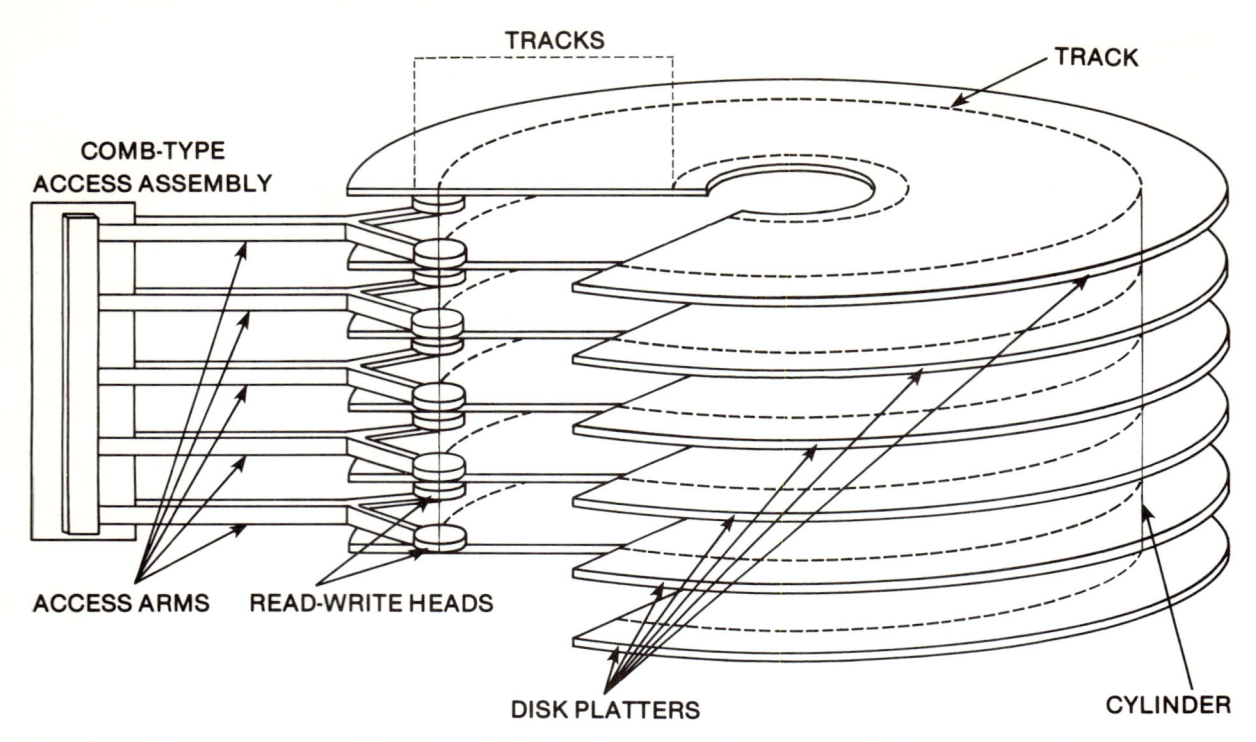

Figure 18.3. Operation of a Magnetic Disk Drive. *Courtesy of International Business Machines Corporation.*

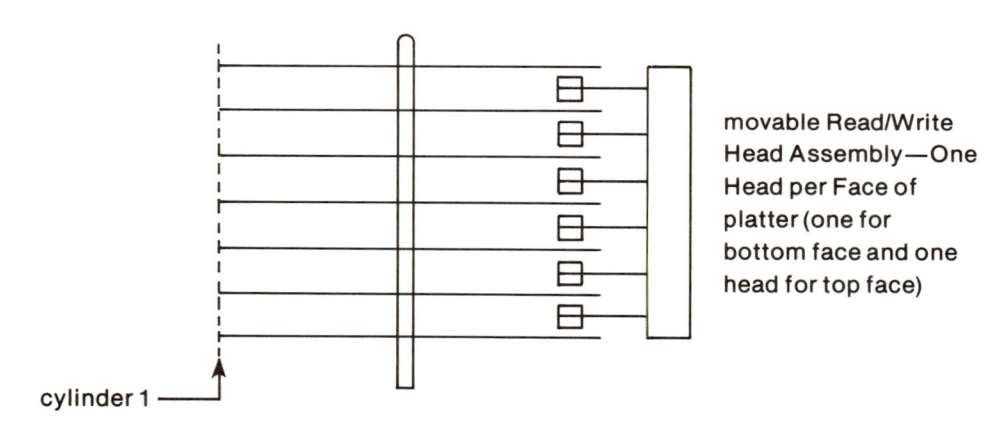

Figure 18.4. Concept of a Cylinder on a Magnetic Disk Pack

Data may be read/written as unblocked logical records or blocked logical records the same as on magnetic tape. Again, JCL functions to describe the construction of a block or physical record in FORTRAN. Interblock gaps are still used to separate physical records on disk. There is no minimum-size record that can be written on disk. If desired, a single-character record can be recorded and read. Disk files are normally retained on disk storage until they are "purged" by the programmer or data base administrator.

18.2 FORTRAN Statements Used with Sequential Magnetic Disk Files

Even though magnetic disk storage provides for the direct access of records, disk is often used to store sequentially organized files. Sequential file organization is still the best design for files with a large volume of activity such as updating with transaction records or when most of the file must be read. Sequential disk files provide on-line access to the file and eliminate certain setup operations for the job, such as mounting and dismounting tape files.

The same FORTRAN statements available for use in processing sequential magnetic tape files can be used in processing sequential disk files. The form and a brief explanation of each statement is as follows:

(1) Formatted READ statement:

READ (unit,n) list
or
READ (unit,n,END = a,ERR = b) list

The file unit used for disk files can be any integer constant or variable (except 5, 6, or 7) that is accepted on the computer system. The formatted READ statement uses a related FORMAT statement to describe the form of the input data fields. All other elements in the statement function the same as with sequential card or tape input files.

(2) Unformatted READ statement:

READ (unit) list
or
READ (unit,END = a,ERR = b) list

The elements of an unformatted READ statement serve the same function as with tape files. Input data is read in binary word format.

(3) Formatted WRITE statement:

WRITE (unit,n) list

The formatted WRITE statement uses a related FORMAT statement to describe the form of the output fields. No carriage control specification is needed in an output record. The elements in the formatted WRITE statement function the same as with tape files.

(4) Unformatted WRITE statement:

WRITE (unit) list

The elements of the unformatted WRITE statement function the same as with unformatted records on tape. Data is transmitted to disk in the same word format as when the fields resided in internal storage.

(5) REWIND statement:

REWIND unit

The REWIND statement allows us to reposition ourselves at the beginning of a file. There is no reflective strip to indicate the beginning of a file on disk. Each disk pack maintains a *directory* of the starting address of each file stored on that disk pack. The REWIND statement for magnetic disk files, however, serves the same purpose as for tape files—to allow us to start processing at the first record in the file. The address of the first record in the file is obtained as the location of the next record to be read or written.

(6) ENDFILE statement:

ENDFILE unit

The ENDFILE statement causes an end-of-file record to be written after the previously recorded record. It is used in conjunction with the "END" option in the READ statement to sense an end-of-file condition.

(7) BACKSPACE statement:

BACKSPACE unit

The BACKSPACE statement allows the programmer to access the previous logical record. As with magnetic tape operations, the BACKSPACE statement is used with read operations on input files only.

The concepts of sequential file processing with records on magnetic disk are the same as for processing sequential files on tape. The records are read and written in a serial manner. To access the tenth record, the first nine must have been previously read. File maintenance with sequential disk files also remains the same. A new disk file must be created as the newly updated master file.

Following is a program to build a sequential file on disk. A blood donor file with blood donor number, blood type, and telephone number is read from cards and created on file unit 3. The file is written without format control.

```
C *** PROGRAM TO BUILD A BLOOD DONOR FILE
C *** ON SEQUENTIAL DISK IN UNFORMATTED FORM
   99    FORMAT (I7,A4,I7)
   10    READ (5,99) NRDONR, TYPEBD, NRTELE
         WRITE (3,99) NRDONR, TYPEBD, NRTELE
         IF (NRDONR .NE. 9999999) GOTO 10
         ENDFILE 3
         STOP
         END
```

18.3 Random File Processing on Magnetic Disk

Files may be organized in a random manner to allow us to go directly to the desired record instead of having to search for it sequentially. Devices that provide us this capability of reading a specific record at random are known as *direct-access devices.* The capacity for locating records directly on these devices exists as a result of being able to assign addresses to storage locations. On magnetic tape, however, no storage address is associated with a given record.

The programmer can consider that his file consists of a given number of records. Each record has an address or index associated with it to make possible the direct reference to a record in the file. In FORTRAN, this index pointer indicates the relative position of the record in the file. The first record has index 1, the second record has index 2, . . . , the last record has index n, where n is the maximum number of records on the file. By using the proper index pointer, any record can be located and read into memory in less than a second.

FORTRAN files which are set up in this manner are said to be in *random organization* and are known as *direct-access files.* These files must be created and processed using direct-access I/O statements which differ slightly from the sequential I/O statements used to create and process sequential files.

ANSI FORTRAN does not provide specifications for random access statements because of the wide variety of statements and techniques used by different compilers. Therefore, this text explains the statements used for IBM 360 and 370 FORTRAN compilers, and WATFIV. Section 18.8 gives the syntax and procedures in using random-access statements for CDC, Honeywell, and Burroughs compilers.

The six direct-access I/O statements in FORTRAN are: (1) DEFINEFILE, (2) formatted random READ, (3) unformatted random READ, (4) formatted random WRITE, (5) unformatted random WRITE, and (6) FIND. We will now discuss each of these statements and look at their syntax construction.

18.4 The DEFINEFILE Statement

The DEFINEFILE statement is used to identify random organized direct-access files in FORTRAN and to specify their characteristics. Any file specified in a DEFINEFILE statement is considered to be a random organized file. One must use the random READ and WRITE statements when these files are used in I/O operations.

In addition to informing the FORTRAN language that a file is to be used as a random-organized, direct-access file, we must indicate the maximum number of records allowed in the file, the size of each record, and whether we are using format statements to describe the data in the records. We must also provide the system with the name of the index pointer to be used when referring to records in the file. All this information is provided in the DEFINEFILE statement whose syntax is given in figure 18.5.

```
┌─────────────────────────────────────────────────────────────────────┐
│                    DEFINEFILE—General Form                          │
├─────────────────────────────────────────────────────────────────────┤
│                                                                     │
│        DEFINEFILE unit (r, s, f, v)                                 │
│                                                                     │
│ where:                                                              │
│                                                                     │
│        unit  —   is an integer constant that represents the file unit.│
│        r     —   is an integer constant specifying the maximum number of records in │
│                  the file.                                          │
│        s     —   is an integer constant specifying the maximum size of each record. │
│                  Record size is measured in bytes or words depending upon specifica-│
│                  tions for f:                                       │
│        f     —   is established from one of the following selections:│
│                  L  —  either with or without format control; maximum record size is │
│                        measured in bytes.                           │
│                  E  —  with format control; maximum record size is measured in │
│                        bytes.                                       │
│                  U  —  without format control; maximum record size is measured in │
│                        words.                                       │
│        v     —   is a nonsubscripted integer variable called an associated variable; it │
│                  is set to a value which points to the next record to be transmitted.│
└─────────────────────────────────────────────────────────────────────┘
```

Figure 18.5. General Notation of the DEFINEFILE Statement

Valid examples of the DEFINEFILE statement are:

DEFINEFILE 2(50,80,L, INDEX)
DEFINE FILE 3(10000,25,U,IREC), 10(10,35,E,KTR)

The first example describes a file on unit 2 to contain a maximum of 50 records. The maximum size of each record is 80 bytes. A byte is equal to one character. The option "L" indicates that the records may be read/written with or without format control, but the record length is measured in bytes (80). The variable, INDEX, is established as the associated variable for the file and can have a range of 1 to 50. The second example defines two files as being type random organization (unit 3 and unit 10). The file on unit 3 is written/read without format control. The file consists of a maximum of 10,000 twenty-five word records. IREC is identified as the associated variable to be used in accessing the records on file unit 3. The file on unit 10 is always written/read with format control. There can be only ten records on the file; each record can be a maximum of 35 characters in length. The associated variable is KTR.

The **f** parameter in the DEFINEFILE statement defines whether the records will be transmitted under format control or nonformat control, and whether the record size is measured in byte or word count. The "L" option indicates that the record size is measured in bytes and that the record may be either read/written under control of a FORMAT statement or under nonformat control. When the "L" option is specified, the programmer has the choice of: (1) reading/writing all records in the file under format control, (2) reading/writing all records in the file under nonformat control, or (3) a mixture of the two forms. If the programmer mixes the forms of the records on the file, he must remember in what form each record was written in order to use the proper form of the READ statement (formatted READ or unformatted READ) in accessing the records. When the "E" option is specified, all records must be written and read under format control. When the "U" option is specified, all records must be written and read without format control.

A direct-access file on disk can have only one associated variable assigned to it. The associated variable given for the file unit in the DEFINEFILE statement must always be used whenever the relative record to be read or written to the file is indicated by an integer variable in the random I/O statements. That is, if INDEX has been identified as the associated variable in the DEFINEFILE statement, only the variable, INDEX, can be used as the associated variable in the related READ, WRITE, and FIND statements. The value in the associated variable is automatically incremented by one to point to the next record in the file after execution of each random READ or WRITE statement.

The DEFINEFILE statement must be present for all FORTRAN files that are organized randomly. Since the DEFINEFILE statement is considered to be a specification statement, it should appear in the program before any executable statement. The DEFINEFILE statement describes the characteristics of a direct-access file and dictates, to a large degree, the form of the random READ and WRITE statements.

18.5 Formatted and Unformatted Random READ Statements

The formatted random READ statement permits the direct reference and reading into memory the record indicated by an integer expression or the associated variable (index pointer). The general form of the formatted random READ statement is given in figure 18.6.

```
                Formatted Random READ Statement—General Form

        READ (unit ' v, n, ERR=b) list

where:

        unit    —   represents the file unit being read.
        '       —   is a necessary separator.
        v       —   is an integer constant, expression, or associated variable used by the
                    system to locate the record to be read.
        n       —   is the number of the format statement used to describe the input
                    record.
        ERR=b   —   b is a statement number to which control is transferred in case of a
                    parity error. This is an optional parameter.
        list    —   is an I/O list.
```

Figure 18.6. General Notation of the Formatted Random READ Statement

Valid examples of the formatted random READ statement are:

```
         READ (9 ' 34, 53, ERR=86) A,B,C
      12 READ (8 ' INDX,87) ACCTNR, UNITS, PRICE, AMT
      76 READ (3 ' 2*L/3,40) X
         READ (10 ' JREC+3,66) I, M, P, Q
```

The formatted random READ uses an integer constant, an expression, or the value in the associated variable to locate the relative record and read it into internal storage. A FORMAT statement is used to describe the external form of the input record being read.

In the example below

>   IREC = 23
>   READ (9 ' IREC, 39) J,K,L,M
>   39 FORMAT (4I5)

the twenty-third record is read and four fields stored at the variables J, K, L, and M.

An integer constant or expression may be used to indicate a specific record to be read. The statement

>   READ (2 ' 20,98) STDNR, GRADES

would read the twentieth record in the file. Whenever this statement is executed, the twentieth record would always be read. The statement

>   READ (8 ' 3*2+1, 95) STDNR, GRADES

would always cause the seventh record in the file to be read. Even though an integer constant expression is used to denote the desired record, the associated variable will always be changed by the system to point to the next record after the desired relative record has been calculated. For example, in the READ statement above, the associated variable would have a value of 8 after the statement had been executed.

The value of the associated variable is also automatically incremented by one when used in the READ statement. Thus, if one wished to read an entire file or a portion of records with ascending position numbers, he would have only to initialize the associated variable to the desired starting location. Following is an example to read the first twenty records on a random file and dump them to the printer.

>   DEFINEFILE 8(50,80,E,NRREC)
>   DIMENSION NAME (5), GRADES (10)
>   98 FORMAT (5A4,10F5.1)
>   97 FORMAT (1H0,10X,5A4,10F7.1)
>   NRREC = 1
>   5 READ (8 ' NRREC, 98) NAME, GRADES
>   WRITE (6,97) NAME, GRADES
>   IF (NRREC .LT. 21) GOTO 5
>   STOP
>   END

A check is made for NRREC to be less than 21 since the associated variable NRREC is incremented by one after each execution of the random READ statement.

Data may also be read from random disk without format control, i.e., in internal word form. The form of the unformatted random READ statement is as follows:

>   READ (unit ' v, ERR = b) list

The programmer supplied elements are the same as the formatted random READ except there is no associated format statement number.

Valid examples of the unformatted random READ statement are:

>   READ (8 ' INDX) A,B,C
>   26 READ (3 ' JREC, ERR = 37) I,J,Z
>   18 READ (2 ' 17) M,N,P
>   READ (9 ' MRCD + 5) K,X,Y

The unformatted random READ statement functions the same as the formatted random READ in all respects except that the transfer of data is in internal word form and not under the control of any FORMAT statement. For data to be transferred without format control, the **U** or **L** option must be specified in the DEFINEFILE statement. The unformatted random READ can only be used on files that have been written without format control (i.e., an unformatted random WRITE).

18.6 The Formatted and Unformatted Random WRITE Statements

The formatted random WRITE statement directs the computer to transfer data from memory to a direct-access file. Data is transferred under control of a related FORMAT statement and is written as the relative disk record as indicated by the integer constant, expression, or associated variable. The general form of the formatted random WRITE statement is given in figure 18.7.

Formatted Random WRITE Statement—General Form

WRITE (unit ' v, n) list

where

unit	—	represents the file unit.
'	—	is a necessary separator.
v	—	represents an integer constant, expression, or associated variable used by the system to indicate the desired record position to be written.
n	—	is the number of the format statement used to describe the output record.
list	—	is an optional I/O list.

Figure 18.7. General Notation of the Formatted Random WRITE Statement

Valid examples of the formatted random WRITE statement are:

16 WRITE (3 ' INDX, 98) A,B,C
WRITE (8 ' 36, 39) I,X,J,Y

The first WRITE statement writes the values of A, B, C to the relative record indicated by INDX on file unit 3. The second WRITE statement writes the values of I, X, J, and Y as the thirty-sixth record on file unit 8. The file unit being written must have been previously defined by a DEFINEFILE statement.

The value in the associated variable is automatically incremented to the next record in the file by the system after execution of each WRITE statement. Thus, if one wished to write an entire file, he would have only to initialize the associated variable to the desired starting location. Following is an example of creating a random file of 50 records with record addresses of 1-50.

```
        DEFINE FILE 8(50,80,E,K)
        DIMENSICN NAME(5), GRADES(10)
  98    FCRMAT (5A4,10F5.1)
        K = 1
  5     READ (5,98,END=200) NAME, GRADES
        WRITE (8 ' K, 98) NAME, GRADES
        GOTC 5
  200   STCP
        END
```

The unformatted random WRITE statement is used to transfer data from memory in word form to a direct-access file. The general form of the unformatted random WRITE statement is as follows:

WRITE (unit ' v) list

The programmer supplied elements are the same as the formatted random WRITE except there is no associated format statement number.

Valid examples of the unformatted random WRITE statement are:

WRITE (3 ' 14) A,B,C
86 WRITE (9 ' JREC) D,E,F

The first example writes the three fields A, B, and C as the fourteenth record on file unit 3. The second example writes the three fields D, E, and F as the relative record indicated by the associated variable JREC on file unit 9.

The unformatted random WRITE statement functions the same as the formatted random WRITE statement in all respects except that the transfer of data is in word form and not under control of any FORMAT statement. For data to be transferred without format control, the "U" or "L" option must be specified in the DEFINEFILE statement.

18.7 The FIND Statement

The FIND statement is used to speed up execution of a program by reducing the time it takes the computer to locate and address a record position on magnetic disk. The function of the FIND statement is to position the read/write heads over the proper track in which a desired record is to be read or written. The general form of the FIND statement is given in figure 18.8.

```
-----------------------------------------------------------------
|                    FIND Statement—General Form                |
-----------------------------------------------------------------
|                                                               |
|       FIND (unit ' v)                                         |
|                                                               |
| where:                                                        |
|                                                               |
|       unit   —   represents the file unit.                    |
|       '      —   is a necessary separator.                    |
|       v      —   is an integer constant, expression, or       |
|                  associated variable used by the              |
|                  system to locate the desired record position.|
```

Figure 18.8. General Notation of the FIND Statement

Valid examples of the FIND statement are:

20 FIND (3 ' 26)
FIND (4 ' IREC + 1)
17 FIND (10 ' INDEX)

The first example positions the read/write heads over the track containing relative record number 26 in the file designated by unit 3. The second example positions the read/write heads over the track containing the relative record indicated by IREC plus 1 for file unit 4. The third example positions the heads over the track containing the relative record indicated by the associated variable, INDEX, for file unit 10.

The FIND statement does not transfer any data to memory; it simply positions the read/write head over the proper track as indicated by the integer constant, expression, or associated variable. A delay is encountered when reading/writing disk records in a random file to allow the mechanical action of moving the heads in or out among the disk platters to locate the proper track. If the read/write heads are positioned at the proper track, a READ or WRITE statement need only to wait for the disk to revolve to the proper record to transfer data. Thus, the FIND statement often allows the programmer to speed up I/O operations by not having to wait for the horizontal delay in positioning the read/write heads over the desired track. The following example illustrates this concept.

```
      DEFINEFILE 8(750,50,U,N)
      DIMENSION REC (50)
      N = 1
    5 READ (8 ' N) REC
      FIND (8 ' N)
      . . .
      . . .
      IF (ISWT .EQ. 0) GOTO 5
```

When control is transferred back to the READ statement, the FIND statement has already caused the read/write heads to be positioned over the proper track of the next record to be read. Data can be transferred as soon as the proper record rotates under the read/write heads. A random READ or WRITE statement immediately following a FIND statement does not provide any benefits in speeding up the accessing of a record.

A Sample Program

The following program illustrates the use of the FORTRAN statements available with random disk files. A disk file of 100 eighty-character records is created on file unit 8 under format control. A student number (within the range from 1-100) is used to specify the address at which the record is written. After the disk file has been created, the records are read beginning at record one and written to printer.

```
           DEFINE FILE 8(100,80,E,NRSTD)
           DIMENSION NAME(5), GRADES(10)
    98     FORMAT (I3,5A4,10F5.1)
    97     FORMAT (5A4,10F5.1)
    96     FORMAT (1H0,I3,5X,5A4,5X,10F7.2)
           NRREC = 0
     3     READ (5,98,END=15) NRSTD,NAME,GRADES
           NRREC = NRREC + 1
           WRITE (8 ' NRSTD, 97) NAME, GRADES
           FIND (8 ' NRSTD)
           IF (NRREC .LE. 100) GOTO 3
    15     NRSTD = 1
    20     READ (8 ' NRSTD, 97) NAME,GRADES
           FIND (8 ' NRSTD)
           WRITE (6, 96) NRSTD, NAME, GRADES
           IF (NRSTD .LT. 100) GOTO 20
           STOP
           END
```

18.8 Nonstandard Language Extensions

Nearly every FORTRAN compiler handles random access files in a different fashion. This section will briefly discuss the conventions, procedures, and statements used for random access files on the CDC 6000, Honeywell 6000, and Burroughs 6700 and 7700 computers.

CDC 6000

Random access files on the CDC 6000 may be read/written to mass storage units (disk, drum, etc.) or to Extended Core Storage (ECS). The subroutine call:

$$\text{CALL OPENMS (parameters)}$$

is used to open a mass storage file and informs the computer that it is a random access file.

Some of the random access statements for direct I/O operations on the CDC 6000 are:

> CALL READMS (parameters)
> CALL READEC (parameters)
> READ MS (parameters)
> CALL WRITEMS (parameters)
> CALL WRITEEC (parameters)
> WRITE MS (parameters)
> WRITE ECS (parameters)

The CALL STINDX statement performs the same function as the FIND statement on IBM compilers.

H6000

On the Honeywell 6000 a subroutine named RANSIZ permits the user to specify the record for a random access file. This subroutine must generally be called before any I/O operation is made to the random file. The general form of this subroutine call is:

$$\text{CALL RANSIZ (unit, recsize, file-indicator)}$$

The "**unit**" parameter is a file unit number, such as 10. The "**recsize**" parameter is the record size in number of words (variables). H6000 random files must have a constant record size that is read or written on each I/O operation. The third parameter, "**file-indicator**", is optional. When this parameter is not supplied, the random file is processed as a standard system format file.

Random files are read with the Random READ statement, general form for which is:

$$\text{READ (unit ' v, ERR = b) list}$$

Random files are written on the H6000 with the Random WRITE statement. Its general form is:

$$\text{WRITE (unit ' v, ERR = b) list}$$

The Random READ and WRITE statements are used with strictly unformatted (binary) files. The data cannot be formatted as with the IBM compilers. Note the ERR = option on the Random WRITE statement to allow an error routine to be executed if a write parity error is encountered. Otherwise, the Random READ and WRITE statements on the H6000 are formed the same as the unformatted Random READ and WRITE statements on the IBM 360 and 370.

B6700 and B7700

Data may be written to and read from a random file in either a formatted or unformatted form. No DEFINEFILE statement or subroutine call is necessary to establish a random access file on the B6700 and 7700 systems.

Random files are read with the Random READ statement. Its general forms are:

READ (unit = v, n, ERR = b) list

or

READ (unit = v, ERR = b) list

The first form is for a formatted read. The second form is for an unformatted read. The "=" is a necessary separator denoting a random READ. The single quote (') may be used in place of the "=" if one so desires.

Random files are written with the Random WRITE statement. Its forms are:

WRITE (unit = v, n, ERR = b) list

or

WRITE (unit = v, ERR = b) list

The first form is for data written under format control. The second form is for an unformatted write. "**unit**" represents the file unit number used. "=" is a necessary separator denoting a Random WRITE. The single quote (') may be used in place of the "=" if desired.

The FIND statement may be used to position a random file at a designated record. The general form of the FIND is:

FIND (unit = v)

where "=" is a necessary separator to indicate a random file. The single quote (') may be used in place of the "=" if one desires.

When a disk file is created on the B6700 and B7700 computers, the operating system will automatically allocate disk storage space for the file. To retain the newly created disk file a LOCK statement must be executed. For example:

LOCK 8

would save the created disk file written on unit 8 and put the file name in the disk directory. If the LOCK statement is not executed, a disk output file will be considered a temporary file and automatically purged at the end of the job.

A temporary disk file may be created on IBM computers by including only a proper Data Definition (DD) JCL card. To retain a created disk file on IBM computers, however, a special utility program called IEFBR14 must be run to allocate the required space and assign a directory name.

Your instructor will assist you in constructing any required JCL statements you might need to use disk files on your computer system.

18.9 REVIEW QUESTIONS

1. What is the primary value of magnetic disk?
2. Random access files must be read in a sequential manner. (True/False)
3. All the tracks in the same position on different platters is called a _____ .
4. Magnetic disk may be used to store sequential files as well as random files. (True/False)
5. The statement used to position a sequential disk file back to the beginning of the file is the _____ .

6. The statement used to define a random access file and its characteristics on an IBM 360/370 is the _____ .

7. When reading/writing to a random access file, an index value must be given to denote the position of a specific record. (True/False)

8. Random access files may be written only under format control. (True/False)

9. The statement used to speed up the access of a particular record by prepositioning the read/write heads over the proper track is the _____ .

18.10 EXERCISES

1. Write a DEFINEFILE statement to identify a random disk file written on file unit 3. The associated variable is to be KPTR. Data is to be transferred under format control only. The maximum number of records is 175 and the maximum size of each record is 110 bytes.

2. Write a DEFINEFILE statement to identify a random disk file written on file unit 10. Data may be transferred with or without format control. The file will contain a maximum of 90 records, each to contain 80 bytes. The variable, JRCD, is the associated variable.

3. Write the READ statement to read the 26th record from a random disk file on file unit 10 and transfer the variables X, Y, and Z without format control.

4. Write a READ statement to read the 35th record from a random disk file on file unit 3 and transfer the variables A, B, and C without format control.

5. Construct a WRITE statement to transfer the values of the variables I, J, K, and L to record position N under format control. Each variable is to be written as a three-digit integer on file unit 9 set up as a random file. Use statement number 99 as the FORMAT statement number.

6. Construct a WRITE statement to transfer the values of the variables A, B, C, and D to record position K-1 (K minus 1) under format control. Each variable is to be written under an F6.2 format specification on file unit 8 set up as a random file. Use statement number 99 as the FORMAT statement number.

7. Write the statement to find record K on file unit 2 which is set up as a random file.

8. Write the statement to find record N on file unit 4 which is set up as a random file.

9. How many records will be read with the following READ?

    READ (4 ' 14, 98) I,J,K,L
    98 FORMAT (I3)

10. How many records will be read with the following READ?

    READ (8 ' 21, 97) W,X,Y,Z
    97 FORMAT (2F6.2, F5.1)

11. How many records will be written with the following WRITE?

    WRITE (8 ' J, 87) I,J,K,L,M
    87 FORMAT (2I5)

12. How many records will be written with the following WRITE?

    WRITE (9 ' K, 96) W,X,Y,Z
    96 FORMAT (F7.1,F6.3)

18.11 PROGRAMMING PROBLEMS

1. Write a program to read a card file containing records of student data as follows:

Card Columns	Field Description
1-20	Name
21-29	Social security number
30-33	Student number (Form is NNNN)
34-39	Class (Form is alphanumeric)
40-42	Section (Form is NNN)
43-44	Blank
45-47	Grade 1 (Form is NNN)
48-50	Grade 2 (Form is NNN)
51-53	Grade 3 (Form is NNN)
54-56	Grade 4 (Form is NNN)
57-59	Grade 5 (Form is NNN)
60-62	Grade 6 (Form is NNN)
63-80	Blank

Create a sequential disk file. Assume that the records are in sequence by name within section within class. Write the input record to disk in the same format specification as read for input. Use a trailer card with a value of 999999999 for SSAN to indicate an end of file on the card input file.

2. Write a program to read the sequential disk file created in problem 1 and compute each student's grade average. Print the student data along with the computed average. Control break on Class. That is, whenever a different class value from the previous value is read, list those student records beginning on a new page.

The printer output specifications are:

Print Positions	Field Description
1-20	Name
26-34	SSAN
39-42	Student number
47-52	Class
57-59	Section
70-72	Grade 1
79-80	Grade 2
86-88	Grade 3
94-96	Grade 4
102-104	Grade 5
110-112	Grade 6
118-123	Average

Include an appropriate report heading and column headings.

3. Write a program to read a card file with records of student data as described in problem 1. Create a random disk file using Student Number as the key (disk address) for each record. Assume a maximum of 300 student records with student numbers from 1-300. Write a disk record that is formatted in the same specifications as the card input record.

4. Write a program to read a card file with only the student numbers to serve as the keys to access the desired records in the random file created in problem 3. The student number

will be read as an I4. Read the proper disk record and then compute each student's grade average. List each record with the newly computed average under the same printer specifications as given in problem 2. Use a trailer card with a value of 9999 for student number to indicate the end of file for the card input file.

5. Read an employee file of 50 card records and load them to sequential disk. The card records are layed out as follows:

Card Columns	Field Description
1-25	Name
26-30	Employee number (Form is NNNNN)
31-32	Age (Form is NN)
33-34	Department (Form is NN)
35-36	Years with company (Form is NN)
37-39	Pay grade (Form is NNN)
40-46	Monthly salary (Form is NNNN.NN)

Write the disk records in an unformatted (binary) form.

6. Read the sequential disk file created in problem 5 and prepare a company roster of all employees with more than 30 years of service with the company.

The printer output specifications are:

Print Positions	Field Description
1-25	Name
31-35	Employee number
41-42	Age
48-49	Department code
55-56	Years with company
62-64	Pay grade code
70-76	Salary

Include an appropriate report heading and column headings.

7. Read a card file containing records with information on the status of the different buildings on a military installation. Each card record will contain the following data:

Card Columns	Field Description
1-4	Building number
5-30	Type of activity/function utilizing the building (alphanumeric)
31-34	Telephone extension of building monitor
35-40	Type structure (alphanumeric)
41-80	Misc. information (alphanumeric)

A trailer card with a value of 9999 for Building Number is used to indicate the end of file. Create a random disk file using the building number as the disk record key address. Write the file in an unformatted (binary) form. Assume a maximum of 500 buildings.

8. Using the unformatted random disk file created in problem 7 and a transaction card file with building number in card columns 1-4, prepare a status report for those buildings

included in the transaction card file. A trailer card with a value of 9999 for Building number is used to indicate the end of the card input file.

The printer output specifications are:

Print Positions	Field Description
1-4	Building number
10-35	Type of activity/function
41-44	Building monitor telephone extension
50-55	Type of structure
61-100	Misc. information

Include an appropriate report and column headings.

Additional FORTRAN Statements and Features 19

19.1 Introduction to Additional FORTRAN Statements and Features

The additional statements and features in ANSI FORTRAN that we have not discussed are included in this chapter to provide thorough coverage of the FORTRAN language. Although the student programmer may find little use for these statements and features, they do serve an important role in many specialized and production applications.

The PAUSE statement permits the programmer to temporarily stop his program, and by so doing, allowing the operator to take specific actions before continuing the program. The ASSIGN and ASSIGNED GOTO statements are additional control statements to selectively transfer control of execution in a program. Object time format specifications provide a great deal of flexibility in the reading of the same data file under different format specifications. These specifications can be provided during the execution of a program and, thus, prevents one from having to change and recompile the program.

There are also several additional nonstandard FORTRAN statements with which you should be familiar. The NAMELIST statement is a convenient way to read and write a list of variables without an accompanying FORMAT statement. It is often used in scientific applications where little I/O is required. The specific READ, PRINT, and PUNCH statements are carryovers from FORTRAN II. These statements are available on most FORTRAN IV compilers and are still used by some programmers.

19.2 The PAUSE Statement

The PAUSE statement is used to *temporarily* stop the program execution, thereby allowing the computer operator time to set a sense switch (a switch on the computer), or mount a magnetic tape or a disk pack. An octal number (consisting of the digits 0-7) is normally included after the keyword to communicate the desired operations to the operator. This coded number is displayed on the operator console along with the word "PAUSE" for viewing by the operator. If the operator is not familiar with the coded number, he can refer to his job "run book" which describes the required operation to take for specific numbers. After the desired operation is accomplished, the operator presses the start button or sends a message that causes the computer to resume execution. Execution of the program then continues at the next statement after the PAUSE.

Figure 19.1 gives the general form of the PAUSE statement.

PAUSE Statement—General Form

PAUSE

or

PAUSE n

where:

    n — is a one- to five-digit octal number which relates to a specific action for the operator to take.

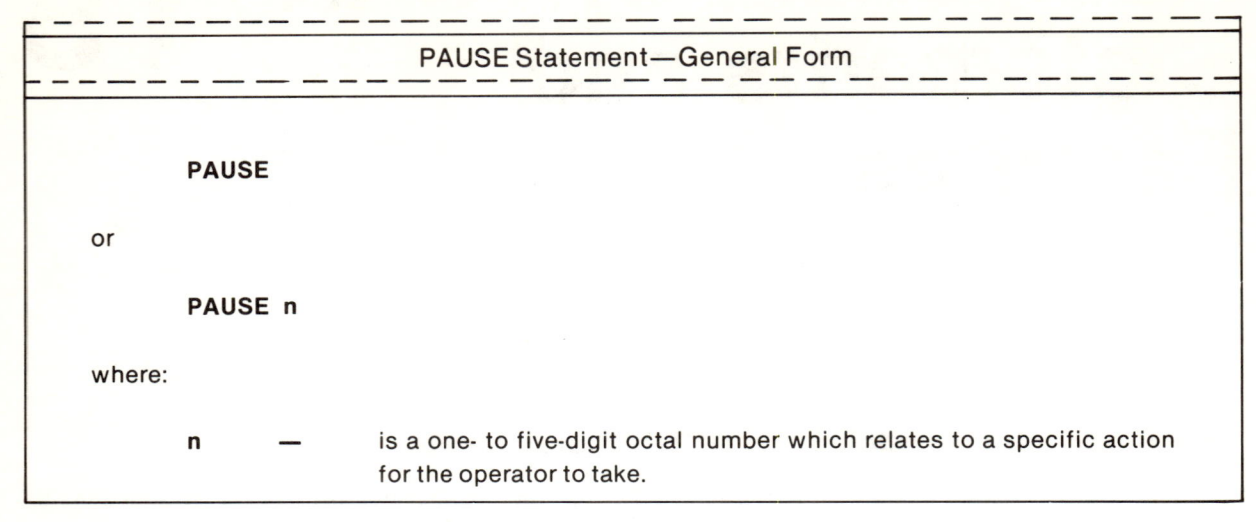

Figure 19.1. General Notation of the PAUSE Statement

Following is an example of the use of the PAUSE statement.

    READ (5,99) NUM,A,B
99 FORMAT (I5,2F5.2)
    IF (NUM .EQ. 1) PAUSE 134

The octal number 134 might mean that the operator is to mount a tape file needed by the program when the variable NUM is equal to one. Some FORTRAN compilers allow a literal constant to be displayed after the word PAUSE to provide a more meaningful message.

Student programmers usually do not require any operator intervention during the running of their job, so the PAUSE statement should not be used in your program. In fact, the WATFIV compiler ignores (skips) any PAUSE statement in a program and continues at the next statement.

19.3 The ASSIGN and Assigned GOTO Statements

The Assigned GOTO statement is very similar to the Computed GOTO; control is transferred to one of several alternative statements in the program. However, control is transferred to a specific statement based on the *statement number* assigned to an integer variable. The ASSIGN statement is used in conjunction with the Assigned GOTO and is used to assign a statement number to the integer variable. Figure 19.2 gives the general form of the ASSIGN statement.

ASSIGN Statement—General Form

ASSIGN n TO ivar

where:

    n — represents the statement number of an executable statement.
    ivar — is an integer variable used to contain the assigned statement number.

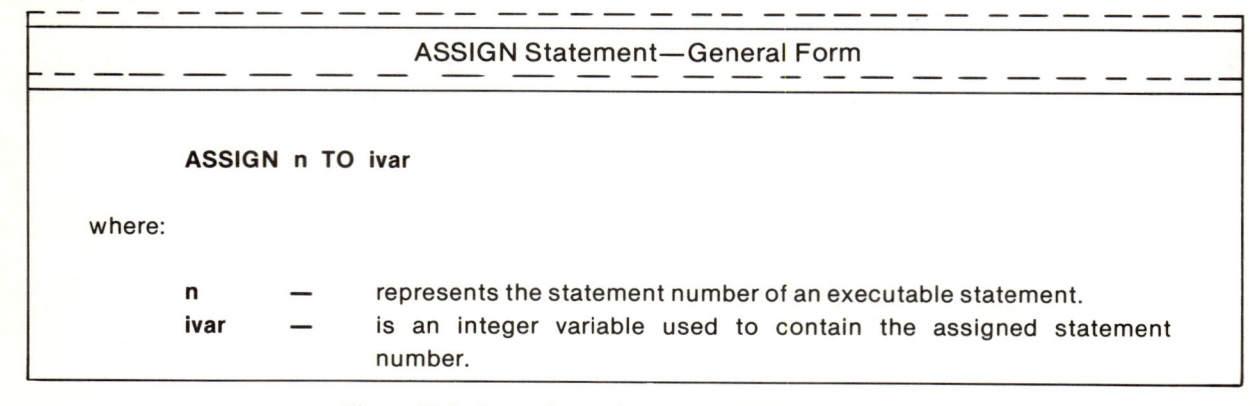

Figure 19.2. General Notation of the ASSIGN Statement

Examples are:

ASSIGN 46 TO K
27 ASSIGN 238 TO NSTMT

The ASSIGN statement must have been executed to assign the desired statement number to the variable before the Assigned GOTO statement is executed. The Assigned GOTO causes a conditional transfer of control to the statement number last assigned to the integer variable by an ASSIGN statement. Figure 19.3 gives the general form of the Assigned GOTO statement.

Assigned GOTO Statement—General Form

GOTO ivar, (n$_1$,n$_2$,. . .,n$_n$)

where:

    ivar — represents an integer variable containing the statement number assigned by the ASSIGN statement.

    (n$_1$,. . ., n$_n$)— are statement numbers of executable statements in the program.

Figure 19.3. General Notation of the Assigned GOTO Statement

Examples are:

GOTO K, (10,5,147)
83 GOTO NSTMT, (20,40,30,10)

Note the syntax requirement of a comma to separate the integer variable and the beginning left parenthesis.

An example of using these two statements is:

ASSIGN 37 TO K

GOTO K, (18,56,37,3)

The statement number 37 is assigned to the variable K. After executing the Assigned GOTO, control is transferred to statement number 37 and will resume processing at that statement. The statement number assigned to the integer variable must be for an executable statement. The position of the statement numbers in the Assigned GOTO is not important, whereas in the Computed GOTO, the position of the statement numbers is important.

We cannot use the arithmetic Assignment statement to assign a statement number. For example:

K = 37
GOTO K, (18,56,37,3)

is invalid. Assigning a statement number with the ASSIGN statement is not the same thing as an arithmetic assignment. The statement number must be assigned with the ASSIGN statement.

Normally, more than one ASSIGN statement is used in conjunction with the Assigned GOTO statement. Suppose we wish to sum two variables if an input code is equal to 1. And we wish to compute the product of the two variables if the input code is equal to 2. (Otherwise, a value of zero is assumed.) A sample routine which uses the ASSIGN and Assigned GOTO statements to perform this logic is:

```
         READ (5,99) K,X,Y
   99    FORMAT (I2,2F5.2)
         RESLT = 0.0
         ASSIGN 30 TO LOC
         IF (K .EQ. 1) ASSIGN 10 TO LOC
         IF (K .EQ. 2) ASSIGN 20 TO LOC
         GOTO LOC, (10,20,30)
   10    RESLT = X + Y
         GOTO 30
   20    RESLT = X * Y
   30    CONTINUE
```

Once the integer variable has been assigned a statement number by the ASSIGN statement, the variable cannot be used in any arithmetic expression. The variable can be used only in an Assigned GOTO statement. The statement number assigned to the integer variable must be one of the statement numbers tested in the Assigned GOTO. If the variable is assigned a statement number other than one of those given in the Assigned GOTO, control continues at the next statement. The next executable statement immediately following the Assigned GOTO should have a statement number. Otherwise it can never be referenced in the program, and a syntax warning message will be given.

The only advantages in the use of the Assigned GOTO over the use of the Computed GOTO are small savings in execution time and storage locations. However, the Computed GOTO is more popular and more in accordance with structured programming. The Computed GOTO is usually simpler and more convenient to use and allows more flexibility, since the value read for an input variable can be used directly by the computer to make the test of which statement to transfer.

19.4 Object Time Format Specifications

Sometimes a user will change the location or even change the form of data fields in input records. Therefore, we must change our program format specifications accordingly to accept the new form and/or location of our input data. When we change our program, we must also recompile it so that the program execution can contain our new changes. It would be nice to have the flexibility to change our format specifications for new forms of data and not have to recompile our program.

What we need is a way to tell the computer our format specifications at execution time. *Object time format specifications* provide us with this capability. Our format specifications are read as alphanumeric data and used to provide the form of our input or even output data with no modification of our existing program. Object time format specifications are also known as *variable formats* or *execution time formats.*

To accomplish object time format specifications we must establish a one-dimensional array to hold our format declarations. The specifications are read from a data card into the established array. The format specifications are punched in the data card with the first character being the beginning left parenthesis. The statement number and the keyword FORMAT are not included in the data card. The entire format specifications that would normally be included within the required pair of parentheses are punched into the data card and read into the array with the alphanumeric (A) format code.

Suppose the input specifications of our input record are:

A punched in card columns 1-5 in the form NN.NN
B punched in card columns 6-9 in the form NN.N
K punched in card columns 13-14 in the form NN

The input specifications, therefore are:

F5.2, F4.1, 3X, I2

These specifications would be included within the normal set of parentheses and punched in a data card as follows:

(F5.2,F4.1,3X,I2)

Now we need to read the specifications from the data card into an array which we will assume is named FARRAY. There are seventeen characters in the format specifications so we must establish FARRAY to be at least five elements. That is, if each element can contain a maximum of four characters, then we need five array elements. It is recommended that we establish the array with greater than five elements, in order to read a longer object time format specifications later under the same array name. So let us establish FARRAY as a twenty-element array which allows us to read an entire card of format specifications. Since the remainder of our data card is blank, there is no harm in having more elements than we need.

Our statements to dimension FARRAY and load the format specifications in it are:

DIMENSION FARRAY(20)
READ (5,99) FARRAY
99 FORMAT (20A4)

Execution of the READ statement would load the specifications from the data card into our array FARRAY which can now be used to provide object time format specifications. Our array FARRAY would have the following values in the array elements:

F(1)	would contain	(F5.
F(2)	would contain	2,F4
F(3)	would contain	.1,3
F(4)	would contain	X,I2
F(5)	would contain	)

and elements 6 through 20 would contain blanks.

To use the format specifications loaded into the array, the array name is substituted for the FORMAT statement number in the READ or WRITE statement. That is, the array name takes the place of the usual format statement number and supplies the format specifications. For example:

array name containing the input format specifications

READ (5, FARRAY) A,B,K

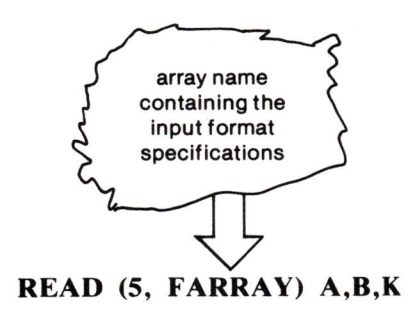

The array name FARRAY contains the input format specifications and tells the computer how to read the three variables A, B, and K.

An array name must be used to store the format specifications, even if the array consists of only one element. The array name must not be subscripted in the READ or WRITE statements when used to refer to object time specifications. For example:

READ (5, FARRAY(I)) A,B,K

would be invalid and cause a syntax error.

Object time format specifications allow us to write programs of a very general nature. They can be quite powerful and timesaving where we use the same program to process data with different format specifications. A program can even be made to selectively modify parts of a format statement by reading in parts of the format statement through specific array elements. Thus, the arrays used to hold our object time format specifications are altered according to the contents of specific array elements.

19.5 Additional Nonstandard FORTRAN Statements (NAMELIST, Specific READ, PRINT, and PUNCH)

The NAMELIST Statement and Operations

Perhaps you have wished that a name could be assigned to a list of variables and the list name used to represent the group of variables. This would certainly provide a shorter means of supplying the list of variables, much to be preferred over the use of an I/O list in READ and WRITE statements that continually refer to the same variables. The NAMELIST statement provides this ability.

The NAMELIST Statement

In addition to representing a group of variables by one name, the NAMELIST statement also provides other advantages. The variable name is printed as a literal title preceding the output value of each variable. This capability greatly simplifies the identification of the output results. The input values for the list of variables are also read without your having to describe the data fields with FORMAT statements. Thus, the NAMELIST statement provides an input and output capability that is virtually format-free.

The NAMELIST statement declares a name to represent a particular list of variable or array names. The list name is formed in the same as other symbolic names, and is enclosed in a pair of slashes. The general form of the NAMELIST statement is given in figure 19.4.

NAMELIST Statement—General Form

NAMELIST /listname$_1$/ varlist$_1$,. . ., /listname$_n$/ varlist$_n$

where:

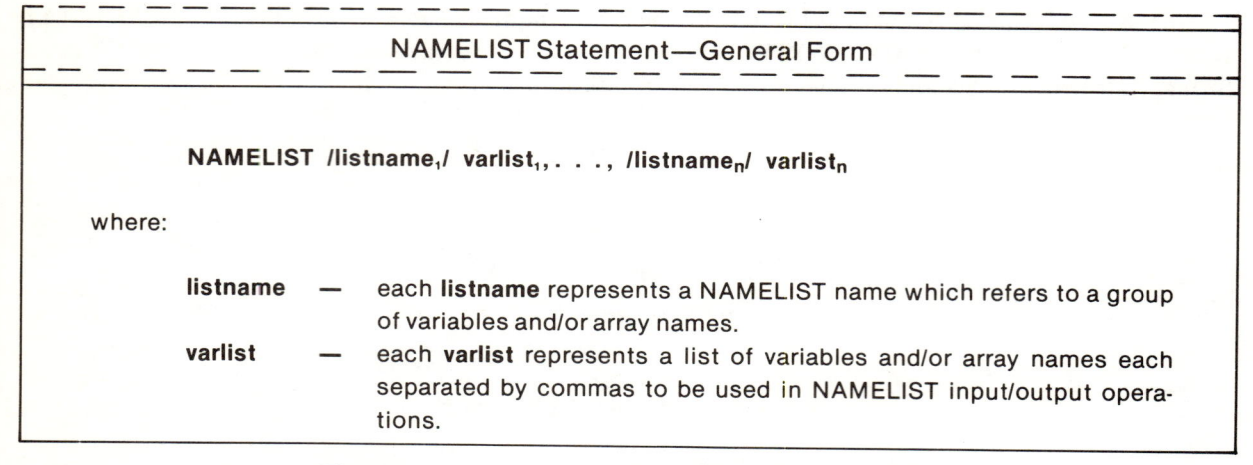

- **listname** — each **listname** represents a NAMELIST name which refers to a group of variables and/or array names.
- **varlist** — each **varlist** represents a list of variables and/or array names each separated by commas to be used in NAMELIST input/output operations.

Figure 19.4. General Notation of the NAMELIST Statement

Examples of the NAMELIST statement follow:

NAMELIST /LIST1/ A,B,C,I,J
NAMELIST /L1/X,Y,Z, /L2/M,N

In the first example the variables A, B, C, I, and J are assigned to the list name of LIST1. In the second example the variables X, Y, and Z are assigned to the list name of L1, while the variables M and N are assigned to the list name of L2.

The NAMELIST statement is a specification statement. It should precede all executable instructions and any statement functions in the program. DIMENSION statements, if used must precede the NAMELIST statement. A single NAMELIST statement (which could include continuation cards) may be used in a program. Or multiple NAMELIST statements may be formed.

The variable or array names in the list may be any symbolic name allowed in FORTRAN. But the variable or array names may not be used for any other I/O operations in the program, other than for NAMELIST use. That is, a variable or array appearing in a NAMELIST list must be used only in a NAMELIST read or write operation, and cannot be used to read or write data under format control. The following paragraphs explain and illustrate how the NAMELIST statement is used in a FORTRAN program.

NAMELIST Input Operations

The syntax of the READ statement is different from the normal construction of the READ statement you have been using. The statement number used to reference a FORMAT statement is replaced by the NAMELIST name. Also, no I/O list is included on the READ statement. The syntax of the READ statement used with NAMELIST input operations is:

READ (unit, listname)

An example of this type of READ is:

READ (5, LIST1)

The input data cards must include the NAMELIST name and the variable names and/or the array elements in the NAMELIST that are being read. Each variable is separated from its corresponding value by an =. An &END must follow the last value to signal the end of the input variables. The NAMELIST name, preceded by an &, must appear at the beginning of the first data card.

The READ statement gives the name of a list which must correspond to the name of a list in the data card. If the list names do not agree, an error message will be given and the program discontinued.

A short FORTRAN program to illustrate the I/O operations of the NAMELIST capability follows.

```
C *** EXAMPLE OF NAMELIST INPUT AND OUTPUT
      DIMENSION K(3)
      NAMELIST /GROUP1/ A,B,K,SUM1,SUM2
      READ (5, GROUP1)
      SUM1 = A + B
      SUM2 = K(1) + K(2) + K(3)
      WRITE (6,GROUP1)
      STOP
      END
```

348   Additional FORTRAN Statements and Features

The input data cards might be:

&GROUP1 A = 15.6, B = 7.83, K(1) = 10, K(2) = 85,
K(3) = −12 &END

When the data cards are read, the input values will be stored at the variables shown on the card, provided that those variable names have been defined as part of the NAMELIST list. All the variables or arrays in the NAMELIST statement do not have to appear in the input data.

The input data must be in a special form in order to be read using a NAMELIST. The rules which pertain to the format of the NAMELIST input data are:

1. The first character in each record to be read must be blank.
2. The second character in the first record must be an & (ampersand), immediately followed by the NAMELIST name.
3. The NAMELIST name must be followed by a blank and must not contain any embedded blanks.
4. This name is followed by data items separated by commas.
5. The form of the data items in an input record is:

   Symbolic name = constant   For example: **AVAL = 13.79**

   The symbolic name may be an array element or a variable name. Subscripts must be integer constants. The form of the constant must be consistent with the type of variable. The constants may be integer, real, literal, complex, or logical. (If constants are logical, they may be in the form: T or .TRUE., and F or .FALSE.).
6. Two forms may be given for the values of arrays. The array name may be specified, followed by the set of constants (each separated by a comma) for the array. The number of constants must be less than or equal to the number of elements in the array. Successive occurrences of the same constant can be represented in the form *i* * constant, where *i* is an integer value specifying the number of times the constant is to be repeated. The second form which may be used to specify array values is to give each array element followed by its respective value.
7. The variable and array names specified in the input data set must appear in the NAMELIST list, but the order is not significant.
8. Each data record must begin with a blank followed by a complete variable, array name, array element, or constant (that is, they cannot be split between cards).
9. Embedded blanks are not permitted in names or constants. Trailing blanks after integers and exponents are treated as zeroes.
10. Within a data record, it is not mandatory to include values for all variables/array names that appear in the NAMELIST statement.
11. The end of a data group is signaled by &END.

Now let us look at the output operations of the NAMELIST.

NAMELIST Output Operations

The syntax of the WRITE statement is also different from that of the normal WRITE statement using format control. The NAMELIST name replaces the statement number used to represent a FORMAT statement. The I/O list is eliminated. The syntax of the WRITE statement used with NAMELIST output operations is:

WRITE (unit, listname)

An example of this type of WRITE is:

WRITE (6, GROUP1)

The output results from our sample NAMELIST program would appear as:

&GROUP1
A = 0.14600000E 02, B = 0.78300000E 01, K = 10, 85, −12, SUM1 = 0.23430000E 02, SUM2 = 0.83000000E 02
&END

The format of the values for NAMELIST output operations is as follows:

1. The first output record will contain only the & (starting in position 2) and the NAMELIST name.
2. Subsequent records will contain the variables and array names followed by their values for all the items in the NAMELIST list.
3. All variables and array names specified in the NAMELIST list and their respective values are output, each according to its type. Real values are written in an exponential form.
4. The last record will contain only &END starting in position 2.

The above description of the NAMELIST statement and operations pertain solely to the IBM 360-370, WATFOR, WATFIV, and B6700 compilers. The control Data 6000-7000, Univac 1108, Honeywell 6000, and Digital Equipment PDP-10 FORTRAN compilers use a dollar sign ($) instead of the & to denote namelist data.

The Specific READ Statement

The specific READ statement is available on many FORTRAN IV compilers to read card files. No unit number is supplied since only punched card files can be read for input. The statement number of an associated FORMAT statement which describes the locations and form of the input fields follows the keyword READ. The variable and array names assigned to each input field follow the FORMAT statement number. Commas are used to separate the FORMAT statement number and each of the variables. The general form of the specific READ statement is given in figure 19.5.

Specific READ Statement—General Form

READ n, list

where:

n	—	is the statement number of the associated FORMAT statement which describes the attributes of the input data.
list	—	is an I/O list which may include one or more variable and/or array names separated by commas.

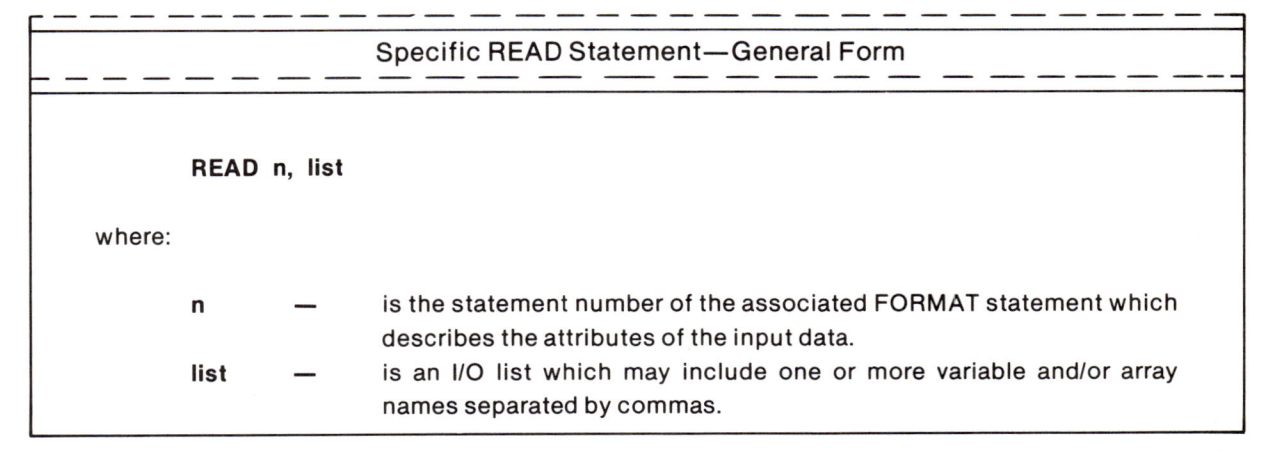

Figure 19.5. General Notation of the Specific READ Statement

Valid examples of the specific READ statement are:

READ 89, A, B, C, I, J
26 READ 132, IDENT, SCORE1, SCORE2

The first example reads the five fields A, B, C, I, and J as described in the FORMAT statement numbered 89. The second example reads the three fields IDENT, SCORE1, and SCORE2 as defined in the FORMAT statement numbered 132.

The PRINT Statement

The PRINT statement is available on many FORTRAN IV compilers to provide the capability for printing data on the line printer. This statement does not include a unit number since only the printer can be used. The statement must have an associated FORMAT statement to describe the output data specifications. An optional list of variables and array names separated by commas follow the associated FORMAT statement number. A comma is used to separate the FORMAT statement number from the first variable, if the I/O list is used. The general form of the PRINT statement is given in figure 19.6.

PRINT Statement—General Form

PRINT n, list

where:

n — is the statement number of the associated FORMAT statement which describes the attributes of the output data.

list — is an optional I/O list which may include one or more variable and/or array names separated by commas.

Figure 19.6. General Notation for the PRINT Statement

Valid examples of the PRINT statement are:

   PRINT 53, S1, S2, S3
87 PRINT 12, TOTAL
   PRINT 174

The first example prints the three fields S1, S2, and S3 according to the output specifications in the FORMAT statement numbered 53. The second example prints the variable TOTAL according to specifications in FORMAT statement 12. The third example prints the data specified in FORMAT statement 174. This last example illustrates the use of the PRINT statement to print only headings which are specified as literal constants inside the FORMAT statement.

The PUNCH Statement

The PUNCH statement is available in many FORTRAN IV compilers to provide the capability for outputting data in the form of punched cards. Like the other specific I/O statements, no unit number is provided since this statement always refers to the card punch. This statement must have an associated FORMAT statement to describe the form of the output fields. An optional list of variables and/or array names, each separated by commas, follow the FORMAT statement number. A comma must separate the FORMAT statement number and the first variable, if present. The general form of the PUNCH statement is given in figure 19.7.

Additional FORTRAN Statements and Features 351

PUNCH Statement—General Form

PUNCH n, list

where:

n — is the statement number of the associated FORMAT statement that describes the attributes of the output data.

list — is an optional I/O list which may include one or more variable and/or array names separated by commas.

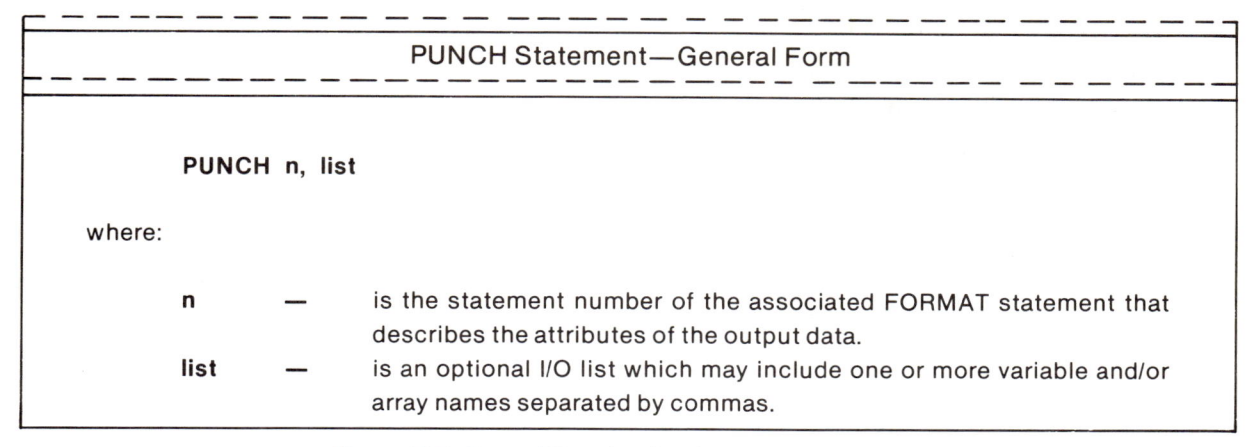

Figure 19.7. General Notation for the PUNCH Statement

Valid examples of the PUNCH statement are:

PUNCH 98, TEST1, TEST2
97 PUNCH 74, XSQ, YSQ, XYSQ

The first example punches the fields TEST1 and TEST2 into a punched card according to the format specifications in statement number 98. The second example punches the fields XSQ, YSQ, and XYSQ into a punched card according to the specifications in the FORMAT statement numbered 74.

19.6 REVIEW QUESTIONS

1. The PAUSE statement is used to permanently terminate the processing of a job. (True/False)
2. The Assigned GOTO statement causes a transfer of control to one of several statements based on the _____ _____ assigned to an integer variable.
3. The ASSIGN statement must be used before the Assigned GOTO to assign a statement number to the integer variable. (True/False)
4. What is the main advantage of using execution time formats?
5. The NAMELIST statement can be used to read data with or without format specifications. (True/False)
6. In addition to the output values the NAMELIST feature also includes the variable name preceding its value. (True/False)
7. The NAMELIST name must be enclosed within a pair of _____ .
8. The single character _____ must be attached to the NAMELIST name on the first input record to indicate that Namelist data is being read on an IBM 360-370 computer.
9. On Namelist input operations the variable/array name followed by an = must be given prior to the input value. (True/False)
10. The specific READ statement can be used to read magnetic tape and disk files as well as card files. (True/False)
11. A comma must be used to separate the keyword READ and the associated FORMAT statement number. (True/False)
12. The PRINT statement is used solely to output values to the line printer. (True/False)
13. The PUNCH statement is used solely to punch output values in a punched card. (True/False)

19.7 EXERCISES

1. Find the errors in the following statements (or group of statements).
   - *a)* PAUSE NOW MOUNT A TAPE
   - *b)* ASSIGN 83 FOR K
   - *c)* GOTO (33, 46, 92, 131, 7)
   - *d)* NAMELIST A,B,C
   - *e)* READ, 99 X,Y,Z
   - *f)* PRINT 37,
   - *g)* PUNCH (7,98) D,E,F
   - *h)* ASSIGN 10.0 TO M
     GOTO M, (20.0,10.0,7.0)

2. Find the errors in the following statements (or group of statements).
   - *a)* PAUSE 573246
   - *b)* ASSIGN 145763 TO K
   - *c)* GOTO K (45,8,17,65)
   - *d)* NAMELIST GROUPA /X,Y,Z/
   - *e)* READ 87 A,B,C
   - *f)* PRINT (6,48) X,Y,Z
   - *g)* PUNCH, 457, D,E,F
   - *h)* J = 7
     GOTO J, (5,19,3,7)

3. Determine which statement number control will be transferred to upon executing the following groups of statements.
   - *a)* ASSIGN 56 TO M
     GOTO M, (31,89,56,5)
   - *b)* ASSIGN 5 TO N
     L = 10
     IF (L/2 .GE.4) ASSIGN 17 TO N
     GOTO N, (4,5,17,90)

4. Determine which statement number control will be transferred to upon executing the following groups of statements.
   - *a)* ASSIGN 13 TO L
     GOTO L, (13,15,8,43)
   - *b)* ASSIGN 7 TO M
     X = 10/3
     IF (X .GT. SQRT(9.0)) ASSIGN 3 TO M
     GOTO M, (7,3)

5. Prepare the Namelist input data formats for the following specifications: A NAMELIST name of LIST2 with the variables A, K, and L containing the values 31.8, 17, and 43, respectively.

6. Prepare the Namelist input data formats for the following specifications: A NAMELIST name of HOLD with the variables M, X, and Z containing the values 14, 89.72, and 56.073, respectively.

19.8 PROGRAMMING PROBLEMS

1. Write a program that reads two values and computes their sum. The input values are read with the variables X and Y under the format specifications 2F6.2. Either variable may contain a negative quantity.

If the sum of the two variables is greater than zero, transfer to a WRITE statement that writes the two variables, their sum, and a message to the effect that the sum is a positive quantity. If the sum of the two variables is equal to zero, transfer to a WRITE statement that writes the variables, their sum, and a message to the effect that the sum is zero. Likewise, transfer to a WRITE statement that writes the variables, their sum, and a message that the sum is negative when the sum is less than zero. Use the ASSIGN and Assigned GOTO statements to accomplish this action.

Print the variable X in positions 5-10, the variable Y in positions 15-20, the computed sum in positions 25-31, and the literal message starting in print position 40. Include appropriate column headings.

2. Write a program that computes the sum and average of ten items. Each item is read from separate data cards. The first five items are punched in card columns 1-5 of each card in the form NNN.N. The remaining five items are punched in card columns 10-14 in the same form. Use object time format specifications to read each group of input values.

Print the input values in the print positions 5-9. Include an appropriate column heading. Print the sum and average on a double-spaced line with appropriate literal constants to identify each value.

3. Write a program that computes the product of the two variables X and Y. The input values are to be read using a NAMELIST data format. The input values and the computed product are to be written with a NAMELIST data format.

4. Write a program to compute the sum of the three variables X, Y, and Z. Use the specific READ statement to read the three variables under the format specifications 3F5.2. Use the PRINT statement to print the three input values and their sum on a double-spaced line under the format specifications 4F8.2. Include appropriate column headings.

Features of the WATFIV Compiler

20

20.1 Introduction to the Importance and Use of the WATFIV Compiler

The WATFIV (pronounced "what five") FORTRAN IV compiler developed by the University of Waterloo, Ontario, Canada in 1969 for the IBM 360, plays an important role in teaching students the FORTRAN language. Hundreds of junior colleges and universities across the nation use this outstanding compiler. The speed of the WATFIV compiler provides computer centers with the potential for giving students faster turnarounds which would, otherwise, not be available with other FORTRAN compilers. The WATFIV compiler also helps the instructor, since students using WATFIV can more readily understand the produced error messages and thus can correct the errors themselves. All in all, the WATFIV compiler can make life easier for everyone concerned with the development of FORTRAN programs.

The two main advantages derived from using the WATFIV compiler are, therefore, summed up as:

1. Its amazing speed in the compilation and execution of small and average-size programs.
2. Its effective and efficient debugging capabilities.

The WATFIV compiler compiles source programs much faster than the other compilers do. This speed of compilation is achieved as a result of two factors. First, the WATFIV compiler is an in-core or in-memory compiler. That is, the compiler stays in memory during the compilation of many WATFIV jobs, whereas, compilations with other FORTRAN compilers require that the compiler be brought back into memory for each FORTRAN job. Thus, only one FORTRAN program can be compiled at a time. Secondly, the WATFIV compiler is designed to take less time in translating a source program into object code. This efficiency in design results in the compilations of WATFIV programs many times faster than program compilations with other compilers.

The WATFIV compiler is superior to other FORTRAN compilers in its debugging abilities. The diagnostic error messages are much more intelligible than those of most compilers, and consequently are a greater help to the struggling student in locating the specific errors. WATFIV provides a clear explanation as to what the error is, and also prints characters surrounding the error. WATFIV also has the ability to detect many kinds of *execution errors* such as an undefined variable, changing the indexing parameters in a DO loop, invalid subscript value, etc., an ability that many compilers do not have. As explained in chapter 14, WATFIV has the capability for identifying the statement it was executing at the point of program blowup.

WATFIV is designed after the IBM "G" FORTRAN IV compiler described in IBM Form C28-6515. However, WATFIV does not include all the features in the IBM "G" compiler and has addition extensions to the "G" compiler. The WATFIV compiler is also written in accord with ANSI standards and will run on ANSI compilers when the standard subset of statements are used.

Following is the control cards setup for a WATFIV job.

$JOB
  FORTRAN source statements are inserted here
$ENTRY
  Data cards, if used, are inserted here

An IBM JOB card is required before the $JOB card. Some installations may require an end-of-file (/*) card after the last data card. The $JOB and $ENTRY cards are punched beginning in column 1 with no embedded blanks. The $ENTRY card is required even if no data cards are used, since this card signifies that execution is ready to begin. No input data card may have a $ in column one, because the dollar-sign symbol is used to indicate a control card.

20.2 WATFIV Extensions to the FORTRAN IV Language

WATFIV contains many features that are extensions to the ANSI standard and IBM "G" level FORTRAN compilers. The more important extensions are discussed in this section.

Format-free Input-output

Perhaps the most important of the additional features is the format-free input/output capability. Thus, the students do not have to worry about the complex data field specifications of the input/output records and can concentrate more on the basic principles of programming concepts.

The WATFIV compiler permits two free-form READ statements that eliminate the need for providing a related FORMAT statement to specify the attributes of the input data.

The form of the specific READ statement to provide format-free I/O in WATFIV is:

READ, list

The form of the general READ statement to provide format-free I/O in WATFIV is:

READ (unit, *, END = a, ERR = b) list

where **unit** may be any valid file unit specified as an integer constant or variable. The * is coded as such to denote the use of a free-form format. The parameters "**END =** " and "**ERR =** " are optional. The parameters **a** and **b** are supplied by the programmer and represent assigned statement numbers when either of the two optional parameters are used; **list** indicates an I/O list of variables and/or array elements.

Valid examples of these two forms of format-free READ statements are:

```
      READ, X, Y, Z
      READ, N, (X(I), I = 1, N)
   16 READ (5, *) D1, D2
    2 READ (5, *, END = 76) N1, N2, N3
```

The input data fields may be punched one per card, or many per card. If multiple fields are punched in the same card, each field must be separated by a comma and/or one or more blanks. The first data field does not have to start in card column one. A data field may not be continued across two cards since the end of a card acts as a delimiter (separator).

Successive cards are read until enough fields have been found to satisfy the requirements of the I/O list. Any fields remaining on the last card read for a particular READ statement will be ignored since the next READ statement executed will cause a new card record to be read.

Data fields may be signed or unsigned and may consist of the following type data: integer, real (in F,

E, or D form), complex, logical, and alphanumeric. The type of input data field must match the type of variable used in the I/O list. Whenever *alphanumeric data* is read under format-free I/O, the characters must be enclosed within a set of *single quotes.* If a quote is required as input in the string, two successive quotes must be punched. Complex values must be enclosed within parentheses and each number separated by a comma. A duplication factor may be given in the data card to avoid punching the same constant many times. For example:

DIMENSION A(50)
READ, A

50 * 0.0

It is permissible to use both the general READ and format-free READ statements in the same program.

Three forms of format-free output statements are available—the PRINT, PUNCH, and WRITE statements.

The form of the format-free PRINT statement is:

PRINT, list

The form of the format-free PUNCH statement is:

PUNCH, list

The form of the format-free WRITE statement is:

WRITE (unit, *) list

where **unit** is a valid file unit specified as an integer constant or variable. The * must be used in place of a format statement number to denote format-free output to the line printer. **list** represents an I/O list of variables and/or array elements.

Valid examples of format-free output statements are:

PRINT, A, B, C
2 PRINT, K
PUNCH, SUM, N
WRITE (6, *) I, J, TOT
WRITE (6, *) (X(J), J = 1,10)

Data output under the format-free I/O statements is formatted as follows:

Type Fields	Format Code Specifications
Integer	I12
Real*4	E16.7
Real*8	D28. 16
Complex*8	'(' E16.7 ' , ' E16.7 ')'
Complex*16	'(' D28. 16 ' , ' D28.16 ')'
Logical	L8
Character * n	An (where n represents the length of the character string)

An additional blank character is included to separate each field. Ten integer data fields can be output to a line; seven real*4 fields can be output to a line. Any combination of type fields can be output on the same line. It is permissible to use both the general WRITE and format-free output statements in the same program.

Expressions in Output Lists

WATFIV permits the use of constants and expressions in output lists. The one resulting value of an expression is printed. Function subprogram references may also be included in the output list. For example:

> PRINT, 12, A + B
> PRINT, "I = ", I
> WRITE (6,*) X, SQRT(X)
> WRITE (6,99) Y**2 + Z − 4, 500.37

The first example illustrates the use of an integer constant and an expression in the output list. Example two illustrates the use of a literal constant and a variable in the list. Example three uses a variable and a built-in function reference in the output list. Example four shows a compound expression and a real constant in the list. The use of constants and expressions in an output list provides a very powerful and flexible way of debugging programs.

One restriction must be observed in forming the output list. The first value in the list must not begin with an open left parenthesis, since this is used as a signal to denote an implied DO. That is, the expression

> PRINT, (K + J)/2

would be invalid. The expression could be written as

> PRINT, +(K + J)/2

to avoid the beginning left parenthesis. Parentheses other than the first character in the I/O list pose no problems. Real values are output in exponential form.

Character Type Variables

A primary weakness in the FORTRAN language lies in the area of its capability for handling input/output character strings. As you know, the character string must be broken up into groups of usually four or less characters. This restriction is understandable because of the intended nature of FORTRAN for scientific applications. WATFIV, however, due to the wide use of FORTRAN for many business applications, has included capabilities for handling character type variables and constants.

Variables which take on character values can be declared in the CHARACTER type statement. Unless specified in the provided length option of the statement, each character type variable has a length of one.

The general form of the CHARACTER type statement is:

> **CHARACTER * s name$_1$*s, . . . , name$_n$*s**

where ***s** is optional and provides the length of the character string contained in the variable. The length option may be from *1 through *255. If the length option is omitted, the default length is one. The **name$_1$, . . . , name$_n$** represent a list of variable and/or array names. As indicated in the general form notation, the length option may be specified after the keyword CHARACTER and/or after each individual name. The length option after each name overrides any length specification given after the keyword. Dimension specifications may be included for array names. Initial data values may also be included after the name by coding the character values as a literal constant and enclosing the value within a set of slashes. This syntax is basically the same for the other Explicit type statements.

Consider the following examples.

```
CHARACTER X, Y*3
CHARACTER * 7 VAL1, VAL2
CHARACTER * 5 R1, S1 * 6/ 'TOTAL = '/
CHARACTER MONTH*3 (12) / 'JAN', 'FEB', 'MAR'/
CHARACTER * 4 A/1HA/,BB/2HBB/,T/ 'ABCDE'/
CHARACTER * 25 NAME
```

In the first example, the variable X is by default assigned a length of one. The variable Y is capable of storing three characters because of the *3. In example two, the length attribute of *7 applies to both variables VAL1 and VAL2 and allows them to store up to seven characters. In the third example, the variable R1 has a length of 5, while the variable S1 has a length of 6. The contents of S1 are also initialized to the characters 'TOTAL = '. In example four, MONTH is declared to be a one-dimensional array of twelve elements; each element is of length three. The first three elements are also initialized to 'JAN', 'FEB', and 'MAR'.

In the fifth example, the three variables A, BB, and T are each declared to be four characters in length. The variable A is initialized to the one symbol 'A'. Since A is of length four, three blanks are added to the right of the constant 'A' to provide an initial value of 'A   '. The same holds true for the variable BB. Two blanks are added to the right to provide the initial value 'BB '. Five characters are coded for the initial value of T. Since the length of T is only four, one character must be dropped. As with the input of alphanumeric data, the leftmost characters in excess of the variable's length are truncated. Thus, the symbol A is dropped and the initial value of T is 'BCDE'. A warning message is always printed whenever initializing constants contain too many characters and some must be truncated. Example five allows a 25-length character string to be stored at the variable NAME.

The character type variables can conserve memory requirements whenever character string processing is needed for individual characters. For example,

```
   CHARACTER * 1 CARD(80)
   READ (5,99) CARD
99 FORMAT (80A1)
```

Only one byte is used to hold each character as opposed to one word when character variables/arrays are not used.

An initially defined character variable or array element must not appear in Blank Common. Otherwise, character type variables and/or arrays may appear in

1. DIMENSION statements
2. COMMON statements
3. NAMELIST statements
4. DATA statements
5. CALL statements
6. FUNCTION references
7. Dummy arguments
8. I/O statements
9. Assignment statements
10. Logical IF statements

WATFIV also permits character data to be assigned to a variable in a CHARACTER Assignment statement. The syntax of this statement is the same as the regular Assignment statement and is:

Character type Variable = Character value

The character value on the right of the = symbol may be either a character variable or a character constant. If the receiving variable to the left of the = symbol is longer than the assigned variable length, blanks are padded in the rightmost positions. If the receiving variable is shorter than the assigned variable length, truncation takes place. A warning message is given whenever the lengths of the two variables are not the same.

Character constants may be assigned to a character variable by coding them as literal constants. For example:

$$\text{CHARA} = \text{'A'}$$
$$\text{STRING} = \text{6HSUM} =$$
$$\text{R(1)} = \text{'SUM OF X + Y IS'}$$

This allows execution time assignment of character values instead of compile time initialization.

Character variables may be used in a Logical IF statement as operands of relational operators provided both operands are of type CHARACTER. For example:

$$\textbf{CHARACTER K, L, KODE(5)}$$
$$\textbf{IF (K .EQ. L) CALL SUBL}$$
$$\textbf{IF (K .EQ. KODE(1)) GOTO 30}$$
$$\textbf{IF (K .NE. 'C') GOTO 40}$$

Thus, character variables, array elements, and constants can be used as either operand in the relational expression. If the operands are not the same length, the shorter operand is padded on the right with blanks to match the length of the longer operand.

Extended Assignment Statement

The extended Assignment statement feature permits a value to be assigned to more than one variable in a single statement. For example:

$$\text{A} = \text{B} = \text{C} = 0.0$$
$$\text{K} = \text{N} = \text{I} = \text{J} = 1$$
$$\text{RESULT} = \text{X} = \text{SQRT(Y)} + 5. * \text{R}$$

The first statement assigns the value of zero to the variables A, B, and C. The second example assigns the value of one to the variables K, N, I, and J. The third example assigns the resulting value of the arithmetic expression to the two variables RESULT and X.

Extreme care must be taken when mixing types of variables in this statement. In the statement

$$\text{A} = \text{I} = \text{W} = 7.5$$

the compiler assigns the values beginning with the rightmost variable and works left. This statement is equivalent to the following three statements:

$$\text{W} = 7.6$$
$$\text{I} = \text{W}$$
$$\text{A} = \text{I}$$

Since I is an integer variable, the value 7 is assigned to I. Thus, A is assigned the value 7.0 rather than 7.6.

Precision may also be lost when an integer value is assigned to a real variable in this type of statement. For example:

$$\text{K} = \text{X} = \text{J} = 872164359$$

J is assigned the value 872164359. If X is a single-precision variable, only about seven significant digits are retained. Thus, X would be assigned the value 8721643, and K would be assigned the value 8721643 also.

Multiple Statements Per Card

This feature permits several FORTRAN statements to be punched on the same card, thus condensing the size of the source deck. For those statements without statement numbers, a semicolon is used to separate successive statements. No semicolon must be placed after the last statement. For example, the card:

$$\text{READ, X, Y; SUM = X + Y; PRINT, X, Y, SUM; STOP; END}$$

is equivalent to the following statements:

```
READ, X, Y
SUM = X + Y
PRINT, X, Y, SUM
STOP
END
```

When statement numbers are used, they must either appear in columns 1-5 as usual, or they must be separated from the FORTRAN statement by a colon. For example, the card:

$$\text{1 READ,N;K = 0; DO 10 I = 1,N; K = K + I; 10:CONTINUE}$$

is equivalent to the following cards:

```
1 READ, N
  K = 0
  DO 10 I = 1,N
  K = K + I
10 CONTINUE
```

Only columns 7-72 of a card can be used since the compiler ignores columns 73-80. The continuation column may still be used to indicate the continuation of the last statement. Comment cards must still be punched in the conventional manner with a C in column one.

Comments on FORTRAN Statements

WATFIV is written to terminate the left-to-right scan by the compiler on a card when the "ZIGAMORPH" character is encountered. Thus, comments may be included after this character to provide the capability for including comments on a FORTRAN statement. For example:

$$\text{AREAC = 3.14 * R ** 2 ¢ COMPUTE AREA OF CIRCLE}$$

where ¢ represents the "ZIGAMORPH" character. The "ZIGAMORPH" symbol is punched as a 12/11/0/7/8/9 multipunched character. This character is not, however, printed on the source listing.

Carriage Control Characters

WATFIV will, without warning, replace invalid carriage control characters with blanks. In addition to the standard set of carriage control characters, WATFIV also permits the use of ' — ' to provide triple spacing.

Transfer Statements as the Object (Range) of the DO

WATFIV allows the last statement (range) of a DO loop to be an Arithmetic IF or a Logical IF. The Logical IF may contain a GOTO statement of any form, a PAUSE, or a STOP statement. In a subprogram the Logical IF may contain a RETURN statement. If the condition expressed in the Logical IF is false, control is transferred back to the beginning of the loop to reexecute the statements in the DO loop. If the condition is true, the action statement on the Logical IF is executed.

Initializing Common Blocks

WATFIV allows the programmer to initialize Labeled Common Blocks in other than BLOCKDATA subprograms. Thus, variables and arrays in Labeled Common may be initialized by the DATA and Explicit type statements in the mainline program. Blank Common may also be initialized with the DATA and/or Explicit type statements.

Error Messages

Error messages may be produced by WATFIV during program compilation and/or execution. Three types of diagnostic messages may appear from the compilation phase. They are:

1. **Extension.** These messages flag the use of language extensions that may not be found in other compilers, such as the use of an implied DO in a DATA statement using arrays, or the use of multiple Assignment statements.
2. **Warning.** These messages flag statements that contain ambiguous syntax, but can nevertheless continue to generate object code. For example, a variable name containing more than six characters is flagged with a warning message, but the first six characters are used as the variable name, and compilation continues. Other examples are "missing END statement" and "this statement should have a statement number."
3. **Error.** These messages flag FORTRAN statements that cannot be translated by the compiler, such as misspelled keywords, missing operator, etc. The error message generally appears after the statement in error, gives a description as to the type of error, and provides information surrounding the error in question.

   Execution error messages are an outstanding feature of the WATFIV compiler that identify programmer blunders and prevent hard-to-detect bugs. Some examples of errors messages that may be produced at execution time are:

   a) Attempt to redefine a DO-loop parameter within the range of the loop.
   b) Format code and data type do not match. For example, reading a real variable with an integer format code.
   c) Attempt to use a subprogram recursively (i.e., calling itself).
   d) Subscript is out of range. That is, it is less than one or greater than the dimension size.
   e) Subscript is undefined. No initial value has been assigned to the subscript variable.
   f) Variable is undefined. For example, coding N = N + 1 where the variable N has never been assigned an initial value.

   Any attempt to print an undefined variable will result in an output value of U's. For example:

   ```
           A = 10.0
           B = 20.5
           SUM = A + B
           WRITE (6,99) A, B, SUMM
        99 FORMAT (1X, F4.1, 5X, F4.1, 5X, F5.1)
   ```

will result in the following values being printed

10.0   20.5   UUUUU

As many U's will be printed as will fill out the format code width specifications. This is a highly useful feature for catching the misspelling of the variables used in the output list or for variables that have never been assigned a value. The U's in the above output result from the spelling of the variable SUMM with two M's instead of SUM.

Since the diagnostic messages are highly descriptive and would also be verbally the same, the available list of WATFIV error messages is not included in this text.

20.3 How WATFIV Differs from the IBM FORTRAN IV "G" Compiler

Features of the IBM "G" compiler not found in WATFIV and major differences are as follows:

1. The concept of the extended range of a DO loop is not supported in WATFIV.
2. The service subprograms named DUMP and PDUMP are not supported in WATFIV.
3. The Debug Facility is not supported in WATFIV.
4. The name of a Labeled COMMON Block must be unique. That is, the name may not also be used as the name of a variable, array, or statement function in WATFIV.
5. The number of continuation cards, as well as the use of operator messages with the STOP and PAUSE statements, are installation options in WATFIV.
6. In WATFIV the character string FORMAT(is a reserved character sequence when used as the first seven characters of a statement. For example,

FORMAT(I) = 7.3

will result in an error message, whereas

A = FORMAT(I)

is legal, assuming that the name FORMAT is an array or Function name.

7. Since WATFIV is a "one-pass" compiler, the ordering of specification statements is highly important. Specification statements referring to variables used in NAMELIST or DEFINEFILE statements must precede the NAMELIST or DEFINEFILE statements. COMMON or EQUIVALENCE statements referring to variables used in data or initializing type statements must precede the DATA or initializing Explicit type statements. A variable may appear in an EQUIVALENCE statement and then in a subsequent Explicit type statement only if the type statement does not change the assumed length of the variable.

WATFIV incompatibilities with the IBM FORTRAN IV G and H level compilers other than those extensions described in section 20.2 are as follows:

1. WATFIV allows any number of contiguous comment cards.
2. Comment cards may be intermixed between continuation cards.
3. Commas are not required between format codes in WATFIV, if their omission does not cause ambiguity.
4. WATFIV does not allow group format repetition or format code repetition field counts to be expressed as a zero.
5. DO loops may be nested to any depth in WATFIV. A maximum of 255 DO loops are allowed in a program unit.
6. Negative values may not be exponentiated (raised to a power).
7. The name SORT may not be used as a variable name. That is, it must be used as a function subprogram reference only.
8. REAL *4 values are printed with a maximum of seven significant digits.

9. The use of the T format code in WATFIV which does a backward tab in an output buffer does not cause the existing characters in the buffer to be blanked out.

Now let us look at the control options that may be included on the WATFIV $JOB control statement and as additional job control cards in a FORTRAN source deck.

20.4 WATFIV Job Control Statements

In a WATFIV job any card containing a "$" in card column 1 is regarded as a job control card. Every WATFIV job must contain a "$JOB" and a "$ENTRY" control card (unless the control card requirements have been changed on your system). The $JOB card must precede your FORTRAN source program. The $ENTRY card must precede your data cards. The $ENTRY card must be included whether data cards are present or not. Other job control cards to cause page eject in source listings, to insert a blank line in the source listing, etc., may also be included in a source deck.

The $JOB card has $JOB punched in columns 1-4, WATFIV in columns 8-13, and an account number beginning in column 16. Various job options may follow this account number. When any of the options are used, a comma must follow the account number. If more than one option is present, a comma is used to separate each option from the previous one. A comment composed of any combination of symbols may follow the first blank column after column 16.

Following are ten job options which may be included on the job card. Any or all of these options may be used. The items in each pair of braces indicate the availabilities, of which you must, in each case, choose one. The underlined option is usually the default value which will be used unless you change it.

Application	*Options*	*Description*
Keypunch Model	KP = $\begin{Bmatrix} \underline{29} \\ 26 \end{Bmatrix}$	The chosen number indicates which keypunch model is used for the source program. The keypunch used determines the encoding of the special characters.
Execution time limit	TIME = n or T = n	This option limits the amount of computer time allowed for execution of the program. **n** is represented by one of the following forms: "**m**"; "**(m,s)**"; or "**(,s)**" where "**m**" denotes the number of minutes and "**s**" seconds.
Execution time limit on number of pages of printed output	PAGES = p or P = p	**p** denotes the maximum number of pages printed before the job is terminated.
Limit of number of lines printed on each page	LINES = n	**n** denotes the maximum number of lines printed per page before a page eject is performed.
Execution time processing with error conditions	RUN = $\begin{Bmatrix} \underline{CHECK} \\ NOCHECK \\ FREE \end{Bmatrix}$	The "RUN =" is optional. See the paragraph on execution time error checking for the CHECK and NOCHECK options. RUN = FREE or simply FREE causes program execution to be attempted even though there are compile time errors.
Printer listing of source program	$\begin{Bmatrix} \underline{LIST} \\ NOLIST \end{Bmatrix}$	This option controls the listing of the source program. "NOLIST" is useful when listings are not desired, especially with long programs.
Printer listing of library subprograms	$\begin{Bmatrix} LIBLIST \\ \underline{NOLIBLIST} \end{Bmatrix}$	This option controls the listing of library subprograms called/invoked in the program.

364   Features of the WATFIV Compiler

Application	Options	Description
Printing of warning messages	{WARN / NOWARN}	This option permits or suppresses the printing of warning messages during program compilation.
Printing of extension messages	{EXTEN / NOEXTEN}	This option permits or suppresses the printing of compiler extension messages during program compilation.
Library mainline program	PGM = name	"name" is the name of a library program to be used as the mainline program.

All 80 columns of the $JOB card may be used. The options may be punched in any order. If a given option is invalid (misspelled, etc.), this option and all options which follow it are ignored. If an option is included more than once, the last specification will be used in the job.

Compile Time Control Cards

The following WATFIV control cards may be used to control output of the source listing produced during program compilation. The "$" must appear in column one.

$PRINTON —All source statements which follow this control card will be listed by the compiler.

$PRINTOFF —All source statements which follow this control card will not be printed by the compiler. This card may be used in conjunction with the $PRINTON control card to selectively turn on and off the printing of source statements by the compiler.

$EJECT —This card causes a page eject to occur in the source listing, and is useful for starting subprograms or special routines on a new page.

$SPACE —This card causes a blank line to appear at its location in the source listing.

$WARN —This card causes compiler WARNING messages to be printed. If NOWARN is set in the $JOB card and the $WARN card is used, this card overrides the option in the $JOB card.

$NOWARN —Compiler WARNING messages will not be printed on the source listing.

$EXTEN —This card causes compiler EXTENSION messages to be printed.

$NOEXTEN —This card caused compiler EXTENSION messages not to be printed on the source listing.

Execution Time Error Checking

Two control cards are available concerning undefined variables during program execution. These two cards are:

$CHECK —This card causes a check to be made during execution to insure that each variable has been assigned a value before it is used in a computation. An error message is produced if the variable is undefined.

$NOCHECK —When this control card is used, no error message is produced if an undefined variable is used in a computation.

Execution Time Tracing of Statements

Two control cards are available to trace the execution of statements during processing.

$ISNON —This control card turns the trace feature on. When this feature is used, statements numbers are listed in the order in which they are executed in the program. This feature is especially useful in debugging to see the flow of execution in a program. At least one executable statement must precede the $ISNON control card.

$ISNOFF   —This card turns the trace feature off. This control card can be used in conjunction with the $ISNON card to limit the trace to specify portions of a program.

20.5 WATFIV-S Features

WATFIV-S is a FORTRAN IV programming language that includes extensions to the WATFIV compiler. The "S" in WATFIV-S stands for "structured" and means that the extensions in WATFIV-S are designed to include features to make it easier to write structured programs. WATFIV-S includes several new statements and block structures such as IF-THEN-ELSE, WHILE-DO, etc., which provide structured control blocks in the WATFIV language. These new extensions, therefore, allow one to develop better top-down, structured programs which are more easily read and understood. Many of these extensions in WATFIV-S are included in a new proposed standard FORTRAN set of specifications (1977).

The new statements in WATFIV-S produce a block structure or simply a block. By this, we mean that a block structure is a group of consecutive FORTRAN statements which begins with a new special statement (such as IF (logical-expression) THEN DO, or WHILE (logical-expression) DO) and ends with a special END statement (such as END IF or END WHILE). These block structures may be wholly contained within another block structure, but must not overlap, just as nested DO loops must not overlap. For example, an IF-THEN-ELSE block could be contained within a WHILE-DO block of statements.

IF-THEN-ELSE : END IF Block

This new block structure is an extension to the Logical IF statement. This block permits more than one statement to be executed if the logical expression being tested is true. Remember, the Logical IF statement permits only a single statement to be given after the logical expression. This certainly makes structured programming awkward in the current FORTRAN language. The new block structure also permits an optional ELSE DO capability to the IF-THEN-ELSE block to include statements to be executed whenever the logical condition is false. The general form of this new block structure is as follows:

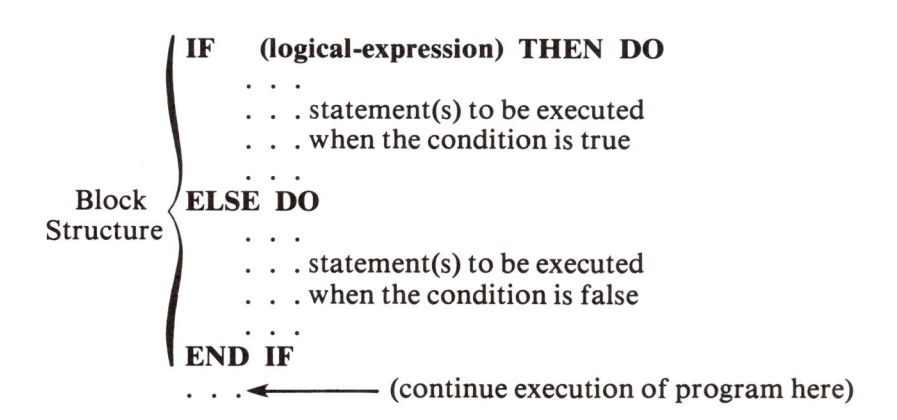

```
                   IF    (logical-expression) THEN DO
                     . . .
                     . . . statement(s) to be executed
                     . . . when the condition is true
                     . . .
         Block      ELSE DO
       Structure      . . .
                     . . . statement(s) to be executed
                     . . . when the condition is false
                     . . .
                   END IF
                     . . . ◄──────── (continue execution of program here)
```

When the logical-expression is true, the statements following the IF (logical-expression) THEN DO statement and preceding the ELSE DO statement are executed. Control of execution is then transferred to the statement following the END IF statement. When the logical expression is false, the statements following the ELSE DO statement and preceding the END IF statement are executed. Control of execution is then transferred to the statement following the END IF statement. Consider the following example.

```
            IF (N .LT. 0) THEN DO
               NNEGSM = NNEGSM + N
               NNEG = NNEG + 1
            ELSE DO
               NPOSSM = NPOSSM + N
               NPOS = NPOS + 1
            END IF
```

This block adds N to a specific counter and also increments another counter by one depending upon whether N is negative or not.

The ELSE DO statement is optional. When this statement is omitted and the logical-expression is false, control of execution is transferred to the statement following the END IF statement. For example:

```
            IF (X .GT. 0.0) THEN DO
               XSUM = XSUM + X
               N = N + 1
            END IF
```

This block adds X to a counter and increments the counter N by one only if the value of X is positive.

The ELSE DO and END IF statements are nonexecutable. If a branch is made to a statement inside the block structure, execution proceeds in a sequential manner with the ELSE DO (if present) and END IF statements being ignored. Control of execution may be transferred to a statement outside the block structure at any time. When this is desired, a GOTO statement is normally included immediately before the ELSE DO or END IF statements. The IF (logical-expression) THEN DO statement cannot be the range statement in a DO loop or the statement given in a regular Logical IF statement.

WHILE-DO : END WHILE Block

This block structure implements the WHILE-DO control structure in structured programming. The WHILE-DO block permits the repeated execution of a group of statements while a given expression is found to be true (that is, until the expression becomes false). The general form of this block structure is:

```
                 ⎛ WHILE (logical-expression) DO
                 ⎪    . . .
     Block       ⎨    . . . statement(s) to be executed
   Structure     ⎪    . . . while the condition is true
                 ⎪    . . .
                 ⎝ END WHILE
                      . . . ←——— (continue execution of program here)
```

If the logical-expression is true, then the statements following the WHILE (logical-expression) DO statement and preceding the END WHILE statement are executed. Control of execution then returns to the WHILE (logical-expression) DO statement and retests the logical-expression. This flow of execution continues repeatedly until the logical-expression becomes true or another test statement inside the block causes a transfer out of the block structure. When the logical-expression is false (which includes the first test), control of execution resumes at the statement following the END WHILE. For example:

```
         99 FORMAT (F5.1)
         98 FORMAT (1H0,F8.2,5X,I3)
            N = 0
            XSUM = 0.0
            WHILE (N .LE. 100) DO
               READ (5,99) X
               IF (X .EQ. -99.9) GOTO 10
               XSUM = XSUM + X
               N = N + 1
            END WHILE
         10 WRITE (6,98) XSUM, N
```

The loop would be reexecuted until 100 cards were read or a value of -99.9 for X is read.

When a WHILE-DO control structure is used in a program, a particular variable is normally identified as the *loop-control variable.* The loop-control variable is that variable whose value is continually tested in the logical-expression. This variable must be correctly initialized prior to entering the WHILE DO block in order for the loop to be executed the correct number of times. The value of the loop-control variable should be updated within each loop execution. Otherwise, the logical-expression may never become false and the loop exit may never occur. The updating of the loop-control variable is usually the last executable statement in the WHILE loop.

The END WHILE statement is a nonexecutable statement. If control of execution is transferred to a statement inside the block structure, the statements are executed in a sequential manner and the END WHILE statement is ignored. Control of execution may be transferred to a statement outside the block structure at any time. The WHILE (logical-expression) DO statement cannot be the range statement in a DO loop or the given statement in a Logical IF.

DO CASE : END CASE Block

The DO CASE block implements the Select case i of $(S_1, S_2, \ldots, S_n)$ control structure. This block structure provides capabilities similar to those of a Computed GOTO statement. The general form of the DO CASE block is:

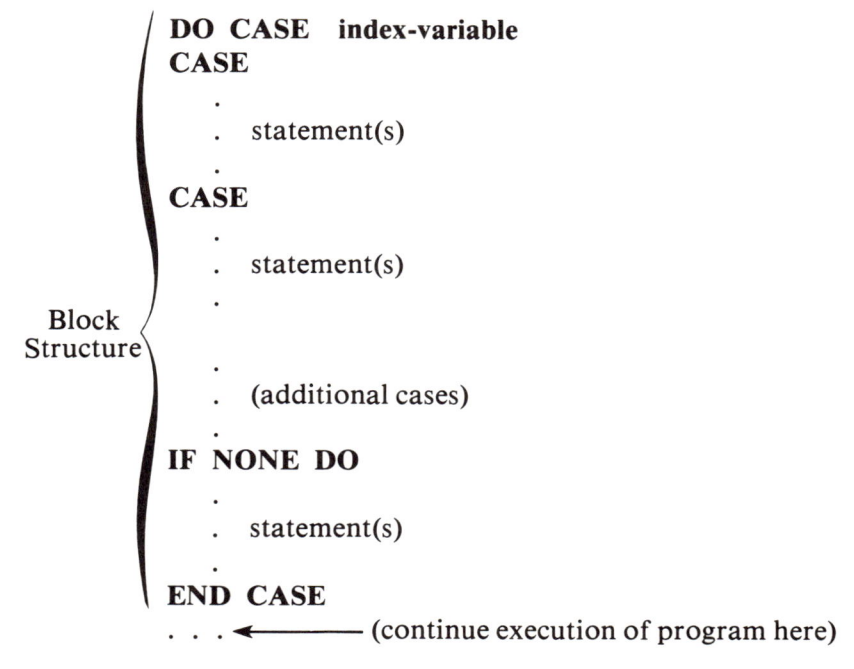

Block Structure:
```
DO CASE   index-variable
CASE
   .
   .  statement(s)
   .
CASE
   .
   .  statement(s)
   .
   .
   .  (additional cases)
   .
IF NONE DO
   .
   .  statement(s)
   .
END CASE
 . . . ←────── (continue execution of program here)
```

The "index-variable" must be an integer variable whose value is used to select the proper CASE and its related statements. That is, if the value of the index-variable is 1, then the statements belonging to the first CASE statement are executed; if the value of the index-variable is 2, then the statements belonging to the second CASE statement are executed; and so on. After the statements belonging to the selected CASE statement are executed, control of execution resumes at the statement immediately following the END CASE statement (unless a GOTO statement was encountered).

If the value of the index-variable is negative, zero, or greater than the number of CASE statements, then the statements following the IF NONE DO statement and preceding the END CASE statement are executed. Control of execution is resumed at the statement following the END CASE statement. The IF NONE DO statement acts like an "else" option to cover all invalid values of the index-variable.

The IF NONE DO statement is optional. If this statement is omitted, control of execution resumes at the next statement after the END CASE whenever the index-variable is negative, zero, or greater than the number of cases. If the IF NONE DO statement is used, it must follow the statements associated with the last case. Only one IF NONE DO statement is allowed in a DO CASE block structure.

An example of the use of a DO CASE is as follows:

```
      99 FORMAT (I2)
      98 FORMAT (1H0,30HINDX IS NOT A 1,2, OR 3, BUT = ,I4)
         READ (5,99) INDX
         DO CASE INDX
            CASE
               I1S = I1S + 1
            CASE
               I2S = I2S + 2
            CASE
               I3S = I3S + 3
            IF NONE DO
               WRITE (6,98) INDX
         END CASE
```

When a value of 1 is read for INDX, a 1 is added to a special counter named I1S. If the value of INDX is 2, a 2 is added to a special counter named IS2. A 3 is added to a special counter named I3S when INDX has a value of 3. If the value of INDX is not a 1, 2, or 3, then a message is written to this effect.

The CASE, IF NONE DO, and END CASE statements are nonexcutable statements. Control may be transferred to a statement inside the block structure from a statement outside the block. If such a transfer of control is made, any CASE statement, the IF NONE DO, and END CASE statements are ignored; thus, control of execution continues in a sequential manner from the statement to which the transfer is made. Control of execution may be transferred from a statement inside the block to a statement outside the block at any time.

The DO CASE statement must not be the range statement in a DO loop or the statement given on a Logical IF statement. The first CASE statement immediately following the DO CASE statement is optional. If the first CASE statement is omitted, correct logic is performed as if it were present. A CASE statement may be followed immediately by another CASE statement. This allows a value of the index-variable to be ignored and control of execution resumes at the statement after the END CASE. A comment may be added after the keyword "CASE" on the CASE statement. Care should be taken in this capability since an " = " symbol in the comments may be taken as an Assignment statement.

EXECUTE Statement

The EXECUTE statement causes a REMOTE BLOCK structure to be executed. The general form of this statement is:

EXECUTE name

where **name** is the name assigned to the REMOTE BLOCK as given in the REMOTE BLOCK statement.

Control of execution resumes at the statement following the EXECUTE statement after execution of the REMOTE BLOCK named in the EXECUTE statement. An EXECUTE statement may be the range statement of a DO loop or it may be the statement given in a Logical IF statement.

REMOTE BLOCK : END BLOCK Structure

This block structure identifies a group of consecutive statements that are preceded by the REMOTE BLOCK statement and terminated by an END BLOCK statement. The REMOTE BLOCK structure can be executed only by using an EXECUTE statement in a program unit. Thus, the use of REMOTE BLOCK structures provides a parameterless, subroutine-like capability within a program unit. Since no arguments or parameter list is used, the current value of a variable is used upon entering a REMOTE BLOCK. Any variable modified within the REMOTE BLOCK retains its new value upon exit from the block.

Each REMOTE BLOCK structure is assigned a symbolic name which is formed from the rules for choosing other FORTRAN variable names. Each REMOTE BLOCK must be assigned a unique name. This name, however, may be the same name as a variable, array, or subprogram used in the same program unit. The general form of the REMOTE BLOCK structure is:

```
            ⎧ REMOTE BLOCK   name
            ⎪    . . .
  Block     ⎨    . . .   statement(s) to be executed
Structure   ⎬    . . .   when the remote block is referenced
            ⎪    . . .
            ⎩ END BLOCK
```

REMOTE BLOCK structures may be located anywhere in a program unit. However, a GOTO or a transfer of control statement must immediately precede the REMOTE BLOCK statement to prevent sequentially "falling into" the block structure. The first executable statement (not belonging to another REMOTE BLOCK structure) following the END BLOCK statement should have a statement number assigned to it. Otherwise, it could never be executed. It is recommended that you place all REMOTE BLOCK structures at the end of a program unit between a STOP or RETURN and the END statement in order to avoid placement errors. A maximum of 255 REMOTE BLOCK structures may be used in a program unit.

The REMOTE BLOCK and END BLOCK statements are nonexecutable. A comment may be included on the END BLOCK statement. Care should be taken to avoid certain special characters in the comment since a " = " symbol may be taken as an Assignment statement.

Consider the following example.

```
   99 FORMAT (1H0,I7)
      I = 7
      EXECUTE X
      WRITE (6,99) XRSLT
      I = 11
      EXECUTE X
      WRITE (6,99) XRSLT
      STOP
      REMOTE BLOCK X
          XRSLT = (I*3) / (I-1)
          END BLOCK OF X
      END
```

A REMOTE BLOCK structure must not be nested (defined) within another REMOTE BLOCK structure. No statement in a REMOTE BLOCK structure may transfer control from within a REMOTE BLOCK to another statement in the program unit (a compile error will occur). Otherwise any WATFIV or WATFIV-S statements and block structures may be used in a REMOTE BLOCK structure. Subprograms may be called/invoked from within a REMOTE BLOCK structure. A REMOTE BLOCK may contain an EXECUTE statement which causes another REMOTE BLOCK structure to be executed. As with subprograms, a REMOTE BLOCK structure must not call a subprogram, execute itself, or execute another REMOTE BLOCK which results in a loop.

A REMOTE BLOCK may be entered only by means of an EXECUTE statement. The BLOCK execution is terminated by executing the statement immediately preceding the END BLOCK statement. Transfers or loops may be performed by statements within the block as long as control of execution is maintained with the block. Any attempt to transfer control to a statement inside a REMOTE BLOCK from a location outside the block will result in an execution error.

WHILE-EXECUTE Statement

The WHILE-EXECUTE statement causes a REMOTE BLOCK structure to be executed repeatedly while a specified condition in a logical expression is true. The general form of the WHILE-EXECUTE statement is:

WHILE (logical-expression) EXECUTE name

If the logical-expression is true, the REMOTE BLOCK identified by "**name**" is executed and control of execution returns to the WHILE statement. When the logical-expression is false, control of execution resumes at the next executable statement after the WHILE. For example:

WHILE (KNT .LE. 50) EXECUTE RB1

The REMOTE BLOCK named RB1 will be repeatedly executed while KNT is less than or equal to 50. If the REMOTE BLOCK named RB1 does not assign a value to KNT which is greater than 50, the program will loop indefinitely (until execution time is exhausted). The WHILE-EXECUTE statement must not be the range statement in a DO loop or the given statement in a Logical IF statement.

AT END DO : END AT END Block

The AT END DO block structure is used to specify the processing to be done when an end-of-file condition is encountered by a READ statement. The general form of this block is:

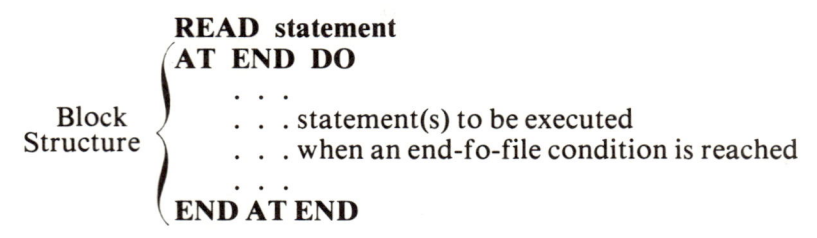

```
                READ statement
               ⎛AT END DO
               ⎜   . . .
     Block     ⎬   . . . statement(s) to be executed
   Structure   ⎜   . . . when an end-fo-file condition is reached
               ⎜   . . .
               ⎝END AT END
```

The AT END DO statement must immediately follow a READ statement.

If an end-of-file condition is encountered by the READ statement, the statements following the AT END DO statement and preceding the END AT END statement are executed. Control of execution then resumes at the statement following the END AT END statement. If an end-of-file condition is not encountered by the READ statement, control of execution is transferred to the statement immediately following the END AT END statement.

The AT END DO and END AT END statements are nonexecutable. Control of execution may be transferred to a statement inside the block from a statement outside the block. When such a transfer is made, the END AT END statement is ignored and execution continues in a sequential manner. Control of execution may be transferred to a statement outside the block from a statement inside the block at anytime. An example is:

```
99 FORMAT (F5.2)
98 FORMAT (1H0,15HEND OF DATA SET)
   READ (5,99) X
   AT END DO
      N = N - 1
      WRITE (6,98)
      GOTO 20
   END AT END
   . . .
   . . .
   . . .
20 CONTINUE
```

When the end-of-file is encountered on unit 5, one will be subtracted from the variable N, an end-of-data-set message is printed, and control of execution is transferred to statement number 20.

The READ statement associated with an AT END DO block cannot be 1) the range statement in a DO loop, 2) the statement given in a Logical IF statement, 3) used with direct access files or "core-to-core" read operations, or 4) used when the "END = " option is included in the READ statement.

WATFIV is indeed a highly versatile, efficient compiler. The rapid compilation for fast turnarounds and the outstanding error diagnostics for student independence in program debugging make this an attractive FORTRAN compiler in the academic institutions.

Appendix A: The 80-Column Punch Card

The most common medium used to input programs and data to the computer in the student environment is the punch card. It contains 80 vertical columns, each of which holds one character. There are twelve horizontal rows on a punch card. A particular character is encoded by punching a hole in the proper rows for the respective column.

Some terms you should know about the punch card are:

Character—A character is a single alphabetic letter, numeric digit, or special character encoded in one vertical column on the card.

Column—A column is one vertical area on the card used to contain one character. Each column is broken down into twelve horizontal rows.

Column 1—Column 1 is the leftmost column in the card, located at the left side.

Column 80—Column 80 is the rightmost column in the card, located at the right side.

Row—A row is one horizontal area across the card. There are twelve different row areas on a card. The top two rows are the 12 and 11 rows. These two rows are not labeled (printed with digits). The remaining ten rows are labeled 0-9.

12 Edge—The 12 edge is the top edge of the card, the edge nearest the 12 row.

9 Edge—The 9 edge is the bottom edge of the card, the edge nearest the 9 row.

Zone rows—The top three rows (12, 11, 0) of the card are known as the zone rows. The top two zone rows are not labeled.

Digit rows—The bottom ten rows (0-9) of the card are known as the digit rows. Note that row 0 is considered to be both a zone and a digit row. The character contained in the column (letter or digit) determines which type of row the 0 row is used as. The ten digit rows have their corresponding digit printed in each of the 80 columns.

Field—A field is one or more adjacent columns used to contain an item of information. For example, the nine-digit SSAN is a field requiring nine adjacent columns to encode this item. If a name field consists of twenty characters, twenty adjacent columns are required to encode this item.

Face—The face of the card is the front side with the printed information.

Interpreting—Interpreting is the translation of the punched code into a form readable by humans. The interpreting of encoded characters is printed at the top of the card. Each character is interpreted at the top of the column in which it is punched. The interpreting is done automatically by the keypunch machine as the card is punched. The print switch must be on, however, to print the characters; otherwise, if the print switch is off, the characters are not interpreted. The interpreted characters along the top edge of the card mean nothing to the computer. These interpreted characters are solely for our benefit, to read the encoded information. The card reader reads the punched holes and translates the information from which holes are punched.

These features of the punch card are illustrated in figure A.1.

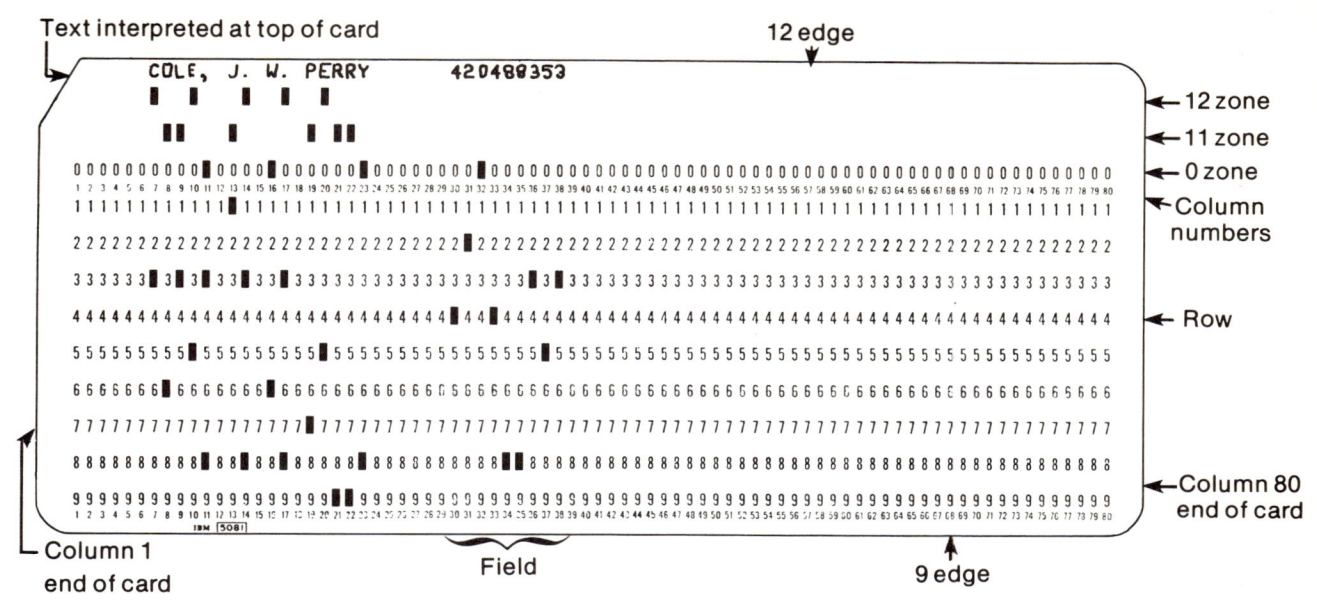

Figure A.1. Features of the 80-column Punch Card

Holes are punched for the various characters in accordance with a special coding scheme. This coding scheme used to encode the different types of characters is as follows:

Numeric digits. A numeric digit is coded in a column by a single hole in the respective row for the desired digit. For example, the digit 2 would be coded by punching a hole in row 2 in a specific card column; the digit 7 would be coded by punching a hole in row 7.

Alphabetic letters. An alphabetic letter is coded by punching two holes in a column. One hole is punched in the zone area (either row 12, 11, or 0) and one hole is punched in the numeric area (one of the rows 1 through 9). For example, the letter A would be represented by a punch in the 12 row and 1 row; the letter M would be coded by a hole in the 11 row and the 4 row.

Special characters. Special characters (characters other than numeric digits and alphabetic letters) are coded as one, two, or three punches. For example, the minus sign (−) is represented by a single hole in row 11. The plus sign (+) is represented by three holes in rows 12, 6, and 8, respectively. The decimal point (.) is coded as three holes in rows 12, 3, and 8, respectively.

It is not necessary to memorize the punched hole code required to represent the different characters in the punch card. The holes are punched into a card by means of a machine called a keypunch. This machine has a keyboard very similar to that of the typewriter. By typing the proper character on the keyboard, the correct holes are punched.

Figure A.2 on the next page illustrates the coding scheme for the various characters punched with an IBM 029 keypunch. Special care must be taken when punching the special characters since they are often encoded by different punch combinations on different computer systems (i.e., the 026 keypunch may be required). Note that all alphabetic letters are capital letters.

The standard 80-column card is 3 1/4" high by 7 3/8" long. Punch cards come in a variety of colors. The color is of no significance to the computer; to the computer operator, however, various colors are used to represent different types of cards. For example, one color might be used to identify the beginning job card. Punch cards come with and without printed fields on them. The printed information is needed in many production environments, and means something to the user. The punch card normally used in the student environment is the standard card without any identifying fields printed on it. A punch card may have a left corner cut off in the upper left edge, or a right corner cut off in the upper right edge. Different types of corner cut cards are used to identify different types of cards, such as the beginning or ending card in a deck of cards.

Punched cards are used extensively because of their flexibility. If we wish to correct a card in a pro-

374 Appendixes

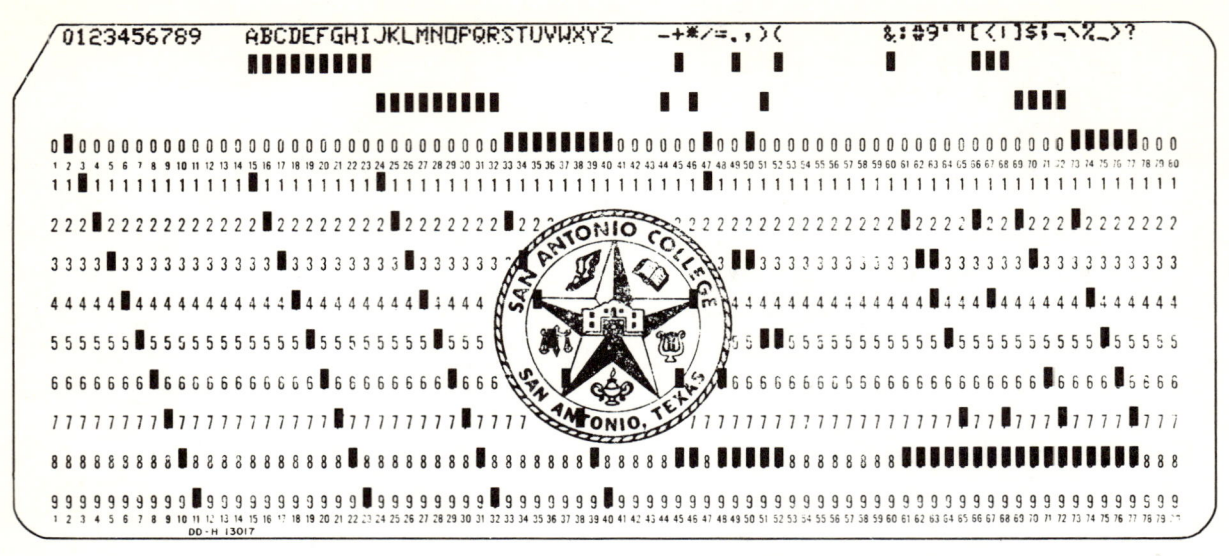

Figure A.2. Coding Scheme of the Characters Available on the 029 Keypunch

gram or data file, we can simply locate the card, repunch it, and replace the corrected card into the deck. Punched cards provide a permanent record which can be used repeatedly. However, punch cards do have their limitations. Their record size is limited to 80 characters. Punched cards cannot be folded, spindled, mutilated, or stapled and are sensitive to high humidity and cold temperatures. They also are a very slow medium for input and output of data when compared to other input/output media.

IBM 029 Keypunch Operations

Appendix B

In a student environment, most FORTRAN jobs are submitted to the computer on punched cards. The FORTRAN statements, JCL, and data cards are all punched onto the 80-column punch card by using a special machine known as a *keypunch machine.* (Some people also refer to this machine as a card punch machine.) Consequently, students using punch cards must learn how to operate the keypunch. Even though your computer center may keypunch your source program, you must still punch your corrections.

There are several types of keypunch machines available for punching the 80-column card. Univac has a very efficient keypunch known as the model 1710. IBM produces several models—the 026, 029, and 129. The 026 model may be used at some installations, especially those having CDC computers. The machine most commonly used is the 029, which is used to keypunch programs for IBM, Honeywell, and Burroughs computers. The main difference between the two machines (026 and 029) is in some of the special characters. The purpose of this Appendix is to provide an overview to the operations of the most frequently used features of the IBM 029 keypunch machine.

The four stations on the keypunch machine are as follows:

1. Input card hopper—in which the blank cards are loaded to be fed into the machine.
2. Punch station—at which a card is positioned for punching.
3. Read station—to which the punched card is advanced after being punched. The read station may be used for duplicating a card or portions of it, in case of errors.
4. Output card stacker—at which the cards are stacked after leaving the read station.

Each punch card usually passed through each of these stations in the machine operation. That is, a card is fed from the input hopper into the punch station where it is punched. The card is then released to the read station and then to the output stacker. However, a card may be duplicated or corrected by inserting it directly into the read station. Figure B.1 presents a picture of the IBM 029 keypunch machine and the location of these four stations. Note that the card hopper is located in the upper right-hand portion of the machine. The read and punch stations are located at the middle of the machine behind the keyboard. The card stacker is located in the upper left-hand portion of the machine.

Figure B.1 also shows the location of other important features of the keypunch. The mainline power switch is a red toggle lever located beneath the keyboard and to the far right. You flip the switch up to turn on the machine and down to turn it off. The keyboard for typing is located towards the right-hand side on the reading board (table top). The other features you will use are the column indicator and backspace button. These are located in the center of the machine above the read station.

The Keyboard. The keyboard is used to type the characters to be punched and to control the movement of the cards through the four stations. The position of the alphabetic characters is identical to their position on the typewriter keyboard. The location of the digits and the special characters are different. The numeric digits are all located on the right side of the keyboard. Since most input data is numeric, the location of all the digits on the right of the keyboard allows a keypunch operator to control the coding sheets with the left hand and to punch the digits with the right hand. The space bar which moves the card forward

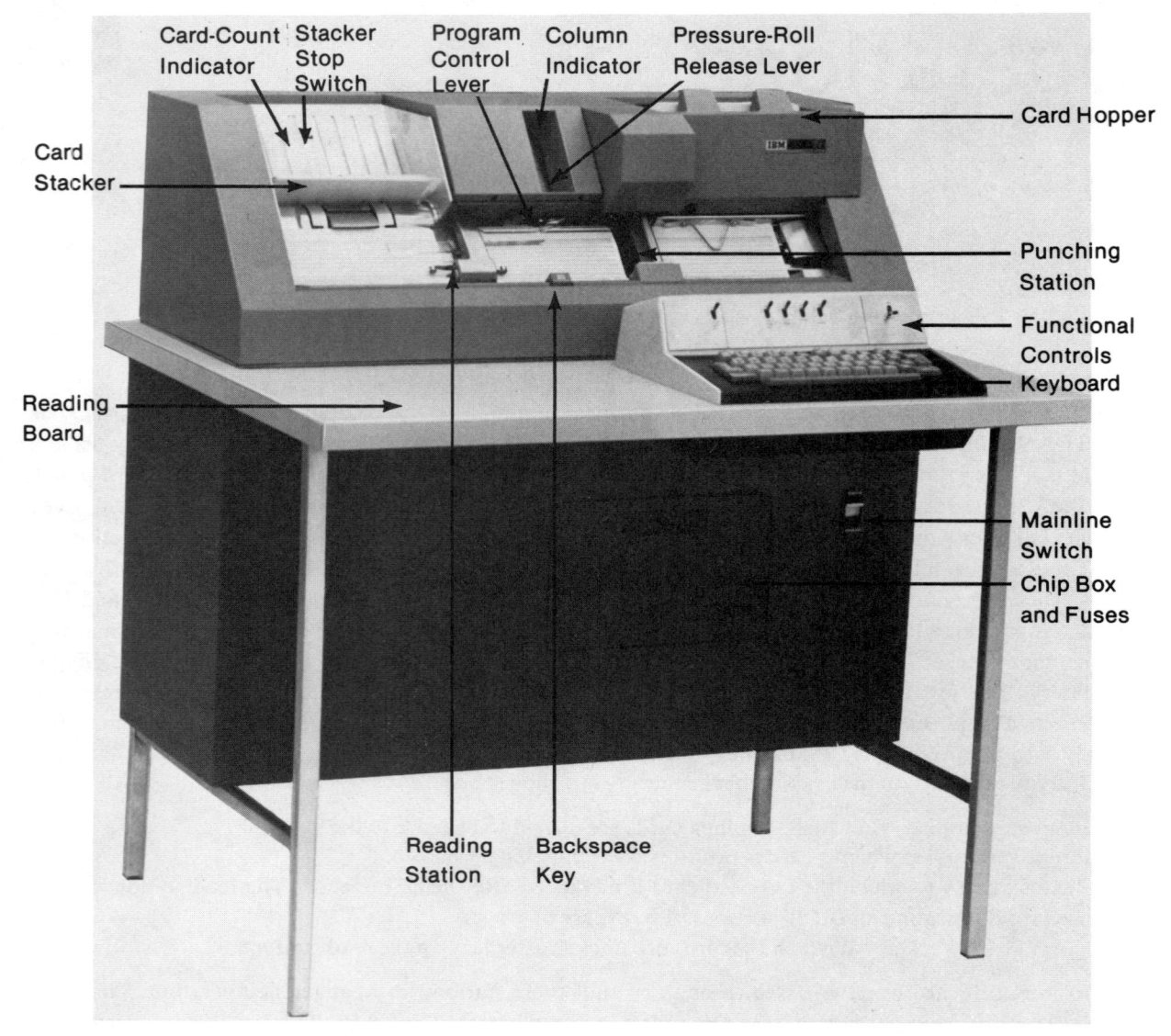

Figure B.1. IBM 029 Keypunch. *Courtesy of International Business Machines Corporation.*

one position each time it is depressed is located at the bottom of the keyboard, as on the typewriter. Study the location of these keys on the keyboard as shown in figure B.2 before you proceed to learn the control keys.

The light or gray-colored keys are for punching characters. The dark or blue-colored keys are control keys for special purposes, such as feeding cards, registering cards at a station, shift selection, etc.

Three main control keys you will use all the time are the feed (FEED), register (REG), and release (REL). These are located towards the far right of the keyboard and are used to control the movement of the cards. The FEED key feeds a card from the card hopper down to the punch station, but normally does not engage it at the station. The REG key must be depressed to register the card at the punch station before you are able to begin typing. If you depress the FEED key twice, the machine will feed two cards from the hopper and automatically register the first one at the punch station. You must never feed more than two cards down to the punch station, otherwise a card jam will occur.

The REL key releases a card to the next station. If a card is in the punch station it is moved to the read station. If the machine is in auto feed mode, the card will be automatically registered at the read station.

Figure B.2. Keyboard on an IBM 029 Keypunch. *Courtesy of International Business Machines Corporation.*

Otherwise you must depress the REG key to register the card at the read station. Again, depressing the REL key followed by the REG key will move the card at the read station into the stacker. If the machine is in auto feed mode, the cards will be automatically registered at the next station when the REL key is depressed.

Notice the NUMERIC shift key at the lower left corner of the keyboard and the ALPHA shift key at the lower right. These are similar to the shift keys on the typewriter. But we have only capital letters on our keyboard, therefore these two keys allow us to shift from alpha mode to numeric, and vice versa. The alphabetic letters, comma, period, and some special characters are in alpha mode. The numeric digits and other special characters (indicated above the letters) are in numeric shift.

The standard mode of the machine is ALPHA shift. To punch the characters indicated on the lower half of the key, you simply type the proper key. But, to punch a digit or some special characters indicated on the upper half of the key, you must depress the NUMERIC shift key. Releasing the NUMERIC key returns the machine to alpha mode. We can place the machine so that it will be in numeric mode all the time by using the program control feature and lowering the "starwheels" control mechanism. Then, when we need alpha mode characters, we must depress the ALPHA shift key. The student seldom uses the program control feature so we will assume the alpha shift mode to be our normal setting. If however, you get special characters or numeric digits when you depress a key, look to see if the "starwheels" control mechanism is engaged. This switch is located above the backspace button and can be disengaged by flipping it in the opposite direction.

The DUP key is located in the center of the top row of keys. When depressed it will duplicate the information in the card at the read station into the same card columns of the card at the punch station. This key gets considerable use in making corrections. You simply insert the card to be corrected into the read station, feed and register a blank card into the punch station, and duplicate the information up to the column in error. Then you can type in the correct character or information and even dup the rest of the card.

If you attempt to depress two keys at the same time, the keyboard will lock and you are unable to type a character. To free the keyboard, depress the ERROR RESET key located on the far left side of the keyboard. Then you can resume typing. The MULT PCH key located at the upper left corner of the keyboard allows you to punch multiple punches into a column on a card. When this key is depressed, the card is locked at the punch station and any number of characters can be typed into the same column. Of course, this usually produces an invalid character, but that is sometimes desired. Releasing the MULT PCH key then causes the card to move to the next column in the punch station.

Switches. Located immediately above the keyboard are six switches. Starting from the right, these switches and functions are:

CLEAR switch—Flipping this switch in an upward motion will clear the cards in both the punch and read stations and stack them in the card stacker. This provides a shortcut way of retrieving a punched card without having to go through the procedure of repeatedly depressing the REL and REG keys. This switch is spring loaded so that it automatically returns to the down position.

LZ PRINT (left-zero print) switch—Is used to print high-order zeroes in a field and should be set on (up).

PRINT switch—Causes the characters to be printed across the top of the card. When turned off, the printing operation is not performed. This switch should always be set on (up).

AUTO FEED switch—Provides for the automatic feeding of cards. If you wish to have the cards continuously fed in an automatic mode, then set this switch up. When you are duplicating individual cards, however, you normally wish to feed and register them individually. Therefore, set the switch off (down).

PROG SEL (Program select) switch—Is used in conjunction with the program control feature. Since the student does not use this feature, it is normally set down. However, the setting of this switch has no effect if the program control feature is not engaged.

AUTO SKIP DUP (auto skip/duplicate) switch—Is also used in conjunction with the program control feature. Thus, the setting of this switch is immaterial.

Column Indicator and Backspace Button—The location of the characters on the card is important. For example, the FORTRAN statement must begin in column 7 or later. When the card is registered at the punch or read stations, it is positioned at card column 1. We could advance the card to column 7 by depressing the space bar six times. However, it is impractical to keep mental track of the current position of the card. So the **column indicator** is used to tell us the card column at which we are positioned.

Above the read station is a window through which can be seen a red pointer and a dial indicating the current position of the card. The dial has hashed marks like a ruler to indicate the current position (card column). The odd positions have a long hash mark and the even positions have a short hash mark, with the indicated position number. Thus, we can read the current position of a card from the dial by noting the column indicated by the red pointer.

The backspace button is a large, protruding, light-colored button located beneath the center of the read station. Each time we depress this button the cards in the read and punch stations are moved backwards one position. Of course, it is impractical to backspace more than a few positions. Simply feed a new card and use the DUP key to help make the desired corrections.

Loading Cards into the Hopper. Many computer centers provide cards for the students. Others expect the students to provide their own. In either case, the student must know how to properly load the cards into the hopper when it is empty.

A spring-loaded pressure plate is provided in the card hopper to force the cards to the front of the hopper. When this plate is pushed to the rear of the hopper, a catch will retain it in this position. The cards can now be inserted into the hopper. You should take extreme care to ensure that the cards are loaded in straight. Pat them down gently to make sure that the top of the cards are even and inserted correctly. Next, depress the flat metal button on top of the spring-loaded plate to cause it to move up against the cards. Should the cards fail to feed, it usually means that the front card is bent at the bottom. Push the plate back and remove the bent card(s).

Clearing a Card Jam. When more than one card is registered at the punch or read station, a card jam occurs. The cards will not move when you type, depress the REL key, or flip the CLEAR switch. Depressing the FEED key will only jam more cards. **DO NOT** attempt to pull the cards out with brute force since this will damage the punch dies and the machine.

The top center portion of the machine which houses the window to the column indicator will fold down. Simply pull upwards and forward on this portion. The metal portion will fold out and lay down against the read station, exposing the column indicator. Located at the lower right of the column indicator is a silver-colored, metal lever called the pressure-roll release lever. By depressing this metal lever towards the back of the machine, the pressure is removed from the punch dies. It usually takes a great deal of force to depress this lever since it has a heavy spring. But fully depress this lever first, then you can pull out the jammed cards without damaging the machine. If you are uncertain of this operation, then contact someone in the computer center who knows how to remove the cards. Remember, do not attempt to pull out the cards physically without fully depressing this lever. If you are unable to clear the card jam, put a note on the keypunch as to the problem and find a new machine.

Summary of Operations. Assuming that cards are in the card hopper, the following rules summarize the operations of the keypunch machine.

1. Make sure the red power switch is on, i.e., in the up position.
2. Press the FEED key.
3. Press the REG (Register) key.
4. Punch the card. Watch the column indicator as to current position of the card to make sure you punch the characters in the desired columns.
5. Press the REL (Release) key to move the card from the punch to the read station.
6. If more cards are to be punched, repeat steps 2-5. Otherwise, flip the CLEAR switch or press the REG and REL keys alternately until the card is placed into the stacker.
7. If the AUTO FEED switch is on (up), you need only to depress the FEED key twice. This will cause the machine to automatically feed and register the cards for you.
8. The NUMERIC shift key must be depressed when punching digits and most special characters.
9. The backspace key may be used to move a card backwards in the stations.
10. Corrections cannot normally be made by backspacing and repunching. A character punched over another character in the same column usually produces an invalid character which cannot be read by the computer.

    Corrections are normally made by releasing the current card at the punch station, duplicating the correct portion of the old card, and punching the correction onto the new card.

Practice is usually the best teacher in learning how to use the keypunch machine. Take time to learn the many control keys and their functions to aid you in using the machine. Punch cards will probably be with us for a long time in data processing. Thus, you should learn how to use the keypunch machine to punch the cards for your various needs. For a complete discussion on the full use and capabilities of the IBM 029 keypunch machine, you should refer to Reference Manual IBM 029 Card Punch, A24-3332-5.

Appendix C Overview to Time-sharing Operations

Timesharing is defined as the sharing of computer time in a real-time, interactive processing mode between the user and computer. ***Real-time*** means the actions take place as they occur, usually within seconds, in an "on-line" mode to the computer. The term, ***on-line,*** refers to a system in which all necessary devices are immediately accessible to the computer. ***Interactive processing*** refers to a direct link between the user and the computer to provide an on-line processing capability. That is, a user, seated at a terminal, enters his program and data, and receives the results, displayed on his terminal, directly from the computer.

In a timesharing environment all intermediate handling of program decks is eliminated. The user can type in his program, compile, and execute it at the terminal with the results coming back immediately to the terminal. Thus, the steps in solving a problem on the computer can be reduced to a matter of minutes or hours instead of being extended over several days or weeks.

The terminals can be located in the same room with the computer or they can be located a considerable distance from the processing computer. They can be scattered across the campus or even around the country to provide an on-line, interactive processing capability on a computer system. However, a timesharing system allows all users, by the use of on-line-terminals, to share time on the processing computer. The computer allots a small amount of time to each user in turn. From the user's point of view, he appears to be the only one using the computer, since most of his time is spent typing in his program or data.

Timesharing Language. In order to create files, compile and execute programs, save or remove files, list the contents of files, etc., a timesharing system must have its own command language. That is, a separate *timesharing language* must be used to tell the computer what operations to perform on a file. It is from this set of commands that we can manipulate our program and data files in an on-line, interactive mode. Before discussing the types of commands normally found on a timesharing system, you need to understand the concepts of a work file and permanent files.

A user can have many permanent program and/or data files stored on disk. A *permanent file* means that the file is retained on disk until some later date when it is finally removed. But when the user works with a file in a timesharing mode, he must normally work with a temporary disk file known as a ***workfile.***

A workfile can be a program file which the user is creating, modifying, compiling, executing, etc. Or a workfile can be a data file that the user is forming. But the user is only allowed one workfile that he can do work on at any given time. To create a new workfile, the old workfile must either be saved as a permanent file or removed (purged) from the system. To work on a permanent file, it must be loaded as a temporary workfile before it is normally available for use. Thus, commands like **GET** or **LOAD** must be used to obtain a permanent file as a workfile. Commands like **SAVE** or **REMOVE** tell the system what to do with a workfile when we have finished working with it.

The operations, commands, and capabilities of a timesharing system vary from computer to computer. The concepts and many types of basic commands, however, are the same. We will, therefore, look at some of the more common types of commands that are found on most timesharing systems. If you have access to a timesharing system, you must consult the user's guide for the specific system that you have.

The first thing you must do is to learn the operation of the available terminal. The terminal may be a typewriter-like device or a cathode-ray tube (CRT) with a visual display screen. You must learn the operation of many of the control buttons used to position the cursor, if a CRT is used, and how to transmit a line of information. The *cursor* is usually a white spot on the screen of a CRT type terminal used to indicate where you are. As you type each character of data, the cursor will move along the screen to indicate your current position.

Transmitting a line of information to the computer is normally done by placing an ETX (end of text) character at the end of the line to be sent to the computer. Then you must move the cursor back to the beginning of the line or information to be transmitted. Finally, you depress a transmit key which causes the data to be sent to the computer. On typewriter devices, a return carriage key is used. The computer will normally respond with a message such as "DONE" or some control character such as a "#" to let you know it received the information.

Now you are ready to learn how to use the terminal in a timesharing mode. You must first sign on by giving your password or usercode, such as:

HELLO COLE/1234

Passwords or *usercodes* are assigned to each user and must be correctly entered for you to gain access to the system. If your sign-on is correct, the computer will respond that you are on line to a specific session of the timesharing system, and give the date and time. You can now proceed to use the timesharing commands to perform your desired operations. When you have finished and want to get off, you normally enter a command, such as **BYE** to sign off and get an accounting of time used.

Most major programming languages such as FORTRAN, COBOL, ALGOL (Algorithmic Language), PL/I (Programming Language One), and BASIC (Beginner's All-purpose Symbolic Instruction Code) are available for the writing and compiling of programs for timesharing systems. Before you can create a program or data file you must inform the system to make you a workfile. The **MAKE** or **CREATE** command is normally used to inform the system that you want to create a new workfile. You must give it a name and specify its type such as DATA, FORTRAN, etc. For example, to make a new workfile named EXER2 to be a FORTRAN type file you might enter:

MAKE EXER2 FORTRAN

The computer will normally respond that you have a workfile named EXER2 which is type FORTRAN.

Now you can proceed to enter your FORTRAN statements (program). You must include a *sequence number* before each statement. The system uses these sequence numbers to keep your statements in a specific order. Also, the sequence numbers are used to delete statements, add new statements, change existing statements, list portions of your workfile, etc. Therefore, these sequence numbers are of the utmost importance to the computer. The sequence numbers are normally kept in columns 73-80 of a statement, but are printed before the information in column 1 for your convenience. You should use some increment, such as by 10's, or even 100's, for the sequence numbers.

For example, say we want to type in a FORTRAN program to compute the sum of two numbers and print the results. Our typed program might be:

```
100        A  =  10.0
200        B  =  20.5
300        C  =  A  +  B
400        WRITE (6,99) A,B,C
500     99 FORMAT (1X,3F7.1)
600        STOP
700        END
```

To list a file or certain portions of a file, a **LIST** command is normally used. For example:

LIST
LIST 100-300, 500

The first example will cause the entire file to be listed. A hyphen is normally used between two sequence numbers to indicate a range of statements to be listed. A comma is normally used to indicate additional individual statements to be listed. Thus, the second example will list the statements with sequence numbers 100 through 300 and 500.

Say we made a mistake, or want to change portions of our program. To correct a statement, we simply type in its corresponding sequence number and the new statement. The computer will automatically replace the old statement with the new one. To add new statements to an existing file, we simply give the new statements a sequence number which will be merged into their proper order. For example, to add a statement between sequence numbers 200 and 300, we would assign a sequence number of 210 or 250 or whatever, as long as it falls between 200 and 300. This is why we assign sequence numbers in increments of 10 or 100, in order for us to be able to insert new statements in a file.

We can delete a statement from our file by typing the word **DELETE** followed by the statement number. For example,

DELETE 100

Or we can delete an entire range of statements by using the hyphen between the beginning sequence number and the ending sequence number. For example,

DELETE 200-400

would delete all statements with sequence numbers in the range of 200 through 400.

Many timesharing systems also allow us to edit our file. That is, we can replace certain characters with new ones, find the statements with certain characters, fix portions of a line, etc. These commands are normally **REPLACE, FIND, FIX, EDIT,** etc. For example,

REPLACE /NUM/ /RNUM/
FIND /FORMAT/

The slash or any special character is used as a delimiter. The first example replaces all occurrences of "NUM" with the new term "RNUM." The second example will find all statements with the character string "FORMAT." Edit-type commands of this nature vary greatly from system to system.

Now, say we are ready to compile and execute our program. Many systems use the command **RUN** to cause the workfile to be compiled and executed. Separate commands such as **COMPILE** and **EXECUTE** are available on many systems for just compiling or executing a program. If a program does not compile, the system will provide syntax error messages back to the terminal. The compute will list each statement in error, followed by the syntax message. You must correct the mistake and recompile the program. You may have your output results displayed back to the terminal or even send them to a line printer in case of voluminous output.

But what happens if you have not finished work on a program and it is time to go to class? You can save your workfile as a permanent file and later retrieve your permanent file to continue working where you left off. To save a workfile you simply say **SAVE**. This will save your workfile as a permanent file with the name you gave it when it was created. Perhaps you retrieved your permanent file as your workfile and want to save it under a new name if you are not sure of the results. Say we want to save it as EXER2TEST. Then you would type:

SAVE AS EXER2TEST

You could have many versions of the same program stored on separate, permanent files. This "eats up" a lot of disk storage. So when you want to delete a permanent file you use commands such as **REMOVE, UNSAVE,** or **PURGE** followed by the permanent file name. For example,

REMOVE EXER2TEST

This would purge the permanent file named EXER2TEST from disk. This means it is no longer available to you. To remove a workfile we simply type **REMOVE**. This will delete the workfile and allow you to create a new one.

Now, how do we retrieve a permanent file from disk and have it loaded as our workfile? The most common commands are **GET** or **LOAD** followed by the file name. Say we wish to load a permanent file named EXER2 to our workfile. We would say:

GET EXER2

or

LOAD EXER2

On some systems you use commands like **OLD** or **NEW** followed by a filename to indicate the retrieval of an existing permanent file or the creation of a new file.

There are many other things we could add about timesharing systems. The method of transmitting data will vary by the type of terminal used. Some systems allow abbreviations to be used for the commands to shorten our typing. The response time varies greatly from system to system. Usually, the larger the processing computer, the faster the response time to your input commands. This appendix is intended solely as an overview to the concepts and operations of a timesharing system. You must consult the user's guide, your instructor, or some other expert to learn the full range of capabilities available on your system. Once you succeed in learning how to use a terminal and the timesharing commands, you should find the direct communication with the computer an enjoyable and a very practical substitute for having to continuously submit a job from a source deck.

Appendix D: The Scientific Notation for Expressing Constants

Mathematicians, engineers, and other personnel engaged in scientific work often represent numeric quantities in a shorthand notation. This shorthand notation is referred to as a *Scientific Notation* for expressing numbers.

A constant is represented as a decimal value followed by a power of ten value such as 1.678×10^{15} or 42.3×10^{-12}. The decimal value is multiplied by the power of ten value to obtain the true value of the constant. The advantage of the scientific notation is the ease and brevity with which very large and very small numbers can be represented. For example $.871265 \times 10^{20}$ takes less space and time to write and is easier to read than

$$87126500000000000000.$$

Likewise, $.1265 \times 10^{-11}$ is shorter and easier to read than

$$.000000000001265$$

The power of ten value is known as the exponent part. Since a power of ten involves moving the decimal point one position, the exponent tells us how many positions left or right the decimal point in the number is to be moved to obtain the true value of the number. If the power of ten is a positive number, then the true value of the constant is found by moving the decimal point that many positions to the right. When the power of ten is a negative number then we have the reciprocal of a positive power. That is: $10^{-2} = \frac{1}{10^2}$. Therefore, a division by the power of ten value takes place. We must move the decimal point that many positions to the left to find the true value of the constant.

In FORTRAN there is no way to raise characters a half line to indicate a superscript or exponent value. We simply drop the value of 10 since we know that the exponent always indicates a power of ten. To indicate the end of the decimal value and the beginning of the exponent, we use the letter E. For example, 1.234E09. Thus, the letter E may be interpreted as "times 10 to the power of."

The exponent value must be a one- or two-digit integer. A sign may precede the integer exponent value to denote the value, e.g., E+10, E−2, E−03, E5. The exponent values from 0 to 9 may be represented by a single digit or as two digits, however one wishes. The maximum and minimum values (range) of the exponent varies by computer and must not be exceeded. For example, the largest positive exponent value on the IBM 360 and 370 is +75 and the smallest negative exponent is −78. The decimal portion of the constant must be formed in accordance with the rules of the basic real constants.

Examples of scientific formed constants with an exponent in FORTRAN are:

147.34E+1	(value of 1473.4)
23.5E7	(value of 235000000.)
.1E05	(value of 10000.)
1.0E0	(value of 1.0)
1.E−06	(value of .000001)
.1E−5	(value of .000001)

The true value of an exponential constant is determined as follows. If the exponent value following the E is positive, we must move the decimal point that many positions to the right. For example,

$$93.0E+06$$

means to move the decimal point 6 places to the right to give us the true value of 93000000. If the exponent value following the E is negative, we must move the decimal point that many positions to the left. For example,

$$1.5E-03$$

means move the decimal point 3 places to the left to give us the true value of .0015. If the exponent value following the E is zero (0 or 00), we do not move the decimal point at all. For example,

$$12.34E+00$$

represents the true value of 12.34.

Numeric input values may also be formed with an exponent. They can be read with the F format code. They may also be read with the E format code which is discussed in chapter 16.

Floating-point values in a program may be written with an exponent by using the E format code. Since some compilers print the output values written with format-free I/O in exponential form, let us briefly discuss this output form.

All output values written in exponential form appear as a normalized number. That is, the decimal number is written as a total fraction with the decimal point appearing before the first (leftmost) digit. The exponent value is adjusted according to represent this operation. A maximum of seven fractional digits are written followed by the exponent. The exponent part consists of the letter E followed by a sign and two digits. A blank is replaced for the sign if the exponent is positive.

Consider the following examples as they will be written by WATFIV and some other compilers using format-free output.

Value	*Output Form*
1.34	0.1340000E 01
.1234	0.1234000E 00
−12.34	−0.1234000E 02
.00123	0.1230000E-02
−.0003456	−0.3456000E-03

For mathematicians and engineers the scientific notation is a very common and widely used form for the expression of numerical constants.

Appendix E

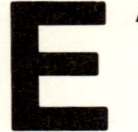

Job Control Language (JCL) Statements for Various Computers

1. JCL for an IBM OS System/360 or System/370 FORTRAN IV "G" Level compile and execute:

```
Card Columns
          1         2         3         4
12345678901234567890123456789012345678901234 5

//JWC JOB (110,318,200,01),'NAME',MSGLEVEL=1
//ST1 EXEC FORTGCLG
//FORT.SYSIN DD *
     (FORTRAN source program deck goes here)
//GO.SYSIN DD *
     (Data cards, if present, go here)
/*
```

NOTE: The format of the JOB card varies by installation.

2. JCL for an IBM DOS S/360 and S/370 Basic FORTRAN IV compile and execute:

```
Card Columns
          1         2
12345678901234567890
// JOB JOBNAME
// OPTION LINK
// EXEC FORTRAN
     (FORTRAN source program deck goes here)
/*
// EXEC LNKEDT
// EXEC
     (Data cards, if present, go here)
/*
/&
```

NOTE: The JOBNAME in the JOB card is a unique name assigned to the job by the programmer.

3. JCL for WATFIV compile and execute on IBM 360 and 370 systems:

```
Card Columns
         1         2         3         4
12345678901234567890123456789012345
//INITLS JOB (110,318,200,01,),'NAME',CLASS=W
$JOB
     (FORTRAN source program deck goes here)
$ENTRY
     (Data cards, if present, go here)
/*
```

NOTE 1: The format of the JOB card varies by installation.
NOTE 2: The end-of-file card (/*) must not be present in the job deck at some installations.

4. JCL for an NCR 200 FORTRAN compile and execute:

```
Card Columns
         1         2         3         4         5         6
123456789012345678901234567890123456789012345678901234567890

        C         NEXTDO FORT/I    ENN      EN1    2B1N032
        C         STOPRD
000000PPGMNAME    FORT/FN
*       P         AUTHORP COLE
*       F         50002P1108001
*       F         66232P01136              060
     (FORTRAN source program deck goes here)
END$
     (Data cards, if present, go here)
END$
```

NOTE: PGMNAME is a unique name assigned by the programmer.

5. JCL for a Honeywell 6000 FORTRAN IV "Y" Level compile and execute:

```
Card Columns
         1         2
123456789012345678901234
$       SNUMB    100
$       IDENT
$       USERID
$       OPTION   FORTRAN
$       LOWLOAD
$       FORTY    NDECK
     (FORTRAN source program deck goes here)
$       EXECUTE
$       DATA     05
     (Data cards, if present, go here)
$       ENDJOB
***EOF
```

NOTE 1: The $IDENT and $USERID cards vary by installation.
NOTE 2: File units 5, 6, and 7 may be used to reference the card reader, line printer, and card punch units respectively. File units 41, 42, and 43 may also be used to reference the card reader, line printer, and card punch units respectively.

6. Work Flow Language (WFL) for Burroughs 6700 and 7700 computer systems FORTRAN compile and execute:

```
Card Columns
          1         2
123456789012345678901234

?JOB "JOBNAME"
?USER Your user code
?BEGIN
?COMPILE PGMNAME FORTRAN
?DATA
FILE   5=CARDS,UNIT=READER
FILE   6=PRTR,UNIT=PRINTER
    (FORTRAN source program deck goes here)
?DATA CARDS
    (Data cards, if present, go here)
?END JOB
```

NOTE 1: The ? in card column 1 represents an invalid punch—normally a 1/2/3.
NOTE 2: The JOBNAME and PGMNAME represent a unique name supplied by the programmer. Your user code is substituted in the ?USER card.

7. JCL for CDC computers using SCOPE 341 operating system (CYBER 74, 7600, etc.) FORTRAN compile and execute:

```
Card Columns
          1         2
1234567890123456789012345678 9

COLE ,T77777,CM100000,P37.
FTN.
LGO.
?
    (FORTRAN source program goes here)
?
    (Data cards, if present, go here)
&
```

NOTE 1: The ? in column 1 represents a 7/8/9 punch.
NOTE 2: The & in column 1 represents a 6/7/8/9 punch.

Comparison of FORTRAN Compilers

Appendix **F**

This appendix compares the statements and features contained in various FORTRAN compilers. A "Yes" is placed in the column if the statement or feature is contained in the version of the FORTRAN compiler identified by the column heading. A "No" is placed in the column if the version of the indicated compiler does not contain the specifications. A "N/A" (not applicable) is placed in the column for those specifications that relate to a specific computer and, thus, not described in ANSI FORTRAN. An "UKN" is placed in those columns for specification which are unknown to the author.

Two standardized versions of FORTRAN compilers have been specified by the American National Standards Institute (ANSI). The first column contains the specifications for the standard Full FORTRAN version described in ANSI FORTRAN, X3.9-1966. The statements and features of ANSI Full FORTRAN are the standard statements and specifications described in this text. The second column contains the specifications for the standard Basic FORTRAN version described in ANSI Basic FORTRAN, X3.10-1966.

Many computer manufacturers have extended the capability of their FORTRAN compilers to provide additional features to ANSI FORTRAN. The remaining columns list those features contained in many of the FORTRAN compilers by the various computer manufacturers, specifications of which the author is familiar with. The IBM Basic FORTRAN IV contains the specifications described in IBM Basic FORTRAN Language C28-6629. The IBM G level FORTRAN compiler is as described in IBM System/360 and System/370 FORTRAN IV Language GC28-6515. Control Data Corporation 6000 series compiler specifications are as contained in CDC FORTRAN Extended Version 4 Reference Manual 60305600E. The Honeywell 6000 series FORTRAN Y compiler specifications are as described in their FORTRAN Manual BJ67. The Burroughs 6700 and 7000 compiler specifications are as described in their FORTRAN Reference Manual 5000458.

FORTRAN Feature	ANSI Standard Full	ANSI Standard Basic	IBM Basic FORTRAN IV	IBM G	WATFOR	WATFIV	CDC 6000	Honey-well 6000 Y	Bur-roughs 6700 7700	Your FORTRAN Compiler
Character Set:										
A-Z, 0-9	Yes	Yes	Yes	Yes	Yes	Yes	Yes	Yes	Yes	
+−*/=.,( )b	Yes	Yes	Yes	Yes	Yes	Yes	Yes	Yes	Yes	
$	Yes	No	Yes	Yes	Yes	Yes	Yes	Yes	Yes	
' (single quote)	No	No	Yes	Yes	Yes	Yes	Yes	Yes	Yes	
" (double quote)	No	No	No	No	No	No	No	Yes	Yes	
&	No	No	No	Yes	Yes	Yes	No	Yes	Yes	
Number of characters	47	46	48	49	49	49	48	51	50	
Variable name	1-6	1-5	1-6	1-6	1-6	1-6	1-7	1-8	1-6	
Statement label	1-99999	1-9999	1-32767	1-99999	1-99999	1-99999	1-99999	1-99999	1-99999	
Continuation lines	19	5	19	19	No limit	No limit	19	19	No limit	
Data types: (constants)										
Integer	Yes	Yes	Yes	Yes	Yes	Yes	Yes	Yes	Yes	
Real	Yes	Yes	Yes	Yes	Yes	Yes	Yes	Yes	Yes	
Double Precision	Yes	No	Yes	Yes	Yes	Yes	Yes	Yes	Yes	
Literal (Hollerith)	Yes	No	No	Yes	Yes	Yes	Yes	Yes	Yes	
Logical	Yes	No	No	Yes	Yes	Yes	Yes	Yes	Yes	
Complex	Yes	No	No	Yes	Yes	Yes	Yes	Yes	Yes	
Hexadecimal	No	No	No	Yes	Yes	Yes	Yes	No	Yes	
Octal	No	No	No	No	No	No	Yes	Yes	Yes	
Real Constant										
Basic real constant followed by E exponent	Yes	Yes	Yes	Yes	Yes	Yes	Yes	Yes	Yes	
Integer constant followed by a decimal exponent	Yes	No	Yes	Yes	No	Yes	No	No	Yes	
$ may be used as a letter in symbolic name	No	No	Yes	Yes	Yes	Yes	No	No	Yes	
Double-precision constant Real constant with "D" in place of "E"	Yes	No	Yes	Yes	No	Yes	Yes	Yes	Yes	
Size of computer word (in bytes)	N/A	N/A	4	4	4	4	10	6	6	
Relational expressions	Yes	No	No	Yes	Yes	Yes	Yes	Yes	Yes	
Logical Operators (.NOT.,.AND., .OR.)	Yes	No	No	Yes	Yes	Yes	Yes	Yes	Yes	
Mixed mode expressions	No	No	Yes	Yes	Yes	Yes	Yes	Yes	Yes	

FORTRAN Feature	ANSI Standard Full	ANSI Standard Basic	IBM Basic FORTRAN IV	IBM G	WATFOR	WATFIV	CDC 6000	Honey-well 6000 Y	Bur-roughs 6700 7700	Your FORTRAN Compiler
Double exponentiation permitted w/o parentheses (i.e., A**B**C)	No	No	Yes	Yes	Yes	Yes	Yes	Yes	Yes	
Statements:										
ASSIGN	Yes	No	No	Yes	Yes	Yes	Yes	Yes	Yes	
BACKSPACE	Yes	Yes	Yes	Yes	Yes	Yes	Yes	Yes	Yes	
BLOCKDATA	Yes	No	No	Yes	Yes	Yes	Yes	Yes	Yes	
CALL	Yes	Yes	Yes	Yes	Yes	Yes	Yes	Yes	Yes	
CHARACTER (type)	No	No	No	No	UKN	Yes	No	Yes	No	
COMMON										
Blank (unlabeled)	Yes	Yes	Yes	Yes	Yes	Yes	Yes	Yes	Yes	
Labeled (named)	Yes	No	No	Yes	Yes	Yes	Yes	Yes	Yes	
COMPLEX (type)	Yes	No	No	Yes	Yes	Yes	Yes	Yes	Yes	
CONTINUE	Yes	Yes	Yes	Yes	Yes	Yes	Yes	Yes	Yes	
DATA	Yes	No	No	Yes	Yes	Yes	Yes	Yes	Yes	
DEFINEFILE	N/A	N/A	Yes	Yes	No	Yes	No	No	No	
DIMENSION	Yes	Yes	Yes	Yes	Yes	Yes	Yes	Yes	Yes	
DO	Yes	Yes	Yes	Yes	Yes	Yes	Yes	Yes	Yes	
DOUBLE PRECISION	Yes	Yes	Yes	Yes	Yes	Yes	Yes	Yes	Yes	
END	Yes	Yes	Yes	Yes	Yes	Yes	Yes	Yes	Yes	
ENDFILE	Yes	Yes	Yes	Yes	Yes	Yes	Yes	Yes	Yes	
ENTRY	No	No	No	Yes	Yes	Yes	Yes	Yes	Yes	
EQUIVALENCE	Yes	Yes	Yes	Yes	Yes	Yes	Yes	Yes	Yes	
EXTERNAL	Yes	No	Yes	Yes	Yes	Yes	Yes	Yes	Yes	
FIND	N/A	N/A	Yes	Yes	No	Yes	No	No	Yes	
FORMAT	Yes	Yes	Yes	Yes	Yes	Yes	Yes	Yes	Yes	
FUNCTION	Yes	Yes	Yes	Yes	Yes	Yes	Yes	Yes	Yes	
GOTO										
Unconditional	Yes	Yes	Yes	Yes	Yes	Yes	Yes	Yes	Yes	
Computed	Yes	Yes	Yes	Yes	Yes	Yes	Yes	Yes	Yes	
Assigned	Yes	No	No	Yes	Yes	Yes	Yes	Yes	Yes	
IF										
Arithmetic	Yes	Yes	Yes	Yes	Yes	Yes	Yes	Yes	Yes	
Logical	Yes	No	No	Yes	Yes	Yes	Yes	Yes	Yes	
IMPLICIT	No	No	No	Yes	Yes	Yes	Yes	Yes	Yes	
INTEGER (type)	Yes	No	Yes	Yes	Yes	Yes	Yes	Yes	Yes	
LOGICAL (type)	Yes	No	No	Yes	Yes	Yes	Yes	Yes	Yes	
NAMELIST	No	No	No	Yes	No	Yes	Yes	Yes	Yes	
PAUSE	Yes	Yes	Yes	Yes	Yes	Yes	Yes	Yes	Yes	
PRINT	No	No	No	Yes	Yes	Yes	Yes	Yes	Yes	
PUNCH	No	No	No	Yes	Yes	Yes	Yes	Yes	Yes	
READ										
formatted	Yes	Yes	Yes	Yes	Yes	Yes	Yes	Yes	Yes	
unformatted	Yes	Yes	Yes	Yes	Yes	Yes	Yes	Yes	Yes	
format-free	No	No	No	No	Yes	Yes	Yes	No	Yes	
specific	No	No	No	Yes	Yes	Yes	Yes	Yes	Yes	
direct access	N/A	N/A	Yes	Yes	No	Yes	Yes	Yes	Yes	
REAL (type)	Yes	No	Yes	Yes	Yes	Yes	Yes	Yes	Yes	
RETURN										
standard	Yes	Yes	Yes	Yes	Yes	Yes	Yes	Yes	Yes	
return n option	No	No	No	Yes	Yes	Yes	Yes	Yes	Yes	
REWIND	Yes	Yes	Yes	Yes	Yes	Yes	Yes	Yes	Yes	

Appendixes

FORTRAN Feature	ANSI Standard Full	ANSI Standard Basic	IBM Basic FORTRAN IV	IBM G	WATFOR	WATFIV	CDC 6000	Honeywell 6000 Y	Burroughs 6700 7700	Your FORTRAN Compiler
STOP	Yes	Yes	Yes	Yes	Yes	Yes	Yes	Yes	Yes	
SUBROUTINE	Yes	Yes	Yes	Yes	Yes	Yes	Yes	Yes	Yes	
WRITE										
formatted	Yes	Yes	Yes	Yes	Yes	Yes	Yes	Yes	Yes	
unformatted	Yes	Yes	Yes	Yes	Yes	Yes	Yes	Yes	Yes	
format-free	No	No	No	No	Yes (PRINT)	Yes	Yes	No	Yes	
direct access	N/A	N/A	Yes	Yes	No	Yes	Yes	Yes	Yes	
Assignment Statement										
var = arith exp	Yes	Yes	Yes	Yes	Yes	Yes	Yes	Yes	Yes	
var = logical exp	Yes	No	No	Yes	Yes	Yes	Yes	Yes	Yes	
extended Asgnmt Stmt. (multiple replacement)	No	No	No	No	Yes	Yes	Yes	No	No	
Character Asgnmt Stmt.	No	No	No	No	Yes	Yes	Yes	Yes	Yes	
DO—extended range	Yes	No	Yes	Yes	No	No	Yes	Yes	Yes	
END = option in READ	No	No	No	Yes	Yes	Yes	Yes	Yes	Yes	
ERR = option in READ	No	No	No	Yes	Yes	Yes	Yes	Yes	Yes	
Memory-to-Memory I/O formatting	No	No	No	No	UKN	Yes	Yes EN-CODE DE-CODE	Yes EN-CODE DE-CODE	Yes Array I/O	
Multiple statements per line	No	No	No	No	No	Yes (;)	Yes ($)	Yes (;)	Yes (;)	
Format Codes:										
A (Alphabetic)	Yes	No	Yes	Yes	Yes	Yes	Yes	Yes	Yes	
C (Character)	No	No	No	No	No	No	No	No	Yes	
D (Dble prec FP)	Yes	No	Yes	Yes	Yes	Yes	Yes	Yes	Yes	
E (Single prec FP)	Yes	Yes	Yes	Yes	Yes	Yes	Yes	Yes	Yes	
F (Single prec FP)	Yes	Yes	Yes	Yes	Yes	Yes	Yes	Yes	Yes	
G (Generalized)	Yes	No	No	Yes	Yes	Yes	Yes	Yes	Yes	
H (Hollerith)	Yes	Yes	Yes	Yes	Yes	Yes	Yes	Yes	Yes	
' (single quote)	No	No	Yes	Yes	Yes	Yes	No(*)	Yes	Yes	
I (Integer)	Yes	Yes	Yes	Yes	Yes	Yes	Yes	Yes	Yes	
J (Integer)	No	No	No	No	No	No	No	No	Yes	
L (Logical)	Yes	No	No	Yes	Yes	Yes	Yes	Yes	Yes	
O (Octal)	No	No	No	No	No	No	Yes	Yes	Yes	
R (Right just. alpha)	No	No	No	No	No	No	Yes	Yes	(see C)	
T (Tab)	No	No	Yes	Yes	Yes	Yes	Yes	Yes	Yes	
X (Blank)	Yes	Yes	Yes	Yes	Yes	Yes	Yes	Yes	Yes	
Z (Hexadecimal)	No	No	No	Yes	Yes	Yes	Yes	No	Yes	
P scale factor	Yes	No	Yes	Yes	Yes	Yes	Yes	Yes	Yes	

FORTRAN Feature	ANSI Standard Full	ANSI Standard Basic	IBM Basic FORTRAN IV	IBM G	WATFOR	WATFIV	CDC 6000	Honeywell 6000 Y	Burroughs 6700 7700	Your FORTRAN Compiler
Carriage Control Opns for printer output (space before printing)	Yes	No	Yes	Yes	Yes	Yes	Yes	Yes	Yes	
blank or 1X : 1 line	Yes	No	Yes	Yes	Yes	Yes	Yes	Yes	Yes	
0 : 2 lines	Yes	No	Yes	Yes	Yes	Yes	Yes	Yes	Yes	
1 : new page	Yes	No	Yes	Yes	Yes	Yes	Yes	Yes	Yes	
+ : no advance	Yes	No	Yes	Yes	Yes	Yes	Yes	Yes	Yes	
Format group repetition levels	2	1	1	2	2	2	2	2	2	
Object time format specifications	Yes	No	No	Yes	Yes	Yes	Yes	Yes	Yes	
Number of array dimensions	3	2	3	7	7	7	3	7	31	
Dimensions may be specified in type statements	Yes	No	Yes	Yes	Yes	Yes	Yes	Yes	Yes	
Dimensions may be specified in COMMON	Yes	No	Yes	Yes	Yes	Yes	Yes	Yes	Yes	
Object time dimensions in subprograms	Yes	No	No	Yes	Yes	Yes	Yes	Yes	Yes	
Generalized subscript form	No	No	No	Yes	Yes	Yes	Yes	Yes	Yes	
Subprogram types:										
FUNCTION	Yes	Yes	Yes	Yes	Yes	Yes	Yes	Yes	Yes	
Statement function	Yes	Yes	Yes	Yes	Yes	Yes	Yes	Yes	Yes	
Subroutine	Yes	Yes	Yes	Yes	Yes	Yes	Yes	Yes	Yes	
Type specification in FUNCTION statement	Yes	No	Yes	Yes	Yes	Yes	Yes	Yes	Yes	
FUNCTION subprogram may define or re-define its dummy arguments	Yes	No	Yes	Yes	Yes	Yes	Yes	Yes	Yes	
Statement function must precede first executable statement	Yes	Yes	Yes	Yes	Yes	Yes	Yes	Yes	Yes	
Actual arguments may be:										
numeric constant	Yes	Yes	Yes	Yes	Yes	Yes	Yes	Yes	Yes	
numeric variable/array	Yes	Yes	Yes	Yes	Yes	Yes	Yes	Yes	Yes	
literal (Hollerith) constant	Yes	No	No	Yes	Yes	Yes	Yes	Yes	Yes	
logical constant	Yes	No	No	Yes	Yes	Yes	Yes	Yes	Yes	
logical variable/array	Yes	No	No	Yes	Yes	Yes	Yes	Yes	Yes	
external subprogram name	Yes	No	Yes	Yes	Yes	Yes	Yes	Yes	Yes	

answers to review questions and odd-numbered exercises

CHAPTER 1

Answers to Review Questions

1. People have mixed emotions about computers. Some like them and think they are great. Others consider them a jinx—always messing things up.
2. It is simply a machine that must be told everything that it does; it has occasional breakdowns or failures; and it can not solve a problem unless the problem can be defined in explicit steps.
3. Speed, accuracy, and economy
4. There are so many application areas. Just a few of them are: Banking, Accounting, Payroll, Tax Returns, Military, Criminology, Hospitals, Numerical Analysis, Universities.
5. Control unit, arithmetic-logic unit, memory unit, input units, and output units.
6. Hollerith 80-column card
7. To display or write data onto specific media
8. Line printer, card punch, magnetic tape unit, magnetic disk unit, paper tape punch, remote visual display units, typewriters, etc.
9. 132
10. A specific location in the memory unit in which an item of data can be stored.
11. To perform calculations and decision-making operations
12. Control unit, arithmetic-logic unit, and memory unit
13. To direct all the other components of the system and to execute the program instructions.
14. A set of instructions called a program is stored in the memory unit. It provides the flexibility needed in general-purpose computers to solve a wide variety of problems. That is, different programs for different problems can be stored and executed.
15. No. To be reexecuted, they must be reloaded.
16. Problem analysis and definition, planning the solution, describing the problem solution, implementing the solution, and documenting the solution.
17. Define the problem, identify the variables involved, determine the relationship between the variables, develop a model, test the model by implementing the solution, and check the results to see if they are correct.
18. To prepare a visual plan for solving the problem and the program logic.
19. It denotes the specific type of computer operation required in the problem solution.
20. Terminal, Processing, Input/Output, Decision, Flowlines, Logic Connector, and Annotation. See the text for the shape of these symbols.
21. Program
22. Compiler
23. FORTRAN—scientific applications, and COBOL—business applications.

24. Job Control Cards
25. This is the stage at which one supplies test data in the execution of the program to test the logic. If the output results are not correct, then the program must be corrected and rerun.
26. It is no longer available for execution unless it is reloaded. In essence, the computer eliminates the program from its list of jobs to be run.
27. To provide an orderly approach to writing programs which are easier to read and to debug.
28. Level one is the statement of the problem—to compute one's paycheck. Level two consists of the main tasks of computing gross pay, deductions, and net pay. Level three shows the subtasks involved under each main task, such as computing regular pay, overtime pay, FICA deductions, withholding tax, other miscellaneous deductions, and netpay.

CHAPTER 2

Answers to Review Questions

1. **FOR**mula **TRAN**slation or **FOR**mula **TRAN**slator
2. 1957
3. Input/Output, Arithmetic Assignment, and Logical and Control
4. 47; 26 letters, 10 digits, and 11 special characters
5. To identify the operation to be performed by the computer
6. As the first word in each statement
7. A known numeric value that represents a fixed quantity
8. Integer and Real
9. Whole number, i.e., one without a fraction
10. Floating-point
11. 7
12. A symbolic name that refers to the contents of a storage location.
13. 1 to 6 characters
14. Alphabetic, alphabetic or numeric
15. I through N
16. A through H, and O through Z
17. +, −, *, /, and **
18. Assign, replace, or store
19. True
20. 1 through 5
21. 1 to 99999
22. True
23. 6
24. Blank or 0 (Zero)
25. 7, 72
26. 73 through 80
27. C, 1
28. True
29. O, I, Z, S, and B
30. They may be included or omitted as one desires. They are normally included to improve readability.

Answers to Odd-numbered Exercises

1.
   1. Integer
   2. Real
   3. Real
   4. Real
   5. Real
   6. Real
   7. Real
   8. Integer
   9. Integer
   10. Real

3.
   1. Exceeds 6 characters
   2. Begins with a numeric digit
   3. Contains a special character
   4. Exceeds 6 characters
   5. Contains a special character
   6. Contains a special character

5.
   1. Real
   2. Integer
   3. Real
   4. Integer
   5. Real
   6. Integer
   7. Real
   8. Integer
   9. Real
   10. Integer

7.
   1. Contains a comma
   2. Sign must precede the number
   3. Exceeds the largest allowed integer value
   4. Contains two decimal points
   5. Sign must be given before decimal point
   6. Contains a special character

CHAPTER 3

Answers to Review Questions

1. To identify the location and form of the output results on a printed line.
2. To describe the function of various routines and statements and what they are doing so that the reader can understand the program.
3. FORMAT
4. To describe the form and length of each data field to be read or written.
5. True
6. To control the movement of the paper
7. False, you must use a Integer (I) format code.
8. STOP
9. True
10. False
11. True
12. False
13. True

Answers to Odd-numbered Exercises

1.
   a) **READ (5,10) HOURS, RATE**   Missing comma
   b) **10 FORMAT (F5.2, 3X, F4.1)**   Missing parenthesis
   c) **GPAY = HOURS * RATE**   Invalid multiplication operator
   d) **WRITE (6, 33) HOURS, RATE, GPAY**   Invalid output unit number and missing parenthesis

Answers to Review Questions and Odd-numbered Exercises 397

3. A with F5.2, I with I3, J with I1, and B with F3.1
5. A should not be read with an I5
7. First, addition because of the parenthesis—(D + E)
   Second, division—C / (D + E)
   Third, multiplication—C / (D + E) * S
   Fourth, subtraction—B − C / (D + E) * C
9. A with F4.1, B with F5.1, C with F6.2, and I with I3

CHAPTER 4

Answers to Review Questions

1. Desk-checking (or bench-checking)
2. FORTRAN statements—to provide the source program. Data—to supply the input values. JCL—to specify which computer resources are needed and used to process the job.
3. False, they are placed after the program.
4. Batch processing involves reading and processing your program with a Job deck of cards. Timesharing involves using an on-line terminal to enter and process your program.
5. It is translated into machine language (object code) which can be executed by the computer.
6. True
7. True
8. Interspersed within the program normally after each statement in error and at the end of the program.
9. Read the diagnostic message very carefully to understand what the compiler is trying to tell you. Look for any information the compiler supplies about the location of the syntax error. Reread the rules for the construction of the statement or the language elements used. Check the syntax error count provided by the compiler. Examine each character around the error very closely.
10. The machine language produced in the compilation phase from the source program is executed by the computer, and the input data processed and output results produced.
11. Perform hand calculations of some simple test values and check these results against the computer results.
12. Play computer. Follow the program manually and describe what the computer would do.
13. Echo check the input data. Write out intermediate results calculated at various points in the program.
14. True
15. True
16. Use a control statement (IF) to test for a special condition such as a special data card used as a trailer card.
17. To cause a branch to another statement in the program.
18. To test various variables for certain conditions and provide a decision-making capability.
19. .LT. less than ($<$); .LE. less than or equal to ($\leqslant$); .GT. greater than ($>$); .GE. greater than or equal to ($\geqslant$); .EQ. equal to (=); .NE. not equal to ($\neq$).
20. The next executable statement after the Logical IF.

Answers to Odd-numbered Exercises

1. a) READ (5,99) X,Y   Parentheses missing
   b) 99 FORMAT (2F5.2)   Statement number missing
   c) WRITE (6, 98) X,Y   Incorrect file unit for WRITE
   d) XYZTOT   = X + Y = Z   Variable too long
   e) 79 FORMAT (3X,F5.2,I2)   Incomplete F format code specifications

    f)   **GOTO 91357**   Statement number too large
    g)   **IF (N .LT. 10) GOTO 30**   Parentheses missing
    h)   **STOP**   Invalid syntax for STOP statement
    i)   **WRITE (6,97) A,B**   Erroneous comma

3.

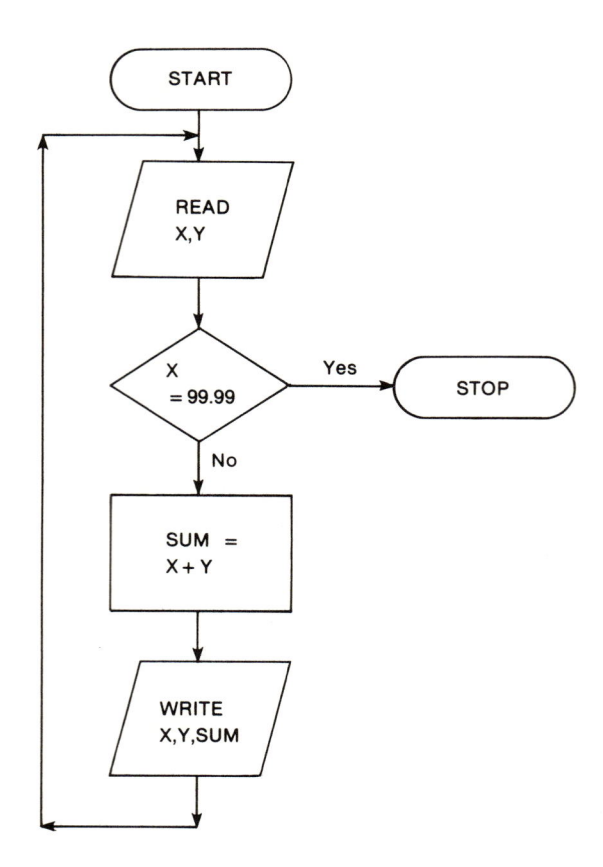

```
                START
                  │
                  ▼
                READ
                X,Y
                  │
       ┌──────────▼──────────┐
       │                     │
       │    X = 99.99    ────┼──── Yes ──→  STOP
       │                     │
       └──────────┬──────────┘
                  │ No
                  ▼
                SUM =
                X + Y
                  │
                  ▼
                WRITE
                X,Y,SUM
                  │
                  └────────→ (loop back to READ)
```

CHAPTER 5

Answers to Review Questions

1. It is generally the most frequently used statement in a program.
2. Variables, constants, and arithmetic operators
3. ** highest
   * / next highest
   + – lowest
4. To override the hierarchy precedence of the arithmetic operators
5. *a)* Two adjacent arithmetic operators: Z/(−3)
   b) Missing multiplication operators: 5*X + 7*Y
   c) Unbalanced parentheses: (A + B*(C − D))
   d) Missing operand for exponentiation operator: 6 − X**2
6. The type of variable to which the resulting value is assigned
7. Integer and real
8. Variable = expression. A = B + C − 10.0
9. Store or to replace
10. Only one

Answers to Odd-numbered Exercises

1.
   a) A = 17.0
   b) K = 3
   c) I = 8
   d) R = 48.4
   e) J = −27
3.
   a) Valid
   b) Invalid. DISTAN = RATE * TIME
   c) Valid
   d) Valid
   e) Invalid. CSQD = A * A + B * B
   f) Invalid. GRSPAY = 210.75
   g) Invalid. C = A * (−B)
   h) Valid
   i) Valid
   j) Invalid. T = A * B / Z

CHAPTER 6

Answers to Review Questions

1. To transmit input data from an external medium to memory and to transmit data from memory to an output medium.
2. Record
3. Field
4. File
5. READ
6. WRITE
7. To specify the form, length, and location of data values to be read or written.
8. Integer (I)
9. Real or Floating-point (F)
10. Blank (X)
11. Width
12. To output data to the line printer
13. True
14. False, they are never executed.
15. True
16. False
17. True
18. True
19. True
20. No
21. False, they are treated as zeroes.
22. Asterisks (*)
23. Blanks
24. True
25. True
26. True
27. False, they are right-justified.

28. Five
29. Location of the field, type of data, size or width of the field, and location of the decimal point.
30. As the first format specification code
31. Same record
32. New or next record

Answers to Odd-numbered Exercises

1.   **READ (5,95) J,K,L**
   95 **FORMAT (I3,2X,I3,1X,I3)**
3.   **WRITE (6,89) J,K,L**
   89 **FORMAT (1H0,2X,I3,4X,I3,2X,I3)**
5. I2
   I4
   I3
   F6.2
   F6.3
   F6.2
   F6.0
7. −567
   ****
   ßßß0 (bbb0)
   b21
   .00
   0.0
   bb49.93
   bb−49.9
   bbbb50.
   ****
   b49.9250
   0.6347
   .635
   b0.6
   b1.
   −0.6347
   b−0.6347
   −.6347
   b−0.634700
9. a) 11
   b) 7
11. Four

CHAPTER 7

Answers to Review Questions

1. A character string preceded by the Hollerith (H) format code which is preceded by an integer constant specifying the number of characters in the string.
2. Report titles, page and column headings, and identifying literals in the data output.
3. False
4. A syntax error occurs

5. False
6. To shorten the format specifications
7. Parentheses
8. Two
9. True
10. Printer Layout Form

Answers to Odd-numbered Exercises

1. *a)*   **WRITE (6,99)**
       99 **FORMAT (1H1,49X,27HSTUDENT REGISTRATION ROSTER)**
   b)   **WRITE (6,98)**
       98 **FORMAT (1H0,15HEMPLOYEE NUMBER,4X,5HHOURS,3X,4HRATE)**
   c)   **WRITE (6,97) X, XSQD**
       97 **FORMAT (1H ,4HX = ,2X,F4.1,4X,8HX − SQD = ,2X,F5.1)**
3. *a)*   37 **FORMAT (2I2, 2(3X,I3))**
   b)   86 **FORMAT (3 (F6.2,I3))**
   c)   97 **FORMAT (5X,2I6, 2(F5.2,2X))**
   d)   94 **FORMAT (2 (I4,2 (I3,2X), 5X))**
   e)   89 **FORMAT (1H0, 3(I3,3X), F8.2)**

CHAPTER 8

Answers to Review Questions

1. Terminating the program, Decision making, Branching/Looping
2. An unconditional branch always causes a transfer of control to the statement number. A condition branch is taken only when the condition is true.
3. Input-bound loops, Counter loops, Iterative loops
4. By some end-of-file card (i.e., JCL or trailer data card)
5. Initialization, increment, and testing
6. When the test condition is met
7. Header card
8. Repeat Until or While DO
9. When the test condition is met
10. .AND., .OR., and .NOT.
11. .AND.
12. .OR.
13. .NOT.—first, .AND.—second, .OR.—last
14. True
15. Three
16. Integer variable
17. Editing

Answers to Odd-numbered Exercises

1. **GOTO 11**
3. *a)*   **IF (C1 .GT. C2 .OR. C1 .GT. C3 .OR. C1 .GT. C4) GOTO 76**
   b)   **IF (R .LE. SUM / 2.) READ (5,99) R**

  c) IF (ROOT .EQ. 0.0) STOP
  d) IF (A .EQ. 10.0 .AND. B .GT. 0.0) GOTO 41
  e) IF ((X .EQ. 0.0 .OR. Y .EQ. 0.0) .AND. Z .NE. 0.0) KTR = KTR + 1
5. *a)* IF (3.5 * A + C) 10,20,30
 b) IF (N / 2 * 2 – N) 12, 33, 12
 c) IF (SUM) 46, 46, 18
7. GOTO (32,47,15,53,27), M

9. | | | |
|---|---|---|
| *a)* | GOTO (2,18,5,162), KNT | Missing comma |
| *b)* | IF (J .LT. 2 * M – 3) GOTO 42 | Missing arithmetic operator |
| *c)* | IF (SUM .LT. 10.0) GOTO 13 | Invalid syntax |
| *d)* | GOTO (3,1,16), KODE | Invalid type variable |
| *e)* | IF (A .EQ. B) C = A | Missing periods on relational operator |
| *f)* | IF (KNT) 10,11,10 | Missing parentheses |
| *g)* | GOTO 19999 | Statement label too large |
| *h)* | IF (N .EQ. 1 .AND. L .GT. 5) STOP | Missing periods on logical operator |

CHAPTER 9

Answers to Review Questions

1. To execute a series of statements one or more times
2. Initialization—to supply an initial value to the index variable. Increment—to supply the increment value to the index variable each time through the loop. Test—to supply the test value against which the index variable is tested at the end of each loop.
3. One
4. When the loop is executed the specified number of times or if a control statement causes a branch out of the loop.
5. Provides greater flexibility
6. It is a rectangle divided into three areas to show the indexing parameters used.
7. True
8. When a transfer is made out of the loop prematurely. If the loop is completed the specified number of times the index variable is undefined.
9. CONTINUE
10. Loops contained within another loop

Answers to Odd-numbered Exercises

1. **DO 56 II = 5,500,5**
3. *a)* 8; 1 through 8
 b) 2; 2 and 5
 c) 3; 4, 6, and 8
5. *a)* 12
 b) 6

7. | | | |
|---|---|---|
| *a)* | DO 20 K = 1, N, 1 | Illegal test variable |
| *b)* | DO 30 I = 1, 10, 1 | Extra comma |
| *c)* | DO 40 ICOUNT = 1,L | Variable too long |
| *d)* | DO 50 J = 1, N, K | Invalid increment parameter |
| *e)* | DO 60 M = 1, 20, 1 | Invalid increment parameter |

CHAPTER 10

Answers to Review Questions

1. A collection of related items stored in adjacent storage locations under a common name.
2. Provides a flexible and powerful tool which makes long, unwieldy problems feasible and easy to solve.
3. DIMENSION
4. Like any other symbolic name, i.e., 1 to 6 characters; the first must be an alphabetic letter.
5. True
6. False, they must be constants.
7. False
8. Subscript
9. True
10. False
11. Subscripted variable, array name, or implied DO loop
12. False
13. True
14. True
15. False
16. Searching
17. Sorting
18. It is given in a DIMENSION statement.
19. Yes
20. Provides a powerful means of indexing one's way through an array.
21. Prior to the executable statements

Answers to Odd-numbered Exercises

1.
   a) **DIMENSION GROUP (12)**
   b) **DIMENSION LISTX (40)**
   c) **DIMENSION LISTA (60), LISTB (50)**
3.
   a) **DIMENSION A(10), B(10)**   Omitted comma
   b) **READ (5,99) (LIST(I),I = 1,10)**   Invalid subscript/index variable
   c) **READ (5,89) (A(I),I = 1,7), (B(I),I = 1,5)**   Missing comma
   d) **WRITE (6,97) (GROUP(J),J = 1,8)**   Missing parenthesis
   e) **WRITE (6,96) (HOLD(2*N),N = 1,K,2)**   Missing comma
5.
   a) **READ (5,99) (ARRAY(I),I = 1,11)**
      99 FORMAT (11F5.2)
   b) **WRITE (6,98) (ARRAY(I),I = 1,10)**
      98 FORMAT (1H0,10F7.2)
7.
   a) 5
   b) 1
   c) 7
9.
   a)   **DO 10 I = 1,50**
            A(I) = 0.0
        10 CONTINUE
   b)   **ISUM = 0**
          DO 10 I = 1,16,2
            ISUM = ISUM + L16(I)
        10 CONTINUE

c)
```
    XHLD = X20(1)
    DO 10 I = 2,20
        X20(I – 1) = X20(I)
10  CONTINUE
    X20(20) = XHLD
```

CHAPTER 11

Answers to Review Questions

1. To provide a more convenient means of handling multiple groups of related items, i.e., a matrixlike arrangement.
2. With two subscripts which provide the row-column coordinates. The first one refers to the desired row element while the second subscript refers to the desired column element.
3. By specifying two levels of dimension in the DIMENSION statement
4. By multiplying the size of each dimension specification by the others.
5. True
6. True
7. When an additional level of dimension is added to a matrix arrangement of related data.

Answers to Odd-numbered Exercises

1. *a)* **DIMENSION MATRIX (7,5)**
   b) **DIMENSION TABLE (20,11)**
   c) **DIMENSION CLASS (6,4,4)**
3. *a)* 6
   b) 10
5. *a)*
```
    READ (5,99)((X(I,J),J = 1,10),I = 1,6)
 99 FORMAT (10F6.2)
```
   b)
```
    WRITE (6,98)((X(I,J),J = 1,8),I = 1,6)
 98 FORMAT (1H0,8F10.2)
```
7. *a)*
```
    DO 20 I = 1,9
        DO 10 J = 1,30
            K(I,J) = 1
10      CONTINUE
20  CONTINUE
```
   b)
```
    NNSUM = 0
    DO 20 I = 1,5,2
        DO 10 J = 1,10
            NNSUM = NNSUM + NN(I,J)
10      CONTINUE
20  CONTINUE
```
   c)
```
    DO 20 J = 1,7
        DO 10 I = 1,50
            IF (D(I,J) .EQ. 76.5) GOTO 777
10      CONTINUE
20  CONTINUE
```

CHAPTER 12

Answers to Review Questions

1. False
2. True
3. True
4. False, the excess leftmost characters are lost.
5. True
6. To override basic default specifications and to make certain requests to the FORTRAN compiler.
7. END
8. To initialize variables/array elements at compile time
9. To override the default type of a variable according to its first letter
10. To equate multiple variable names to the same storage location
11. Prior to the executable statements

Answers to Odd-numbered Exercises

1. *a)* (4A4,A1)
   b) (A2)
   c) (2A3) or (A4,A2)
   d) (3A4,A2)
   e) (4A4,A2)
3. DATA A,X,M / 3.75, 8.6, 12 /
   or
   DATA A / 3.75 /,X / 8.6 / , M / 12 /
5. INTEGER A, B, X(100)
7. REAL I,J,INTRST, LIST2(7)
9. EQUIVALENCE (L,L1,L2,LX)
11. *a)* DATA A / 5.5 /, B / 6.3/ — Slashes should be around constants, not variables
    b) REAL M,N
       INTEGER X,Y — Must be two separate statements
    c) INTEGER XXARAY (20) — Array name too long
    d) INTEGER YARRAY (100) — ANS FORTRAN does not permit initialization of variables or arrays on type statements
    e) REAL I,J,K — Initialization not allowed
    f) REAL NSUM, NTOTAL — Invalid syntax, keyword EXPLICIT does not exist
    g) EQUIVALENCE (A,AX,ZA) — Equated variables must be enclosed in parentheses

CHAPTER 13

Answers to Review Questions

1. Reduction in amount of required work, incorporation of modular programming technique, ease in debugging, simplification in the division of work, reduction in memory required, use of library subprograms which are tested and debugged.
2. By the SUBROUTINE statement
3. *a)* Subprogram (SUBROUTINE or FUNCTION) statement
   b) At least one executable statement

c) RETURN
   d) END
4. RETURN
5. By using its name in the CALL statement
6. Arguments in a parameter list
7. The next executable statement after the CALL
8. There is no minimum. No arguments must be passed.
9. Number, order, type, and length
10. To assign variables/arrays to a common block of memory to be shared between programs units.
11. Blank (unlabeled) and Labeled (named)
12. To allow one to initialize at compile time those variables/array elements assigned to Labeled Common areas.
13. No

Answers to Odd-numbered Exercises

1. *a)* Unequal number of arguments in the two parameter lists.
   b) The first argument in the CALL statements does not agree in type with the first argument in the SUBROUTINE statement.
   c) The name of the subroutine in the CALL is not the same name in the SUBROUTINE statement.
3. *a)* **SUBROUTINE PAGE**
   b) **SUBROUTINE PAGELN (LINCNT)**
5. *a)* **CALL SUBRTN**
   b) **CALL SUB1 (LINCNT)**
7. *a)* 5
   b) 14
   c) 42
   d) 101
9. **BLOCKDATA**
   COMMON /GROUP1/ A,X,J
   DATA A /5.3/, X /8.7/, J/3/
   END

CHAPTER 14

Answers to Review Questions

1. True
2. False, software development.
3. True
4. Programs easier to read and comprehend, reduction in program logic errors, increased programmer productivity, and reduction in program maintenance.
5. Sequence, If-Then, If-Then-Else, While-Do, Repeat-Until, and Select-case
6. False
7. True
8. True
9. False
10. True
11. Logical and system errors

12. Desk checking
13. Echo check
14. Checkpoint
15. Exponent overflow and underflow, divide by zero, invalid operator, invalid subscript, and various I/O errors.
16. Failure to initialize required variables, failure to check for input errors and all possible conditions, misspelled variables, order of computations, integer division, truncation errors, failure to read the input data correctly, and failure to check for system-caused errors.

CHAPTER 15

Answers to Review Questions

1. When a single calculated value is needed
2. By the FUNCTION statement
3. By using its name in an arithmetic expression such as in an Assignment or IF statement.
4. One
5. One
6. Via the Function subprogram name
7. The type of the Function subprogram name
8. By the Assignment statement
9. Back to the same statement that invoked the Function subprogram
10. Number, order, type, and length
11. To perform complex calculations that are repeated many times in a program.
12. Yes, at least one.
13. A statement function is a one statement subprogram belonging to the program in which it appears. A Function subprogram is a separate program unit.
14. After the specification statements and before the first executable statement.
15. By using its name in an arithmetic expression, i.e., the same as a Function subprogram.
16. Back to the same statement that invoked the statement function.
17. To identify subprogram names passed as arguments in a parameter list.
18. To provide the dimension specifications of an array at execution time.
19. To provide multiple entry points into a Function or Subroutine subprogram.

Answers to Odd-numbered Exercises

1.   a)   FUNCTION FUN1 (A,I,P)
    b)   FUNCTION FY (Y)
    c)   FUNCTION CALC (X,Y,TOTAL)
3.   a)   Invalid argument
    b)   Name used as both a variable and a Function subprogram
    c)   Invalid number of arguments
    d)   Omission of the closing parenthesis
    e)   No argument list
    f)   Invalid type of argument
5.   a)   3.0
    b)   9.0
    c)   27.0

7. *a)* Subroutine
   b) Function
   c) Function
   d) Subroutine

CHAPTER 16

Answers to Review Questions

1. Numeric fields represented in exponential form
2. Integer or real constant followed by an exponential value. The exponential value consists of the letter E followed by a one- or two-digit integer constant representing a power of ten. A minus sign may precede the constant when needed to represent a negative power of ten. A plus sign to represent a positive exponent is optional.
3. A total fractional number consisting of the number of digits as given in the precision portion of the format code, followed by the letter E, followed by a space for a positive exponent or a minus sign for a negative exponent, and ending with a two-digit integer constant for the exponent value.
4. By the constant having more than 7 digits or by using the letter D in the exponential portion of the constant.
5. DOUBLE PRECISION
6. Double precision numeric values expressed in exponential form
7. .TRUE. and .FALSE.
8. LOGICAL
9. L
10. Real and imaginary
11. Two real constants separated by a comma and enclosed within a set of parentheses.
12. COMPLEX
13. F, D, or E. But two format codes must always be used for each complex variable.
14. Numeric fields expressed in decimal (F) or exponential (E) form.
15. True
16. To override and move the position of the decimal point in the data.
17. The letter P preceding the format code and a positive or negative integer constant preceding the letter P to specify which way to move the decimal point.

Answers to Odd-numbered Exercises

1. *a)* E6.1
   b) E6.0
   c) E7.2
   d) E3.0
3. *a)* b̸0.123Eb̸02
   b) −0.12345Eb̸02
   c) b̸b̸0.123450Eb̸03
   d) b̸b̸b̸0.12346Eb̸03
   e) b̸b̸−0.12346Eb̸02
   f) b̸−0.1200E−02

5. a) .TRUE.
   b) .TRUE.
   c) .TRUE.
   d) .TRUE.
   e) .FALSE.
7. a) ␣␣99.␣␣␣␣
   b) ␣0.10␣␣␣␣
   c) ␣0.146E−07
   d) ␣−57.4␣␣␣␣
9. a) 2PF4.2
   b) −1PF6.5
   c) 3PF5.1
   d) −6PF2.0

CHAPTER 17

Answers to Review Questions

1. Compact storage of large volumes of data, economic storage, and fast I/O speeds.
2. 2400
3. File or data reel
4. Read/write head
5. Load-point marker (or BOT marker)
6. True
7. True
8. End-of-file or tape mark
9. Block
10. Interblock gap.
11. True
12. False
13. True
14. True
15. REWIND
16. ENDFILE
17. BACKSPACE
18. False
19. Master file
20. Transaction file
21 File Maintenance (or File Update)
22. True
23. Add, delete, and change

Answers to Odd-numbered Exercises

1. a) **READ (11,98) X,Y,Z**
   b) **READ (11) X,Y,Z**

3. a) **WRITE (11,96) M,N,O**
   b) **WRITE (11) M,N,O**
5. a) **REWIND 12**
   b) **ENDFILE NTAPE**
   c) **BACKSPACE 3**

CHAPTER 18

Answers to Review Questions

1. To provide on-line storage and direct access to a particular record.
2. False
3. Cylinder
4. True
5. REWIND
6. DEFINEFILE
7. True
8. False
9. FIND

Answers to Odd-numbered Exercises

1. **DEFINE FILE 3 (175,110,E,KPTR)**
3. **READ (10 ' 26) X,Y,Z**
5. **WRITE (9 ' N,99) I,J,K,L**
7. **FIND (2 ' K)**
9. 4
11. 3

CHAPTER 19

Answers to Review Questions

1. False
2. Statement number
3. True
4. One can define the formats of input/output values at execution time.
5. False
6. True
7. Slashes
8. Ampersand (&)
9. True
10. False
11. False, to separate the format statement number and the first variable.
12. True
13. True

Answers to Odd-numbered Exercises

1. a) Invalid syntax following keyword PAUSE
   b) Word FOR should be TO
   c) Omitted integer variable and comma after keyword GOTO
   d) Missing NAMELIST name within slashes
   e) Comma should go after 99 and not after the keyword READ
   f) No comma is needed after the format statement number 37
   g) The parentheses, file unit number 7 and comma after the 7 should be omitted. A comma must be inserted after the 98.
   h) The constants must be integer statement numbers
3. a) 56
   b) 17
5. &LIST2 A = 31.8, K = 17, L = 43 &END

Index

Index

A

A (Alphanumeric) format code, 207
Abnormal termination, 49, 102
ABS function, 275, 276
Access organization of file
  random (direct), 324, 328
  sequential, 85, 313, 326
Accumulated totals logic, 80
Actual argument, 233
Addition operation in file maintenance, 319
Addition operator, 31, 74
Address, in storage, 3, 28
Address, record in Random I/O, 328
Adjustable array dimensions, 283
AINT function, 276
Algebraic order, 31, 73
ALGOL language, 381
Algorithm, 7, 39
ALOG function, 275, 276
ALOG10 function, 276
Alphanumeric, definition of, 207
Alphanumeric data, 207-10
Alphanumeric (A) format code, 207
AMAX0 function, 276
AMAX1 function, 275, 276
AMIN0 function, 276
AMIN1 function, 275, 276
AMOD function, 276
Ampersand (&) to denote Namelist data, 348-49
Ampersand (&) to denote statement label in multiple return option, 247
Analog computer, 4
.AND. logical operator, 128
Annotation flowchart symbol, 11
ANSI FORTRAN, 22, 25, 328, 389
ARCOS function, 277
Argument
  actual, 233-36
  dummy, 233-36
  input, 233
  input/output, 233
  output, 233
Argument list, 233
Arguments, 230, 233-36
Arithmetic Assignment statement
  explanation of, 46, 73, 77-78
  for counting and accumulating totals, 80
  general form, 47
  in sequence, 79-80
Arithmetic expression
  definition of, 73
  evaluation, 31, 46, 73, 75
  integer mode, 76
  mixed mode, 80-81
  real (floating point) mode, 76
  rules for evaluation, 73-77
Arithmetic IF statement
  explanation of, 130-31
  general form, 130
Arithmetic-logic unit, 2, 3
Arithmetic operation, unary, 74
Arithmetic operations, 31, 73
Arithmetic operator, 31, 73
Arithmetic operators' precedence, 31, 74
Array
  definition of, 167
  input operations, 170-74, 191-93
  manipulations, 176-83, 194-98, 199-202
  one-dimensional, 166-67
  output operations, 174-76, 193-94
  two-dimensional, 189-98
  three-dimensional, 198-99
Array concepts, 166-68, 189-91
Array declaration, 168-69, 190, 198
Array element, 166
Array name, 169
ARSIN function, 277
ASSIGN statement
  explanation of, 342-44
  general form, 342
Assigned GOTO statement
  explanation of, 342-44
  general form, 343
Assignment (=) operator, 23, 31-32, 46, 69, 77-78
Assignment statement
  explanation of, 46, 73, 77-81
  general form, 47
Assignment statement for storing
  alphanumeric value, 211
  complex value, 299
  double-precision value, 293
  integer value, 46, 77-81
  logical value, 295-96
  real value, 46, 77-81
Associated variable for random I/O, 329-33
Asterisk (*) to denote comment in CDC compilers, 36, 119
Asterisk (*) to denote dummy argument in multiple return, 247
Asterisks to denote improper field width, 49
AT END DO block in WATFIV-S, 370-71
ATAN function, 277
ATAN2 function, 277
Auxiliary devices, 2

B

BACKSPACE statement
  explanation of, 316, 327
  general form, 317
Basic elements of FORTRAN language, 24
BASIC language, 381
Batch Processing, 59
BCD, 311
Bench checking, 58, 264-65
Binary digit (BIT), 303-5
Binary (unformatted) I/O, 314-15
Binary number, 303-5
BIT (binary digit), 303-5
Blank character (space) in statements, 35, 110
Blank Common, 240-42
Blank (X) format code, 48, 62, 87
Block, tape, 312
BLOCKDATA statement, explanation of, 246-47
BLOCKDATA subprogram, 246-47
Blocked records, 312-13
Blocking factor, 312
Blocks, structured in WATFIV-S, 365-71
Boolean constant, 289
Bottom-up approach to programming, 18
Branching
  explanation of, 65, 122-23
  flowchart symbol, 10
  three-way, 130-31
Bug, program, 16, 51, 264

413

Index

Built-in functions, 274-77
Byte, 304, 329

C

C (for comment card), 34, 36
C (Character) format code, 219
CALL EXIT statement, 50, 122
CALL statement
   explanation of, 232-33
   general form, 233
Calling program, 229
Card
   continuation, 33
   data, 40-41, 58, 67
   FORTRAN, 34, 57, 58
   header, 125
   Job control, 58
   punch, 3, 40, 58, 372-74
   sentinal, 66
   source statement, 58
   trailer, 66
Card deck, 85
Card punch, 57
Card reader, 3, 4
Carriage control character, 48, 87, 110, 360
Central processing unit, 2
Change operation in file maintenance, 319
Character count for H format code, 109
Character data (alphanumeric), 207-12
Character (C) format code, 219
Character set for Standard FORTRAN, 25, 36, 390
Character string, 109-10
Character type variable, 357-59
Check
   desk (bench), 57, 264-65
   echo, 64, 265
   edit, 133, 319
   parity, 320
Checkpoint message, 265, 266
COBOL language, 14, 381
Code
   machine, 60
   object, 16, 60
   source, 14-16, 35, 60
Coding, 14, 35
Coding form/sheet, 14, 32, 33, 35, 57
Collating sequence, 212
Column headings, 109-10
Column order, 196
Column vector, 189
Comment card, 34, 36, 43
COMMON
   blank (unlabeled), 240-42
   equivalence in, 244-46
   labeled (named), 242-44
   name, 243
COMMON statement
   explanation of, 240-46
   general form, 243
Comparing alphanumeric data, 211-12, 219-20

Compilation errors, 61-64
Compilation of program, 15, 60-61
Compile and Go, 59
Compile phase, 60-61
Compile time initialization of variables, 213-14
Compiler
   diagnostic messages, 16, 61
   FORTRAN, 15, 60
   operation, 15, 60
Compiling, 15, 60
COMPLEX
   assignment, 299
   constants, 298
   data, 297-98
   expressions, 299
   function subprogram, 299-300
   I/O, 299
   variables, 298
Complex number, 297-98
COMPLEX type statement
   explanation of, 297-300
   general form, 298
Compound condition (relational), 128-29
Computed GOTO flowchart symbol, 132
Computed GOTO statement
   explanation of, 131-32
   general form, 131
Computer
   advantages of, 1, 6
   analog, 4
   components of, 2
   definition of, 5
   digital, 5
   general-purpose, 4
   special-purpose, 4
   units, 2
   uses of, 1-2
Computer instruction, 4, 5, 14, 39
Computer program
   a tool, 19
   general description, 4, 14, 39
Computer system, 2
Computer word, 26, 208, 303-5
Conditional branch/transfer, 122
Conditional statements
   Arithmetic IF, 130-31
   Assigned GOTO, 342-44
   Computed GOTO, 131-32
   Logical IF, 68-70
Conditional transfer, 70, 122-23
Connector flowchart symbol, 10
Consecutive slashes, 97
Constants
   complex, 298
   double-precision, 292
   hexadecimal, 303-4
   Hollerith, 109
   integer, 26
   literal, 109, 118
   logical, 295
   numeric, 25
   octal, 305
   real (floating-point), 27

Continuation card, 33, 36
Continuation character, 33
Continuation column, 33
CONTINUE statement
   explanation of, 150
   general form, 150
Control, carriage for printer, 48, 87, 110, 360
Control card, 125
Control language (JCL), 58, 266, 386-88
Control statements, 40, 122
Control unit, 2
COS function, 277
COSH function, 277
COTAN function, 277
Counter logic, 80, 100, 124
Counter loops, 80, 123-25, 142
CPU, 2
Creation of variable names, 29-31, 41, 44
Cylinder on disk, 324, 326

D

D exponent, 292
D (double-precision) format code, 294
Data card, 40-41, 58, 67, 85
Data reel, 310
DATA statement
   explanation of, 213-14
   general form, 213
DATA types
   alphanumeric, 207-10
   complex, 297-300
   double-precision, 292-94
   exponential, 289-91
   floating-point, 45, 215-16
   generalized, 300-301
   hexadecimal, 303-4
   Hollerith, 109
   integer, 45, 215
   literal, 109, 118
   logical, 295-97
   octal, 305
   real, 45, 215-16
Debugging, 16, 64, 267
Decimal position in data, 90, 302
Decimal scaling, 302
Decision, flowcharting symbol, 9, 70
Decision making, 32, 70-122
Deck
   job, 58, 59
   source, 57
Declaration of arrays, 168-69, 190, 198
Declaration statements
   COMMON, 243
   COMPLEX, 298
   DATA, 213
   DIMENSION, 168-69
   DOUBLEPRECISION, 293
   EQUIVALENCE, 216
   EXTERNAL, 282
   FORMAT, 45, 49
   FUNCTION, 273
   IMPLICIT, 221

INTEGER, 215
LOGICAL, 295
placement in program, 217
REAL, 215
SUBROUTINE, 232
Default typing of variables, 30
Defined variable, 77
DEFINEFILE statement
 explanation of, 328-30
 general form, 329
Deletion operation in file maintenance, 319
Delimiter, 44
Density, tape, 313
Desk checking, 57, 64, 264-65
Detailed flowcharting level, 12
Diagnostic message, 61
Digital computer, 5
DIM function, 276
DIMENSION
 for one-dimension arrays, 167-69
 for two-dimension arrays, 190-91
 for three-dimension arrays, 198
 variable/object time, 283
DIMENSION statement
 explanation of, 168-69, 190, 198, 283
 general form, 168
Direct access, 324, 328
Disk, magnetic, 324-36
Disk I/O flowchart symbol, 9
Disk operations, magnetic, 324-28
Division
 integer, 76
 operator, 31, 74
DO CASE block in WATFIV-S, 367-68
DO loop
 exit, 144, 148
 flowchart symbol, 145
 implied, 173-76
 index variable, 142-44
 nested, 142
 parameters, 142-44
 range of, 144
 rules of, 147-49, 153-55
DO statement
 explanation of, 142-46
 general form, 143
Document output flowchart symbol, 9
Documentation, 17-18
Dollar sign ($) as letter in variable name, 36, 221
Double quote (") for literals, 36, 119
Double-precision
 assignment, 293
 built-in functions, 294
 constants, 292
 expressions, 293
 function subprogram, 294-95
 I/O, 294
 mode, 27, 292-94
 variables, 292-93
Doubleprecision (D) format code, 294
DOUBLEPRECISION type statement
 explanation of, 292-94

general form, 293
Doubleword, 221, 292, 306
Dummy argument, 233

E

E exponent, 289-91
E (Exponential) format code, 290
EBCDIC, 311
Echo check of input data, 64, 265
Edit processing, 133, 319
Efficiency in multidimensional array operations, 199-202
Element, array, 166
Elements of FORTRAN language, 24-32
END = option in READ, 102
End of job
 abnormal, 49
 normal, 49, 122
END statement
 explanation of, 50-51
 general form, 51
ENDFILE statement
 explanation of, 316, 327
 general form, 316
Endless loop, 126
End-of-file condition, 12, 102
End-of-file record, 312
End-of-reel marker, 312
End-of-tape marker, 312
Entry, multiple to subprograms, 283-85
ENTRY statement
 explanation of, 283-85
 general form, 284
.EQ. relational operator, 65, 69
Equals sign (assignment operator), 23, 31-32, 46, 69, 77-78
Equipment, peripheral, 2
Equivalence in COMMON blocks, 244-46
EQUIVALENCE statement
 explanation of, 216-17
 general form, 216
ERR = option in READ, 320
Error
 compilation, 16, 61-63
 execution, 16, 64, 264, 354
 logical, 15, 64, 264-66
 message, 61-63, 361
 syntax, 15, 61
 system, 264, 266
Exchange operation in sorting, 180-81
Executable statement, 23, 65, 68, 70
EXECUTE statement in WATFIV-S, 368-69
Execution
 error, 16, 64, 264, 354
 of a program, 40, 64
 phase, 60-61
Execution time format specification, 344-46
Exit from a loop, 65-66, 125, 144, 148
EXP function, 276
Explicit-type statements
 COMPLEX, 298

DOUBLEPRECISION, 293
INTEGER, 215
LOGICAL, 295
REAL, 215-16
Exponent
 with D format code, 292-94
 with E format code, 289-91
 part, 290
 range of, 290
Exponential (E) format code, 290
Exponential notation, 289, 384-85
Exponentiation operation, 74
Exponentiation operator, 31, 74
Expression
 arithmetic, 31, 46, 73
 complex, 299
 double-precision, 293
 floating-point mode, 76
 integer mode, 76
 logical, 70, 128-29, 296
 mixed mode, 80-81
 real mode, 76
 relational, 69-70
 use of parentheses, 31, 74, 129
Expressions in output list, 103, 357
Extended Assignment statement, 359
Extended range of the DO, 155
EXTERNAL statement
 explanation of, 280-83
 general form, 282
EXTERNAL subprogram, 281-82

F

F (floating-point) format code
 explanation of, 90-92
 general form, 90
 input, 90-91
 output, 48, 49, 91-92
Factorial, 287
.FALSE. logical constant, 295
Field, definition of, 85
Field width, 45
Fields I/O, 87-89
File
 definition of, 85
 random (direct) access, 324, 328
 sequential access, 85
File maintenance, 318-20
File reel, 310
File unit, 44, 53
File update, 318-20
FIND statement
 explanation of, 333-34
 general form, 333
Finite loop, 127
Fixed-point number, 26
Flagging symbol to denote syntax error, 61
FLOAT function, 275, 276
Floating-point (F) format code, 90-92
Floating-point number, 27
Flowchart
 conventions, 7
 definition of, 7

Index

levels, 12
rules, 11-12
symbols, 8-11, 70, 145, 238
Flowcharting symbols
  annotation, 11
  connector, 10
  decision, 9, 70
  flowlines, 10
  input/output, 8
  iteration (DO), 145
  predefined process, 238
  processing, 8
  terminal, 8
Flowlines flowcharting symbol, 10
Form
  coding, 14, 32, 33, 35, 57
  printer layout, 41, 95
Formal argument, 233
Format code, 45, 86, 92
Format code, repetition of, 46, 112
Format code, rescan of, 88, 95, 314
FORMAT codes
  A, 207
  C, 219
  D, 294
  E, 290
  F, 45, 90-92
  G, 300
  H, 109-10
  I, 45, 89-90
  J, 103
  L, 296
  O, 305
  P, 302
  R, 220
  T, 117
  X, 48
  Z, 304
  ' (single quote), 117
Format-free alphanumeric data in WATFIV, 356
Format-free I/O, 45, 52, 355-57
Format-free READ, 45, 52, 355-56
Format specifications, 45-49, 92
FORMAT statement
  explanation of, 45-49
  general form, 46
  placement in program, 45, 218
  rules in use, 86-89
Formatted READ
  Random
    explanation of, 330-32
    general form, 331
  Sequential
    explanation of, 43-45
    general form, 44
Formatted WRITE
  Random
    explanation of, 332-33
    general form, 332
  Sequential
    explanation of, 47-49, 313
    general form, 48, 327
FORTRAN
  character set, 24, 25
  coding form/sheet, 14, 32-33, 57

compiler, 22, 60, 389-93
constants, 25
job, 57, 59
keywords, 24, 25
language, 14, 22
program, 23-24, 35, 57-58
statement coding, 14, 34-36, 57
variables, 28, 36
Fullword, 57, 59, 303
Function subprogram
  argument of, 270
  built-in functions, 274-77
  complex, 299-300
  declaring, 269
  double-precision, 294-95
  evaluation, precedence, 271
  integer type, 273
  library, 274-77
  logical, 297
  name, 269
  real type, 273
  reference, 270
  value returned by, 270
FUNCTION subprogram statement
  explanation of, 269-74
  general form, 273

G

G (generalized) format code, 300
Gap, interblock, 312, 326
.GE. relational operator, 69
General flowcharting level, 12
General form notation, 44-45
General purpose computer, 4
Generalized (G) format code, 300
GOTO statement
  Assigned
    explanation of, 342-44
    general form, 343
  Computed
    explanation of, 131-32
    general form, 131
  Unconditional
    explanation of, 65-68, 122-23
    general form, 68
Group format repetition, 112-13
.GT. relational operator, 69

H

H (Hollerith) format code, 109
H (Hollerith) format code
  character count, 109
  continued string between two cards, 110
  for printer carriage control, 48
  string, 109-11
Halfword, 221
Hardware, 2
Header card, 125
Headings
  column, 109-10
  report, 109-10
Hexadecimal constants, 303-4
Hexadecimal digit, 303
Hexadecimal (Z) format code, 304
Hierarchy chart, 19, 261-62

High-order position, 26
High-order zeroes, 41
Hollerith constant, 109
Hollerith (H) format code, 109

I

I (Integer) format code
  explanation of, 89-90
  field width, 45, 89
  general form, 89
  input, 45, 89
  output, 48, 90
IABS function, 275, 276
Identifiers, 24
IDIM function, 276
IF statements
  Arithmetic
    explanation of, 130-31
    general form, 130
  Logical
    explanation of, 65-69, 127-29
    general form, 69
IF-THEN control structure, 253, 257-58, 365-66
IF-THEN-ELSE control structure, 254, 258-59, 365-66
IFIX function, 275, 276
Imaginary part of complex number, 297-98
IMPLICIT statement
  explanation of, 220-21
  general form, 221
Implied DO in I/O list, 173-76, 192-94
Increment parameter of a DO, 142-44
Index of DO loop
  initial value, 144
  preservation of, 148
Index pointer for random access, 328
Index variable, 123, 142, 144
Indexing parameters, 142-44
Infinite loop, 126
Initialization
  of compile time variables, 213-14
  of DO index variable, 144
  of indexing parameters, 144
  of variables, 78
Initialization (DATA) statement, 213-14
Inner loops, 151, 153
Input argument, 233
Input-bound loops, 66, 123
Input data cards/records, 40, 41
Input devices, 2, 3
Input flowcharting symbol, 8
Input list, 44
Input list rules, 87-89
Input operations
  for alphanumeric data, 208-9
  for complex data, 299
  for floating-point data, 92-95
  for generalized data, 300
  for hexadecimal data, 304
  for integer data, 92-95
  for logical data, 297
  for Namelist data, 347-48
  for octal data, 305
  for one-dimensional arrays, 170-74

for two-dimensional arrays, 191-93
format-free, 45, 52, 355
scale factor, 302
unformatted, 314-15
Input unit, 2
Input/output, definition of, 85
Input/output flowcharting symbol, 8, 9
Input/output list rules, 87-89
Instruction, program, 2, 4, 5, 14, 39, 58, 61
INT function, 276
Integer division, 76, 79
Integer (I) format code, 89
Integer mode arithmetic/expression, 76-77
Integer type
   constants, 26
   expressions, 76
   variables, 30
INTEGER type statement
   explanation of, 215
   general form, 215
Interactive processing, 380
Interblock gap, 312, 326
Interchange operation in sorting, 180-82, 197-98
Intermediate output method of debugging, 64, 365
Internal data representation, 212
Internal program documentation, 17-18, 43
Internal sorting, 180
Invalid subscript error, 170, 267
I/O, definition of, 85
I/O list rules, 87-89
ISIGN function, 276
Iteration flowchart symbol, 145
Iterative loops, 125-26

J

J format code, 103
JCL, 58
JOB card, 58, 59
Job control card, 15, 58
Job control language, 58, 386-88
Job deck, 58, 59
Job setup for subprograms, 239
Justification
   left, 33, 208
   right, 33, 49

K

Key, 313
Keypunch machine, 41, 57, 375
Keypunch operation, 57, 375-79
Keywords, 24-25, 36

L

L (Logical) format code, 296
Label, statement location in card, 33
Labeled COMMON, 242-44
Language
   job control, 58, 386-88
   machine, 15, 61

object, 16, 60
source, 15, 35, 60
Layout design of input record, 41, 93
.LE. relational operator, 69
Left justified, 33, 208
Level subscript, 198, 199
Library functions, 274-77
Line counter logic, 100
Line printer, 3, 4
Linear array, 166
List
   argument, 233
   I/O, 44, 48
   linear, 166
   parameter, 233
Literal
   constant, 109
   data, 109-11
   format code (H), 109
Load-point marker, 310
LOCK statement, 336
Logic, 6, 7, 39
Logic chart, 7
Logic connector flowcharting symbol, 10
Logic error, 16, 64, 264-66
Logical
   assignment, 295-96
   constants, 295
   data, 296-97
   expressions, 68, 296
   function subprograms, 297
   I/O, 296-97
   variables, 295-96
Logical errors, 16, 64
Logical (L) format code, 296
Logical IF statement
   explanation, 65-71, 127-29
   general form, 69
Logical operations, 128, 296
Logical operators, 128
Logical record, 85, 313
LOGICAL type statement
   explanation of, 295-97
   general form, 295
Loop
   DO, 142-61
   exit from, 144
   explanation of, 65
   flowchart symbol, 145
   implied DO, 173-76
   logic, 40, 66, 123-26
   nested, 151-52
Looping, 65, 123, 142
Loops
   counter, 80, 123-25, 142
   general, 40, 65
   input-bound, 66, 123
   iterative, 125-26

M

Machine language, 14-15, 22
Machine language (code)
   transforming statements into, 15, 60
Machine reel, 310
Magnetic disk, 324-36

Magnetic tape, 310-20
Main program, 228
Mainline program, 228
Maintenance
   file, 318-20
   program, 16-17
Manipulating array items, 176-80, 194-98, 199-202
Master file, 318-20
Matrix, 189
MAX0 function, 276
MAX1 function, 276
Mean, 187-88
Median, 288
Memory, 3, 28
Memory contents, 3
Memory dump, 266
Memory unit, 2, 3
Message
   diagnostic, 15-16, 61, 361
   error, 16, 61, 266, 361
   syntax (compilation), 15-16, 61, 361
   warning, 61, 361
MIN0 function, 276
MIN1 function, 276
Mixed mode expression, 80-81
MOD function, 275, 276
Mode, 27, 75-78, 80-81
Model for problem solution, 6
Modification to program, 16-17
Modular programming, 252, 261
Multiple assignment/replacement, 81, 359
Multiple common blocks, 242-44
Multiple returns in subroutine, 247
Multiple statements per card, 36, 360
Multiple subprogram entry, 283-85
Multiplication operator, 31, 74
Multiprogramming, 58

N

Name
   array, 169
   common, 243
   function, 269
   namelist, 346
   subroutine, 230
   symbolic, 3
   variable, 28
Named (labeled) Common, 242-44
Namelist input, 347-48
Namelist output, 348-49
Namelist statement
   explanation of, 346-49
   general form, 346
.NE. relational operator, 69
Negation, 74
Nested DO loops, 151-52
Nested loops, 151
Nonexecutable statement, 23, 214
Normalization of a number, 291, 385
.NOT. logical operator, 128
Number
   complex, 298
   double precision, 292

file unit, 44, 53
floating-point, 27
integer, 26
real, 27
statement, 33
Numeric data
　input, 92-95
　output, 93-97

O

O (Octal) format code, 305
Object code/program, 16, 17, 60-61
Object time dimensions, 283
Object time format specifications, 344-46
Octal digit, 50, 305
Octal digit string, 50, 305
Octal (O) format code, 305
Octal input, 305
Octal output, 305
Off-page connector flowcharting symbol, 10
One-dimensional arrays, 166-83
Operand, 73
Operation
　arithmetic, 31, 73-74
　assignment, 31-32, 69, 77-78
　logical, 128, 296
　relational, 32, 65, 69
Operator
　addition, 31, 74
　arithmetic, 31, 73-74
　assignment, 31, 32, 77-78
　division, 31, 74
　exponentiation, 31, 74
　logical, 128
　multiplication, 31, 74
　precedence of execution, 31, 69, 74, 128
　relational, 32, 69
　subtraction, 31, 74
　unary (negation), 74
OR. logical operator, 128
Order
　sequence of computation, 31
　sequence of execution in a program, 40
　sequence of specification statements in program, 217
Outer loops, 151
Output argument, 233
Output devices, 2, 3
Output flowchart symbol, 8
Output list rules, 87-89
Output operation
　for alphanumeric data, 209-10
　for floating-point data, 95-97
　for hexadecimal data, 304
　for integer data, 95-97
　for literal data, 110-17
　for logical data, 297
　for namelist data, 348-49
　for octal data, 305
　for one-dimensional arrays, 174-76
　for two-dimensional arrays, 193-94

format-free, 52, 356
scale factor, 303
unformatted, 315
Overflow error, 266

P

P scale factor, 302
Paging, 200
Paired interchange sort, 180-82
Paper spacing control, 48, 87, 110, 360
Parameter list, 233
Parentheses in expressions
　nested, 74
　use of, 31, 74, 129
Parity check, 320
Parity (even/odd) mode, 321
Password, 381
Patch, program, 16
PAUSE statement
　explanation of, 341-42
　general form, 342
Peripheral equipment, 2
Phase
　compile, 15, 60-61
　execution, 17, 60-61
Physical record, 312
Planning of program design, 6, 64
PL/I language, 381
Preassigned data values, 213-14
Precedence rules
　arithmetic operators, 31, 46, 73, 75
　function reference, 271
　logical operators, 128
　relational operators, 65, 69
Precision, 27
Precision, limit of
　double, 27
　single, 27
Predefined process flowchart symbol, 238
Preparing program for execution, 57-61
PRINT statement
　explanation of, 350
　general form, 350
Printer carriage control character, 48, 87, 110, 360
Printer layout form, 41-42, 95
Priority of operators, 31, 46, 65, 69, 74, 129, 271
Problem
　definition, 5-6
　solving, 5-6, 39-40
Processing flowchart symbol, 8
Processing unit, central, 2
Program
　coding, 35, 39
　coding sheet, 32
　compilation, 15, 60-61
　definition of, 4, 14
　design, 18, 19, 263-64
　documentation, 17-18
　execution, 17, 40
　flowchart, 7-13
　implementation, 17
　instruction, 2, 4, 5, 14, 39, 58, 61

logic, 6-7, 39
modifications, 17
object, 16, 60-61
source, 60
stored, 5
termination of, 49, 102
testing, 15
unit, 229
Program identification/sequence number, 34
Programmer, 5, 14-16
Programming, 19, 51
Punch card, 3, 40, 58, 372-74
PUNCH statement
　explanation of, 350-51
　general form, 351
Punched card flowchart symbol, 9

Q

Quote, double (") format code, 36, 119
Quote, single (') format code, 118-19

R

R (Right-justified) format code, 220
Random, (direct) access, 324, 328
Random READ statement, 330-32
Random WRITE statement, 332-33
Range of the DO, 143-44
Range statement of the DO, 144
READ
　format free/free-field, 45, 52, 355
　formatted sequential, 43-45
　Namelist, 347-48
　random, 330-32
　specific, 349-50
　unformatted sequential, 314-15
READ statement
　formatted sequential
　　explanation of, 43-45, 327
　　general form, 44, 327
　random (formatted/unformatted)
　　explanation of, 330-32
　　general form, 330, 331
　unformatted sequential
　　explanation of, 314-15, 327
　　general form, 314, 327
Real mode arithmetic/expression, 76-77
Real part of complex number, 297-98
Real type
　built-in functions, 274-77
　constants, 27
　declaration, 215
　expression, 76
　variable, 30
REAL type statement
　explanation of, 215-16
　general form, 215
Real-time processing, 324, 380
RECORD
　definition of, 85
　format, 14
　input, 43-46, 85-86
　layout, 41, 93
　logical, 85, 313
　output, 47-49, 85-86

Index

physical, 312
size, 314
Record selection with slash, 97-99
Reference to a statement, 33
Relational expression, 69
Relational operators, 32, 65, 69
REMOTE BLOCK in WATFIV-S, 369-70
Repeat-UNTIL control structure, 254, 259-60
Repeat UNTIL loops, 125-26, 254, 259-60
Repetition of format codes
  group, 112-13
  simple, 46
Report headings, 109-10
Rescan of format specifications, 88-89, 95, 314
RETURN statement
  explanation of, 231, 232
  general form, 232
RETURN to alternate points in program, 247
REWIND statement
  explanation of, 315-16, 327
  general form, 316
Right justified, 33, 49
Right-justified (R) format code, 220
Rounding of fractions, 91-92
Round-off error/truncation, 79
Row order, 196
ROW vector, 166, 189
Rules
  arithmetic expressions, 73-77
  DO loop, 147-49
  efficiency in two-dimensional arrays, 202
  flowchart, 11-12
  format, 86-87
  input/output list, 87-89
  nested DO loops, 153-56
  nested implied DO's, 192
  rounding of fractions, 91-92
  slash, 97-98

S

Scalar variable, 166
Scale factor (P) format specifications, 302-3
Scientific approach to problem solving, 6
Scientific notation, 27, 289, 384-85
Searching operations with arrays, 178-80, 196-97
Segment, 262
Segmentation, 200, 262
SELECT-CASE control structure, 255, 260, 367-68
Selection of variable names, 29, 31, 44
Selection with interchange sort, 183
Semicolon (;) to separate multiple statements per card, 36, 360
Semidetailed level of flowcharting, 12
Sentinal card, 66
Sequence control structure, 253, 257
Sequence of execution in a program, 40

Sequence number in timesharing files, 381
Sequential access, 67, 85, 313
Sequential file, 85, 313, 326
Sequential file maintenance, 318-20
Sequential search, 178, 196-97
Serial access, 67, 85, 313
Sign, unary, 74
SIGN function, 276
SIN function, 274, 277
Single quote (') format code, 36, 118-19
Single-line spacing, 48, 87
Single-precision mode, 27
SINH function, 277
Skipped fields, 94
Slash, use of in format specifications, 97-100, 111-12
Software, 5, 251
Sort keys, 318
Sorting
  ascending, 180
  definition of, 180
  descending, 180
  external, 180
  internal, 180
  paired exchange method, 180-82
  selection with interchange method, 183, 197
Source deck, 57
Source language, 15
Source program, 23, 60
Source program listing, 60
Source statements, 23, 57, 58
Special characters, 25, 36, 373
Special purpose computer, 4
Specific READ statement, 349-50
Specifications, format, 45-49, 92
Specification statements, 23, 214, 246
SQRT function, 275-76
Standard deviation, 188
Statement
  executable, 23, 65, 68, 70
  nonexecutable, 23
  order of specification, 58
  placement within program, 217
  specification, 23
Statement function definition, 278
Statement function subprogram, 277-80
Statement label, 33, 36
Statement location in punched card, 33-34
Statement number, 33, 36
Statements, general form of:
  Arithmetic assignment, 47
  Arithmetic IF, 130
  ASSIGN, 342
  Assigned GOTO, 343
  BACKSPACE, 317
  BLOCKDATA, 246
  CALL, 233
  COMMON, 243
  COMPLEX, 298
  Computed GOTO, 131
  CONTINUE, 150
  DATA, 213

  DEFINEFILE, 329
  DIMENSION, 168
  DO, 143
  DOUBLEPRECISION, 293
  END, 51
  ENDFILE, 316
  ENTRY, 284
  EQUIVALENCE, 216
  EXTERNAL, 282
  FIND, 333
  FORMAT, 46
  FUNCTION, 273
  GOTO (unconditional), 68
  IMPLICIT, 221
  INTEGER, 215
  LOCK, 336
  LOGICAL, 295
  Logical IF, 69
  PAUSE, 342
  PRINT, 350
  PUNCH, 351
  READ, 44, 330, 349
  REAL, 215
  RETURN, 232
  REWIND, 315
  STOP, 50
  SUBROUTINE, 232
  WRITE, 48
STOP statement
  explanation of, 49-50
  general form, 50
Storage
  address, 3, 28
  disk, 324-26
  internal, 3
  location, 3, 26, 28, 208
  tape, 310-13
Stored program, 5
String, 109
Structured programming, 18-19, 64, 134, 144, 251-63
Subprogram
  actual arguments, 233-36
  advantages, 231
  arguments, 280
  array dimensions, 283
  defining, 230, 269
  defintion of, 228
  entries, multiple, 283-85
  flowchart symbol, 238
  function, 269-73
  input arguments, 233
  output arguments, 233
  statement function, 277-80
  subroutine, 228-33
  types of, 228, 273
SUBROUTINE statement
  explanation of, 230, 232
  general form, 232
SUBROUTINE subprogram, 228-39
Subscript, 166, 169-70
Subscript expression, 170-184
Subscripted variable, 166-67
Subtraction operator, 31, 74
Symbolic name, 3, 28, 169

Symbols, flowcharting, 8-11, 70, 145, 238
Syntax
   definition of, 61
   error, 15, 61-64
   location of message, 61-63
   message, 61-63
System error, 264, 266

T

T (Tabulation) format code, 117-18
Table, 189
TAN function, 277
TANH function, 277
Tape I/O flowchart symbol, 9
Tape mark, 312
Tape operations, magnetic, 310-20
Terminal devices, 59
Terminal flowchart symbol, 8
Terminal statement of DO loop, 144
Termination of program
   abnormal, 49, 102
   normal, 49, 122
Test parameter of a DO, 142-44
Testing, 16-17, 137
Three-dimensional arrays, 198-99
Three-way branch, 130
Time
   compile, 15, 60-61
   execution, 17, 60-61
Timesharing, 59, 380-83
Top down program design, 18, 255, 260-63
Totals logic, 80
Trace of program execution, 64, 364
Track, on disk, 324
Trailer card, 66, 102
Transaction file, 318-20
Transfer instruction/operation, 65, 67-68, 122-23
.TRUE. logical constant, 295
Truncation errors, 79
Truncation, integer, 77, 79
Truth value, 289, 295
Two-dimensional arrays, 189-98
Two-way branch, 137-38
Type
   of constants, 26-28
   of variables, 30-31

Type declaration statements, 214-16, 292-300

U

Unary minus sign/operator, 74
Unblocked records, 312-13
Unconditional branch/transfer, 65, 67-68, 122-23
Unconditional GOTO statement
   explanation of, 65-68, 122-23
   general form, 68
Undefined variable, 77, 126-27
Underflow error, 266
Unformatted (binary) input, 314-15
Unformatted (binary) output, 315
Unformatted READ statement
   random, 330-32
   sequential, 314-15
Unformatted WRITE statement
   random, 332-33
   sequential, 315
Unlabeled (blank) COMMON, 240-42
Update, file, 318-20
Usercode, 381

V

Variable
   assigning value to, 28
   definition of, 24, 28
   dimension in subprogram, 283
   index in DO loop, 142-44
   maximum number of characters, 28, 36
   name, 28, 36, 77
   subscripted, 166-67
   undefined/defined, 77, 126-27
Variable dimensions in subprograms, 283
Variable formats, 344-46
Vector, 166, 189

W

Warning messages, 61
WATFIV compiler, 22, 51, 61, 62, 354-71
WATFIV job control statements, 363, 387
WATFIV-S features, 365-71

WATFOR compiler, 22, 51, 61, 62
While-DO control structure, 254, 259, 366-67
WHILE EXECUTE statement in WATFIV-S, 370
While loop, 125-26, 254, 259
Whole number, 26
Width of a field, 45
Word, computer, 26, 208, 303-5
WRITE
   format-free, 45, 52, 356-57
   formatted sequential, 47, 313, 327
   general, 47
   Namelist, 348-49
   random (formatted/unformatted), 332-33
   unformatted sequential, 315, 327
WRITE statement
   formatted sequential
      explanation of, 47, 313, 327
      general form, 48, 327
   random (formatted/unformatted)
      explanation of, 332-33
      general form, 332
   unformatted sequential
      explanation of, 315, 327
      general form, 315, 327

X

X (blank) format code, 48, 62, 87

Z

Z (Hexadecimal) format code, 304
Zero, division by, 266
Zero, leading (H/O), 41
Zigamorph character, 360